A SOURCE BOOK
in
MATHEMATICS

Newton

A SOURCE BOOK
in
MATHEMATICS

By

DAVID EUGENE SMITH

DOVER PUBLICATIONS, INC.

NEW YORK

This Dover edition, first published in 1959, is an
unabridged republication of the first edition, originally
published in 1929 by McGraw-Hill Book Co., Inc.

International Standard Book Number: 0-486-64690-4
Library of Congress Catalog Card Number: 59-14227

Manufactured in the United States of America
Dover Publications, Inc.
31 East Second Street,
Mineola, N.Y. 11501

A SOURCE BOOK IN MATHEMATICS

Author's Preface

The purpose of a source book is to supply teachers and students with a selection of excerpts from the works of the makers of the subject considered. The purpose of supplying such excerpts is to stimulate the study of the various branches of this subject—in the present case, the subject of mathematics. By knowing the beginnings of these branches, the reader is encouraged to follow the growth of the science, to see how it has developed, to appreciate more clearly its present status, and thus to see its future possibilities.

It need hardly be said that the preparation of a source book has many difficulties. In this particular case, one of these lies in the fact that the general plan allows for no sources before the advent of printing or after the close of the nineteenth century. On the one hand, this eliminates most of mathematics before the invention of the calculus and modern geometry; while on the other hand, it excludes all recent activities in this field. The latter fact is not of great consequence for the large majority of readers, but the former is more serious for all who seek the sources of elementary mathematics. It is to be hoped that the success of the series will permit of a volume devoted to this important phase of the development of the science.

In the selection of material in the four and a half centuries closing with the year 1900, it is desirable to touch upon a wide range of interests. In no other way can any source book be made to meet the needs, the interests, and the tastes of a wide range of readers. To make selections from the field, however, is to neglect many more sources than can possibly be selected. It would be an easy thing for anyone to name a hundred excerpts that he would wish to see, and to eliminate selections in which he has no special interest. Some may naturally seek for more light on our symbols, but Professor Cajori's recent work furnishes this with a satisfactory approach to completeness. Others may wish for a worthy treatment of algebraic equations, but Matthiessen's *Grundzüge* contains such a wealth of material as to render the undertaking unnecessary. The extensive field of number theory will appeal to many readers, but the monumental work of Professor Dickson, while not a source book in the ordinary sense of the term, satisfies most of the needs in this respect. Consideration must always be given to the demands of readers, and naturally these demands change as the literature of the history of mathematics becomes more extensive. Furthermore, the possibility of finding source material that is stated succinctly enough for purposes of quotation has to be considered, and also that of finding material that is not so ultra-technical as to serve no useful purpose for any considerable number of readers. Such are a few of the many difficulties which will naturally occur to everyone and which will explain some of the reasons which compel all source books to be matters of legitimate compromise.

Although no single department of "the science venerable" can or should be distinct from any other, and although the general trend is strongly in the direction of unity of both purpose and method, it will still serve to assist the reader if his attention is called to the rough classification set forth in the Contents.

The selections in the field of Number vary in content from the first steps in printed arithmetic, through the development of a few selected number systems, to the early phases of number theory. It seems proper, also, to consider the mechanics of computation in the early stages of the subject, extending the topic to include even as late a theory as nomography. There remains, of course, a large field that is untouched, but this is a necessary condition in each branch.

The field of Algebra is arbitrarily bounded. Part of the articles classified under Number might have been included here, but such questions of classification are of little moment in a work of this nature. In general the articles relate to equations, symbolism, and series, and include such topics as imaginary roots, the early methods of solving the cubic and biquadratic algebraic equations and numerical equations of higher degree, and the Fundamental Theorem of Algebra. Trigonometry, which is partly algebraic, has been considered briefly under Geometry. Probability, which is even more algebraic, is treated by itself, and is given somewhat more space than would have been allowed were it not for the present interest in the subject in connection with statistics.

The field of Geometry is naturally concerned chiefly with the rise of the modern branches. The amount of available material is such that in some cases merely a single important theorem or statement of purpose has been all that could be included. The topics range from the contributions of such sixteenth-century writers as Fermat, Desargues, Pascal, and Descartes, to a few of those who, in the nineteenth century, revived the study of the subject and developed various forms of modern geometry.

The majority of the selections thus far mentioned have been as non-technical as possible. In the field of Probability, however, it has been found necessary to take a step beyond the elementary bounds if the selections are to serve the purposes of those who have a special interest in the subject.

The fields of the Calculus, Function Theory, Quaternions, and the general range of Mathematics belong to a region so extensive as to permit of relatively limited attention. It is essential that certain early sources of the Calculus should be considered, and that some attention should be given to such important advances as relate to the commutative law in Quaternions and Ausdehnungs-lehre, but most readers in such special branches as are now the subject of research in our universities will have at hand the material relating to the origins of their particular subjects. The limits of this work would not, in any case, permit of an extensive offering of extracts from such sources.

It should be stated that all the translations in this work have been contributed without other reward than the satisfaction of assisting students and teachers in knowing the sources of certain phases of mathematics. Like the editor and the advisory committee, those who have prepared the articles have given their services gratuitously. Special mention should, however, be made of the unusual interest taken by a few who have devoted much time to assisting the editor and committee in the somewhat difficult labor of securing and assembling the material. Those to whom they are particularly indebted for assistance beyond the preparation of special articles are Professor Lao G. Simons, head of the department of mathematics in Hunter College, Professor Jekuthiel Ginsburg, of the Yeshiva College, Professor Vera Sanford of Western Reserve University, and Professor Helen M.

Walker, of Teachers College, Columbia University. To Professor Sanford special thanks are due for her generous sacrifice of time and effort in the reading of the proofs during the editor's prolonged absence abroad.

The advisory committee, consisting of Professors Raymond Clare Archibald of Brown University, Professor Florian Cajori of the University of California, and Professor Leonard Eugene Dickson of the University of Chicago, have all contributed of their time and knowledge in the selection of topics and in the securing of competent translators. Without their aid the labor of preparing this work would have been too great a burden to have been assumed by the editor.

In the text and the accompanying notes, the remarks of the translators, elucidating the text or supplying historical notes of value to the reader, are inclosed in brackets [ ]. To these contributors, also, are due slight variations in symbolism and in the spelling of proper names, it being felt that they should give the final decision in such relatively unimportant matters.

DAVID EUGENE SMITH.

NEW YORK,
September, 1929.

Contents

CONTENTS

III. THE FIELD OF GEOMETRY

IV. THE FIELD OF PROBABILITY

A SOURCE BOOK IN MATHEMATICS

SOURCE BOOK IN MATHEMATICS

I. FIELD OF NUMBER

THE FIRST PRINTED ARITHMETIC

TREVISO, ITALY, 1478

(Translated from the Italian by Professor David Eugene Smith, Teachers College, Columbia University, New York City.)

Although it may justly be said that mere computation and its simple applications in the lives of most people are not a part of the science of mathematics, it seems proper that, in a source book of this kind, some little attention should be given to its status in the early days of printing. For this reason, these extracts are selected from the first book on arithmetic to appear from the newly established presses of the Renaissance period.[1] The author of the work is unknown, and there is even some question as to the publisher, although he seems to have been one Manzolo or Manzolino. It is a source in the chronological rather than the material sense, since the matter which it contains had apparently but little influence upon the other early writers on arithmetic. The work is in the Venetian dialect and is exceedingly rare.[2] The copy from which this translation was made is in the library of George A. Plimpton of New York City. As with many other *incunabula*, the book has no title. It simply begins with the words, *Incommincia vna practica molto bona et vtilez a ciascbaduno chi vuole vxare larte dela mercbadantia. chiamata vulgarmente larte de labbacbo.* It was published at Treviso, a city not far to the north of Venice, and the colophon has the words "At Treviso, on the 10th day of December, 1478."

Here beginneth a Practica, very helpful to all who have to do with that commercial art commonly known as the abacus.

I have often been asked by certain youths in whom I have much interest, and who look forward to mercantile pursuits, to put into writing the fundamental principles of arithmetic, commonly

---

[1] For the most part, these selections are taken from an article by this translator which appeared in *Isis*, Vol. VI (3), pp. 311–331, 1924, and are here published by permission of the editor. For a more extended account of the book, the reader is referred to this periodical.

[2] A critical study of it from the bibliographical standpoint was made by Prince Boncompagni in the *Atti dell' Accademia Pontificia de' Nuovi Lincei*, tomo XVI, 1862–1863.

called the abacus.   Therefore, being impelled by my affection for them, and by the value of the subject, I have to the best of my small ability undertaken to satisfy them in some slight degree, to the end that their laudable desires may bear useful fruit.   Therefore in the name of God I take for my subject this work in algorism, and proceed as follows:

All things which have existed since the beginning of time have owed their origin to number.   Furthermore, such as now exist are subject to its laws, and therefore in all domains of knowledge this Practica is necessary.   To enter into the subject, the reader must first know the basis of our science. Number is a multitude brought together or assembled from several units, and always from two at least, as in the case of 2, which is the first and the smallest number.   Unity is that by virtue of which anything is said to be one.   Furthermore be it known that there are three kinds of numbers, of which the first is called a simple number, the second an article, and the third a composite or mixed number.   A simple number is one that contains no tens, and it is represented by a single figure, like i, 2, 3, etc.   An article is a number that is exactly divisible by ten, like i0, 20, 30 and similar numbers.   A mixed number is one that exceeds ten but that cannot be divided by ten without a remainder, such as ii, i2, i3, etc.   Furthermore be it known that there are five fundamental operations which must be understood in the Practica, viz., numeration, addition, subtraction, multiplication, and division. Of these we shall first treat of numeration, and then of the others in order.

Numeration is the representation of numbers by figures.   This is done by means of ten letters or figures, as here shown.   .i., .2.,

.3., .4., .5., .6., .7., .8., .9., .0.. Of these the first figure, i, is not called a number but the source of number. The tenth figure, 0, is called cipher or "nulla," *i. e.*, the figure of nothing, since by itself it has no value, although when joined with others it increases their value. Furthermore you should note that when you find a figure by itself its value cannot exceed nine, *i. e.*, 9; and from that figure on, if you wish to express a number you must use at least two figures, thus: ten is expressed by i0, eleven by ii, and so on. And this can be understood from the following figures.[1]

..Thousands of millions	..Hundreds of millions	..Tens of millions	..Millions	..Hundreds of thousands	..Tens of thousands	..Thousands	..Hundreds	..Tens	..Units
									i
									2
									3
									4
									5
									6
									7
									8
									9
								i	0
							i	2	0
						i	2	3	0
					i	2	3	4	0
				i	2	3	4	5	0
			i	2	3	4	5	6	0
		i	2	3	4	5	6	7	0
	i	2	3	4	5	6	7	8	0
i	2	3	4	5	6	7	8	9	0
2	3	4	5	6	7	8	9	0	0
3	4	5	6	7	8	9	0	0	0
4	5	6	7	8	9	0	0	0	0
5	6	7	8	9	0	0	0	0	0
6	7	8	9	0	0	0	0	0	0
7	8	9	0	0	0	0	0	0	0
8	9	0	0	0	0	0	0	0	0
9	0	0	0	0	0	0	0	0	0

---

[1] The figure 1 was not always in the early fonts of type, the letter "i" being then used in its stead.

To understand the figures it is necessary to have well in mind the following table:[1]

i times i makes i	i times i0 makes i0
i times 2 makes 2	2 times i0 makes 20
i times 3 makes 3	3 times i0 makes 30
i times 4 makes 4	4 times i0 makes 40
i times 5 makes 5	5 times i0 makes 50
i times 6 makes 6	6 times i0 makes 60
i times 7 makes 7	7 times i0 makes 70
i times 8 makes 8	8 times i0 makes 80
i times 9 makes 9	9 times i0 makes 90
i times 0 makes 0	0 times i0 makes 0

And to understand the preceding table it is necessary to observe that the words written at the top[2] give the names of the places occupied by the figures beneath. For example, below 'units' are the figures designating units, below 'tens' are the tens, below 'hundreds' are the hundreds, and so on. Hence if we take each figure by its own name, and multiply this by its place value, we shall have its true value. For instance, if we multiply i, which is beneath the word 'units,' by its place,—that is, by units,—we shall have 'i time i gives i,' meaning that we have one unit. Again, if we take the 2 which is found in the same column, and multiply by its place, we shall have 'i time 2 gives 2,' meaning that we have two units,. . . and so on for the other figures found in this column. . . This rule applies to the various other figures, each of which is to be multiplied by its place value.

And this suffices for a statement concerning the 'act'[3] of numeration.

Having now considered the first operation, viz. numeration, let us proceed to the other four, which are addition, subtraction, multiplication, and division. To differentiate between these operations it is well to note that each has a characteristic word, as follows:

---

[1] [The table continues from "i times i00 makes i00" to "0 times i00 makes 0."]

[2] [That is, the numeration table shown on p. 3.]

[3] [The fundamental operations, which the author calls "acts" (atti) went by various names. The medieval Latin writers called them "species," a word that appears in *The Crafte of Nombryng*, the oldest English manuscript on arithmetic, where the author speaks of "7 spices or partes of this craft." This word, in one form or another, is also found in various languages. The Italians used both 'atti' and 'passioni.']

Addition has the word   *and,*
Subtraction has the word   *from,*
Multiplication has the word *times,*
Division has the word   *in.*

It should also be noticed that in taking two numbers, since at least two are necessary in each operation, there may be determined by these numbers any one of the above named operations. Furthermore each operation gives rise to a different number, with the exception that 2 times 2 gives the same result as 2 and 2, since each is 4.   Taking, then, 3 and 9 we have:

Addition:	3 and	9 make	i2
Subtraction:	3 from	9 leaves	6
Multiplication:	3 times	9 makes	27
Division:	3 in	9 gives	3

We thus see how the different operations with their distinctive words lead to different results.

In order to understand the second operation, addition, it is necessary to know that this is the union of several numbers, at least of two, in a single one, to the end that we may know the sum arising from this increase.   It is also to be understood that, in the operation of adding, two numbers at least are necessary, namely the number to which we add the other, which should be the larger, and the number which is to be added, which should be the smaller. Thus we always add the smaller number to the larger, a more convenient plan than to follow the contrary order, although the latter is possible, the result being the same in either case.   For example, if we add 2 to 8 the sum is i0, and the same result is obtained by adding 8 to 2.   Therefore if we wish to add one number to another we write the larger one above and the smaller one below, placing the figures in convenient order, *i. e.,* the units under units, tens under tens, hundreds under hundreds, etc.   We always begin to add with the lowest order, which is of least value. Therefore if we wish to add 38 and 59 we write the numbers thus:

59
38
Sum 97

We then say, '8 and 9 make i7,' writing 7 in the column which was added, and carrying the i (for when there are two figures in one place we always write the one of the lower order and carry the other to the next higher place).   This i we now add to 3, making 4, and this to the 5, making 9, which is written in the column from which it is derived. The two together make 97.

The proof of this work consists in subtracting either addend from the sum, the remainder being the other. Since subtraction proves addition, and addition proves subtraction, I leave the method of proof until the latter topic is studied, when the proof of each operation by the other will be understood.

Besides this proof there is another. If you wish to check the sum by casting out nines, add the units, paying no attention to 9 or 0, but always considering each as nothing. And whenever the sum exceeds 9, subtract 9, and consider the remainder as the sum. Then the number arising from the sum will equal the sum of the numbers arising from the addends. For example, suppose that you wish to prove the following sum:

.59.
.38.
Sum   .97. | 7

The excess of nines in 59 is 5; 5 and 3 are 8; 8 and 8 are 16; subtract 9 and 7 remains. Write this after the sum, separated by a bar. The excess of nines in 97 is 7, and the excess of nines in 7 equals 7, since neither contains 9. In this way it is possible to prove the result of any addition of abstract numbers or of those having no reference to money, measure, or weight. I shall show you another plan of proof according to the nature of the case. If you have to add 816 and 1916,[1] arrange the numbers as follows:

1916
816
Sum   2732

Since the sum of 6 and 6 is 12, write the 2 and carry the 1. Then add this 1 to that which follows to the left, saying, '1 and 1 are 2, and the other 1 makes 3.' Write this 3 in the proper place, and add 8 and 9. The sum of this 8 and 9 is 17, the 7 being written and the 1 carried to the other 1, making 2, which is written in the proper place, the sum being now complete. If you wish to prove by 9 arrange the work thus:

1916
816
The sum   2732 | 5

---

[1] [From now on, the figure 1 will be used in the translation instead of the letter 'i' which appears always in the original.]

You may now effect the proof by beginning with the upper number, saying '1 and 1 are 2, and 6 are 8, and 8 are 16. Subtract 9, and 7 remains. The 7 and 1 are 8, and 6 are 14. Subtract 9, and 5 remains,' which should be written after the sum, separated by a bar. Look now for the excess of nines in the sum: 2 and 7 are 9, the excess being 0; 3 and 2 are 5, so that the result is correct.[1]

Having now considered the second operation of the Practica of arithmetic, namely the operation of addition, the reader should give attention to the third, namely the operation of subtraction. Therefore I say that the operation of subtraction is nothing else than this: that of two numbers we are to find how much difference there is from the less to the greater, to the end that we may know this difference. For example, take 3 from 9 and there remains 6. It is necessary that there should be two numbers in subtraction, the number from which we subtract and the number which is subtracted from it.

The number from which the other is subtracted is written above, and the number which is subtracted below, in convenient order, viz., units under units and tens under tens, and so on. If we then wish to subtract one number of any order from another we shall find that the number from which we are to subtract is equal to it, or greater, or less. If it is equal, as in the case of 8 and 8, the remainder is 0, which 0 we write underneath in the proper column. If the number from which we subtract is greater, then take away the number of units in the smaller number, writing the remainder below, as in the case of 3 from 9, where the remainder is 6. If, however, the number is less, since we cannot take a greater number from a less one, take the complement of the larger number with respect to 10, and to this add the other, but with this condition: that you add one to the next left-hand figure. And be very careful that whenever you take a larger number from a smaller, using the complement, you remember the condition above mentioned. Take now an example: Subtract 348 from 452, arranging the work thus:

$$452$$
$$348$$
Remainder 104

First we have to take a greater number from a less, and then an equal from an equal, and third, a less from a greater. We proceed

---

[1] [The addition of larger numbers and of the compound numbers like 916 lire 14 soldi plus 1945 lire 15 soldi are now considered.]

as follows: We cannot take 8 from 2, but 2 is the complement of 8 with respect to 10, and this we add to the other 2 which is above the 8, thus: 2 and 2 make 4, which we write beneath the 8 for the remainder.   There is, however, this condition, that to the figure following the 8 (viz., to 4), we add 1, making it 5.   Then 5 from 5, which is an equal, leaves 0, which 0 we write beneath.

Then 3 from 4, which is a less from a greater, is 1, which 1 we write under the 3, so that the remainder is 104.

If we wish to prove this result, add the number subtracted to the remainder, and the result will be the number from which we subtracted.   We may arrange the work as follows:

$$\begin{array}{c|c} 452 & 2 \\ 348 & 6 \\ \hline 104 & 5 \\ \hline 452 & \end{array}$$

Now add, 4 and 8 are 12; write 2 under the 4 and carry 1; then 1 and 4 are 5; write this 5 under the 0; then add 1 and 3, making 4, and write this 4 under the 1, and the work checks.   Thus is found that which was promised you, as you can see[1]. . .

Having now explained the third operation, namely that of subtraction, the reader should give attention to the fourth, namely that of multiplication.   To understand this it is necessary to know that to multiply one number by itself or by another is to find from two given numbers a third number which contains one of these numbers as many times as there are units in the other.   For example, 2 times 4 are 8, and 8 contains 4 as many times as there are units in 2, so that 8 contains 4 in itself twice.   Also the 8 contains 2 as many times as there are units in 4, and 4 has in itself four units, so that 8 contains 2 four times.   It should be well understood that in multiplication two numbers are necessary, namely the multiplying number and the number multiplied, and also that the multiplying number may itself be the number multiplied, and vice versa, the result being the same in both cases. Nevertheless usage and practice demand that the smaller number shall be taken as the multiplying number, and not the larger. Thus we should say, 2 times 4 makes 8, and not 4 times 2 makes 8,

---

[1] [The author now gives a further proof of subtraction by the casting out of nines, after which he devotes about six or seven pages to checks on subtraction and to the subtraction of lire, soldi, grossi, pizoli, and the like.]

although the results are the same. Now not to speak at too great length I say in brief, but sufficiently for the purposes of a Practica, that there are three methods of multiplication, viz., by the tables, cross multiplication, and the chess-board plan. These three methods I will explain to you as briefly as I am able. But before I give you a rule or any method, it is necessary that you commit to memory the following statements, without which no one can understand all of this operation of multiplication[1]...

I have now given you to learn by heart all the statements needed in the Practica of arithmetic, without which no one is able to master the Art. We should not complain, however, at having to learn these things by heart in order to acquire readiness; for I assure you that these things which I have set forth are necessary to any one who would be proficient in this art, and no one can get along with less. Those facts which are to be learned besides these are valuable, but they are not necessary.

Having learned by heart all of the above facts, the pupil

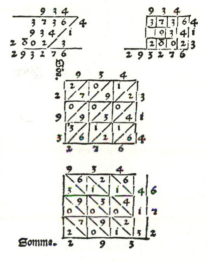

may with zeal begin to multiply by the table. This operation arises when the multiplier is a simple number, and the number multiplied has at least two figures, but as many more as we wish. And that we may more easily understand this operation we shall call the first figure toward the right, units; the second toward the left, tens, and the third shall be called hundreds. This being understood, attend to the rule of working by the table, which is as follows: First multiply together the units of the multiplier and

---

[1] [The author now gives the multiplication table, omitting all duplications like 3 × 2 after 2 × 3 has been given, but extending for "those who are of scholarly tastes" the table to include multiples of 12, 20, 24, 32 and 36, as needed in the monetary systems used by merchants of the time.]

the number multiplied. If from this multiplication you get a simple number, write it under its proper place; if an article, write a 0 and reserve the tens to add to the product of the tens; but if a mixed number is found, write its units in the proper place, and save the tens to add to the product of the tens, proceeding in the same way with all the other orders. Then multiply together the units of the multiplier with the tens; then with the hundreds, and so on in regular order[1]...

In order to understand the fourth operation, viz., division, three things are to be observed, viz., what is meant by division; second, how many numbers are necessary in division; third, which of these numbers is the greater. As to the first I say that division is the operation of finding, from two given numbers, a third number, which is contained as many times in the greater number as unity is contained in the less number. You will find this number when you see how many times the less number is contained in the greater. Suppose, for example, that we have to divide 8 by 2; here 2 is contained 4 times in 8, so we say that 4 is the quotient demanded. Also, divide 8 by 4. Here the 4 is contained 2 times in 8, so that 2 is the quotient demanded.

Second, it is to be noticed that three numbers are necessary in division,—the number to be divided, the divisor, and the quotient, as you have understood from the example above given, where 2 is the divisor, 8 the number to be divided, and 4 the quotient. From this is derived the knowledge of the third thing which is to be noted, that the number which is to be divided is always greater than, or at least is equal to, the divisor. When the numbers are equal the quotient is always 1.

Now to speak briefly, it is sufficient in practice to say that there are two ways of dividing,—by the table and the galley method. In this operation you should begin with the figure of highest value, that is by the one which is found at the left, proceeding thence to the right. If you can divide by the table you will be able to divide by the galley method, and it is well, for brevity, to avoid the latter when you can. Therefore this is the method of dividing by the table: See how many times your divisor is found in the first

---

[1] [The author now gives an example in multiplying by a one-figure number, proving the work by casting out nines. He then gives a proof by casting out sevens, after which he sets forth various methods of multiplication, such as that of the chessboard, that of the quadrilateral, or the one known by the name of gelosia, all of which were in common use at the time.]

left-hand figure, if it is contained in it, and write the quotient beneath it. If it is not so contained, consider this figure as tens and take together with it the following figure; then, finding the quotient write it beneath the smaller of the two figures. If there is any remainder, consider this as tens, and add it to the next number to the right, and see how many times your divisor is found in these two figures, writing the quotient under the units. In this same way proceed with the rest of the figures to the right. And when you have exhausted them all, having set down the quotient, write the remainder at the right, separated by a bar; and if the remainder is 0, place it where I have said. In the name of God I propose the first example, so attend well.

Divide 7624 ducats into two parts, viz. by 2, arranging your work as follows:

$$\text{The divisor} \quad .2. \quad 7624 \mid 0 \text{ the remainder}[1]$$
$$\text{The quotient} \quad \quad 3812 \mid$$

The operations which I have set forth above being understood, it is necessary to take up the method and the rules of using them. The rule you must now study is the rule of the three things. Therefore that you may have occasion to sharpen your understanding in the four operations above mentioned,—addition, subtraction, multiplication, and division,—I shall compare them. As a carpenter (wishing to do well in his profession) needs to have his tools very sharp, and to know what tools to use first, and what next to use, &c., to the end that he may have honor from his work, so it is in the work of this Practica. Before you take the rule of the three things it is necessary that you should be very skilled in the operations which have been set forth in addition, subtraction, multiplication, and division, so that you may enter enthusiastically into your work. Furthermore, that the rule of the three things, which is of utmost importance in this art, may be at your command, you must have at hand this tool of the operations, so that you can begin your labors without spoiling your instruments and without failing. Thus will your labors command high praise.

---

[1] [The author now devotes twelve quarto pages to completing the explanation of division, which shows the degree of difficulty which the subject then offered. The rest of the text is devoted largely to the solution of mercantile problems by the Rule of Three. The preliminary statement and four problems will suffice to show the nature of the work. The subject is treated much more fully in *Isis*, Vol. VI (3), pp. 311–331, 1924.]

The rule of the three things is this: that you should multiply the thing which you wish to know, by that which is not like it, and divide by the other. And the quotient which arises will be of the nature of the thing which has no term like it. And the divisor will always be dissimilar (in weight, in measure, or in other difference) to the thing which we wish to know.

In setting forth this rule, note first that in every case which comes under it there are only two things of different nature, of which one is named twice,—by two different numbers,—and the other thing is named once, by one number alone. For example:

If 1 lira of saffron is worth 7 lire of pizoli, what will 25 lire of this same saffron be worth? Here are not mentioned together both saffron and money, but the saffron is mentioned twice by two different numbers, 1 and 25; and the money is mentioned once, by the one number 7. So this is not called the rule of three things because there are three things of different nature, for one thing is mentioned twice.

. . . .

Three merchants have invested their money in a partnership, whom to make the problem clearer I will mention by name. The first was called Piero, the second Polo, and the third Zuanne. Piero put in 112 ducats, Polo 200 ducats, and Zuanne 142 ducats. At the end of a certain period they found that they had gained 563 ducàts. Required to know how much falls to each man so that no one shall be cheated.

. . . .

There are two merchants of whom the one has cloth worth 22 soldi a yard, but who holds it in barter at 27 soldi. The other has wool which is worth in the country 19 lire per hundredweight. Required to know how much he must ask per hundredweight in barter so that he may not be cheated.

. . . .

The Holy Father sent a courier from Rome to Venice, commanding him that he should reach Venice in 7 days. And the most illustrious Signoria of Venice also sent another courier to Rome, who should reach Rome in 9 days. And from Rome to Venice is 250 miles. It happened that by order of these lords the couriers started on their journeys at the same time. It is required to find in how many days they will meet.

. . . .

What availeth virtue to him who does not labor? Nothing. At Treviso, on the 10th day of December, 1478.

RECORDE

On "The Declaration of the Profit of Arithmeticke"

(Selected by Professor David Eugene Smith, Teachers College, Columbia University, New York City.)

Robert Recorde (c. 1510–1558), a student and later a private teacher at both Oxford and Cambridge, wrote several works on mathematics. His arithmetic, *The Grovnd of Artes*, was not the first one published in England, but it was by far the most influential of the early books upon the subject as far as the English-speaking peoples are concerned. This is not because of its catechetic style, although it doubtless influenced other writers to adopt this form of textbook instruction, but rather because through its subject matter and style of problems it set a standard that has been followed until comparatively recent times. On the principle that a source book should touch at least lightly upon the elementary branches, "The declaration of the profit of Arithmeticke" is here set forth. The exact date of the first edition is uncertain, but it was about 1540 to 1542. Although a number of the early editions are available in the library of George A. Plimpton of New York City, it has been thought best to select one which represents the results of Recorde's influence for a full century,—that of 1646. As the title page says, this was "afterward augmented by M. John Dee," the promoter of the first English edition of Euclid; "enlarged—By John Mellis," and "diligently perused, corrected, illustrated and enlarged by R. C.", and its tables "diligently calculated by Rv: Hartwell, Philomathemat." It therefore represents the best efforts of the teaching profession for a hundred years.

The following is an extract from Recorde's preface:

TO THE LOVING Readers,
The Preface of Mr. *Robert Record*

Sore oft times have I lamented with my self the unfortunate condition of England, seeing so many great Clerks to arise in sundry other parts of the world, and so few to appear in this our Nation: whereas for pregnancy of naturall wit (I think) few Nations do excell Englishmen: But I cannot impute the cause to any other thing, then to be contempt, or misregard of learning. For as Englishmen are inferiour to no men in mother wit, so they passe all men in vain pleasures, to which they may attain with great pain and labour: and are as slack to any never so great

13

commodity; if there hang of it any painfull study or travelsome labour.

Howbeit, yet all men are not of that sort, though the most part be, the more pity it is: but of them that are so glad, not onely with painfull study, and studious pain to attain learning, but also with as great study and pain to communicate their learning to other, and make all England (if it might be) partakers of the same; the most part are such, that unneath they can support their own necessary charges, so that they are not able to bear any charges in doing of that good, that else they desire to do.

But a greater cause of lamentation is this, that when learned men have taken pains to do things for the aid of the unlearned, scarce they shall be allowed for their wel-doing, but derided and scorned, and so utterly discouraged to take in hand any like enterprise again.

The following is "The declaration of the profit of Arithmeticke" and constitutes the first ten pages of the text. It may be said to represent the influence of this text upon establishing for a long period what educators at present speak of as "the objectives" of elementary arithmetic.

A Dialogue between the Master and the Scholar: *teaching the* Art *and use of Arithmetick with Pen.*
The Scholar speaketh.

SIR, such is your authority in mine estimation, that I am content to consent to your saying, and to receive it as truth, though I see none other reason that doth lead me thereunto: whereas else in mine own conceit it appeareth but vain, to bestow any time privately in learning of that thing, that every childe may, and doth learn at all times and hours, when he doth any thing himself alone, and much more when he talketh or reasoneth with others.

Master. Lo, this is the fashion and chance of all them that seek to defend their blinde ignorance, that when they think they have made strong reason for themselves, then have they proved quite contrary. For if numbring be so common (as you grant it to be) that no man can do anything alone, and much lesse talk or bargain with other, but he shall still have to do with number: this proveth not number to be contemptible and vile, but rather right excellent and of high reputation, sith it is the ground of all mens affairs, in that without it no tale can be told, no communication without it can be continued, no bargaining without it can duely be ended, or no business that man hath, justly completed. These commodi-

ties, if there were none other, are sufficient to approve the worthi-
nesse of number. But there are other innumerable, farre passing
all these, which declare number to exceed all praise. Wherefore
in all great works are Clerks so much desired? Wherefore are
Auditors so richly fed? What causeth Geometricians so highly
to be enhaunsed? Why are Astronomers so greatly advanced?
Because that by number such things they finde, which else would
farre excell mans minde.

Scholar. Verily, sir, if it bee so, that these men by numbring,
their cunning do attain, at whose great works most men do wonder,
then I see well I was much deceived, and numbring is a more
cunning thing then I took it to be.

Master. If number were so vile a thing as you did esteem it,
then need it not to be used so much in mens communication.
Exclude number, and answer to this question: How many years
old are you?

Scholar. Mum.

Master. How many dayes in a weeke? How many weeks in
a year? What lands hath your Father? How many men doth
hee keep? How long is it since you came from him to me?

Scholar. Mum.

Master. So that if number want, you answer all by Mummes:
How many miles to London?

Scholar. A poak full of plums.

Master. Why, thus you may see, what rule number beareth,
and that if number bee lacking it maketh men dumb, so that to
most questions they must answer Mum.

Scholar. This is the cause, sir, that I judged it so vile, because
it is so common in talking every while: Nor plenty is not dainty,
as the common saying is.

Master. No, nor store is no sore, perceive you this? The more
common that the thing is, being needfully required, the better is
the thing, and the more to be desired. But in numbring, as
some of it is light and plain, so the most part is difficult, and not
easie to attain. The easier part serveth all men in common, and
the other requireth some learning. Wherefore as without numbring a man can do almost nothing, so with the help of it, you may
attain to all things.

Scholar. Yes, sir, why then it were best to learn the Art of
numbring, first of all other learning, and then a man need learn
no more, if all other come with it.

Master.   Nay not so: but if it be first learned, then shall a man be able (I mean) to learn, perceive, and attain to other Sciences; which without it he could never get.

Scholar.   I perceive by your former words, that Astronomy and Geometry depend much on the help of numbring: but that other Sciences, as Musick, Physick, Law, Grammer, and such like, have any help of Arithmetick, I perceive not.

Master.   I may perceive your great Clerk-linesse by the ordering of your Sciences: but I will let that passe now, because it toucheth not the matter that I intend, and I will shew you how Arithmetick doth profit in all these somewhat grosly, according to your small understanding, omitting other reasons more substantiall.

First (as you reckon them) Musick hath not onely great help of Arithmetick, but is made, and hath his perfectnesse of it: for all Musick standeth by number and proportion: And in Physick, beside the calculation of criticall dayes, with other things, which I omit, how can any man judge the pulse rightly, that is ignorant of the proportion of numbers?

And so for the Law, it is plain, that the man that is ignorant of Arithmetick, is neither meet to be a Judge, neither an Advocate, nor yet a Proctor.   For how can hee well understand another mans cause, appertaining to distribution of goods, or other debts, or of summes of money, if he be ignorant of Arithmetick?   This oftentimes causeth right to bee hindered, when the Judge either delighteth not to hear of a matter that hee perceiveth not, or cannot judge for lack of understanding: this commeth by ignorance of Arithmetick.

Now, as for Grammer, me thinketh you would not doubt in what it needeth number, sith you have learned that Nouns of all sorts, Pronouns, Verbs, and Participles are distinct diversly by numbers: besides the variety of Nouns of Numbers, and Adverbs. And if you take away number from Grammer, then is all the quantity of Syllables lost.   And many other ways doth number help Grammer.   Whereby were all kindes of Meeters found and made? was it not by number?

But how needfull Arithmetick is to all parts of Philosophy, they may soon see, that do read either Aristotle, Plato, or any other Philosophers writings.   For all their examples almost, and their probations, depend of Arithmetick.   It is the saying of Aristotle, that hee that is ignorant of Arithmetick, is meet for no Science.

And **Plato** his Master wrote a little sentence over his Schoolhouse door, Let none enter in hither (quoth he) that is ignorant of Geometry. Seeing hee would have all his Scholars expert in Geometry, much rather he would the same in Arithmetick, without which Geometry cannot stand.

And how needfull Arithmetick is to Divinity, it appeareth, seeing so many Doctors gather so great mysteries out of number, and so much do write of it. And if I should go about to write all the commodities of Arithmetick in civill acts, as in governance of Common-weales in time of peace, and in due provision & order of Armies, in time of war, for numbering of the Host, summing of their wages, provision of victuals, viewing of Artillery, with other Armour; beside the cunningest point of all, for casting of ground, for encamping of men, with such other like: And how many wayes also Arithmetick is conducible for all private Weales, of Lords and all Possessioners, of Merchants, and all other occupiers, and generally for all estates of men, besides Auditors, Treasurers, Receivers, Stewards, Bailiffes, and such like, whose Offices without Arithmetick are nothing: If I should (I say) particularly repeat all such commodities of the noble Science of Arithmetick, it were enough to make a very great book.

Scholar. No, no sir, you shall not need: For I doubt not, but this, that you have said, were enough to perswade any man to think this Art to be right excellent and good, and so necessary for man, that (as I think now) so much as a man lacketh of it, so much hee lacketh of his sense and wit.

Master. What, are you so farre changed since, by hearing these few commodities in generall: by likelihood you would be farre changed if you knew all the particular Commodities.

Scholar. I beseech you Sir, reserve those Commodities that rest yet behinde unto their place more convenient: and if yee will bee so good as to utter at this time this excellent treasure, so that I may be somewhat inriched thereby, if ever I shall be able, I will requite your pain.

Master. I am very glad of your request, and will do it speedily, sith that to learn it you bee so ready.

Scholar. And I to your authority my wit do subdue; whatsoever you say, I take it for true.

Master. That is too much; and meet for no man to bee beleeved in all things, without shewing of reason. Though I might of my Scholar some credence require, yet except I shew reason, I do it not desire. But now sith you are so earnestly set this Art to attaine, best it is to omit no time, lest some other passion coole this great heat, and then you leave off before you see the end.

Scholar. Though many there bee so unconstant of mind, that flitter and turn with every winde, which often begin, and never come to the end, I am none of this sort, as I trust you partly know. For by my good will what I once begin, till I have it fully ended, I would never blin.

Master. So have I found you hitherto indeed, and I trust you will increase rather then go back. For, better it were never to assay, then to shrink and flie in the mid way: But I trust you will not do so; therefore tell mee briefly: What call you the Science that you desire so greatly.

Scholar. Why sir, you know.

Master. That maketh no matter, I would hear whether you know, and therefore I ask you. For great rebuke it were to have studied a Science, and yet cannot tell how it is named.

Scholar. Some call it Arsemetrick, and some Augrime.

Master. And what do these names betoken?

Scholar. That, if it please you, of you would I learn.

Master. Both names are corruptly written: Arsemetrick for Arithmetick, as the Greeks call it, and Augrime for Algorisme, as the Arabians found it: which both betoken the Science of Numbring: for Arithmos in Greek is called Number: and of it commeth Arithmetick, the Art of Numbring. So that Arithmetick is a Science or Art teaching the manner and use of Numbring: This Art may be wrought diversly, with Pen or with Counters. But I will first shew you the working with the Pen, and then the other in order.

Scholar. This I will remember. But how many things are to bee learned to attain this Art fully?

Master. There are reckoned commonly seven parts or works of it.

Numeration, Addition, Subtraction, Multiplication, Division, Progression, and Extraction of roots: to these some men adde Duplication, Triplation, and Mediation. But as for these three last they are contained under the other seven. For Duplication, and Triplation are contained under Multiplication; as it shall appear in their place: And Mediation is contained under Division, as I will declare in his place also.

Scholar. Yet then there remain the first seven kinds of Numbring.

Master.   So there doth: Howbeit if I shall speak exactly of the parts of Numbring, I must make but five of them: for Progression is a compound operation of Addition, Multiplication and Division. And so is the Extractions of roots.   But it is no harme to name them as kindes severall, seeing they appear to have some severall working.   For it forceth not so much to contend for the number of them, as for the due knowledge and practising of them.

Scholar.   Then you will that I shall name them as seven kindes distinct.   But now I desire you to instruct mee in the use of each of them.

Master.   So I will, but it must be done in order: for you may not learn the last so soon as the first, but you must learn them in that order, as I did rehearse them, if you will learn them speedily, and well.

Scholar.   Even as you please.   Then to begin; Numeration is the first in order: what shall I do with it?

Master.   First, you must know what the thing is, and then after learn the use of the same.

STEVIN

On Decimal Fractions

(Translated from the French by Professor Vera Sanford, Western
Reserve University, Cleveland, Ohio.)

The invention of the decimal fraction cannot be assigned to any single
individual. Pellos (1492) used a decimal point to set off one, two, or three
places in the dividend when the divisor was a multiple of 10, 100, or 1000.
Adam Reise (1522) printed a table of square roots in which values to three
places were computed for the irrationals. Most important of all, Rudolff
(1530) used the symbol as a decimal point in a compound interest table.[1]

The first person to discuss the theory of decimal fractions and their arithmetic was Simon Stevin (c.1548–c.1620), a native of Bruges and a firm supporter of
William the Silent in the struggle of the Low Countries against Spain. Stevin
was tutor to Maurice of Nassau, served as quartermaster general in the
Dutch army, and acted as commissioner of certain public works, especially
of the dikes. He is reported to have been the first to adapt the principles of
commercial bookkeeping to national accounts, and his studies in hydraulics
resulted in theorems which foreshadowed the integral calculus.

Stevin's work on decimal fractions was published in 1585, two editions
appearing in that year—one in Flemish with the title *La Thiende*, the other in
French with the title *La Disme*.

The translation that follows was made from *Les Oeuvres Mathematiques de
Simon Stevin*, edited by Girard and published in Leyden in 1634.[2]

La Disme

Teaching how all Computations that are met in Business may be performed
by Integers alone without the aid of Fractions
Written first in Flemish and now done into French
by
Simon Stevin of Bruges

To Astrologers, Surveyors, Measurers of Tapestry, Gaugers, Stereometers in
General, Mint-masters, and to All Merchants Simon Stevin Sends Greeting

A person who contrasts the small size of this book with your
greatness, my most honorable sirs to whom it is dedicated, will

---

[1] These instances are discussed with facsimiles of the cases in point in "The Invention
of the Decimal Fraction," by David Eugene Smith, *Teachers College Bulletin*, First Series.
No. 5.

[2] A facsimile of the original edition with an introduction by the late Father Bosmans was
printed by the Société des Bibliophiles Anversois in Antwerp, in 1924, with the title *La
"Thiende" de Simon Stevin*.

think my idea absurd, especially if he imagines that the size of this volume bears the same ratio to human ignorance that its usefulness has to men of your outstanding ability; but, in so doing, he will have compared the extreme terms of the proportion which may not be done. Let him rather compare the third term with the fourth.

What is it that is here propounded? Some wonderful invention? Hardly that, but a thing so simple that it scarce deserves the name invention; for it is as if some stupid country lout chanced upon great treasure without using any skill in the finding. If anyone thinks that, in expounding the usefulness of decimal numbers, I am boasting of my cleverness in devising them, he shows without doubt that he has neither the judgment nor the intelligence to distinguish simple things from difficult, or else that he is jealous of a thing that is for the common good. However this may be, I shall not fail to mention the usefulness of these numbers even in the face of this man's empty calumny. But, just as the mariner who has found by chance an unknown isle, may declare all its riches to the king, as, for instance, its having beautiful fruits, pleasant plains, precious minerals, etc., without its being mputed to him as conceit; so may I speak freely of the great usefulness of this invention, a usefulness greater than I think any of you anticipates, without constantly priding myself on my achievements.

As your daily experience, Messieurs, makes you sufficiently aware of the usefulness of number, which is the subject of *La Disme*, it will not be necessary to say many words with reference to this. The astrologer[1] knows that, by computation, using tables of declinations, the pilot may describe the true latitude and longitude of a place and that by such means every point upon the earth's surface may be located. But as the sweet is never without the bitter, the labor of such computations cannot be disguised, for they involve tedious multiplications and divisions of sexagesimal fractions,[2] degrees, minutes, seconds, thirds, etc. The surveyor

---

[1] [This is used for "astrologer" and for "astronomer" as well.]

[2] [Fractions whose denominators were the powers of sixty. They were not restricted to the measurement of time or angles but were used by scientists and mathematicians in all sorts of computations. They afforded a convenient way of expressing the approximate root of an equation,—in one case, for instance a root is given to the tenth sexagesimal,—but although their use facilitated the comparing of one number with another and although they were well suited to addition and subtraction, multiplication, division, and square root were difficult and were frequently performed by tables.]

knows the great benefit which the world receives from his science by which it avoids many disputes concerning the unknown areas of land.   And he who deals in large matters, cannot be ignorant of the tiresome multiplications of rods, feet, and inches[1] the one by the other, which often give rise to error tending to the injury of one of the parties, and to the ruin of the reputation of the sur- veyor.   So too, with mint-masters, merchants, etc., each in his own business.   The more important these calculations are, and the more laborious their execution, so much the greater is this dis- covery of decimal numbers which does away with all these diffi- culties.   To speak briefly, *La Disme* teaches how all computations of the type of the four principles of arithmetic—addition, sub- traction, multiplication and division—may be performed by whole numbers with as much ease as in counter-reckoning.[2]

If by these means, time may be saved which would otherwise be lost, if work may be avoided, as well as disputes, mistakes, lawsuits, and other mischances commonly joined thereto, I willingly submit *La Disme* to your consideration.   Someone may raise the point that many inventions which seem good at first sight are of no effect when one wishes to use them, and as often happens, new methods good in a few minor cases are worthless in more important ones.   No such doubt exists in this instance, for we have shown this method to expert surveyors in Holland and they have abandoned the devices which they have invented to lighten the work of their computations and now use this one to their great satisfaction.   The same satisfaction will come to each of you, my most honorable sirs, who will do as they have done.

ARGUMENT

La Disme consists of two parts,—definitions and operations. In the first part, the first definition explains what decimal numbers[3] are, the second, third, and fourth explain the meaning of the terms unit,[4] prime, second, etc., and the other decimal numbers.

---

[1] [Here Stevin uses the units *verge, pied, doigt.*]

[2] [That is, reckoning with jetons or counters, a method of reckoning that was still in vogue in Stevin's time.]

[3] [In this translation, the words "decimal numbers" are used where the literal translation would be "the numbers of *La Disme.*"]

[4] [In the Flemish version, Stevin uses the word *Begbin* and in the French one, *Commencement.*]

In the operations, four propositions show the addition, subtraction, multiplication, and division of decimal numbers. The order of these topics may be succinctly represented in a table.

La Disme has two divisions......
- Definitions.......
  - Decimals
  - Unit
  - Prime, Second, etc.
  - Decimal Numbers
- Operations.......
  - Addition
  - Subtraction
  - Multiplication
  - Division

At the end of this discussion, there will be added an appendix setting forth the use of decimal numbers in real problems.

THE FIRST DIVISION OF LA DISME
OF DEFINITIONS
Definition I

Decimal numbers[1] are a kind of arithmetic based on the idea of the progression by tens, making use of the ordinary Arabic numerals, in which any number may be written and by which all computations that are met in business may be performed by integers alone without the aid of fractions.

Explanation

Let the number one thousand one hundred eleven be written in Arabic numerals 1111, in which form it appears that each 1 is the tenth part of the next higher figure. Similarly, in the number 2378, each unit of the 8 is the tenth part of each unit of the 7, and so for all the others. But since it is convenient that the things which we study have names, and since this type of computation is based solely upon the idea of the progression by tens[2] as will be seen in our later discussion, we may properly speak of this treatise as *La Disme* and we shall see that by it we may perform all the computations we meet in business without the aid of fractions.

---

[1] [Disme est une espece d'arithmetique.]

[2] [Disme: "tithe," later the word was contracted into dîme. Earlier forms in English use are *dyme* and *dessime*. Disme came into the language when Stevin's work was translated in 1608. It was used as a noun meaning a tenth and as a synonym for decimal arithmetic; also as a verb, to divide into tenths.]

Definition II

Any given number is called the *unit* and has the sign ⓪.

Explanation

In the number three hundred sixty four, for example, we call the three hundred sixty four units and write the number 364⓪. Similarly for other cases.

Definition III

The tenth part of a unit is called a *Prime*, and has the sign ①, and the tenth of a prime is called a *Second*, and has the sign ②. Similarly for each tenth part of the unit of the next higher figure.

Explanation

Thus 3①7②5③9④ is 3 primes, 7 seconds, 3 thirds, 9 fourths, and we might continue this indefinitely.   It is evident from the definition that the latter numbers are $\frac{3}{10}$, $\frac{7}{100}$, $\frac{5}{1000}$, $\frac{9}{10,000}$, and that this number is $\frac{3759}{10,000}$. Likewise 8⓪9①3②7③ has the value $8\frac{9}{10}$, $\frac{3}{100}$, $\frac{7}{1000}$, or $8\frac{937}{1000}$.   And so for other numbers. We must also realize that in these numbers we use no fractions and that the number under each sign except the "unit" never exceeds the 9.   For instance, we do not write 7①12② but 8①2② instead, for it has the same value.

Definition IV

The numbers of the 2nd and 3d definitions are called *Decimal Numbers*.

The End of the Definitions[1]

---

[1] [These same names and symbols are given a more general application in Stevin's other works. The following discussion is from his work on the subject in *l'Arithmétique* where geometric progressions, or geometric numbers play a prominent part.

He says, "When the ancients realized the value of progressions of the sort where the first term multiplied by itself gives the second term..., they saw that it would be necessary to choose meaningful names for these numbers so that they might the more readily distinguish them.   Thus they called the first term *Prime* which we denote by ①, the next *Second* which we write as ② etc. "For example,

①2②4③ 8④16...
①3②9③27④81...
. . . .

"We intend that the *Unit* of a quantity shall mean something distinct from the first quantity or prime. Any arithmetic number or radical which one uses in algebraic computation as 6 or $\sqrt{3}$ or $2 + \sqrt{3}$..., we will call

The Second Division of La Disme
Of Operations

Proposition I.—To add decimal numbers. Given three decimal numbers, 27⓪8①4②7③, 37⓪6①7②5③, 875⓪7①8②2③.
Required to find their sum.

Construction.—Arrange the numbers as in the accompanying figure, adding them in the usual manner of adding integers. This (by the first problem of *l'Arithmé-tique*[1]) gives the sum 941304,[1] which, as the signs above the numbers show[2] is 941⓪3①0②4③. And this is the sum required.

```
      ⓪①②③
    2 7 8 4 7
    3 7 6 7 5
  8 7 5 7 8 2
  _____
  9 4 1 3 0 4
```

Proof.—By the third definition of this book, the given number 27⓪8①4②7③ is $278\frac{8}{10}$, $\frac{4}{100}$, $\frac{7}{1000}$, or $27847\frac{}{1000}$. Similarly, the 37⓪6①7②5③ is $37675\frac{}{1000}$, and the 875⓪7①8②3③ is $875783\frac{}{1000}$. These three numbers $27847\frac{}{1000}$, $37675\frac{}{1000}$, $875783\frac{}{1000}$ added, according to the 10th problem of *l'Arithmétique*, make $941304\frac{}{1000}$, but 941⓪3①0②4③ has this same value, and is therefore the true sum which was to be shown.

Conclusion.—Having been given decimal numbers to add, we have found their sum which was to be done.

Note.—If, in the numbers in question, some figure of the natural order be lacking, fill its place with a zero. For example, in the numbers 8⓪5①6② and 5⓪7② where the second lacks a figure of order prime, insert 0① and take 5⓪0①7② as the given number and add as before. This note applies to the three following propositions also.

```
    ⓪①②
    8 5 6
    5 0 7
    _____
  1 3 6 3
```

the *Unit* and we will give it the symbol ⓪: but this symbol shall be used only when the arithmetic number or radical is not denominate. (quand les nombres Arithmetiques ou radicaux ne sont pas absoluement descripts)."

Stevin writes denominate numbers as 1 hour 3①5②, 5 degrees 4①18②, 2790 verges 5①9②. He later notes (*l'Arithmétique*, p. 8) that Bombelli has used this symbolism also except for the ⓪. In Bombelli's *Algebra* (1572) the symbols ⌣1⌣, ⌣2⌣, ⌣3⌣... are used for the powers of the unknown quantity just as Stevin used his ①, ②, ③,..., i. e., a specialized form of the geometric progression.

The names for these quantities except that of the unit, are easily traced to the *pars minuta prima*, etc., of the sexagesimals.]

[1] [*La Pratiqve D'Arithmetiqve De Simon Stevin De Brvges, Leyden*, 1585.]

[2] [Stevin has three ways of writing these numbers, depending upon the exigencies of the case: 27⓪8①4②7③, $\begin{matrix}⓪①②③\\27\ 8\ 4\ 7.\end{matrix}$, $\begin{matrix}27\ 8\ 4\ 7.\\⓪①②③\end{matrix}$]

Proposition II.—To subtract decimal numbers.

Given the number 237⓪5①7②8③ from which the number 59⓪7①4②9③ is to be subtracted.

Required to find the remainder.

Construction.—Place the numbers in order as in the adjoining figure, subtracting after the usual manner of subtracting integers (by the 2nd problem of *l'Arithmétique*). There remains 177829 which, as indicated by the signs above the numbers, is 177⓪8①2②9③; and this is the remainder required.

$$
\begin{array}{r}
⓪①②③ \\
2\ 3\ 7\ 5\ 7\ 8 \\
5\ 9\ 7\ 4\ 9 \\
\hline
1\ 7\ 7\ 8\ 2\ 9 \\
\end{array}
$$

Proof.—By the third definition of *la Disme*, the 237⓪ 5①7②8③ is $2375\frac{5}{10}$, $\frac{7}{100}$, $\frac{8}{1000}$ or $237^{578}\frac{}{1000}$. And, by the same reasoning, the 59⓪7①4②9③ is $5974^{9}\frac{}{1000}$; subtracting this from $237^{578}\frac{}{1000}$, according to the tenth problem of *l'Arithmétique*, leaves $1778^{29}\frac{}{1000}$. But the aforesaid 177⓪8①2②9③ has this same value and is, therefore, the true remainder, which was to be proved.

Conclusion.—Having been given a decimal number and a similar number which is to be subtracted from it, we have found the remainder which was to be done.

Proposition III.—To multiply decimal numbers.

Given the number 32⓪5①7② and the multiplier 89⓪4①6②. Required to find their product.

Construction.—Place the numbers in order and multiply in the ordinary way of multiplying whole numbers (by the third problem of *l'Arithmétique*). This gives the product 29137122. To find what this is, add the last two signs of the given numbers, the one ② and the other ② also, which together are ④. We say, then, that the sign of the last figure of the product will be ④. Once this is established, all the signs are known on account of their continuous order. Therefore, 2913⓪7①1②2③2④ is the required product.

$$
\begin{array}{r}
⓪①② \\
3\ 2\ 5\ 7 \\
8\ 9\ 4\ 6 \\
\hline
1\ 9\ 5\ 4\ 2 \\
1\ 3\ 0\ 2\ 8 \\
2\ 9\ 3\ 1\ 3 \\
2\ 6\ 0\ 5\ 6 \\
\hline
2\ 9\ 1\ 3\ 7\ 1\ 2\ 2 \\
⓪①②③④ \\
\end{array}
$$

Proof.—As appears by the third definition of *La Disme*, the given number 32⓪5①7② is $32\frac{5}{10}$, $\frac{7}{100}$, or $325^{7}\frac{}{100}$, and likewise the multiplier 89⓪4①6② is $894^{6}\frac{}{100}$. Multiplying the aforesaid $325^{7}\frac{}{100}$ by this number gives the product $29137122\frac{}{10,000}$ (by the twelfth problem of *l'Arithmétique*). But the aforesaid product 2913⓪7①1②2③2④ has this value and is, therefore, the true product which we were required to prove. We will now explain why second multiplied by second gives the product fourths, which is the sum of their signs, and why fourth by fifth gives the product ninth and why unit by third gives the product third and so forth.

Let us take for example $\frac{2}{10}$ and $\frac{3}{100}$ which, by the definitions of *La Disme*, are the values of 2① and 3②. Their product is $\frac{6}{1000}$, which by the third defintion given above is 6③. Hence, multiplying prime by second gives a product in thirds, that is, a number whose sign is the sum of the given signs.

Conclusion.—Having been given a decimal number to multiply and the multiplier we have found the product, which was to be done.

NOTE.—If the last sign of the multiplicand is not equal to the last sign of the multiplier, for example, 3④7⑤6⑥ and 5①4②, proceed as above. The placing of the figures will appear as here shown.

```
            ④⑤⑥
          3  7  8
            5  4②
         ───────────
          1 5 1 2
        1 8 9 0
         ───────────
        2 0 4 1 2
        ④⑤⑥⑦⑧
```

Proposition IV.—To divide decimal numbers.
Given 3⓪4①4②3③5④2⑤ to be divided by 9①6②.
Required to find their quotient.

Construction.—Omitting their signs, divide the given numbers in the ordinary way of dividing whole numbers by the fourth problem of *l'Arithmétique*. This gives the quotient 3587. To determine the signs, subtract the last sign of the divisor ② from the last sign of the dividend ⑤, leaving ③ as the sign of the last digit of the quotient. Once this is determined, all the other signs are known because of their continuous order. 3⓪5①8②7③ is, therefore, the quotient required.

```
   1
   18
   5164
   7687      ⓪①②③
   344352  (3 5 8 7
   96666
   999
```

Proof.—By the third definition of *La Disme* the dividend 3⓪4①4②3③5④2⑤ is 3 $\frac{4}{10}$, $\frac{4}{100}$, $\frac{3}{1000}$, $\frac{5}{10,000}$ $\frac{2}{100,000}$ or $344,35\frac{2}{100,000}$. The divisor 9①6② is $\frac{9}{10}$, $\frac{6}{100}$ or $9\frac{6}{100}$. By the thirteenth problem of *l'Arithmétique*, the quotient of these numbers is $358\frac{7}{1000}$. The aforesaid 3⓪5①8②7③ is therefore the true quotient which was to be shown.

Conclusion.—Having been given a decimal number to be divided and the divisor, we have found the quotient, which was to be done.

Note I.—If the signs of the divisor be higher than the signs of the dividend, add to the dividend as many zeros as may be necessary. For example, in dividing 7② by 4⑤, I place zeros after the 7 and divide as above getting the quotient 1750.

It sometimes happens that the quotient cannot be expressed by whole numbers as in the case of 4② divided by 3④. Here. it appears that the quotient will be infinitely many threes with always one third in addition. In such a case, we may approach as near to the real quotient as the problem requires and omit the remainder. It is true indeed that 13⓪3①3② or 13⓪3①3②3⅓③ is the exact result, but in this work we propose to use whole numbers only, and, moreover, we notice that in business one does not take account of the thousandth part of a maille[1] or of a grain. Omissions such as these are made by the principal Geometricians and Arithmeticians even in computations of great consequence. Ptolemy and Jehan de Montroyal, for instance, did not make up their tables with the utmost accuracy that could be reached with mixed numbers, for in view of the purpose of these tables, approximation is more useful than perfection.

> 3̸2̸   ⓪
> 7̸0̸0̸0̸ (175 0
> 4̸4̸4̸4̸
>
> 1̸1̸1̸   ⓪①②
> 4̸0̸0̸0000 (1 3 3 3
> 3̸3̸3̸3̸

Note II.—Decimal numbers may be used in the extraction of roots. For example, to find the square root of 5②2③9④, work according to the ordinary method of extracting square root and the root will be 2①3②. The last sign of the root is always one half the last sign of the given number. If, however, the last sign is odd, add (a zero in place of) the next sign and extract the root of the resulting number as above. By a similar method with cube root, the third of the last sign of the given number will be the sign of ther oot—similarly for all other roots.

> 1̸
> 5̸2̸9̸
> 2, 3
> ‾‾‾
> 4̸2̸9̸

The End of La Disme

Appendix

Decimal numbers have been described above. We now come to their applications and in the following six articles we shall show how all computations which arise in business may be performed by them. We shall begin with the computations of surveying as this subject was the first one mentioned in the introduction.

---

[1] [¼ ounce.]

Article One

Of the Computations of Surveying

When decimal numbers are used in surveying, the verge[1] is
called the unit, and it is divided into ten equal parts or primes,
each prime is divided into seconds and, if smaller units are required,
the seconds into thirds and so on so far as may be necessary. For
the purposes of surveying the divisions into seconds are sufficiently
small but in matters that require greater accuracy as in the measuring of lead roofs, thicknesses etc., one may need to use the thirds.
Many surveyors, however, do not use the verge, but a chain
three, four, or five verges long, and a cross-staff[2] with its shaft
marked in five or six feet divided into inches. These men may
follow the same practise here substituting five or six primes with
their seconds. They should use these markings of the cross-staff
without regard to the number of feet and inches that the verge
contains in that locality, and add, subtract, multiply, and divide
the resulting numbers as in the preceding examples. Suppose,
for instance, four areas[3] are to be added: 345⓪7①8②, and 872⓪-
5①3②, 615⓪4①8②, and 956⓪8①6②. Add these as in the
first proposition of *La Disme*. This gives the sum ⓪①②
2790 verges 5 primes, 9 seconds. This number, 3 4 5 7 2
divided by the number of verges in an arpent[4] gives 8 7 2 5 3
the number of arpents required. To find the number 6 1 5 4 8
of small divisions in 5 primes 9 seconds, look on the 9 5 6 8 6
other side of the verge to see how many feet and inches 2 7 9 0 5 0
match with them; but this is a thing which the surveyor must do but once, *i. e.*, at the end of the account which he

---

[1] [The word verge was used for a measuring rod of that length.]

[2] [The cross-staff was a piece of wood mounted at its mid-point perpendicular
to a shaft and free to move along this shaft to positions parallel to the first one.
To use this instrument, the observer adjusts it so the lines of sight from the
ends of the staff to the tip of the cross-piece coincide with the end-points of the
line to be measured, and the distances are computed from one measurement and
similar triangles. When neither point of the required line is accessible, the
distance is computed from two observations at known distances from each
other.]

[3] [Stevin's area units evidently proceed directly from his decimal scheme.
Thus the prime of the area unit is one tenth of the unit itself, not a square
whose side is the prime of the linear unit.]

[4] [The arpent was the common unit of area. It varied from about 3000 to
5100 square meters, according to the locality.]

gives to the proprietaries, and often not then, as the majority think it useless to mention the smaller units.

Secondly, to subtract 32⓪5①7② from 57⓪3①2②, work according to the second proposition of *La Disme*, and there remains 24 verges, 7 primes, 5 seconds.

⓪①②
5 7 3 2
3 2 5 7
———
2 4 7 5

Thirdly, to multiply 8⓪7①3② by 7⓪5①4② (these might be the sides of a rectangle or quadrangle), proceed as above according to the third proposition of *La Disme*, getting the product, or area, 65 verges 8 primes etc.

⓪①②
8 7 3
7 5 4
———
3 4 9 2
4 3 6 5
6 1 1 1
———
6 5 8 2 4 2
⓪①②③ ④

Fourthly, suppose that the rectangle *ABCD* has the side *AD* 26⓪3①. From what point on *AB* should a line be drawn parallel to *AD* to cut off a rectangle of area 367⓪6①? Divide 367⓪6① by 26⓪3① according to the fourth proposition of *La Disme* getting the quotient 13⓪9①7② which is the required distance *AE*. If greater accuracy is desired, this division may be carried further, although such accuracy does not seem necessary. The proofs of these problems are given above in the propositions to which we have referred.

1
2̷2̷
7̷6̷
2̷5̷0̷8
4̷6̷3̷1
1̷0̷4̷7̷3̷9   ⓪①②
3̷6̷7̷6̷0̷0̷ (13 9 7
2̷6̷3̷3̷3̷3̷
2̷6̷6̷6̷
2̷2̷

Article Two

Of the Computations of the Measuring of Tapestry

The aûne[1] is the unit of the measurer of tapestry. The blank side of the aûne should be divided into ten equal parts each of which is 1 prime, just as is done in the case of the verge of the surveyor. Then each prime is divided into ten equal parts, each 1 second etc. It is not necessary to discuss the use of this measure as examples of it would be similar in every respect to those given in the article on surveying.

---

[1] [About 46 inches.]

Article Three
Of Computations Used in Gauging and in the Measuring of all Casks
. . . .

Article Four
Of Computations of Volume Measurement in General

It is true, indeed, that gauging is stereometry that is the science of measuring volumes, but all stereometry is not gauging. We therefore distinguish them in this treatise. The stereometer who uses the method of *La Disme*, should mark the customary measure of his town, whether this be the verge or the aûne, with the decimal divisions described in the first and second articles above.

Let us suppose that he is to find the volume of a rectangular column whose length is 3①2②, breadth 2①4②, 2⓪3①5②. He should multiply the length by the breadth (according to the 4th proposition of *La Disme*) and this product by the height, getting as his result 1①8②4④8⑤.

```
        3 2
        2 4
      ─────
      1 2 8
        6 4
      ─────
        7 6 8
        2 3 5
    ─────────
      3 8 4 0
    2 3 0 4
  1 5 3 6
  ───────────
  1 8 0 4 8 0
  ①②③④⑤⑥
```

NOTE.—Someone who is ignorant of the fundamentals of stereometry—for it is such a man that we are addressing now—may wonder why we say that the volume of the above column is but 1① etc., for it contains more than 180 cubes of side 1 prime. He should realize that a cubic verge is not 10 but 1000 cubes of side 1 prime. Similarly, 1 prime of the volume unit is 100 cubes each of side 1 prime. The like is well known to surveyors for when one says 2 verges 3 feet of earth, he does not mean 2 verges 3 feet square, but 2 verges and (counting 12 feet to the verge) 36 feet square.

If however the question had been how many cubes of side 1 prime are in the above column, the result would have been changed to conform to this requirement, bearing in mind that each prime of volume units is 100 cubes of side 1 prime, and each second is 10 such cubes. If the tenth part of the verge is the greatest measure that the stereometer intends using, he should call it the unit and proceed as above.

Article Five

Of Astronomical Computations

The ancient astronomers who divided the circle into 360 degrees, saw that computations with these and their fractional parts were too laborious. They divided the degree, therefore, into sub-multiples and these again into the same number of equal parts, and in order to always work with integers, they chose for this division the sexagesimal progression for sixty is a number commensurate with many whole numbers, to wit with 1, 2, 3, 4, 5, 6, 10, 12, 15, 20, 30. If we may trust to experience, however, and we say this with all reverence for the past, the decimal and not the sexagesimal is the most convenient of all the progressions that exist potentially in nature. Thus, we would call the 360 degrees the unit and we would divide the degree into ten equal parts, or primes, and the prime in turn into ten parts and so forth, as has been done several times above. Having once agreed upon this division, we might describe the easy methods of adding, subtracting, multiplying, and dividing these numbers, but as this does not differ from the preceding propositions such a recital would be but a waste of time. We therefore let those examples illustrate this article. Moreover we would use this division of the degree in all astronomical tables and we hope to publish one such[1] in our own Flemish language which is the richest, the most ornate, and the most perfect of all languages. Of its exquisite uniqueness, we contemplate a fuller proof than the brief one which Pierre and Jehan made in the *Bewysconst* or *Dialectique*[2] which was recently published.

---

[1] [Stevin did not make this promise good, however, for in his work on astronomy, the *Wiscontige Gedachtenissen* (1608), he keeps to the old partition of the degree. Father Bosmans is of the opinion that this was due to the tremendous labor involved, but he also points out that the errors incident to converting readings to the decimal system for computation and then shifting back again would be greater than those involved in the mere computation with sexagesimals.]

[2] [*Dialectike Ofte Bewysconst. Leerende van allen saeken recht ende constelick Oirdeelen; Oock openende den wech tot de alderiepste verborgentheden der Nature-ren. Bescbreven int Neerdytach door Simon Stevin van Brugghe. Tot Leyden by Christoffel Plantijn* M.D.LXXXV.

In this book, which was an imitation of Cicero's *Tusculan Disputations*, Pierre and Jehan discuss the beauties of the Flemish tongue. (First edition, pp. 141–166.)

The more complete proof comes as a digression at the end of the preface of Stevin's *Begbinselen der Weegbconst*, 1596.]

Article Six

Of the Computations of Mint-masters, Merchants, and in General of All States

To summarize this article, we might say that all measures—linear, liquid, dry, and monetary—may be divided decimally, and that each large unit may be called the unit. Thus the marc is the unit of weight for gold and silver; the livre, for other common weights; and the livre gros in Flanders, the livre esterlain in England, and the ducat in Spain, are the units of money in those countries. In the case of the money, the highest symbol (and the lowest denomination) of the marc would be the fourth, for the prime weighs half the Es of Antwerp. The third suffices for the highest symbol of the livre gros, for the third is less than the fourth of a penny.

Instead of the demi-livre, once, demi-once, esterlain, grain etc. the subdivisions of the weights should be the 5, 3, 2, 1 of each sign, that is to say, after a livre would follow a weight of 5 primes (or ½ lb.) then 3 primes, then 2, then 1. Similar parts of other weights would have the 5 and the other multiples of the division following.

We think it essential that each subdivision should be named prime, second, third etc., whatever sort of measure it may be, for it is evident to us that second multiplied by third gives the product fifth (2 & 3 make 5 as was said above), and third divided by second gives the quotient prime, facts which cannot be shown so neatly by other names. But when one wishes to name them so as to distinguish the systems of measure, as we say demi-aûne, demi-livre, demi-pinte, etc., we may call them prime of marc, second of marc, second of livre, second of aûne, etc.

As examples of this, let us suppose that 1 marc of gold is worth 36 lb. 5①3②, how much is 8 marcs 3③5②4③ worth? Multiply 3653 by 8354 getting the product 305 lb. 1①7②1③ which is the required solution.[1] As for the 6④2⑤, these are of no account here.

Again, take the case of 2 aûnes 3① (of cloth) which cost 3 lb. 2①5②, what would be the cost of 7 aûnes 5①3②? To find this, multiply the last of the given numbers by the second and divide this product by the third according to the usual custom, that is to

---

[1] [Stevin is not consistent in his approximation of results. When he computes his interest tables, he says he considers $\frac{10}{101}$ an extra unit 'for it is more than one half.']

say 753 by 325 which gives 244725.   This divided by 23 gives the quotient and solution 10 lb. 6①4②.

We might give examples of all the common rules of arithmetic that pertain to business, as the rules of partnership, interest, exchange etc., and show how they may be carried out by integers alone and also how they may be performed by easy operations with counters, but as these may be deduced from the preceding, we shall not elaborate them here.   We might also show by comparison with vexing problems with fractions the great difference in ease between working with ordinary numbers and with decimal numbers, but we omit this in the interest of brevity.

Finally, we must speak of one difference between the sixth article and the five preceding articles, namely that any individual may make the divisions set forth in the five articles, but this is not the case in the last article where the results must be accepted by every one as being good and lawful.   In view of the great usefulness of the decimal division, it would be a praiseworthy thing if the people would urge having this put into effect so that in addition to the common divisions of measures, weights, and money that now exist, the state would declare the decimal division of the large units legitimate to the end that he who wished might use them. It would further this cause also, if all new money should be based on this system of primes, seconds, thirds, etc.   If this is not put into operation so soon as we might wish, we have the consolation that it will be of use to posterity, for it is certain that if men of the future are like men of the past, they will not always be neglectful of a thing of such great value.

Secondly, it is not the most discouraging thing to know that men may free themselves from great labor at any hour they wish.

Lastly, though the sixth article may not go into effect for some time, individuals may always use the five preceding articles indeed it is evident that some are already in operation.

The End of the Appendix

DEDEKIND

On Irrational Numbers

(Translated from the German by the Late Professor Wooster Woodruff
Beman, University of Michigan, Ann Arbor, Michigan. Selection
made and edited by Professor Vera Sanford, Western
Reserve University, Cleveland, Ohio)

Julius Wilhelm Richard Dedekind (1831-1916) studied at Göttingen and
later taught in Zürich and Braunschweig. His essay *Stetigkeit und irra-
tionale Zablen* published in 1872 was the outcome of researches begun in
Zürich in 1858 when Dedekind, teaching differential calculus for the first
time, became increasingly conscious of the need for a scientific discussion of the
concept of continuity.

This work is included in *Essays on the Theory of Numbers*, by Richard
Dedekind, translated by the late Professor Wooster Woodruff Beman (Open
Court Publishing Company, Chicago, 1901). The extract here given is on
pages 6 to 24 of this translation and is reproduced by the consent of the
publishers.

The author begins with a statement of three properties of rational numbers
and of the three corresponding properties of the points on a straight line.
These properties are as follows:

For Numbers

I. If $a > b$, and $b > c$, then $a > c$.

II. If a, c are two different numbers, there are infinitely many different
numbers lying between a, c.

III. If a is any definite number, then all numbers of the system R fall into
two classes, A_1 and A_2, each of which contains infinitely many individuals;
the first class A_1 comprises all numbers a_1 that are $< a$, the second class A_2
comprises all numbers a_2 that are $> a$; the number a itself may be assigned
at pleasure to the first or second class, being respectively the greatest number
of the first class or the least of the second.

For the Points on a Line

I. If p lies to the right of q, and q to the right of r, then p lies to the right of r;
and we say that q lies between the points p and r.

II. If p, r are two different points, then there always exist infinitely many
points that lie between p and r.

III. If p is a definite point in L, then all points in L fall into two classes, P_1
and P_2 each of which contains infinitely many individuals; the first class P_1
contains all the points p_1 that lie to the left of p, and the second class P_2
contains all the points p_2 that lie to the right of p; the point p itself may be
assigned at pleasure to the first or second class.   In every case, the separation

of the straight line L into the two classes or portions P_1, P_2 is of such a character that every point of the first class P_1 lies to the left of every point of the second class P_2.

III.

CONTINUITY OF THE STRAIGHT LINE.

Of the greatest importance, however, is the fact that in the straight line L there are infinitely many points which correspond to no rational number. If the point p corresponds to the rational number a, then, as is well known, the length op is commensurable with the invariable unit of measure used in the construction, i. e., there exists a third length, a so-called common measure, of which these two lengths are integral multiples. But the ancient Greeks already knew and had demonstrated that there are lengths incommensurable with a given unit of length, e. g., the diagonal of the square whose side is the unit of length. If we lay off such a length from the point o upon the line we obtain an end-point which corresponds to no rational number. Since further it can be easily shown that there are infinitely many lengths which are incommensurable with the unit of length, we may affirm: The straight line L is infinitely richer in point-individuals than the domain R of rational numbers in number individuals.

If now, as is our desire, we try to follow up arithmetically all phenomena in the straight line, the domain of rational numbers is insufficient and it becomes absolutely necessary that the instrument R constructed by the creation of the rational numbers be essentially improved by the creation of new numbers such that the domain of numbers shall gain the same completeness, or as we may say at once, the same *continuity*, as the straight line.

The previous considerations are so familiar and well known to all that many will regard their repetition quite superfluous. Still I regarded this recapitulation as necessary to prepare properly for the main question. For, the way in which the irrational numbers are usually introduced is based directly upon the conception of extensive magnitudes—which itself is nowhere carefully defined—and explains number as the result of measuring such a magnitude by another of the same kind.[1] Instead of this I demand that arithmetic shall be developed out of itself.

---

[1] The apparent advantage of the generality of this definition of number disappears as soon as we consider complex numbers. According to my view, on the other hand, the notion of the ratio between two numbers of the same kind can be clearly developed only after the introduction of irrational numbers.

That such comparisons with non-arithmetic notions have furnished the immediate occasion for the extension of the number-concept may, in a general way, be granted (though this was certainly not the case in the introduction of complex numbers); but this surely is no sufficient ground for introducing these foreign notions into arithmetic, the science of numbers. Just as negative and fractional rational numbers are formed by a new creation, and as the laws of operating with these numbers must and can be reduced to the laws of operating with positive integers, so we must endeavor completely to define irrational numbers by means of the rational numbers alone. The question only remains how to do this.

The above comparison of the domain R of rational numbers with a straight line has led to the recognition of the existence of gaps, of a certain incompleteness or discontinuity of the former, while we ascribe to the straight line completeness, absence of gaps, or continuity. In what then does this continuity consist? Everything must depend on the answer to this question, and only through it shall we obtain a scientific basis for the investigation of *all* continuous domains. By vague remarks upon the unbroken connection in the smallest parts obviously nothing is gained; the problem is to indicate a precise characteristic of continuity that can serve as the basis for valid deductions. For a long time I pondered over this in vain, but finally I found what I was seeking. This discovery will, perhaps, be differently estimated by different people; the majority may find its substance very commonplace. It consists of the following. In the preceding section attention was called to the fact that every point p of the straight line produces a separation of the same into two portions such that every point of one portion lies to the left of every point of the other. I find the essence of continuity in the converse, i. e., in the following principle:

"If all points of the straight line fall into two classes such that every point of the first class lies to the left of every point of the second class, then there exists one and only one point which produces this division of all points into two classes, this severing of the straight line into two portions."

As already said I think I shall not err in assuming that every one will at once grant the truth of this statement; the majority of my readers will be very much disappointed in learning that by this commonplace remark the secret of continuity is to be revealed. To this I may say that I am glad if every one finds the above principle so obvious and so in harmony with his own ideas of a line; for I am utterly unable to adduce any proof of its correctness, nor has any one the power. The assumption of this property of the line is nothing else than an axiom by which we attribute to the line its continuity, by which we find continuity in the line. If space has at all a real existence it is *not* necessary for it to be continuous; many of its properties would remain the same even were it discontinuous. And if we knew for certain that space was discontinuous there would be nothing to prevent us, in case we so desired, from filling up its gaps, in thought, and thus making it continuous; this filling up would consist in a creation of new point-individuals and would have to be effected in accordance with the above principle.

IV.

CREATION OF IRRATIONAL NUMBERS.

From the last remarks it is sufficiently obvious how the discontinuous domain R of rational numbers may be rendered complete so as to form a continuous domain. In Section I it was pointed out that every rational number a effects a separation of the system R into two classes such that every number a_1 of the first class A_1 is less than every number a_2 of the second class A_2; the number a is either the greatest number of the class A_1 or the least number of the class A_2. If now any separation of the system R into two classes A_1, A_2, is given which possesses only *this* characteristic property that every number a_1 in A_1 is less than every number a_2 in A_2, then for brevity we shall call such a separation a *cut* [Schnitt] and designate it by (A_1, A_2). We can then say that every rational number a produces one cut or, strictly speaking, two cuts, which, however, we shall not look upon as essentially different; this cut possesses, *besides*, the property that either among the numbers of the first class there exists a greatest or among the numbers of the second class a least number. And conversely, if a cut possesses this property, then it is produced by this greatest or least rational number.

But it is easy to show that there exist infinitely many cuts not produced by rational numbers. The following example suggests itself most readily.

Let D be a positive integer but not the square of an integer, then there exists a positive integer λ such that

$$\lambda^2 < D < (\lambda + 1)^2.$$

If we assign to the second class A_2, every positive rational number a_2 whose square is $> D$, to the first class A_1 all other rational numbers a_1, this separation forms a cut (A_1, A_2), i. e., every number a_1 is less than every number a_2. For if $a_1 = 0$, or is negative, then on that ground a_1 is less than any number a_2, because, by definition, this last is positive; if a_1 is positive, then is its square $\leq D$, and hence a_1 is less than any positive number a_2 whose square is $> D$.

But this cut is produced by no rational number. To demonstrate this it must be shown first of all that there exists no rational number whose square $= D$. Although this is known from the first elements of the theory of numbers, still the following indirect proof may find place here. If there exist a rational number whose square $= D$, then there exist two positive integers t, u, that satisfy the equation

$$t^2 - Du^2 = 0,$$

and we may assume that u is the *least* positive integer possessing the property that its square, by multiplication by D, may be converted into the square of an integer t. Since evidently

$$\lambda u < t < (\lambda+1)u,$$

the number $u' = t - \lambda u$ is a positive integer certainly *less* than u. If further we put

$$t' = Du - \lambda t,$$

t' is likewise a positive integer, and we have

$$t'^2 - Du'^2 = (\lambda^2 - D)(t^2 - Du^2) = 0,$$

which is contrary to the assumption respecting u.

Hence the square of every rational number x is either $< D$ or $> D$. From this it easily follows that there is neither in the class A_1 a greatest, nor in the class A_2 a least number. For if we put

$$y = \frac{x(x^2 + 3D)}{3x^2 + D},$$

we have

$$y - x = \frac{2x(D - x^2)}{3x^2 + D}$$

and

$$y^2 - D = \frac{(x^2 - D)^3}{(3x^2 + D)^2}.$$

If in this we assume x to be a positive number from the class A_1, then $x^2 < D$, and hence $y > x$ and $y^2 < D$. Therefore y likewise belongs to the class A_1. But if we assume x to be a number from the class A_2, then $x^2 > D$, and hence $y < x$, $y > 0$, and $y^2 > D$. Therefore y likewise belongs to the class A_2. This cut is therefore produced by no rational number.

In this property that not all cuts are produced by rational numbers consists the incompleteness or discontinuity of the domain R of all rational numbers.

Whenever, then, we have to do with a cut (A_1, A_2) produced by no rational number, we create a new, an *irrational* number a, which we regard as completely defined by this cut (A_1, A_2); we shall say that the number a corresponds to this cut, or that it produces this cut. From now on, therefore, to every definite cut there corresponds a definite rational or irrational number, and we regard two numbers as *different* or *unequal* always and only when they correspond to essentially different cuts.

In order to obtain a basis for the orderly arrangement of all *real*, i. e., of all rational and irrational numbers we must investigate the relation between any two cuts (A_1, A_2) and (B_1, B_2) produced by any two numbers a and β. Obviously a cut (A_1, A_2) i ˙ ˜ ̃ completely when one of the two classes, e. g., the first A_1 is because the second A_2 consists of all rational numbers not cou. tained in A_1, and the characteristic property of such a first class lies in this that if the number a_1 is contained in it, it also contains all numbers less than a_1. If now we compare two such first classes A_1, B_1 with each other, it may happen

1. That they are perfectly identical, i. e., that every number contained in A_1 is also contained in B_1, and that every number contained in B_1 is also contained in A_1. In this case A_2 is necessarily identical with B_2, and the two cuts are perfectly identical, which we denote in symbols by $a = \beta$ or $\beta = a$.

But if the two classes A_1, B_1 are not identical, then there exists in the one, e. g., in A_1, a number $a'_1 = b'_2$ not contained in the other B_1 and consequently found in B_2; hence all numbers b_1 contained in B_1 are certainly less than this number $a'_1 = b'_2$ and therefore all numbers b_1 are contained in A_1.

2. If now this number a'_1 is the only one in A_1 that is not contained in B_1, then is every other number a_1 contained in A_1 also contained in B_1 and is consequently $< a'_1$, i. e., a'_1 is the greatest among all the numbers a_1, hence the cut (A_1, A_2) is produced by the rational number $a = a'_1 = b'_2$. Concerning the other cut (B_1, B_2) we know already that all numbers b_1 in B_1 are also contained in A_1 and are less than the number $a'_1 = b'_2$ which is

contained in B_2; every other number b_2 contained in B_2 must, however, be greater than b'_2, for otherwise it would be less than a'_1, therefore contained in A_1 and hence in B_1; hence b'_2 is the least among all numbers contained in B_2, and consequently the cut (B_1, B_2) is produced by the same rational number $\beta = b'_2 = a'_1 = \alpha$. The two cuts are then only unessentially different.

3. If, however, there exist in A_1 at least two different numbers $a'_1 = b'_2$ and $a''_1 = b''_2$, which are not contained in B_1, then there exist infinitely many of them, because all the infinitely many numbers lying between a'_1 and a''_1 are obviously contained in A_1 (Section I, II) but not in B_1. In this case we say that the numbers α and β corresponding to these two essentially different cuts (A_1, A_2) and (B_1, B_2) are *different*, and further that α is *greater* than β, that β is *less* than α, which we express in symbols by $\alpha > \beta$ as well as $\beta < \alpha$. It is to be noticed that this definition coincides completely with the one given earlier, when α, β are rational.

The remaining possible cases are these:

4. If there exists in B_1 one and only one number $b'_1 = a'_2$, that is not contained in A_1 then the two cuts (A_1, A_2) and (B_1, B_2) are only unessentially different and they are produced by one and the same rational number $\alpha = a'_2 = b'_1 = \beta$.

5. But if there are in B_1 at least two numbers which are not contained in A_1, then $\beta > \alpha$, $\alpha < \beta$.

As this exhausts the possible cases, it follows that of two different numbers one is necessarily the greater, the other the less, which gives two possibilities. A third case is impossible. This was indeed involved in the use of the *comparative* (greater, less) to designate the relation between α, β; but this use has only now been justified. In just such investigations one needs to exercise the greatest care so that even with the best intention to be honest he shall not, through a hasty choice of expressions borrowed from other notions already developed, allow himself to be led into the use of inadmissible transfers from one domain to the other.

If now we consider again somewhat carefully the case $\alpha > \beta$ it is obvious that the less number β, if rational, certainly belongs to the class A_1; for since there is in A_1 a number $a'_1 = b'_2$ which belongs to the class B_2, it follows that the number β, whether the greatest number in B_1 or the least in B_2 is certainly $\leq a'_1$ and hence contained in A_1. Likewise it is obvious from $\alpha > \beta$ that the greater number α, if rational, certainly belongs to the class B_2, because $\alpha \geq a'_1$. Combining these two considerations we get the following result: If a cut is produced by the number a then any rational number belongs to the class A_1 or to the class A_2 according as it is less or greater than a; if the number a is itself rational it may belong to either class.

From this we obtain finally the following: If $a > \beta$, i. e., if there are infinitely many numbers in A_1 not contained in B_1 then there are infinitely many such numbers that at the same time are different from a and from β; every such rational number c is $< a$, because it is contained in A_1 and at the same time it is $> \beta$ because contained in β_2.

V.

CONTINUITY OF THE DOMAIN OF REAL NUMBERS.

In consequence of the distinctions just established the system $\mathfrak{R}$ of all real numbers forms a well-arranged domain of one dimension; this is to mean merely that the following laws prevail:

I. If $a > \beta$, and $\beta > \gamma$, then is also $a > \gamma$. We shall say that the number β lies between a and γ.

II. If a, γ are any two different numbers, then there exist infinitely many different numbers β lying between a, γ.

III. If a is any definite number then all numbers of the system $\mathfrak{R}$ fall into two classes $\mathfrak{U}_1$ and $\mathfrak{U}_2$ each of which contains infinitely many individuals; the first class $\mathfrak{U}_1$ comprises all the numbers a_1 that are less than a, the second $\mathfrak{U}_2$ comprises all the numbers a_2 that are greater than a; the number a itself may be assigned at pleasure to the first class or to the second, and it is respectively the greatest of the first or the least of the second class. In each case the separation of the system $\mathfrak{R}$ into the two classes $\mathfrak{U}_1$, $\mathfrak{U}_2$ is such that every number of the first class $\mathfrak{U}_1$ is smaller than every number of the second class $\mathfrak{U}_2$ and we say that this separation is produced by the number a.

For brevity and in order not to weary the reader I suppress the proofs of these theorems which follow immediately from the definitions of the previous section.

Beside these properties, however, the domain $\mathfrak{R}$ possesses also *continuity*; i. e., the following theorem is true:

IV. If the system $\mathfrak{R}$ of all real numbers breaks up into two classes $\mathfrak{U}_1$, $\mathfrak{U}_2$ such that every number a_1 of the class $\mathfrak{U}_1$ is less than every number a_2 of the class $\mathfrak{U}_2$ then there exists one and only one number a by which this separation is produced.

Proof. By the separation or the cut of $\Re$ into $\mathfrak{U}_1$ and $\mathfrak{U}_2$ we obtain at the same time a cut (A_1, A_2) of the system R of all rational numbers which is defined by this that A_1 contains all rational numbers of the class $\mathfrak{U}_1$ and A_2 all other rational numbers, i. e., all rational numbers of the class $\mathfrak{U}_2$. Let a be the perfectly definite number which produces this cut (A_1, A_2). If β is any number different from a, there are always infinitely many rational numbers c lying between a and β. If $\beta < a$, then $c < a$; hence c belongs to the class A_1 and consequently also to the class $\mathfrak{U}_1$, and since at the same time $\beta < c$ then β also belongs to the same class $\mathfrak{U}_1$, because every number in $\mathfrak{U}_2$ is greater than every number c in $\mathfrak{U}_1$. But if $\beta > a$, then is $c > a$; hence c belongs to the class A_2 and consequently also to the class $\mathfrak{U}_2$, and since at the same time $\beta > c$, then β also belongs to the same class $\mathfrak{U}_2$, because every number in $\mathfrak{U}_1$ is less than every number c in $\mathfrak{U}_2$. Hence every number β different from a belongs to the class $\mathfrak{U}_1$ or to the class $\mathfrak{U}_2$ according as $\beta < a$ or $\beta > a$; consequently a itself is either the greatest number in $\mathfrak{U}_1$ or the least number in $\mathfrak{U}_2$, i. e., a is one and obviously the only number by which the separation of $\Re$ into the classes $\mathfrak{U}_1$, $\mathfrak{U}_2$ is produced. Which was to be proved.

VI.

OPERATIONS WITH REAL NUMBERS.

To reduce any operation with two real numbers a, β to operations with rational numbers, it is only necessary from the cuts (A_1, A_2), (B_1, B_2) produced by the numbers a and β in the system R to define the cut (C_1, C_2) which is to correspond to the result of the operation, γ. I confine myself here to the discussion of the simplest case, that of addition.

If c is any rational number, we put it into the class C_1, provided there are two numbers one a_1 in A_1 and one b_1 in B_1 such that their sum $a_1 + b_1 \geqq c$; all other rational numbers shall be put into the class C_2. This separation of all rational numbers into the two classes C_1, C_2 evidently forms a cut, since every number c_1 in C_1 is less than every number c_2 in C_2. If both a and β are rational, then every number c_1 contained in C_1 is $\leqq a + \beta$, because $a_1 \leqq a$, $b_1 \leqq \beta$, and therefore $a_1 + b_1 \leqq a + \beta$; further, if there were contained in C_2 a number $c_2 < a + \beta$, hence $a + \beta = c_2 + p$, where p is a positive rational number, then we should have

$$c_2 = (a - \tfrac{1}{2}p) + (\beta - \tfrac{1}{2}p),$$

which contradicts the definition of the number c_2, because $a - \frac{1}{2}p$ is a number in A_1, and $\beta - \frac{1}{2}p$ a number in B_1; consequently every number c_2 contained in C_2 is $\geqq a + \beta$. Therefore in this case the cut (C_1, C_2) is produced by the sum $a + \beta$. Thus we shall not violate the definition which holds in the arithmetic of rational numbers if in all cases we understand by the sum $a + \beta$ of any two real numbers a, β that number γ by which the cut (C_1, C_2) is produced. Further, if only one of the two numbers a, β is rational, e. g., a, it is easy to see that it makes no difference with the sum $\gamma = a + \beta$ whether the number a is put into the class A_1 or into the class A_2.

Just as addition is defined, so can the other operations of the so-called elementary arithmetic be defined, viz., the formation of differences, products, quotients, powers, roots, logarithms, and in this way we arrive at real proofs of theorems (as, e. g., $\sqrt{2} \cdot \sqrt{3} = \sqrt{6}$), which to the best of my knowledge have never been established before. The excessive length that is to be feared in the definitions of the more complicated operations is partly inherent in the nature of the subject but can for the most part be avoided. Very useful in this connection is the notion of an *interval*, i. e., a system A of rational numbers possessing the following characteristic property: if a and a' are numbers of the system A, then are all rational numbers lying between a and a' contained in A. The system R of all rational numbers, and also the two classes of any cut are intervals. If there exist a rational number a_1 which is less and a rational number a_2 which is greater than every number of the interval A, then A is called a finite interval; there then exist infinitely many numbers in the same condition as a_1 and infinitely many in the same condition as a_2; the whole domain R breaks up into three parts A_1, A, A_2 and there enter two perfectly definite rational or irrational numbers a_1, a_2 which may be called respectively the lower and upper (or the less and greater) *limits* of the interval; the lower limit a_1 is determined by the cut for which the system A_1 forms the first class and the upper a_2 by the cut for which the system A_2 forms the second class. Of every rational or irrational number a lying between a_1 and a_2 it may be said that it lies *within* the interval A. If all numbers of an interval A are also numbers of an interval B, then A is called a portion of B.

Still lengthier considerations seem to loom up when we attempt to adapt the numerous theorems of the arithmetic of rational numbers (as, e. g., the theorem $(a+b)c=ac+bc$) to any real numbers. This, however, is not the case. It is easy to see that it all reduces to showing that the arithmetic operations possess a certain continuity. What I mean by this statement may be expressed in the form of a general theorem:

"If the number λ is the result of an operation performed on the numbers a, β, γ, . . . and λ lies within the interval L, then intervals A, B, C, . . . can be taken within which lie the numbers a, β, γ, . . . such that the result of the same operation in which the numbers a, β, γ, . . . are replaced by arbitrary numbers of the intervals A, B, C, . . . is always a number lying within the interval L." The forbidding clumsiness, however, which marks the statement of such a theorem convinces us that something must be brought in as an aid to expression; this is, in fact, attained in the most satisfactory way by introducing the ideas of *variable magnitudes, functions, limiting values*, and it would be best to base the definitions of even the simplest arithmetic operations upon these ideas, a matter which, however, cannot be carried further here.

WALLIS

On Imaginary Numbers

(Selected from the English Version, by Professor David Eugene Smith, Teachers College, Columbia University, New York City)

John Wallis (1616–1703), Savilian professor of geometry at Oxford (1649–1703), contemporary of Newton (see also p. 217), was the first to make any considerable contribution to the geometric treatment of imaginary numbers. This appeared in his *Algebra* (1673), cap. LXVI (Vol. II, p. 286), of the Latin edition. The following extract is from his English translation:

CHAP. LXVI.[1]

Of Negative Squares, *and their* Imaginary Roots *in Algebra*.

We have before had occasion (in the Solution of some Quadratick and Cubick Equations) to make mention of Negative Squares, and Imaginary Roots, (as contradistinguished to what they call Real Roots, whether Affirmative or Negative:) But referred the fuller consideration of them to this place.

These *Imaginary* Quantities (as they are commonly called) arising from the *Supposed* Root of a Negative Square, (when they happen,) are reputed to imply that the Case proposed is Impossible.

And so indeed it is, as to the first and strict notion of what is ⸺ ⸢posed. For it is not possible, that any Number (Negative or ⸺ native) Multiplied into itself, can produce (for instance) −4. ⸺ e that Like Signs (whether + or −) will produce +; and ⸺ efore not −4.

⸺ ut it is also Impossible, that any Quantity (though not a ⸺ posed Square) can be *Negative*. Since that it is not possible ⸺ any *Magnitude* can be *Less than Nothing*, or any *Number* ⸺ er *than None*.

⸺ et[2] is not that Supposition (of Negative Quantities,) either ⸺ useful or Absurd; when rightly understood. And though, as to the bare Algebraick Notation, it import a Quantity less than nothing: Yet, when it comes to a Physical Application, it denotes as Real a Quantity as if the Sign were +; but to be interpreted in a contrary sense.

---

[1] [Page 264.]   [2] [Page 265.]

As for instance: Supposing a man to have advanced or moved forward, (from A to B,) 5 Yards; and then to retreat (from B to C) 2 Yards: If it be asked, how much he had Advanced (upon the whole march) when at C? or how many Yards he is now Forwarder than when he was at A? I find (by Subducting 2 from 5,) that he is Advanced 3 Yards. (Because $+5-2=+3$.)

But if, having Advanced 5 Yards to B, he thence Retreat 8 Yards to D; and it be then asked, How much he is Advanced when at D, or how much Forwarder than when he was at A: I say -3 Yards. (Because $+5-8=-3$.) That is to say, he is advanced 3 Yards less than nothing.

Which in propriety of Speech, cannot be, (since there cannot be less than nothing.) And therefore as to the Line AB *Forward*, the case is Impossible.

But if (contrary to the Supposition,) the Line from A, be continued *Backward*, we shall find D, 3 Yards *Behind* A. (Which was presumed to be *Before* it.)

And thus to say, he is *Advanced* $-$ 3 Yards; is but what we should say (in ordinary form of Speech), he is *Retreated* 3 Yards; or he wants 3 Yards of being so Forward as he was at A.

Which doth not only answer Negatively to the Question asked. That he is not (as was supposed,) Advanced at all: But tells moreover, he is so far from being advanced, (as was supposed) that he is Retreated 3 Yards; or that he is at D, more Backward by 3 Yards, than he was at A.

And consequently $-$ 3, doth as truly design the Point D; as $+$ 3 designed the Point C. Not Forward, as was supposed; but Backward, from A.

So that $+$ 3, signifies 3 Yards Forward; and $-$ 3, signifies 3 Yards Backward: But still in the same Streight Line. And each designs (at least in the same Infinite Line,) one Single Point: And but one. And thus it is in all Lateral Equations; as having but one Single Root.

Now what is admitted in Lines, must on the same Reason, be allowed in Plains also.

As for instance: Supposing that in one Place, we Gain from the Sea, 30 Acres, but Lose in another Place, 20 Acres: If it be now asked, How many Acres we have gained upon the whole: The

Answer is, 10 Acres, or + 10. (Because of 30−20=10.) Or, which is all one 1600 Square Perches. (For the *English* Acre being Equal to a Plain of 40 Perches in length, and 4 in breadth, whose Area is 160; 10 Acres will be 1600 Square Perches.) Which if it lye in a Square Form, the Side of that Square will be 40 Perches in length; or (admitting of a Negative Root,) − 40.

But if then in a Third place, we lose 20 Acres more; and the same Question be again asked, How much we have gained in the whole; the Answer must be − 10 Acres. (Because 30−20−20=−10.) That is to say, The Gain is 10 Acres less than nothing. Which is the same as to say, there is a Loss of 10 Acres: or of 1600 Square Perches.

And hitherto, there is no new Difficulty arising, nor any other Impossibility than what we met with before, (in supposing a Negative Quantity, or somewhat Less than nothing:) Save only that $\sqrt{1600}$ is ambiguous; and may be + 40, or − 40. And from such Ambiguity it is, that Quadratick Equations admit of Two Roots.

But now (supposing this Negative Plain, − 1600 Perches, to be in the form of a Square;) must not this Supposed Square be supposed to have a Side? And if so, What shall this Side be?

We[1] cannot say it is 40, nor that it is −40. (Because either of these Multiplyed into itself, will make + 1600; not −1600).

But thus rather, that it is $\sqrt{-1600}$, (the Supposed Root of a Negative Square;) or (which is Equivalent thereunto) $10\sqrt{-16}$, or $20\sqrt{-4}$, or $40\sqrt{-1}$.

Where $\sqrt{}$ implies a Mean Proportional between a Positive and a Negative Quantity. For like as $\sqrt{bc}$ signifies a Mean Proportional between $+b$ and $+c$; or between $-b$, and $-c$; (either of which, by Multiplication, makes $+bc$:) So doth $\sqrt{-bc}$ signify a Mean Proportional between $+b$ and $-c$, or between $-b$ and $+c$; either of which being Multiplied, makes $-bc$. And this as to Algebraick consideration, is the true notion of such Imaginary Root, $\sqrt{-bc}$.

CHAP. LXVII.

The same Exemplified in Geometry.

What hath been already said of $\sqrt{-bc}$ in Algebra, (as a Mean Proportional between a Positive and a Negative Quantity:) may be thus Exemplified in Geometry.

---

[1] [Page 266].

If (for instance,) Forward from A, I take A B $= +b$; and Forward from thence, BC $= +c$; (making AC $= +$AB$+$BC$= +b+c$, the Diameter of a Circle:) Then is the Sine, or Mean Proportional BP$= \sqrt{+bc}$.

But if Backward from A, I take AB $= -b$; and then Forward from that B, BC$=+c$; (making AC $= -$AB$+$BC$= -b+c$, the Diameter of the Circle:) Then is the Tangent or Mean Proportional BP$= \sqrt{-bc}$.

So that where $\sqrt{+b}$ c signifies a Sine; $\sqrt{-b}$ c shall signify a Tangent, to the same Arch (of the same Circle,) AP, from the same Point P, to the same Diameter AC.

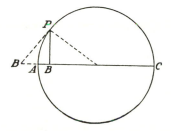
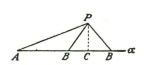

Suppose now (for further Illustration,) A Triangle standing on the Line AC (of indefinite length;) whose one Leg AP$=20$ is given; together with (the Angle PAB, and consequently) the Height PC$=12$; and the length of the other Leg PB$=15$: By which we are to find the length of the Base AB.

'Tis manifest that the Square of AP being 400; and of PC, 144; their Difference 256 ($= 400-144$) is the Square of AC.

And therefore AC($= \sqrt{256}$) $= +16$, or -16; Forward or Backward according as we please to take the Affirmative or Negative Root. But we will here take the Affirmative.

Then, because the Square of PB is 225; and of PC, 144; their Difference 81, is the Square of CB. And therefore CB $= \sqrt{81}$; which is indifferently, $+9$ or -9: And may therefore be taken Forward or Backward from C. Which gives a Double value for the length of AB; to wit, AB$=16+9=25$, or AB$=16-9=7$. Both Affirmative. (But if we should take, Backward from A, AC$=-16$; AB$=-16+9=-7$, and AB$=-16-9=-25$. Both Negative.)

Suppose[1] again, AP$=15$, PC$=12$, (and therefore AC$=\sqrt{\ }: 225 -144 \ : \ =\sqrt{81}=9$,) PB$=20$ (and therefore BC$=\sqrt{\ }:400-144:$

---

[1] [Part 267.]

$=\sqrt{256}=+16$, or -16:) Then is $AB=9+16=25$, or AB $=9-16=-7$. The one Affirmative, the other Negative. (The same values would be, but with contrary Signs, if we take $AC=$ $\sqrt{81}=-9$: That is, $AB=-9+16=+7$, $AB=-9-16=-25$.)

In all which cases, the Point B is found, (if not Forward, at least Backward,) in the Line AC, as the Question supposeth.

And of this nature, are those Quadratick Equations, whose Roots are Real, (whether Affirmative or Negative, or partly the one, partly the other;) without any other Impossibility than (what is incident also to Lateral Equations,) that the Roots (one or both) may be Negative Quantities.

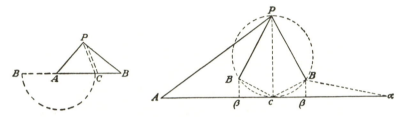

But if we shall Suppose, $AP=20$, $PB=12$, $PC=15$, (and therefore $AC=\sqrt{175}$:) When we come to Subtract as before, the Square of PC (225,) out of the Square PB (144,) to find the Square of BC, we find that cannot be done without a Negative Remainder, $144-225=-81$.

So that the Square of BC is (indeed) the Difference of the Squares of PB, PC; but a defective Deference; (that of PC proving the greater, which was supposed the Lesser; and the Triangle PBC, Rectangled, not as was supposed at C, but at B:) And therefore $BC=\sqrt{-81}$.

Which gives indeed (as before) a double value of AB, $\sqrt{175}$, $+\sqrt{-81}$, and $\sqrt{175}$, $-\sqrt{-81}$: But such as requires a new Impossibility in Algebra, (which in Lateral Equations doth not happen;) not that of a Negative Root, or a Quantity less than nothing; (as before,) but the Root of a Negative Square. Which in strictness of speech, cannot be: since that no Real Root (Affirmative or Negative,) being Multiplied into itself, will make a Negative Square.

This Impossibility in *Algebra*, argues an Impossibility of the case proposed in Geometry; and that the Point B cannot be had, (as was supposed,) in the Line AC, however produced (forward or backward,) from A.

Yet are there Two Points designed (out of that Line, but) in the same Plain; to either of which, if we draw the Lines AB, BP, we have a Triangle; whose Sides AP, PB, are such as were required: And the Angle PAC, and Altitude PC, (above AC, though not above AB,) such as was proposed; And the Difference of Squares of PB, PC, is that of CB.

And like as in the first case, the Two values of AB (which are both Affirmative,) make the double of AC, (16+9, +16−9, =16+16=32:) So here, $\sqrt{175}+\sqrt{-81}$, $+\sqrt{175}-\sqrt{-81}$, $=2\sqrt{175}$.

And (in the Figure,) though not the Two Lines themselves, AB, AB, (as in the First case, where they lay in the Line AC;) yet the Ground-lines on which they stand, Aβ, Aβ, are Equal to the Double of AC: That is, if to either of those AB, we join Bα, equal to the other of them, and with the fame Declivity; ACα (the Distance of Aα) will be a Streight Line equal to the double of AC; as is AC α in the First case.

The greatest difference is this; That in the first Case, the Points B, B, lying in the Line AC, the Lines AB, AB, are the fame with their Ground-Lines, but not so in this last case, where BB are so raised above β β (the respective Points in their Ground-Lines, over which they stand,) as to make the case feasible; (that is, so much as is the versed Sine of CB to the Diameter PC:) But in both ACα (the Ground-Line of ABα) is Equal to the Double of AC.

So that, whereas in case of Negative Roots, we are to say, The Point B cannot be found, so as is supposed in AC Forward, but Backward from A it may in the same Line: We must here say, in case of a Negative Square, the Point B cannot be found so as was supposed, in the Line AC; but Above that Line it may in the same Plain.

This I have the more largely insisted on, because the Notion (I think) is new; and this, the plainest Declaration that at present I can think of, to explicate what we commonly call the *Imaginary Roots* of Quadratick Equations. For such are these.

For instance; The Two Roots of this Equation, $aa-2a\sqrt{175}+256=0$; are $a=\sqrt{175}+\sqrt{-81}$, and $a=\sqrt{175}-\sqrt{-81}$. (Which are the values of AB in the last case.) For if from 175 (the Square of half the Coefficient,) we Subduct the Absolute Quantity 256, the Remainder is −81; the Root of which, Added to, and Subducted from, the half Coefficient,) makes $\sqrt{175}\pm\sqrt{-81}$: Which are therefore the Two Roots of that Equation. In the same manner as in the Equation $a^2 - 32\ a + 175 = 0$; if from 256 (the Square of Half 32,) we Subduct 175, the Remainder is $+81$; whose Root $\sqrt{81} = 9$, Added to and Subducted from, 16 (the half Coefficient,) makes 16 ± 9; which are the values of A B in the First case.

CHAP. LXVIII.

The Geometrical Construction accommodated hereunto.

In the former Chapter, we have shewed what in Geometry answers to the Root of a Negative Square in *Algebra*.

I shall now shew some Geometrical Effections, answering to the Resolution of such Quadratick Equations whose Roots may have (what we call) *Imaginary* values, arising from such Negative Squares.

The natural Construction of this Equation $aa \mp ba + a = 0$; is this. The Coefficient b being the Sum of Two Quantities, whose Rectangle is a, the Absolute Quantity: This cannot be more naturally expressed, in Magnitudes, than by making b ($= Aa$) the Diameter of a Circle, and $\sqrt{a}$ ($= BS$) a Right Sine or Ordinate thereunto. (For it is one of the most known Properties of a Circle, that the Sine or Ordinate is a mean Proportional between the Two Segments of the Diameter.) And because BS (of the same length,) may be taken indifferently on either side of CT, we have therefore, in the Diameter, two Points B, B, (answering to SS in the Semicircumference,) either of which divide the Diameter into AB, Ba, the Two Roots desired. (Both Affirmative, or both Negative, according as in the Equation we have $-ba$, or $+ba$.) And as BS increaseth, so B approacheth (on either Side) to C; and CB (the Co-Sine, or Semi-difference of Roots,) decreaseth.

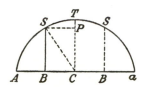

But because the Sine BS can never be greater than CT the Semidiameter: Therefore, whenever $\sqrt{a}$ is greater than $\tfrac{1}{2}b$; the Case according to this construction is Impossible.

1. The Geometrical Effection, therefore answering to this Equation, $a\ a \mp b\ a + a = 0$, (so as to take in both cases at once, Possible and Impossible; that is, whether $\tfrac{1}{4}b\ b$ be or be not less than a;) may be this.

On[1] $ACa = b$, bisected in C, erect a Perpendicular $CP = \sqrt{a}$. And taking $PB = \tfrac{1}{2}b$, make (on whether Side you please of CP,)

---

[1] [Page 269 on this p. 52, a has been used for the æ of Wallis.]

PBC, a Rectangled Triangle. Whose Right Angle will therefore be at C or B, according as PB or PC is bigger; and accordingly, BC a Sine or a Tangent, (to the Radius PB,) terminated in PC.

The Streight Lines AB, B a, are the two values of a. Both Affirmative if (in the Equation,) it be $- ba$: Both Negative, if $+b\ a$. Which values be (what we call) *Real*, if the Right-Angle be at C: But *Imaginary* if at B.

In both cases (whether the Right Angle be at C or B,) the Point B may indifferently be taken on either side of PC, in a like Position. And the Two Points B, B, are those which the Equation designs.

In the former case; ABa is a Streight Line, and the same with ACa.

In the latter; ABa makes at B, such an Angle, as that ACa is the distance α Aa; and is the Ground-line, on which if ABa be Ichnographically projected, B falls on β, the point just under it.

And therefore, if (in the Problem which produceth this Equation) ABa were supposed to be a Streight Line; or the Point B, in the Line ACa; or the same with β; or that ACa be Equal to the Aggregate of AB+Ba; or any thing which doth imply any of these: This Construction shews that Case (so understood) to be Impossible; but how it may be qualified, so as to become possible.

The difference between this Impossibility, and that incident to a Lateral Equation, is this. When in a Lateral Equation, we are reduced to a Negative value; it is as much as to say the Point B demanded, cannot be had (in the Line AC proposed,) Forward from A, as is presumed: But backward from A it may, at such a distance Behind it. But when in a Quadratick Equation, we be reduced, (not to a Negative value; wherein it communicates with the Lateral; but) to (what is wont to be called) an Imaginary value; it is as much as to say, The Point B cannot be had in the Line AC, as was presumed; but, out of that Line it may (in the same Plain;) at such a distance Above it.

The other form of Quadratick Equations, $aa \mp ba - x = 0$; is naturally thus Effected. Taking CA, or CP, $= \frac{1}{2}b$; and PB

$= \sqrt{x}$; containing a Right Angle at P. The Hypothenuse, BC continued, will cut the Circle PAa, in Aa. And the two Roots desired, are AB, Ba; between which the Tangent PB is a mean Proportional, and Aa their Difference. But one of them is to be understood Affirmative, the other Negative. (Because if AB be Forward, Ba is Backward; if that be Backward, this Forward.) To wit, +AB, −Ba, if we have (in the Equation) +ba; or −AB, +Ba, if −ba.

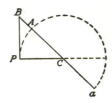

But this Construction belongs not properly to this place: Because in this form of Equation, we are never reduced to these Imaginary values. For PB, of whatever length, may be a Tangent to that Circle.

WESSEL

On Complex Numbers

(Translated from the Danish by Professor Martin A. Nordgaard, St. Olaf College, Northfield, Minnesota.)

Caspar Wessel (1745–1818) was a Norwegian surveyor. In 1797, he read a paper upon the graphic representation of complex numbers. The paper was printed in 1798 and appeared in the memoirs of the Royal Academy of Denmark in 1799. This paper may be said to have been the first noteworthy attempt at the modern method. Within a few years thereafter, numerous other attempts were made, all leading to similar results (see Smith, *History of Mathematics*, Vol. II, pp. 263–267). Wessel's work attracted little attention at the time and was almost unknown until the French translation appeared in 1897. The present translation of certain essential passages is made from the original Danish.

On the Analytical Representation of Direction; an Attempt,[1]

Applied Chiefly to the Solution of Plane and Spherical Polygons (By Caspar Wessel, Surveyor.)

This present attempt deals with the question, how may we represent direction analytically; that is, how shall we express right lines so that in a single equation involving one unknown line and others known, both the length and the direction of the unknown line may be expressed.

To help answer this question I base my work on two propositions which to me seem undeniable. The first one is: changes in direction which can be effected by algebraic operations shall be indicated by their signs. And the second: direction is not a subject for algebra except in so far as it can be changed by algebraic operations. But since these cannot change direction (at least, as

---

[1] [In recent histories of mathematics, there have come about very misleading translations into English of Wessel's title word "forsög" as "essay on, etc." This possibility comes from the word "essai" used in the French translation of Wessel's memoir, the French word meaning both an attempt or endeavor, and a treatise (essay.) Wessel's word "forsög" can only mean *attempt* or *endeavor*.]

commonly explained) except to its opposite, that is, from positive to negative, or *vice versa*, these two are the only directions it should be possible to designate, by present methods; for the other directions the problem should be unsolvable.   And I suppose this is the reason no one has taken up the matter.[1]   It has undoubtedly been considered impermissible to change anything in the accepted explanation of these operations.

And to this we do not object so long as the explanation deals only with quantities in general.   But when in certain cases the nature of the quantities dealt with seems to call for more precise definitions of these operations and these can be used to advantage, it ought not to be considered impermissible to offer modifications. For as we pass from arithmetic to geometric analysis, or from operations with abstract numbers to those with right lines, we meet with quantities that have the same relations to one another as numbers, surely; but they also have many more.   If we now give these operations a wider meaning, and do not as hitherto limit their use to right lines of the same or opposite direction; but if we extend somewhat our hitherto narrow concept of them so that it becomes applicable not only to the same cases as before, but also to infinitely many more; I say, if we take this liberty, but do not violate the accepted rules of operations, we shall not contravene the first law of numbers.   We only extend it, adapt it to the nature of the quantities considered, and observe the rule of method which demands that we by degrees make a difficult principle intelligible.

It is not an unreasonable demand that operations used in geometry be taken in a wider meaning than that given to them in arithmetic.   And one will readily admit that in this way it should be possible to produce an infinite number of variations in the directions of lines.   Doing this we shall accomplish, as will be proved later, not only that all impossible operations can be avoided —and we shall have light on the paradoxical statement that at times the possible must be tried by impossible means—, but also that the direction of all lines in the same plane can be expressed as analytically as their lengths without burdening the mind with new signs or new rules.   There is no question that the general validity of geometric propositions is frequently seen with greater ease if direction can be indicated analytically and governed by alge-

---

[1] Unless it be Magister Gilbert, in Halle, whose prize memoir on *Calculus Situs* possibly contains an explanation of this subject.

braic rules than when it is represented by a figure, and that only in certain cases. Therefore it seems not only permissible, but actually profitable, to make use of operations that apply to other lines than the equal (those of the same direction) and the opposite. On that account my aim in the following chapters will be:

I. First, to define the rules for such operations;
II. Next, to demonstrate their application when the lines are in the same plane, by two examples;
III. To define the direction of lines lying in different planes by a new method of operation, which is not algebraic;
IV. By means of this method to solve plane and spherical polygons;
V. Finally, to derive in the same manner the ordinary formulas of spherical trigonometry.

These will be the chief topics of this treatise. The occasion for its being was my seeking a method whereby I could avoid the impossible operations; and when I had found this, I applied it to convince myself of the universality of certain well-known formulas. The Honorable Mr. Tetens, Councillor-of-state, was kind enough to read through these first investigations. It is due to the encouragement, counsel, and guidance of this distinguished savant that this paper is minus some of its first imperfections and that it has been deemed worthy to be included among the publications of the Royal Academy.

A Method Whereby from Given Right Lines to Form Other Right Lines by Algebraic Operations; and How to Designate Their Directions and Signs

Certain homogeneous quantities have the property that if they are placed together, they increase or diminish one another only as increments or decrements.

There are others which in the same situation effect changes in one another in innumerable other ways. To this class belong right lines.

Thus the distance of a point from a plane may be changed in innumerable ways by the point describing a more or less inclined right line outside the plane.

For, if this line is perpendicular to the axis of the plane, that is, if the path of the point makes a right angle with the axis, the point remains in a plane parallel to the given plane, and its path has no effect on its distance from the plane.

If the described line is indirect, that is, if it makes an oblique angle with the axis of the plane, it will add to or subtract from the distance by a length less than its own; it can increase or diminish the distance in innumerable ways.

If it is direct, that is, in line with the distance, it will increase or diminish the same by its whole length; in the first case it is positive, in the second, negative.

Thus, all the right lines which can be described by a point are, in respect to their effects upon the distance of a given point from a plane outside the point, either direct or indirect or perpendicular[1] according as they add to or subtract from the distance the whole, a part, or nothing, of their own lengths.

Since a quantity is called absolute if its value is given as immediate and not in relation to another quantity, we may in the preceding definitions call the distance the absolute line; and the share of the relative line in lengthening or shortening the absolute line may be called the "effect" of the relative line.

There are other quantities besides right lines among which such relations exist. It would therefore not be a valueless task to explain these relations in general, and to incorporate their general concept in an explanation on operations. But I have accepted the advice of men of judgment, that in this paper both the nature of the contents and plainness of exposition demand that the reader be not burdened here with concepts so abstract. I shall consequently make use of geometric explanation only. These follow.

§1

Two right lines are added if we unite them in such a way that the second line begins where the first one ends, and then pass a right line from the first to the last point of the united lines. This line is the sum of the united lines.

For example, if a point moves forward three feet and backward two feet, the sum of these two paths is not the first three and the last two feet combined; the sum is one foot forward. For this path, described by the same point, gives the same effect as both the other paths.

---

[1] "Indifferent" would be a more fitting name were it not so unfamiliar to our ears.

Similarly, if one side of a triangle extends from a to b and the other from b to c, the third one from a to c shall be called the sum. We shall represent it by $ab + bc$, so that ac and $ab + bc$ have the same meaning; or $ac = ab + bc = -ba + bc$, if ba is the opposite of ab. If the added lines are direct, this definition is in complete agreement with the one ordinarily given. If they are indirect, we do not contravene the analogy by calling a right line the sum of two other right lines united, as it gives the same effect as these. Nor is the meaning I have attached to the symbol $+$ so very unusual; for in the expression $ab + \dfrac{ba}{2} = \dfrac{1}{2}ab$ it is seen that $\dfrac{ba}{2}$ is not a part of the sum. We may therefore set $ab + bc = ac$ without, on that account, thinking of bc as a part of ac; $ab + bc$ is only the symbol representing ac.

§2

If we wish to add more than two right lines we follow the same procedure. They are united by attaching the terminal point of the first to the initial point of the second and the terminal point of this one to the initial point of the third, etc. Then we pass a right line from the point where the first one begins to the point where the last one ends; and this we call their sum.

The order in which these lines are taken is immaterial; for no matter where a point describes a right line within three planes at right angles to one another, this line has the same effect on the distances of the point from each of the planes. Consequently any one of the added lines contributes equally much to the determination of the position of the last point of the sum whether it have first, last, or any other place in the sequence. Consequently, too, the order in the addition of right lines is immaterial. The sum will always be the same; for the first point is supposed to be given and the last point always assumes the same position.

So that in this case, too, the sum may be represented by the added lines connected with one another by the symbol $+$. In a quadrilateral, for example, if the first side is drawn from a to b, the second from b to c, the third from c to d, but the fourth from a to d, then we may write: $ad = ab + bc + cd$.

§3

If the sum of several lengths, breadths and heights is equal to zero, then is the sum of the lengths, the sum of the breadths, and the sum of the heights each equal to zero.

§4

It shall be possible in every case to form the product of two right lines from one of its factors in the same manner as the other factor is formed from the positive or absolute line set equal to unity. That is:

Firstly, the factors shall have such a direction that they both can be placed in the same plane with the positive unit.

Secondly, as regards length, the product shall be to one factor as the other factor is to the unit. And,

Finally, if we give the positive unit, the factors, and the product a common origin, the product shall, as regards its direction, lie in the plane of the unit and the factors and diverge from the one factor as many degrees, and on the same side, as the other factor diverges from the unit, so that the direction angle of the product, or its divergence from the positive unit, becomes equal to the sum of the direction angles of the factors.

§5

Let $+1$ designate the positive rectilinear unit and $+\epsilon$ a certain other unit perpendicular to the positive unit and having the same origin; then the direction angle of $+1$ will be equal to $0°$, that of -1 to $180°$, that of $+\epsilon$ to $90°$, and that of $-\epsilon$ to $-90°$ or $270°$. By the rule that the direction angle of the product shall equal the sum of the angles of the factors, we have: $(+1)(+1) = +1$; $(+1)(-1) = -1$; $(-1)(-1) = +1$; $(+1)(+\epsilon) = +\epsilon$; $(+1)(-\epsilon) = -\epsilon$; $(-1)(+\epsilon) = -\epsilon$; $(-1)(-\epsilon) = +\epsilon$; $(+\epsilon)(+\epsilon) = -1$; $(+\epsilon)(-\epsilon) = +1$; $(-\epsilon)(-\epsilon) = -1$.

From this it is seen that ϵ is equal to $\sqrt{-1}$; and the divergence of the product is determined such that not any of the common rules of operation are contravened.

§6

The cosine of a circle arc beginning at the terminal point of the radius $+1$ is that part of the radius, or of its opposite, which begins at the center and ends in the perpendicular dropped from the terminal point of the arc. The sine of the arc is drawn perpendicular to the cosine from its end point to the end point of the arc.

Thus, according to §5, the sine of a right angle is equal to $\sqrt{-1}$. Set $\sqrt{-1} = \epsilon$. Let v be any angle, and let sin v represent a right line of the same length as the sine of the angle v, positive, if the measure of the angle terminates in the first semi-circumference, but negative, if in the second. Then it follows from §§4 and 5 that $\epsilon \sin v$ expresses the sine of the angle v in respect to both direction and extent. . . .

§7

In agreement with §§1 and 6, the radius which begins at the center and diverges from the absolute or positive unit by angle v is equal to $\cos v + \epsilon \sin v$. But, according to §4, the product of the two factors, of which one diverges from the unit by angle v and the other by angle u, shall diverge from the unit by angle $v + u$. So that if the right line $\cos v + \epsilon \sin v$ is multiplied by the right line $\cos u + \epsilon \sin u$, the product is a right line whose direction angle is $v + u$. Therefore, by §§1 and 6, we may represent the product by $\cos (v + u) + \epsilon \sin (v + u)$.

§8

The product $(\cos v + \epsilon \sin v)(\cos u + \epsilon \sin u)$, or $\cos (v + u) + \epsilon \sin (v + u)$, can be expressed in still another way, namely, by adding into one sum the partial products that result when each of the added lines whose sum constitutes one factor is multiplied by each of those whose sum constitutes the other. Thus, if we use the known trigonometric formulas

$$\cos (v + u) = \cos v \cos u - \sin v \sin u,$$
$$\sin (v + u) = \cos v \sin u + \cos u \sin v,$$

we shall have this form:

$$(\cos v + \epsilon \sin v)(\cos u + \epsilon \sin u) = \cos v \cos u - \sin u$$
$$+ \epsilon(\cos v \sin u + \cos u \sin v).$$

For the above two formulas can be shown, without great difficulty, to hold good for all cases,—be one or both of the angles acute or obtuse, positive or negative. In consequence, the propositions derived from these two formulas also possess universality.

§9

By §7 $\cos v + \epsilon \sin v$ is the radius of a circle whose length is equal to unity and whose divergence from $\cos 0°$ is the angle v. It follows that $r \cos v + r\epsilon \sin v$ represents a right line whose length is r and whose direction angle is v. For if the sides of a right angled triangle increase in length r times, the hypotenuse increases r times; but the angle remains the same. However, by §1, the sum of the sides is equal to the hypotenuse; hence,

$$r \cos v + r\epsilon \sin v = r(\cos v + \epsilon \sin v).$$

This is therefore a general expression for every right line which lies in the same plane with the lines cos 0° and ϵ sin 90°, has the length r, and diverges from cos 0° by v degrees.

§10

If a, b, c denote direct lines of any length, positive or negative, and the two indirect lines $a + \epsilon b$ and $c + \epsilon d$ lie in the same plane with the absolute unit, their product can be found, even when their divergences from the absolute unit are unknown. For we need only to multiply each of the added lines that constitute one sum by each of the lines of the other and add these products; this sum is the required product both in respect to extent and direction: so that $(a + \epsilon b)(c + \epsilon d) = ac - bd + \epsilon(ad + bc)$.

Proof.—Let the length of the line $a + \epsilon b$ be A, and its divergence from the absolute unit be v degrees; also let the length of $c + \epsilon d$ be C, and its divergence be u. Then, by §9, $a + \epsilon b = A \cos v + B\epsilon \sin v$, and $c + \epsilon d = C \cos u + C\epsilon \sin u$. Thus $a = A \cos v$, $b = A \sin v$, $c = C \cos u$, $d = C \sin u$ (§3). But, by §4, $(a + \epsilon b)(c + \epsilon d) = AC[\cos (v + u) + \epsilon \sin (v + u)] = Ac[\cos v \cos u - \sin v \sin u + \epsilon(\cos v \sin u + \cos u \sin v)]$ (§8). Consequently, if instead of $A\,C \cos v \cos u$ we write ac, and for $A\,C \sin v \sin u$ write bd, etc., we shall derive the relation we set out to prove.

It follows that, although the added lines of the sum are not all direct, we need make no exception in the known rule on which the theory of equations and the theory of integral functions and their simple divisors are based, namely, that if two sums are to be multiplied, then must each of the added quantities in one be multiplied by each of the added quantities in the other. It is, therefore, certain that if an equation deals with right lines and its root has the form $a + \epsilon b$, then an indirect line is represented. Now, if we should want to multiply together right lines which do not both lie in the same plane with the absolute unit, this rule would have to be put aside. That is the reason why the multiplication of such lines is omitted here. Another way of representing changes of direction is taken up later, in §§24–35.

The quotient multiplied by the divisor shall equal the dividend. We need no proof that these lines must lie in the same plane with the absolute unit, as that follows directly from the definition in §4. It is easily seen also that the quotient must diverge from the absolute unit by angle $v - u$, if the dividend diverges from the same unit by angle v and the divisor by angle u.

Suppose, for example, that we are to divide $A(\cos v + \epsilon \sin v)$ by $B(\cos u + \epsilon \sin u)$. The quotient is

$$\frac{A}{B}[\cos \ (v - u) + \epsilon \sin (v - u)] \text{ since}$$

$$\frac{A}{B}[\cos (v - u) + \epsilon \sin (v - u)] \times B(\cos u + \epsilon \sin u)$$

$$= A (\cos v + \epsilon \sin v),$$

by §7. That is, since $\frac{A}{B}[\cos \ (v - u) + \epsilon \sin (v - u)]$ multiplied by the divisor $B(\cos u + \epsilon \sin u)$ equals the dividend $A(\cos v + \epsilon \sin v)$, then $\frac{A}{B}[\cos \ (v - u) + \epsilon \ \sin (v - u)]$ must be that required quotient. . . .

§12

If a, b, c, and d are direct lines, and the indirect lines $a + \epsilon b$ and $c + \epsilon d$ are in the same plane with the absolute unit: then

$$\frac{1}{c + \epsilon d} = \frac{c - \epsilon d}{c^2 + d^2}, \text{ and the quotient } \frac{a + \epsilon b}{c + \epsilon d} = (a + \epsilon b) \cdot \frac{1}{c + \epsilon d}$$

$$= (a + \epsilon b) \cdot \frac{c - \epsilon d}{c^2 + d^2} = [ac + bd + \epsilon(bc - ad)] : (c^2 + d^2).$$

For by §9 we may set $a + \epsilon b = A(\cos v + \epsilon \sin v)$, and

$$c + \epsilon d = C(\cos u + \epsilon \sin u),$$

so that

$$c - \epsilon d = C(\cos u - \epsilon \sin u), \text{ by §3.}$$

Since

$$(c + \epsilon d)(c - \epsilon d) = c^2 + d^2 = C^2, \text{ by §10,}$$

then

$$\frac{c - \epsilon d}{c^2 + d^2} = \frac{1}{C}(\cos u - \epsilon \sin u), \text{ by §10;}$$

or

$$\frac{c - \epsilon d}{c^2 + d^2} = \frac{1}{C}[\cos (-u) + \epsilon \sin (-u)] = \frac{1}{c + \epsilon d}, \text{ by §11.}$$

Multiplying by $a + \epsilon b = A(\cos v + \epsilon \sin v)$, gives

$$(a + \epsilon b) \cdot \frac{c - \epsilon d}{c^2 + d^2} = \frac{A}{C}[\cos (v - u) + \epsilon \sin (v - u)] = \frac{a + \epsilon b}{c + \epsilon d} \text{ by §11.}$$

Indirect quantities of this class have also this in common with direct, that i᷄ ˙   dividend is a sum of several quantities, then each of these ded by the divisor, gives a quotient, and the sum of these   tute the required quotient.

§13

If m is an integer, then $\cos \dfrac{v}{m} + \epsilon \sin \dfrac{v}{m}$ multiplied by itself m times gives the power $\cos v + \epsilon \sin v$ (§7); therefore we have:

$$(\cos v + \epsilon \sin v)^{\frac{1}{m}} = \cos \frac{v}{m} + \epsilon \sin \frac{v}{m}.$$

But, according to §11,

$$\cos \left(-\frac{v}{m}\right) + \epsilon \sin \left(-\frac{v}{m}\right) = \frac{1}{\cos \dfrac{v}{m} + \epsilon \sin \dfrac{v}{m}} =$$

$$\frac{1}{(\cos v + \epsilon \sin v)^{\frac{1}{m}}} = (\cos v + \epsilon \sin v)^{-\frac{1}{m}}.$$

Consequently, whether m is positive or negative, it is always true that

$$\cos \frac{v}{m} + \epsilon \sin \frac{v}{m} = (\cos v + \epsilon \sin v)^{\frac{1}{m}}.$$

Therefore, if both m and n are integers, we have;

$$(\cos v + \epsilon \sin v)^{\frac{n}{m}} = \cos \frac{n}{m}v + \epsilon \sin \frac{n}{m}v.$$

In this way we find the value of such expressions as $\sqrt[n]{b + c\sqrt{-1}}$ or $\sqrt[m]{a} \sqrt[n]{b + c\sqrt{-1}}$. For example, $\sqrt[3]{4\sqrt{3} + 4\sqrt{-1}}$ denotes a right line whose length is 2 and whose angle with the absolute unit is 10°.

§14

If two angles have equal sines and equal cosines their difference is 0, or ∓ 4 right angles, or a multiple of ± 4 right angles; and conversely, if the difference between two angles is 0 or ± 4 right angles taken once or several times, then their sines as well as their cosines are equal.

§15

If m is an integer and π is equal to 360°, then $(\cos v + \epsilon \sin v)^{\frac{1}{m}}$ has only the following m different values:

$$\cos v + \epsilon \sin v, \; \cos \frac{\pi + v}{m} + \epsilon \sin \frac{\pi + v}{m}, \; \cos \frac{2\pi + v}{m} + \epsilon \sin \frac{2\pi + v}{m} ...,$$

$$\cos \frac{(m-1)\pi + v}{m} + \epsilon \sin \frac{(m-1)\pi + v}{m};$$

for the numbers by which π is multiplied in the preceding series are in the arithmetical progression 1, 2, 3, 4,...$m - 1$. Consequently the sum of every two of them is m, if the one is as far from 1 as the other is from $m - 1$; and if their number is not even, then the middle one taken two times equals m. Therefore if $\dfrac{(m-n)\pi + v}{m}$

is added to $\dfrac{(m - u)\pi + v}{m}$, and the latter is as far from $\dfrac{\pi + v}{m}$, in the series, as $\dfrac{(m - n)\pi + v}{m}$ is from $\dfrac{(m - 1)\pi + v}{m}$, then

the sum is equal to $\dfrac{2m - u - n}{m}\pi + \dfrac{2v}{m} = \pi + \dfrac{2v}{m}$. But adding

$\dfrac{(m - n)\pi}{m}$ is equivalent to subtracting $\dfrac{(m - n)(-\pi)}{m}$; and since

the difference is π, $\dfrac{(m - n)(-\pi) + v}{m}$ has the same cosine and sine

as $\dfrac{(m - n)\pi + v}{m}$. Hence $(-\pi)$ gives no values not given by $+\pi$.

However, none of these values are equal; for the difference between any two angles of the series is always less than π and never equal to 0. Nor will any more values result if the series is continued; for then the new angles will be $\pi + \dfrac{v}{m}, \pi + \dfrac{\pi + v}{m}, \pi + \dfrac{2\pi + v}{m}$,

etc., and according to §14 the values of the sines and cosines of these will be the same as in the angles we already have. There can be no angle outside of the series; for then π would not be multiplied in the numerator by an integer, and the angles multiplied by m would not produce any angle which subtracted from v gives 0, or $\pm\pi$, or a multiple of $\mp\pi$; consequently the mth power of the cosine and sine of such angles could not equal $\cos v + \epsilon \sin v$.

§16

Without knowing the angle which the indirect line $1 + x$ makes with the absolute, we may find, if the length of x is less than 1,

the power $(1 + x)^m = 1 + \dfrac{mx}{1} + \dfrac{m}{1} \cdot \dfrac{m - 1}{2} x^2 + $ etc. If this series

is arranged according to the powers of m, it has the same value and is changed into the form

$$1 + \frac{ml}{1} + \frac{m^2 l^2}{1} + \frac{m^3 l^3}{1.2.3} + \text{etc.},$$

where

$$l = x - \frac{x^2}{2} + \frac{x^3}{3} - \frac{x^4}{4} + \text{etc.},$$

and is a sum of a direct and a perpendicular line.   If we call the direct line a and the perpendicular $b\sqrt{-1}$, then b is the smallest measure of the angle which $1 + x$ makes with $+1$.   If we set

$$1 + \frac{1}{1} + \frac{1}{1.2} + \frac{1}{1.2.3} + \text{etc.} = e,$$

then

$$(1 + x)^m, \text{ or } 1 + \frac{ml}{1} + \frac{m^2l^2}{1.2} + \frac{m^3l^3}{1.2.3} + \text{etc.,}$$

may be represented by $e^{ma + mb\sqrt{-1}}$; that is, $(1 + x)^m$ has the length e^{m4} and a direction angle whose measure is mb, assuming m to be either positive or negative.   Lines lying in the same plane may thus have their direction expressed in still another way, namely, by the aid of the natural logarithms.   I shall produce complete proofs for these statements at another time, if privileged to do so.   Now, that I have rendered an account of my plan for finding the sums, products, quotients, and powers of right lines, I shall next give a couple of examples illustrating the use of this method.

PASCAL

On the Arithmetic Triangle

(Translated from the French by Anna Savitsky, A. M., Columbia University,
New York City.)

Although Pascal (see p. 165) was not the originator of the arithmetic triangle, such an arrangement of numbers having been anticipated, his name has
been linked with the triangle by his development of its properties, and by the
applications which he made of these properties. The historical interest of
the work is to be found, perhaps, in its bearing on probability discussions and
on the early developments of the binomial theorem. Since Pascal's contributions to the theory of probability are considered elsewhere in this Source Book,
passages pertinent to that theory are omitted in the present translation.
Other omissions, also, are necessarily made with great freedom. The original
article is found in the works of Pascal, the latest edition of which was edited
by Léon Brunschvicg and Pierre Boutroux (Paris, 1908).

Treatise on the Arithmetic Triangle
Definitions

I designate as *an arithmetic triangle* a figure whose construction
is as follows:

I draw from any point, G, two lines perpendicular to each other,
GV, $G\zeta$,[1] in each of which I take as many equal and[2] continuous
parts as I please, beginning at G, which I name 1, 2, 3, 4, etc.; and
these numbers are *the indices*[3] of the divisions of the lines.

Then I join the points of the first division in each of the two lines
by another line that forms a triangle of which it is *the base*.

I join in this manner the two points of the second division by
another line that forms a second triangle of which it is *the base*.

And joining in this manner all the points of division which have
the same index I form with them as many *triangles* and *bases*.

I draw through each of the points of division lines parallel to
the sides, which by their intersections form small squares that I
call *cells*.

---

[1] [The editor of the French edition uses ξ instead of ζ by mistake.]

[2] [Pascal employs the words "continües" and "contigües" interchangeably.
In the translation, they have been rendered literally.]

[3] [The term used is "exposans."]

The cells which lie between two parallels going from left to right are called *cells of the same parallel rank*, like the cells G, σ, π, etc., or ϕ, ψ, θ, etc.

And those which lie between two lines going from the top downward are called *cells of the same perpendicular rank*, like the cells G, ϕ, A, D, etc., and also σ, ψ, B, etc.

Those which are crossed diagonally by the same base are called *cells of the same base*, like the following: D, B, θ, λ, or A, ψ, π.

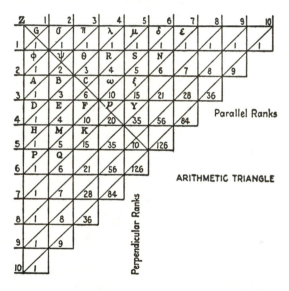

The cells of the same base equally distant from its ends are called *reciprocals*, as E, R, and B, θ, because the index of the parallel rank of the one is the same as the index of the perpendicular rank of the other, as is apparent in the example where E is in the second perpendicular rank and in the fourth parallel, and its reciprocal R is in the second parallel rank and reciprocally in the fourth perpendicular; and it is quite easy to show that those cells which have their indices reciprocally equal are in the same base and equally distant from its extremities.

It is also quite easy to show that the index of the perpendicular rank of any cell whatsoever, added to the index of its parallel rank, exceeds by unity the index of its base.

For example, cell F is in the third perpendicular rank, and in the fourth parallel rank, and in the sixth base; and the two indices of the ranks $3 + 4$ exceed by unity the index of the base 6, which arises from the fact that the two sides of the triangle are divided into an equal number of parts; but this is rather understood than demonstrated.

The above statement is equivalent to saying that each base contains one cell more than the preceding base, and each as many as the number of units in its index; thus, the second $\phi\sigma$ has two cells, the third $A\psi\pi$ has three of them, etc.

Now the numbers which are placed in each cell are found by this method:

The number of the first cell, which is in the right angle, is arbitrary; but when that has been decided upon, all the others necessarily follow; and for this reason, it is called the *generator* of the triangle. Each of the others is determined by this one rule:

The number of each cell is equal to that of the cell which precedes it in its perpendicular rank, added to that of the cell which precedes it in its parallel rank. Thus, the cell F, that is, the number of the cell F, is equal to the cell C, plus the cell E; and likewise for the others.

From these facts there arise several consequences. Below are the principal ones, in which I consider those triangles whose generator is unity; but what is said of them will apply to all others.

Corollary 1.—In every arithmetic triangle, all the cells of the first parallel rank and of the first perpendicular rank are equal to the generator.

For, by the construction of the triangle, each cell is equal to that of the cell which precedes it in its perpendicular rank, added to that which precedes it in its parallel rank. Now the cells of the first parallel rank have no cells which precede them in their perpendicular ranks, nor those of the first perpendicular rank in their parallel ranks; consequently they are all equal to each other and thus equal to the generating first number.

Thus ϕ equals G + zero, that is, ϕ equals G.

Likewise A equals ϕ + zero, that is, ϕ.

Likewise σ equals G + zero, and π equals σ + zero.

And likewise for the others.

Corollary 2.—In every arithmetic triangle, each cell is equal to the sum of all those of the preceding parallel rank, comprising the cells from its perpendicular rank to the first, inclusively.

Consider any cell ω: I assert that it is equal to $R + \theta + \psi + \phi$, which are cells of the parallel rank above, from the perpendicular rank of ω to the first perpendicular rank.

This is evident by defining the cells, merely, in terms of the cells from which they are formed.

For ω equals $R + C$.

$$\overbrace{\theta + B}$$

$$\overbrace{\psi + A}$$

ϕ, for A and ϕ are equal to each other by the preceding.

Hence ω equals $R + \theta + \psi + \phi$.

Corollary 3.—In every arithmetic triangle, each cell is equal to the sum of all those of the preceding perpendicular rank, comprising the cells from its parallel rank to the first, inclusively.

Consider any cell C: I assert that it is equal to $B + \psi + \sigma$, which are the cells of the preceding perpendicular rank, from the parallel rank of the cell C to the first parallel rank.

This appears likewise by the very definition of the cells.

For C equals $B + \theta$.

$$\overbrace{\psi + \pi}$$

σ, for π equals σ by the first (corollary).

Hence C equals $B + \psi + \sigma$.

Corollary 4.—In every arithmetic triangle, each cell diminished by unity is equal to the sum of all those which are included between its perpendicular rank and its parallel rank, exclusively.

Consider any cell ξ: I assert that $\xi - g$ equals $R + \theta + \psi + \phi + \lambda + \pi + \sigma + G$, which are all the numbers included between the rank $\xi\omega CBA$ and the rank $\xi S\mu$, exclusively.

This appears in like manner from the definition.

For ξ equals $\lambda + R + \omega$.

$$\overbrace{\pi + \theta + C}$$

$$\overbrace{\theta + \psi + B}$$

$$\overbrace{G + \phi + A}$$

$$\overbrace{G.}$$

Hence ξ equals $\lambda + R + \pi + \theta + \sigma + \psi + G + \phi + G$.

Note.—I have said in the statement: *each cell diminished by unity*, because unity is the generator; but if it were another number, it would be necessary to say: *each cell diminished by the generating number.*

Corollary 5.—In every arithmetic triangle, each cell is equal to its reciprocal.

For in the second base $\phi\sigma$, it is evident that the two reciprocal cells ϕ, σ, are equal to each other and to G.

In the third $A\psi\pi$, it is likewise seen that the reciprocals π, A, are equal to each other and to G.

In the fourth, it is seen that the extremes D, λ, are again equal to each other and to G.

And those between the two are evidently equal, since B equals $A + \psi$, and θ equals $\psi + \pi$; now $\pi + \psi$ are equal to $A + \psi$, as has been shown; hence, etc.

Likewise it can be shown in all the other bases that the reciprocals are equal, because the extremes are always equal to G, and the rest can always be defined by their equals in the preceding base which are reciprocal to each other.

Corollary 6.—In every arithmetic triangle, a parallel rank and a perpendicular one which have the same index are composed of cells which are respectively equal to each other.

For they are composed of reciprocal cells.

Thus, the second perpendicular rank $\sigma\psi BEMQ$ is exactly equal to the second parallel rank $\phi\psi\theta RSN$.

Corollary 7.—In every arithmetic triangle, the sum of the cells of each base is twice those of the preceding base.

Consider any base $DB\theta\lambda$. I assert that the sum of its cells is double the sum of the cells of the preceding base $A\psi\pi$.

For extremes........................... $D,$ $\lambda,$

are equal to the extremes.................... $A,$ $\pi,$
and each of the others..................... $B,$ $\theta,$

is equal to two of the other base............. $A + \psi,$ $\psi + \pi,$

Hence $D + \lambda + B + \theta$ equal $2A + 2\psi + 2\pi$.

The same thing may be demonstrated for all the others.

Corollary 8.—In every arithmetic triangle, the sum of the cells of each base is a number of the[1] geometric progression which begins with unity, and whose order is the same as the index of the base.

For the first base is unity.

The second is twice the first, hence it is 2.

---

[1] [The term used, "double progression," refers to a geometric progression.]

The third is twice the second, hence it is 4.

And so on to infinity.

NOTE.—If the generator were not unity, but another number like 3, the same thing would be true; however, one should not take the numbers of the geometric progression beginning with unity, that is, 1, 2, 4, 8, 16, etc., but those of another geometric progression beginning with the generator 3, as, 3, 6, 12, 24, 48, etc.

Corollary 9.—In every arithmetic triangle, each base diminished by unity is equal to the sum of all the preceding ones.

For this is a property of the double (geometric) progression.

NOTE.—If the generator were other than unity, it would be necessary to say: *each base diminished by the generator.*

Corollary 10.—In every arithmetic triangle, the sum of a many continuous cells as desired of a base, beginning at one end, is equal to as many cells of the preceding base, taking as many again *less* one.

Let the sum of as many cells as desired of the base $D\lambda$ be taken: for example, the first three $D + B + \theta$.

I assert that it is equal to the sum of the first three cells of the preceding base $A + \psi + \pi$, adding the first two of the same base $A + \psi$.

For D. B. θ.

equals A. $A + \psi$. $\psi + \pi$.

Hence $D + B + \theta$ equals $2A + 2\psi + \pi$.

Definition.—I designate as *cells of the dividend* those which are crossed diagonally by the line which bisects the right angle, as G, ψ, C, ρ, etc.

Corollary 11.—Every cell of the dividend is twice that which precedes it in its parallel or perpendicular rank.

Consider a cell of the dividend C. I assert that it is twice θ, and also twice B.

For C equals $\theta + B$, and θ equals B, by Corollary 5.

NOTE.—All these corollaries are on the subject of the equalities which are encountered in the arithmetic triangle. Now we shall consider those relating to proportions; and for these, the following proposition is fundamental.

Corollary 12.—In every arithmetic triangle, if two cells are contiguous in the same base, the upper is to the lower as the number of cells from the upper to the top of the base is to the number of those from the lower to the bottom, inclusive.

Consider any two contiguous cells of the same base, *E, C*: I assert that:

E is to	*C* as 2		is to 3	
lower,	upper,	because there are two cells from *E* to the bottom, that is, *E, H;*	because there are three cells from *C* to the top, that is, *C,* *R, μ.*	

Although this proposition has an infinite number of cases, I will give a rather short demonstration, assuming two lemmas.

Lemma 1: which is self-evident, that this proportion is met with in the second base; for it is apparent that ϕ is to σ as 1 is to 1.

Lemma 2: that if this proportion is found in any base, it will necessarily be found in the following base.

From which it will be seen that this proportion is necessarily in all the bases: for it is in the second base by the first lemma; hence by the second, it is in the third base, hence in the fourth, and so on to infinity.

It is then necessary only to prove the second lemma in this way. If this proportion is met with in any base, as in the fourth $D\lambda$, that is, if *D* is to *B* as 1 is to 3, and *B* is to θ as 2 is to 2, and θ is to λ as 3 is to 1, etc., I say that the same proportion will be found in the following base $H\mu$, and that, for example, *E* is to *C* as 2 is to 3.

For *D* is to *B* as 1 is to 3, by the hypothesis. Hence

$$D + B \text{ is to } B \text{ as } 1 + 3 \text{ is to } 3.$$

$$E \quad \text{is to } B \text{ as} \quad 4 \quad \text{is to } 3.$$

In the same way *B* is to θ as 2 is to 2, by the hypothesis. Hence

$$B + \theta \text{ is to } B \text{ as } 2 + 2 \text{ is to } 4.$$

$$C \quad \text{is to } B \text{ as} \quad 4 \quad \text{is to } 2.$$

But

$$B \quad \text{is to } E \text{ as} \quad 3 \quad \text{is to } 4.$$

Hence by the[1] mixed proportion, *C* is to *E* as 3 is to 2: Which was to be proved.

The same may be demonstrated in all the rest, since this proof is based only on the assumption that the proportion occurs in the

---

[1] [The term used is "proportion troublée."]

preceding base, and that each cell is equal to its preceding plus the one above it, which is true in all cases.

Corollary 13.—In every arithmetic triangle, if two cells are continuous in the same perpendicular rank, the lower is to the upper as the index of the base of the upper is to the index of its parallel rank.

Consider any two cells in the same perpendicular rank, F, C. I assert that F is to C as   5   is to 3
       the lower,   the upper,   index of   the index of the parallel
                              base of C,   rank of C.

For E is to C as 2 is to 3.

Hence

$$E + C \text{ is to } C \text{ as } 2 + 3 \text{ is to } 3.$$

$$F \quad \text{is to } C \text{ as } 5 \quad \text{is to } 3.$$

Corollary 14.—In every arithmetic triangle, if two cells are continuous in the same parallel rank, the greater is to the preceding one as the index of the base of the preceding is to the index of its perpendicular rank.

Consider two cells in the same parallel rank, F, E. I assert that
     F is to E as   5   is to   2
the greater,   the preceding,   index of the   index of the perpendicular
                             base of E,   pendicular rank of
                                         E.

For E is to C as 2 is to 3.

Hence

$$E + C \text{ is to } E \text{ as } 2 + 3 \text{ is to } 2.$$

$$F \quad \text{is to } E \text{ as } \quad 5 \quad \text{is to } 2.$$

Corollary 15.—In every arithmetic triangle, the sum of the cells of any parallel rank is to the last cell of the rank as the index of the triangle is to the index of the rank.

Consider any triangle, for example, the fourth $GD\lambda$: I assert that for any rank which one takes in it, like the second parallel rank, the sum of its cells, that is $\phi + \psi + \theta$, is to θ as 4 is to 2. For $\phi + \psi + \theta$ equals C, and C is to θ as 4 is to 2, by Corollary 13.

Corollary 16.—In every arithmetic triangle, any parallel rank is to the rank below as the index of the rank below is to the number of its cells.

Consider any triangle, for example the fifth μGH: I assert that, whatever rank one may choose in it, for example the third, the sum of its cells is to the sum of those of the fourth, that is

$A + B + C$ is to $D + E$ as 4, the index of the fourth rank, is to 2, which is the index of the number of its cells, for it contains 2 of them.

For $A + B + C$ equals F, and $D + E$ equals M.

Now F is to M as 4 is to 2, by Corollary 12.

Note.—It may also be stated in this manner: *Every parallel rank is to the rank below as the index of the rank below is to the index of the triangle minus the index of the rank above.*

For the index of a triangle, minus the index of one of its ranks, is always equal to the number of cells contained in the rank below.

Corollary 17.—In every arithmetic triangle, any cell whatever added to all those of its perpendicular rank is to the same cell added to all those of its parallel rank as the number of cells taken in each rank.

Consider any cell B: I assert that $B + \psi + \sigma$ is to $B + A$ as 3 is to 2.

I say 3, because there are three cells added in the antecedent, and 2, because there are two of them in the consequent.

For $B + \psi + \sigma$ equals C, by Corollary 3, and $B + A$ equals E, by Corollary 2.

Now C is to E as 3 is to 2, by Corollary 12.

Corollary 18.—In every arithmetic triangle, two parallel ranks equally distant from the ends are to each other as the number of their cells.

Consider any triangle $GV\zeta$, and two of its ranks equally distant from the ends, as the sixth $P + Q$, and the second $\phi + \psi + \theta + R + S + N$: I assert that the sum of the cells of the one is to the sum of the cells of the other as the number of cells of the first is to the number of cells of the second.

For, by Corollary 6, the second parallel rank $\phi\psi\theta RSN$ is the same as the second perpendicular rank $\sigma\psi BEMQ$, for which we have demonstrated this proportion.

Note.—It may also be stated: *In every arithmetic triangle, two parallel ranks, whose indices added together exceed by unity the index of the triangle, are to each other inversely as their indices.*

For it is the same thing as that which has just been stated.

Final Corollary.—In every arithmetic triangle, if two cells in the dividend are continuous, the lower is to the upper taken four times as the index of the base of the upper is to a number greater (than the base) by unity.

Consider two cells of the dividend ρ, C: I assert that ρ is to $4C$ as 5, the index of the base of C, is to 6.

For ρ is twice ω, and C twice θ; hence 4θ equal $2C$.

Hence 4θ is to C as 2 is to 1.

Now ρ is to $4C$ as ω is to 4θ,

or by a ratio composed of................. ω to C + C to 4θ

by the preceding corollaries............... 5 to 3   1 to 2

                                                     or   3 to 6

                                              5   to   6

Hence ρ is to $4C$ as 5 is to 6.   Which was to be proved.

NOTE.—Thence many other proportions may be drawn that I have passed over, because they may be easily deduced, and those who would like to apply themselves to it will perhaps find some, more elegant than these which I could present.[1]

APPLICATION OF THE ARITHMETIC TRIANGLE

To Find the Powers of Binomials and[2] Apotomes

If it is proposed to find a certain power, like the fourth degree, of a binomial whose first term is A and the other unity, that is to say, if it is required to find the fourth power of $A + 1$, take the fifth base of the arithmetic triangle, namely, the one whose index 5 is greater by unity than 4, the exponent of the proposed order. The cells of this fifth base are 1, 4, 6, 4, 1; the first number, 1, is to be taken as the coefficient of A to the proposed degree, that is, of A^4; then take the second number of the base, which is 4, as the coefficient of A to the next lower degree, that is to say, of A^3, and take the following number of the base, namely 6, as the coefficient of A to the lower degree, namely, of A^2 and the next number of the base, namely 4, as the coefficient of A to the lower degree,

---

[1] [At this point, Pascal establishes a theorem which would be stated in modern notations as follows: The cell in the n-th parallel and r-th perpendicular ranks contains the number

$$\frac{n(n + 1)\ldots(n + r - 2)}{(r - 1)!}$$

He then indicates applications of the arithmetic triangle in the theory of combinations, and in the elementary analysis of questions of mathematical probability suggested by games of chance. All of this material is omitted in the present translation.]

[2] [By "apotome," Pascal means a binomial which is the difference between two terms.]

namely, of the root A, and take the last number of the base, 1, as the absolute number; thus we obtain: $1A^4 + 4A^3 + 6A^2 + 4A + 1$, which is the fourth (square-square) power of the binomial $A + 1$. So that if A (which represents any number) is unity, and thus the binomial $A + 1$ becomes 2, this power $1A^4 + 4A^3 + 6A^2 + 4A + 1$, now becomes $1.1^4 + 4.1^3 + 6.1^2 + 4.1 + 1$.

That is, one times the fourth power of A, which is unity   1
> Four times the cube of 1, that is............ 4
> Six times the square of 1, that is............ 6
> Four times unity, that is.................... 4
> Plus unity................................. 1
> Which added together make................ $\overline{16}$

And indeed, the fourth power of 2 is 16.

If A is another number, like 4, and thus the binomial $A + 1$ is 5, then its fourth power will always be, in accordance with this method,

$$1A^4 + 4A^3 + 6A^2 + 4A + 1$$

which now means,

$$1.4^4 + 4.4^3 + 6.4^2 + 4.4 + 1.$$

That is to say, one times the fourth power of 4, namely 256
> Four times the cube of 4, namely............ 256
> Six times the square of 4.................... 96
> Four times the root 4...................... 16
> Plus unity................................. 1

whose sum.. $\overline{625}$

produces the fourth power of 5: and indeed, the fourth power of 5 is 625.

Likewise for other examples.

If it is desired to find the same degree of the binomial $A + 2$, take the same expression $1A^4 + 4A^3 + 6A^2 + 4A + 1$, and then write the four numbers, 2, 4, 8, 16, which are the first four degrees of 2, under each of the numbers of the base, omitting the first, in this way

$$1A^4 + 4A^3 + 6A^2 + 4A^1 + 1$$
$$2 \qquad 4 \qquad 8 \qquad 16$$

and multiply the numbers which correspond to each other

$$1A^4 + 4A^3 + 6A^2 + 4A^1 + 1$$
$$2 \qquad 4 \qquad 8 \qquad 16$$

in this way $\overline{1A^4 + 8A^3 + 24A^2 + 32A^1 + 16}$

Thus the fourth power of the binomial $A + 2$ is obtained; if A is unity, the fourth power will be as follows:

One times the fourth power of A, which is unity   1
Eight times the cube of unity................   8
24, 1^2................................... 24
32, 1..................................... 32
Plus the fourth power of ı.................. 16
Whose sum................................... $\overline{81}$

is the fourth power of 3.   And indeed, 81 is the fourth power of 3.

If A is 2, then $A + 2$ is 4, and its fourth power will be

One times the fourth power of A, or of 2,
    namely................................ 16
8, 2^3..................................... 64
24, 2^2.................................... 96
32, 2..................................... 64
Plus the fourth power of 2................. 16
whose sum................................... $\overline{256}$

is the fourth power of 4.

In the same way, the fourth power of $A + 3$ can be found, by writing likewise

$$A^4 + \quad 4A^3 + \quad 6A^2 + \quad 4A + \quad 1$$

and below, the numbers,

$$\frac{\qquad 3 \qquad\qquad 9 \qquad\qquad 27 \qquad\qquad 81 \qquad}{1A^4 + 12A^3 + 54A^2 + 108A + 81}$$

which are the first four degrees of 3; and by multiplying the corresponding numbers, we obtain the fourth power of $A + 3$.

And so on to infinity.   If in place of the fourth power, the square-cube, or the fifth degree, is desired, take the sixth base and apply it as I have described in the case of the fifth; and likewise for all the other degrees.

In the same way, the powers of the apotomes $A - 1$, $A - 2$, etc., may be found.   The method is wholly similar, and differs only in the matter of signs, for the signs $+$ and $-$ always alternate, and the sign $+$ is always first.

Thus the fourth power of $A - 1$ may be found in this way. The fourth power of $A + 1$ is, according to the preceding rule, $1A^4 + 4A^3 + 6A^2 + 4A + 1$.   Hence, by changing the signs in the way described, we obtain $1A - 4A^3 + 6A^2 - 4A + 1$. Thus the cube of $A - 2$ is likewise found.   For the cube of $A + 2$, by the preceding rule, is $A^3 + 6A^2 + 12A + 8$.   Hence the cube of $A - 2$ is found by changing the signs, $A^3 - 6A^2 +$ $12A - 8$.   And so on to infinity.

I am not giving a demonstration of all this, because others have already treated it, like Hérigogne; besides, the matter is self-evident.

BOMBELLI AND CATALDI

On Continued Fractions

(Translated from the Italian by Professor Vera Sanford, Western Reserve University, Cleveland, Ohio.)

The study of continued fractions seems to have arisen in connection with the problem of finding the approximate values of the square roots of numbers that are not perfect squares. Various methods of finding such roots had been advanced[1] at an earlier period, but, in general, their operation was difficult and clumsy.

The first mathematician to make use of the concept of continued fractions was Rafael Bombelli (born *c.* 1530). Little is known of his career, but his contribution to mathematics was the writing of a work which has been characterized as "the most teachable and the most systematic treatment of algebra that had appeared in Italy up to that time."[2] The title was *L'Algebra parte maggiore dell' arimetica divisa in tre libri* and the work was published in Bologna in 1572 and brought out in a second edition in that same city in 1579 under the title *L' Algebra Opera*, the editions being identical except for the title pages and the dedicatory letter. This algebra was noteworthy for its treatment of the cubic and biquadratic equations. The selection here given appears on pages 35 to 37 of the edition of 1579.

Method of Forming Fractions in the Extraction of Roots

Many methods of forming fractions have been given in the works of other authors; the one attacking and accusing another without due cause (in my opinion) for they are all looking to the same end. It is indeed true that one method may be briefer than another, but it is enough that all are at hand and the one that is the most easy will without doubt be accepted by men and be put in use without casting aspersions on another method. Thus it may happen that today I may teach a rule which may be more acceptable than those given in the past, but if another should be discovered later and if one of them should be found to be more vague and if another should be found to be more easy, this [latter] would then be accepted at once and mine would be discarded; for as the saying goes, experience is our master and

---

[1] See Smith, D. E., *History of Mathematics,* Vol. II, pp. 144, and 253, Boston, Massachusetts, 1925.

[2] *Ibid.*, Vol. I, p. 301.

the result praises the workman. In short, I shall set forth the method which is the most pleasing to me today and it will rest in men's judgment to appraise what they see: mean while I shall continue my discourse going now to the discussion itself.

Let us first assume that if we wish to find the approximate root[1] of 13 that this will be 3 with 4 left over. This remainder should be divided by 6 (double the 3 given above) which gives $\frac{2}{3}$. This is the first fraction which is to be added to the 3, making $3\frac{2}{3}$ which is the approximate root of 13. Since the square of this number is $13\frac{4}{9}$, it is $\frac{4}{9}$ too large, and if one wishes a closer approximation, the 6 which is the double of the 3 should be added to the fraction $\frac{2}{3}$, giving $6\frac{2}{3}$, and this number should be divided into the 4 which is the difference between 13 and 9. The result is $\frac{3}{5}$ which, added to the 3 makes $3\frac{3}{5}$. This is a closer approximation to the root of 13, for its square is $12\frac{24}{25}$, which is closer than that of the $3\frac{2}{3}$.[2] But if I wish a closer approximation, I add this fraction to the 6 making $6\frac{3}{5}$, divide 4 by this, obtaining $\frac{20}{33}$. This should be added to the 3 as was done above, making $3\frac{20}{33}$. This is a closer approximation for its square is $13\frac{4}{1089}$, which is $\frac{4}{4089}$ too large. If I wish a closer approximation, I divide 4 by $6\frac{20}{33}$, obtaining $\frac{109}{180}$, [and] add this to 3, obtaining $3\frac{109}{180}$. This is much closer than before for its square is

---

[1] [Bombelli's term *latus* was a popular one based on the concept of a square root as the side of a square of given area. In this translation, however, the term *root* will be used because of its greater significance.]

[2] [In modern notation, this would, of course be written as: $3 + \dfrac{4}{6 + \dfrac{4}{6}}$.

Bombelli gives no hint as to the reasons for the success of this method, nor does he tell how he discovered it.]

$13\frac{1}{32400}$, which is $\frac{1}{32400}$ too large. If I wish to continue this even further, I divide 4 by $6\frac{109}{180}$ obtaining $\frac{729}{1189}$, which is the root of $13\frac{4}{1413721}$, which is $\frac{4}{1413721}$ too large, and this process may be carried to within an imperceptible difference. Care should be taken, however, in the formation of these fractions in the many cases when the number whose root is to be found falls just short of being a perfect square (as 8, for example). In this case, since 4 is the largest square number, and since 4 is also the remainder, the fraction becomes $\frac{4}{4}$ which is equal to 1. Adding this to 2 gives 3, whose square is 9. Subtracting the number 8 whose root is required from this number, 1 remains. This should be divided by 6, the double of the 3 giving $\frac{1}{6}$. Subtracting this from the 3 gives $2\frac{5}{6}$ as the approximate root of 8. The square of this number is $8\frac{1}{36}$,[1] which is $\frac{1}{36}$ too large. If a closer approximation is desired, add the $2\frac{5}{6}$ to the 3 getting $5\frac{5}{6}$, and divide 1 by this as was done above, giving $\frac{6}{35}$, which should be subtracted from 3 leaving $2\frac{29}{35}$. This will be a nearer root. If a still closer approximation is desired, divide 1 by $5\frac{29}{35}$. Proceed (as was done above) as close as any one may desire.

. . . .

Pietro Antonio Cataldi[2] (1548–1626) was professor of mathematics and astronomy at Florence, Perugia, and Bologna. He was the author of works on arithmetic, theory of numbers, and geometry and also wrote treatises on topics in algebra. He seems to have been the first to develop a symbolism for continued fractions, and this appears in an essay with the title *Trattato del modo brevissimo Di trouare la Radice quadra delli numeri, Et Regole da approssimarsi di continuo al vero nelle Radici de'numeri non quadrati, con le cause et inuentioni loro, Et anco il modo di pigliarne la Radice cuba, appli-*

---

[1] [Here Bombelli gives $\frac{1}{119}$, evidently a misprint.]

[2] Sometimes given as Cattaldi.

cando il tutto alle Operationi Militari & altre. Bologna, 1613. The selection here given appears on page 70.

Let us now proceed to the consideration of another method of finding roots continuing by adding row on row (*di mano in mano*) to the denominator of the fraction, which finally yields a fraction equal to the fraction of the preceding rule. But for greater convenience, I shall assume a number whose root may be easily taken and I shall assume that the first part of the root is an integer. Then let 18 be the proposed number, and if I assume that the first root is 4. & $\frac{2}{8}$, that is $4\frac{1}{4}$, this will be in excess by $\frac{1}{16}$ which is the square of the fraction $\frac{1}{4}$. The second root will be found by the above mentioned method to be 4. & $\frac{2}{8}$. & $\frac{1}{4}$ which is 4. & $\frac{8}{33}$, which is $\frac{2}{1089}$ too small. This arises from multiplying the entire fraction $\frac{8}{33}$ by $\frac{1}{132}$ in which the whole fraction is less than the $\frac{1}{4}$ which is the added fraction.[1]

Let the root of 18. be

$$4. \& \frac{2}{8}. \& \frac{2}{8}. \& \frac{2}{8}$$

The total fraction added is

$$\begin{array}{ccc} \dfrac{33}{136} & \dfrac{8}{33} & \\[2mm] 1089 & & 1088 \\[2mm] \dfrac{1}{136} & & \\ \end{array}$$

$$\dfrac{2}{8\frac{1}{4}}$$

$$\dfrac{8}{33} \qquad \dfrac{1}{136 \times 33} \times \dfrac{33 \times 1}{136}$$

makes

$$\dfrac{1}{18,496}$$

That is, 4. & $\frac{2}{8}$. & $\frac{8}{33}$

$$\dfrac{66}{272}$$

[1] [The work which follows appears in a column at the side of the page, and is rearranged in this translation.]

which is

$$4\frac{33}{136}$$

Squaring, $\frac{1089}{18496}$.

$$17\frac{16}{17} \qquad 1088$$

The square is $18\frac{1}{18496}$, which is too large by $\frac{1}{18496}$.

Be it noted that in the printing when proceeding hurriedly, it is not possible to form fractions and fractions of fractions conveniently in this form, as for instance in the case of

$$4. \ \& \ \frac{2}{8}.$$

$$\& \ \frac{2}{8}.$$

$$\& \ \frac{2}{8}$$

as we are forcing ourselves to do in this example, but we may denote all of them by adopting this device: $4. \ \& \ \frac{2}{8.} \ \& \ \frac{2}{8.} \ \& \ \frac{2}{8.}$ letting a period by the 8 in the denominator of each fraction mean that the following fraction is a fraction of the denominator.

· · · · · · · · · ·

I shall find the third fraction by the above mentioned method to be $4. \ \& \ \frac{2}{8.} \ \& \ \frac{2}{8.} \ \& \ \frac{2}{8.}$, or as I might say $4. \ \& \ \frac{2}{8.} \ \& \ \frac{2}{8.} \ \& \ \frac{1}{4}$, which is $4. \ \& \ \frac{2}{8} \ \& \ \frac{8}{33}$, or $4. \ \& \ \dfrac{1}{4 + \frac{4}{33}}$, which is $4. \ \& \ \frac{33}{136}$, which will be in excess since $\frac{33}{136}$. the whole fraction is greater than $\frac{8}{33}$. the added fraction. The excess of the square over 18 is $\frac{1}{18496}$ which arises from multiplying $\frac{33}{136}$, the whole fraction by $\frac{1}{136 \times 33}$ in which the whole fraction is greater than the $\frac{8}{33}$ which is the added fraction."[1]

---

[1] [Cataldi continues this work until he reaches the fifteenth fraction.]

JACQUES (I) BERNOULLI

ON THE "BERNOULLI NUMBERS"

(Translated from the Latin by Professor Jekuthiel Ginsburg, Yeshiva College, New York City.)

Of the various special kinds of numbers used in analysis, there is hardly a species that is so important and so generally applicable as the Bernoulli Numbers. Their numerous properties and applications have caused the creation of an extensive literature on the subject which still continues to attract the attention of scholars. The first statement of the properties of these numbers was given to the world by their inventor Jacques (1) Bernoulli (1654-1705) in his posthumously printed work, *Ars Conjectandi* (Basel, 1713), pages 95 to 98. These pages are here translated.

The excerpt is interesting from more than one point of view. First, we witness in it the first stroke of genius that caused ripples in human thought that have not died out even to the present day. Second, the memoir is as fresh and vigorous today as when it was written; in fact, it could be used even now as a popular exposition of the simpler properties of the Bernoulli Numbers. Third, the text reveals the personal touch, the unbounded enthusiasm of the author over the power of the numbers later called by his name. His remark that the results of Bullialdus's enormous treatise could, by means of his numbers, be compressed in less than one page, is both striking and illuminating. Nor is the element of puzzle and mystery lacking. Regardless of the fact that the discovery is more than 200 years old, mathematicians have not been able as yet to find by what process Bernoulli derived the properties of his numbers which he gives in these pages. They can readily be derived by various modern methods, but how did he derive them with the means at his disposal? It is also interesting to compare his criticism of Wallis's use of incomplete induction with his own use of the same imperfect tool. In short, in the compass of three printed pages we get not only information about the invention but also glimpses of the person of the great master.

We will observe here in passing that, many [scholars] engaged in the contemplation of figurate numbers (among them Faulhaber[1]

---

[1] [Johann Faulhaber, a successful teacher of mathematics in Ulm, was born there on May 5, 1580, and died there in 1635 (D. E. Smith, *History of Mathematics*, Vol. I, p. 418). With the help of his friend and protector Johann Remmelin he published a number of mathematical works. In his *Mysterium Arithmeticum*, 1615, he discussed the properties of figurate numbers. Bernoulli possibly refers to this work of his. Faulhaber also developed formulas for Σn^c from $c = 1$ to $c = 17$ (Tropfke, *Geschichte der Elementar Mathematik*, Vol. VI, p. 22).]

and Remmelin of Ulm, Wallis, Mercator,[1] in his *Logarithmo-technia* and others) but I do not know of one who gave a general and scientific proof of this property.[2]

Wallis in his *Arithmetica Infinitorum* investigated by means of induction the ratios that the series of squares, cubes, and other powers of natural numbers have to the series of terms each equal to the greatest term. This he put in the foundation of his method. His next step was to establish 176 properties of trigonal, pyramidal, and other figurate numbers, but it would have been better and more fitting to the nature of the subject if the process would have been reversed and he would have first given a discussion of figurate numbers, demonstrated in a general and accurate way, and only then have proceeded with the investigation of the sums of powers of the natural numbers. Even disregarding the fact that the method of induction is not sufficiently scientific and, moreover, requires special work for every new series; it is a method of common judgment that the simpler and more primitive things should precede others. Such are the figurate numbers as related to the powers, since they are formed by addition, while the others are formed by multiplication; chiefly, however, because the series of figurate numbers, supplied with the corresponding zeros[3] have a submultiple ratio to the series of equals.[4] In case of powers (when

---

[1] [Nicolaus Mercator was born near Cismar in Holstein, c. 1620, and died in Paris in February, 1687. His *Arithmotechnia sive methodus construendi logarithmos nova accurata et facilis...* was published in London in 1678 (Smith, *l. c.*, I, 434). Bernoulli fails to mention Oughtred who pointed out the correspondence between the binominal coefficients and the figurate numbers, as did also Nicolo Tartaglia, Pascal, and others.]

[2] [The property refers to the method of finding the nth term and the sum of n terms in a series of figurate numbers.]

[3] [The number of zeros to put in the triangle of figurate numbers to make it look like a square

1	0	0	0	0	0	0
1	1	0	0	0	0	0
1	2	1	0	0	0	0
1	3	3	1	0	0	0
1	4	6	4	1	0	0
1	5	10	10	5	1	0
1	6	15	20	15	6	1
1	7	21	35	35	21	7

Bernoulli counts each zero as a term. Thus the sum of the terms of the third column is $0 + 0 + 1 + 3 + 6 + 10 + 15 + 21$.]

[4] [That is taking for example column three in the preceding footnote, the ratio of the sum of any number of terms beginning with two zeros to the sum of

the number of terms is finite) this does not hold without some excess or defect no matter how many zeros be added. With the known sums of the figurate numbers it is not difficult to derive the sums of the powers. I will show briefly how it is done.

Let the series of natural numbers 1, 2, 3, 4, 5, etc. up to n be given, and let it be required to find their sum, the sum of the squares, cubes, etc. Since in the table of combinations the general term of the second column is $n - 1$ and the sum of all terms, that is, all $n - 1$, or $\int \overline{n - 1}$ in consequence of above is[2]

$$\frac{n.n - 1}{1.2} = \frac{nn - n}{2}.$$

The sum $\int \overline{n - 1}$ or

$$\int n - \int 1 = \frac{nn - n}{n}.$$

Therefore

$$\int n = \frac{nn - n}{2} + \int 1.$$

But $\int 1$ (the sum of all units) $= n$. Therefore the sum of all n or

$$\int n = \frac{nn - n}{2} + n = \tfrac{1}{2}nn + \tfrac{1}{2}n.$$

A term of the third column is generally taken to be

$$\frac{n - 1.n - 2}{1.2} = \frac{nn - 3n + 2}{2}$$

---

a series of terms each equal to the last term of the first series will be $\frac{1}{3}$. Thus,

$$\frac{0 + 0 + 1 + 3}{3 + 3 + 3 + 3} = \frac{4}{12} = \frac{1}{3}; \quad \frac{0 + 0 + 1 + 3 + 6 + 10}{10 + 10 + 10 + 10 + 10 + 10} = \frac{1}{3};$$

$$\frac{0 + 0 + 1 + 3 + 6 + 10 + 15}{15 + 15 + 15 + 15 + 15 + 15 + 15} = \frac{1}{3}, \text{ etc.}$$

In the fourth column we get ¼, and in the fifth we have ⅕. In every case the first series is a submultiple of the second, which he calls the series of equal terms.]

[2] [I. e., the sum of $0 + 1 + 2 + \ldots + n - 1$ is as was stated above $\frac{1}{2}$ of $(n - 1) + (n - 1) + (n - 1) \ldots (n$ times) since the ratio

$$\frac{0 + 1 + 2 + \ldots + (n - 1)}{(n - 1) + (n - 1) \ldots + (n - 1)} = \frac{1}{2}. \quad \text{Hence} \frac{s}{n(n - 1)} = \frac{1}{2} \therefore s = \frac{n(n - 1)}{1.2}.$$

Throughout the work Bernoulli uses the old form of s, our present integral sign ($\int$) where we would now use Σ. His usage has been followed in the translation. He also writes n.n $- 1$ where we would write n(n $- 1$) and he expresses equality by the sign $\propto$ but in this translation the sign $=$ will be used. His use of nn instead of n² should also be noted.]

and the sum of all terms $\left(\text{that is, of all } \dfrac{nn - 3n + 2}{2}\right)$ is

$$\frac{n.n - 1.n - 2}{1.2.3} = \frac{n^3 - 3nn + 2n}{6}$$

PARS SECUNDA.   97

$$\infty \; \frac{n^4 - 6n^3 + 11nn - 6n}{24}, \text{ erit, utique } \sqrt{\frac{n^3 - 6nn + 11^{n-6}}{6}}, \text{ hoc eft,}$$

$\int \frac{1}{6} n^3 - \int nn + \int \frac{1}{6} n - \int 1 \; \infty \; \frac{n^4 - 6n^3 + 11nn - 6n}{24}$, indeque $\int \frac{1}{6} n^3 \; \infty$
$\frac{n^4 - 6n^3 + 11nn - 6n}{24} + \int nn - \int \frac{1}{6} n + \int 1.$ Et quoniam per modo inventa $\int nn \; \infty \; \frac{1}{3} n^3 + \frac{1}{2} nn + \frac{1}{6} n$; nec non $\int \frac{1}{6} n$ five $\frac{1}{6} \int n \; \infty \; \frac{1}{12} nn + \frac{1}{12} n$, & $\int 1 \; \infty \; n$; hinc factâ horum fubftitutione emerget $\int \frac{1}{6} n^3 \; \infty$
$\frac{n^4 - 6n^3 + 11nn - 6n}{24} + \frac{1}{3} n^3 + \frac{1}{2} nn + \frac{1}{6} n - \frac{1}{12} nn - \frac{1}{12} n + n \; \infty$
$\frac{1}{24} n^4 + \frac{1}{12} n^3 + \frac{1}{24} nn$, ejusque proin fextuplum $\int n^3$ (fumma cuborum) $\infty \; \frac{1}{4} n^4 + \frac{1}{2} n^3 + \frac{1}{4} nn$. Atque fic porrò ad altiores gradatim poteftates pergere, levique negotio fequentem adornare laterculum licet:

Summæ Poteftatum.

$\int n \; \infty \; \frac{1}{2} nn + \frac{1}{2} n.$

$\int nn \; \infty \; \frac{1}{3} n^3 + \frac{1}{2} nn + \frac{1}{6} n.$

$\int n^3 \; \infty \; \frac{1}{4} n^4 + \frac{1}{2} n^3 + \frac{1}{4} nn.$

$\int n^4 \; \infty \; \frac{1}{5} n^5 + \frac{1}{2} n^4 + \frac{1}{3} n^3 \; * - \frac{1}{30} n.$

$\int n^5 \; \infty \; \frac{1}{6} n^6 + \frac{1}{2} n^5 + \frac{5}{12} n^4 \; * - \frac{1}{12} nn.$

$\int n^6 \; \infty \; \frac{1}{7} n^7 + \frac{1}{2} n^6 + \frac{1}{2} n^5 \; * - \frac{1}{6} n^3 \; * + \frac{1}{42} n.$

$\int n^7 \; \infty \; \frac{1}{8} n^8 + \frac{1}{2} n^7 + \frac{7}{12} n^6 \; * - \frac{7}{24} n^4 \; * + \frac{1}{12} nn.$

$\int n^8 \; \infty \; \frac{1}{9} n^9 + \frac{1}{2} n^8 + \frac{2}{3} n^7 \; * - \frac{7}{15} n^5 \; * + \frac{2}{9} n^3 \; * - \frac{1}{30} n.$

$\int n^9 \; \infty \; \frac{1}{10} n^{10} + \frac{1}{2} n^9 + \frac{3}{4} n^8 \; * - \frac{7}{10} n^6 \; * + \frac{1}{2} n^4 \; * - \frac{1}{12} nn.$

$\int n^{10} \; \infty \; \frac{1}{11} n^{11} + \frac{1}{2} n^{10} + \frac{5}{6} n^9 \; * - 1 \, n^7 \; * + 1 \, n^5 \; * - \frac{1}{2} n^3 \; * + \frac{5}{66} n.$

Quin imò qui legem progreffionis inibi attentius infpexerit, eundem etiam continuare poterit abfq; his ratiociniorum ambagibus: Sumtâ enim c pro poteftatis cujuslibet exponente, fit fumma omnium n^c feu

$$\int n^c \; \infty \; \frac{1}{c+1} n^{c+1} + \frac{1}{2} n^c + \frac{c}{2} An^{c-1} + \frac{c.c-1.c-2}{2.3.4} Bn^{c-3} +$$
$$\frac{c.c-1.c-2.c-3.c-4}{2.3.4.5.6} Cn^{c-5} + \frac{c.c-1.c-2.c-3.c-4.c-5.c-6}{2.3.4.5.6.7.8} Dn^{c-7} \ldots$$

& ita deinceps, exponentem poteftatis ipfius n continuè minuendo binario, quoufque perveniatur ad n vel nn. Literæ capitales A, B, C, D &c. ordine denotant coëfficientes ultimorum terminorum pro $\int nn$, $\int n^4$, $\int n^6$, $\int n^8$ &c. nempe A $\infty \; \frac{1}{6}$, B $\infty - \frac{1}{30}$

N

We will have then that

$$\int \frac{n^2 - 3n + 2}{2}$$

or

$$\int \frac{1}{2} \, ^2 nn - \int \frac{3}{2} n + \int 1 = \frac{n^3 - 3nn + 2n}{6}$$

and

$$\int \tfrac{1}{2}nn = \frac{n^3 - 3nn + 2n}{6} + \int \tfrac{3}{2}n - \int 1;$$

but

$$\int \tfrac{3}{2}n = \tfrac{3}{2}\int n = \tfrac{3}{4}nn + \tfrac{3}{4}n$$

and

$$\int 1 = n.$$

Substituting, we have

$$\int \tfrac{1}{2}nn = \frac{n^3 - 3nn + 2n}{6} + \frac{3nn + 3n}{4} - n = \tfrac{1}{6}n^3 + \tfrac{1}{4}nn + \tfrac{1}{12}n,$$

of which the double $\int nn$ (the sum of the squares of all n) $= \tfrac{1}{3}n^3 + \tfrac{1}{2}nn + \tfrac{1}{6}n.$

A term of the fourth column is generally

$$\frac{n - 1.n - 2.n - 3}{1.2.3} = \frac{n^3 - 6nn + 11n - 6}{6},$$

and the sum of all terms is

$$\frac{n.n - 1.n - 2.n - 3}{1.2.3.4} = \frac{n^4 - 6n^3 + 11nn - 6n}{24}.$$

It must certainly be that

$$\int \frac{n^3 - 6nn + 11n - 6}{6};$$

that is

$$\int \tfrac{1}{6}n^3 - \int nn + \int \tfrac{11}{6}n - \int 1 = \frac{n^4 - 6n^3 + 11nn - 6n}{24}.$$

Hence

$$\int \tfrac{1}{6}n^3 = \frac{n^4 - 6n^3 + 11nn - 6n}{24} + \int nn - \int \tfrac{11}{6}n + \int 1.$$

And before it was found that $\int nn = \tfrac{1}{3}n^3 + \tfrac{1}{2}nn + \tfrac{1}{6}n$, $\int \tfrac{11}{6}n$ or $\tfrac{11}{6}\int n = \tfrac{11}{12}nn + \tfrac{11}{12}n$, and $\int 1 = n$.
When all substitutions are made, the following results:

$$\int \tfrac{1}{6}n^3 = \frac{n^4 - 6n^3 + 11nn - 6n}{24} + \tfrac{1}{3}n^3 + \tfrac{1}{2}nn + \tfrac{1}{6}n - \tfrac{11}{12}nn$$

$$- \tfrac{11}{12}n + n$$

$$= \tfrac{1}{24}n^4 + \tfrac{1}{12}n^3 + \tfrac{1}{24}nn;$$

or, multiplying by 6,

$$\int n^3 = \tfrac{1}{4}n^4 + \tfrac{1}{2}n^3 + \tfrac{1}{4}nn.$$

Thus we can step by step reach higher and higher powers and with slight effort form the following table:

Sum of Powers

$$\int n = \tfrac{1}{2}nn + \tfrac{1}{2}n,$$
$$\int nn = \tfrac{1}{3}n^3 + \tfrac{1}{2}nn + \tfrac{1}{6}n,$$
$$\int n^3 = \tfrac{1}{4}n^4 + \tfrac{1}{2}n^3 + \tfrac{1}{4}nn$$
$$\int n^4 = \tfrac{1}{5}n^5 + \tfrac{1}{2}n^4 + \tfrac{1}{3}n^3 - \tfrac{1}{30}n,$$
$$\int n^5 = \tfrac{1}{6}n^6 + \tfrac{1}{2}n^5 + \tfrac{5}{12}n^4 - \tfrac{1}{12}nn,$$
$$\int n^6 = \tfrac{1}{7}n^7 + \tfrac{1}{2}n^6 + \tfrac{1}{2}n^5 - \tfrac{1}{6}n^3 + \tfrac{1}{42}n,$$
$$\int n^7 = \tfrac{1}{8}n^8 + \tfrac{1}{2}n^7 + \tfrac{7}{12}n^6 - \tfrac{7}{24}n^4 + \tfrac{1}{12}nn,$$
$$\int n^8 = \tfrac{1}{9}n^9 + \tfrac{1}{2}n^8 + \tfrac{2}{3}n^7 - \tfrac{7}{15}n^5 + \tfrac{2}{9}n^3 - \tfrac{1}{30}n,$$
$$\int n^9 = \tfrac{1}{10}n^{10} + \tfrac{1}{2}n^9 + \tfrac{3}{4}n^8 - \tfrac{7}{10}n^6 + \tfrac{1}{2}n^4 - \tfrac{3}{20}nn,$$
$$\int n^{10} = \tfrac{1}{11}n^{11} + \tfrac{1}{2}n^{10} + \tfrac{5}{6}n^9 - 1n^7 + 1n^5 - \tfrac{1}{2}n^3 + \tfrac{5}{66}n.$$

Whoever will examine the series as to their regularity may be able to continue the table. Taking c to be the power of any exponent, the sum of all n^c or

$$\int n^c = \frac{1}{c+1} n^{c+1} + \frac{1}{2} n^c + \frac{c}{2} An^{c-1} + \frac{c.c - 1.c - 2}{2.3.4} Bn^{c-3}$$
$$+ \frac{c.c - 1.c - 2.e - 3.c - 4}{2.3.4.5.6} Cn^{c-5}$$
$$+ \frac{c.c - 1.c - 2.c - 3.c - 4.c - 5.c - 6}{2.3.4.5.6.7.8} Dn^{c-7},$$

and so on, the exponents of n continually decreasing by 2 until n or nn is reached. The capital letters A, B, C, D denote in order the coefficients of the last terms in the expressions for $\int nn, \int n^4, \int n^6, \int n^8$ etc., namely, A is equal to $\tfrac{1}{6}$, B is equal to $-\tfrac{1}{30}$, C is equal to $\tfrac{1}{42}$, D is equal to $-\tfrac{1}{30}$.

These coefficients are such that each one completes the others in the same expression to unity. Thus D must have the value $-\tfrac{1}{30}$ because $\tfrac{1}{9} + \tfrac{1}{2} + \tfrac{2}{3} - \tfrac{7}{15} + \tfrac{2}{9} + (+ D) - \tfrac{1}{30} = 1$.

With the help of this table it took me less than half of a quarter of an hour to find that the tenth powers of the first 1000 numbers being added together will yield the sum

$$91,409,924,241,424,243,424,241,924,242,500$$

From this it will become clear how useless was the work of Ismael Bullialdus[1] spent on the compilation of his voluminous *Arithmetica Infinitorum* in which he did nothing more than compute with immense labor the sums of the first six powers, which is only a part of what we have accomplished in the space of a single page.

---

[1] [The title of Bullialdus's (1605–1694) work is *Opus novum ad arithmeticum infinitorum*. It was published in Paris 1682 and consists of six parts.]

EULER

Proof that Every Integer is a Sum of Four Squares

(Translated from the Latin by Professor E. T. Bell, California Institute of Technology, Pasadena, California.)

Léonard (Leonhard) Euler (1707–1783), a pupil of Jean (I) Bernoulli, was not only one of the greatest mathematicians and astronomers of his century, but he was also versed in theology, medicine, botany, physics, mechanics, chemistry, and the Oriental as well as the modern languages. He was a voluminous writer, and there was hardly a branch of mathematics to which he did not contribute. The selection here translated serves to illustrate his method of attacking a problem in the theory of numbers. It is taken from his *Commentationes Aritbmeticae Collectæ*, Petropoli, 1849, edited by P. H. Fuss and N. Fuss (Vol. I, pp. 543–546) but appeared earlier in the *Acta Eruditorum* (p. 193, Leipzig, 1773) and the *Acta Petrop.*, (p. 48, I. II., 1775. Exhib. Sept. 21, 1772). In preparing the article, the effort has been made to give a free translation that shall clearly convey Euler's meaning, in preference to following too closely the rather poor Latin of the day. Of his two proofs for the exceptional case of $n = 2$, only the simpler one has been given. From the modern point of view, the proof of the theorem is not very satisfactory, but it serves to illustrate the theory of numbers of the eighteenth century.

Lemma.—The product of two numbers, each of which is a sum of four squares, may always be expressed as a sum of four squares.

Let such a product be

$$(a^2 + b^2 + c^2 + d^2)(\alpha^2 + \beta^2 + \gamma^2 + \delta^2).$$

Write

$$
\begin{aligned}
A &= a\alpha + b\beta + c\gamma + d\delta, \\
B &= a\beta - b\alpha - c\delta + d\gamma, \\
C &= a\gamma + b\delta - c\alpha - d\beta, \\
D &= a\delta - b\gamma + c\beta - d\alpha.
\end{aligned}
$$

Then

$$A^2 + B^2 + C^2 + D^2 = (a^2 + b^2 + c^2 + d^2)(\alpha^2 + \beta^2 + \gamma^2 + \delta^2),$$

since obviously the cross products in A^2, B^2, C^2, D^2 cancel.

Theorem 1.—*If N is a divisor of a sum of four squares, say of $p^2 + q^2 + r^2 + s^2$, no one of which is divisible by N, then N is the sum of four squares.*

It will first be shown that each of the four roots p, q, r, s may be chosen less than $\frac{1}{2}N$.[1]

I. Let n be the quotient on dividing the sum of four squares by N, so that $Nn = p^2 + q^2 + r^2 + s^2$. Then we may write

$$p = a + n\alpha, \quad q = b + n\beta, \quad r = c + n\gamma, \quad s = d + n\delta,$$

where each remainder a, b, c, d does not exceed $\frac{1}{2}n$ in absolute value.[2]

Hence

$$a^2 + b^2 + c^2 + d^2 < n^2.$$

II. By substituting the above values of p, q, r, s in

$$Nn = p^2 + q^2 + r^2 + s^2,$$

we get

$$Nn = a^2 + b^2 + c^2 + d^2 + 2n(a\alpha + b\beta + c\gamma + d\delta) + n^2(\alpha^2 + \beta^2 + \gamma^2 + \delta^2);$$

whence it follows that n must be a divisor of $a^2 + b^2 + c^2 + d^2$.

Put

$$a^2 + b^2 + c^2 + d^2 = nn'.$$

Then $n > n'$, or $n' < n$. By division we get

$$N = n' + 2A + n(\alpha^2 + \beta^2 + \gamma^? + \delta^2).$$

III. Multiply now by n'. Then, since

$$nn' = a^2 + b^2 + c^2 + d^2,$$

we have, by the Lemma,

$$nn'(\alpha^2 + \beta^2 + \gamma^2 + \delta^2) = A^2 + B^2 + C^2 + D^2,$$

Combining this with the preceding equation we find

$$Nn' = n'^2 + 2n'A + A^2 + B^2 + C^2 + D^2,$$

and therefore

$$(n' + A)^2 + B^2 + C^2 + D^2 = Nn'.$$

IV. By repeating the foregoing argument we obtain a decreasing sequence of integers Nn', Nn'', etc., and hence finally we reach $N.1$ and its expression as a sum of four squares.

---

[1] [It is to be observed in the following proof that n is different from 2; this case is tacitly ignored until the so-called corollary, following the proof, which disposes of the exceptional case implicit in the argument as presented.]

[2] [But see the preceding footnote. The condition as to absolute values safeguards the assertion above, but it does not take care of all possibilities in the proof which immediately follows, unless, as with Euler, and as indicated in the preceding footnote, we attend to the corollary.]

LEONHARD EULER

(*Facing page 92.*)

Corollary.—To dispose of the apparent exception, let p, q, r, s be odd numbers and n an even number. Then, since

$$Nn = p^2 + q^2 + r^2 + s^2,$$

we have

$$\tfrac{1}{2}Nn = \left(\frac{p+q}{2}\right)^2 + \left(\frac{p-q}{2}\right)^2 + \left(\frac{r+s}{2}\right)^2 + \left(\frac{r-s}{2}\right)^2,$$

and the four squares on the right are integers. A like reduction may be performed so long as the roots of all the squares are odd. Thus the exception when $n = 2$ disappears.

THEOREM 2.—*If N is prime, not only 4 squares not divisible by N, can be found in an infinity of ways, whose sum is divisible by N, but also 3 squares.*

For, with respect to N, all numbers are of one or other of the N forms

$$\lambda N,\ \lambda N + 1,\ \lambda N + 2,\ \lambda N + 3, \ldots, \lambda N + N - 1.$$

Disregard the first form, λN, which contains all the multiples of N. There remain $N - 1$ forms, and we observe that the square of a number of the form $\lambda N + 1$, likewise the square of a number of the form $\lambda N + N - 1$, belongs to the same form $\lambda N + 1$. Similarly the square of a number of either form $\lambda N + 2$, $\lambda N + N - 2$ is of the form $\lambda N + 4$; and so on. Thus the squares of all numbers not of the form λN are comprised in the $\frac{1}{2}(N - 1)$ forms

$$\lambda N + 1,\quad \lambda N + 4,\quad \lambda N + 9,\quad \text{etc.,}$$

which will be called forms of the first class, and will be denoted by

$$\lambda N + a,\quad \lambda N + b,\quad \lambda N + c,\quad \lambda N + d, \text{ etc.,}$$

so that a, b, c, d,...denote the squares 1, 4, 9, 16,...or, if these exceed N, their residues on division by N. The remaining $\frac{1}{2}(N - 1)$ forms will be denoted by

$$\lambda N + \alpha,\ \lambda N + \beta,\ \lambda N + \gamma, \text{ etc.,}$$

which will be called forms of the second class. It is easy to prove the following three properties concerning these classes.[1]

I. The product of two numbers of the first class is again contained in the first class, since evidently $\lambda N + ab$ is in the first class. If $ab > N$, the residue of ab on division by N is to be understood.

---

[1] [These merely are the well known elementary properties of quadratic residues, which, since the time of Gauss, are phrased more briefly in modern terminology. The like applies to the proof given presently.]

II. Numbers of the first class a, b, c, d, etc., multiplied into any numbers of the second class α, β, γ, δ, etc., give products in the second class.

III. A product of two numbers in the second class, say $\alpha\beta$, falls into the first class.

We shall now proceed to the proof of Theorem 2, by means of a contradiction.

Suppose then that there are no three squares, not all divisible by N, whose sum is divisible by N. Then, so much the more, there are no two such squares. Hence it follows at once that the form $\lambda N - a$, or what amounts to the same, $\lambda N + (N - a)$, cannot occur in the first class. For, if there were a square of the form $\lambda N - a$, the sum of this and $\lambda N + a$ would be divisible by N, contrary to hypothesis. Hence the form $\lambda N - a$ is necessarily in the second class; the numbers -1, -4, -9, etc., are among those of the set α, β, γ, δ, etc. Let f by any number of the first class, so that there exist squares of the form $\lambda N + f$. If to one of these be added a square of the form $\lambda N + 1$, the sum of the two will have the form $\lambda N + f + 1$. Now if there were squares of the form $\lambda N - f - 1$, there would exist a sum of three squares divisible by N. Since this is denied, the form $\lambda N - f - 1$ is not contained in the first class, and hence it is in the second. But in the second class there appear the numbers -1 and $-f - 1$, and hence, by III above, their product $f + 1$ is in the first class. In the same way it may be shown that the numbers

$$f + 2, \quad f + 3, \quad f + 4, \text{ etc.,}$$

must occur in the first class. Hence, taking $f = 1$, we see that all the numbers

$$\lambda N + 1, \quad \lambda N + 2, \quad \lambda N + 3, \text{ etc.,}$$

occur in the first class, and therefore that there are none left for the second class. But, by the same reasoning, we see that the numbers -1, $-f - 1$, $-f - 2$, etc., occur in the second class, and hence all forms are in the second class. This obviously is a contradiction. It follows therefore that it is false that there are not three squares whose sum is divisible by N. Hence there are indeed three squares, and much more therefore four squares, of the prescribed kind whose sum is divisible by N.

Corollary.—From this theorem, combined with the preceding, it follows obviously that every number is a sum of four or fewer squares.

EULER

Use of the Letter e to Represent 2.718...

(Selections Translated by Professor Florian Cajori, University of California, Berkeley, California.)

Prominent among the mathematicians who have contributed notations which have met with general adoption is the Swiss Leonhard Euler (1707–1783). One of his suggestions, made when he was a young man of twenty or twenty-one, at the court in St. Petersburg, was the use of the letter e to stand for 2.718..., the base of the natural system of logarithms. It occurs in a manuscript of Euler entitled "Meditation upon Experiments made recently on the firing of Canon" (Meditatio in Experimenta explosione tormentorum nuper instituta). The manuscript was first printed in 1862 in Euler's *Opera postuma mathematica et physica*, Petropoli, 1862, edited by P. H. Fuss and N. Fuss (Vol. II, p. 800–804). In this article, seven experiments are cited, which were performed between Aug. 21 and Sept. 2, 1727. These dates, and the word "recently" (nuper) in the title, would indicate that the article was written in 1727 or 1728. In it the letter e occurs sixteen times to represent 2.718... From page 800, we translate the following:

Let c designate the diameter of a globe [spherical projectile], in scruples of Rhenish feet,[1] $m:n$ the ratio of the specific gravity of the globe to the specific gravity of the air or the medium in which the globe moves, let t seconds be the length of time of the globe in air, let also the required height to which the body rises be x. For the number whose logarithm is unity, let e be written, which is 2,7182817...whose logarithm[2] according to Vlacq is 0, 4342944. Also let N indicate the number of degrees of an arc, whose tangent is:

$$\sqrt{e^{\frac{3nx}{4mc}} - 1},$$

the sinus totus [or radius] $= 1$. The required altitude x may be obtained from the following equation:

$$t = \frac{m\sqrt{c}}{447650\sqrt{3n(m-n)}} \left(125N - 7162 \log. \left(\sqrt{e^{\frac{3nx}{4mc}}} - \sqrt{e^{\frac{nx}{4mc}} - 1} \right) \right).$$

That the analysis may proceed more easily, let us call $\sqrt{e^{\frac{3nx}{4mc}} - 1}$

---

[1] [Rhenish foot = 1000 scruples.]

[2] [That is, logarithm to the base 10.]

$= y$, then N will be the number of degrees of the arc whose tangent is y, . . .

In a letter of Nov. 25, 1731, addressed to Goldbach[1] (first published in 1843), Euler solves the differential equation

$$dz - 2zdv + \frac{zdv}{v} = \frac{dv}{v}, \text{ thus:}$$

This multiplied by e^{lv-2v}, or what is the same, by $e^{-2v}v$ (e denotes that number, whose hyperbolic logarithm is $= 1$), becomes

$$e^{-2v}vdz - 2e^{-2v}zvdv + e^{-2v}zdv = e^{-2v}dv,$$

which, integrated, gives

$$e^{-2v}vz = \text{Const.} - \frac{1}{2}e^{-2v}$$

or

$$2vz + 1 = ae^{2v} \ldots$$

The earliest occurrence *in print* of the letter e for 2.718... is in Euler's *Mechanica*, 1736. It is found in Vol. I, page 68, and in other places, as well as in Vol. II, page 251, and on many of the 200 pages following. We quote, in translation, from Vol. I, page 68, where c means the velocity of the point under consideration:

Corollary II

171. Although in the foregoing equation the force p does not occur, its direction still remains, which depends upon the ratio of the elements dx and dy. Given therefore the direction of the force which moves the point and the curve along which the point moves, one can, from these data alone, derive the velocity of the point at any place. For there will be $\dfrac{dc}{c} = \dfrac{dyds}{zdx}$ or $c = e^{\int \frac{dyds}{zdx}}$, where e denotes the number whose hyperbolic logarithm is 1.

The use of the letter e, affected by imaginary exponents, in analytica expressions that were new to mathematics, occurs in a dissertation of Euler's entitled "On the sums of reciprocal series arising from the Powers of the natural Numbers" (De summis serierum reciprocarum ex potestatibus numerorum naturalium ortarum).[2] He lets s denote a circular arc and develops $\sin s$ into the now familiar infinite series. On page 177 he gives without explanation the exponential expression for $\sin s$, and the fundamental limit for e^z in the following passage:

---

[1] *Correspondance mathématique et physique de quelques célèbres géomètres du XVIII*ᵐᵉ *siècle.* Par P. H. Fuss, St. Pétersbourg, 1843, Tome I, p. 58.
[2] *Miscellanea Berolinensia*, p. 172, Vol. VII, Berlin, 1743.

Hence I am now able to write down all the roots or factors of the following infinite expression

$$S - \frac{S^3}{1.2.3} + \frac{S^5}{1.2.3.4.5} - \frac{S^7}{1.2.3...7} + \frac{S^9}{1.2.3...9} - \&c.$$

Indeed that expression is equivalent to this $\dfrac{e^{s\sqrt{-1}} - e^{-s\sqrt{-1}}}{2\sqrt{-1}}$, e denoting the number whose logarithm is $= 1$, and, since $e^z = \left(1 + \dfrac{z}{n}\right)^n$, when n emerges an infinite number, the given infinite expression is reduced to this:

$$\frac{\left(1 + \dfrac{s\sqrt{-1}}{n}\right)^n - \left(1 - \dfrac{s\sqrt{-1}}{n}\right)^n}{2\sqrt{-1}} \quad \ldots$$

More systematic development is found in Euler's *Introductio in analysin infinitorum*, Vol. I, Lausannæ, 1748. We quote from § 138, in which the letter i is an infinitely great number:

...Substituting gives

$$\cos. v = \frac{\left(1 + \dfrac{v\sqrt{-1}}{i}\right)^i + \left(1 - \dfrac{v\sqrt{-1}}{i}\right)^i}{2}$$

and

$$\sin. v = \frac{\left(1 + \dfrac{v\sqrt{-1}}{i}\right)^i - \left(1 - \dfrac{v\sqrt{-1}}{i}\right)^i}{2\sqrt{-1}}$$

In the preceding chapter we saw that

$$\left(1 + \frac{z}{i}\right)^i = e^z$$

e denoting the base of hyperbolic logarithms; writing for z, first $+v\sqrt{-1}$, then $-v\sqrt{-1}$, there will be

$$\cos. v = \frac{e^{+v\sqrt{-1}} + e^{-v\sqrt{-1}}}{2}$$

and

$$\sin. v = \frac{e^{+v\sqrt{-1}} - e^{-v\sqrt{-1}}}{2\sqrt{-1}}$$

From these it is perceived how imaginary exponential quantities are reduced to the sine and cosine of real arcs. For, there is

$$e^{+v\sqrt{-1}} = \cos. v + \sqrt{-1}. \sin. v$$
$$e^{-v\sqrt{-1}} = \cos. v - \sqrt{-1}. \sin. v.$$

If in the formula for $e^{+v\sqrt{-1}}$ one substitutes π and v, there results the famous formula $e^{\pi\sqrt{-1}} = -1$, indicating the strange interrelation of π and e. Euler states this relation in the logarithmic form and generalized, in his paper, "De la Controverse entre Mrs. Leibnitz & Bernoulli sur les logarithmes des nombres negatifs et imaginaires," *Histoire de l'academie royale des sciences et belles lettres*, année 1749, Berlin, 1751, where on page 168 he refers to:

...this formula $\cos \varphi + \sqrt{-1}.\sin \varphi$, all logarithms of which are included in this general formula

$$l(\cos \varphi + \sqrt{-1}.\sin \varphi) = (\varphi + p\pi)\sqrt{-1},$$

p indicating any even integer, either affirmative, or negative, or even zero. From this we derive...

$$l - 1 = (1 + p)\pi\sqrt{-1} = q\pi\sqrt{-1},$$

taking q to mark any odd integer. One has therefore:

$$l - 1 = \pm\pi\sqrt{-1}; \pm3\pi\sqrt{-1}; \pm5\pi\sqrt{-1}; \pm7\pi\sqrt{-1}; \&c.$$

HERMITE

On the Transcendence of e

(Translated from the French by Dr. Laura Guggenbühl, Hunter College, New York City.)

Charles Hermite (1822–1901) was one of the best-known writers upon the function theory in the second half of the nineteenth century. He was a professor in the Ecole Polytechnique, an honorary professor in the University of Paris, and a member of the Académie des Sciences. His memoir on the transcendence of e was published in 1873. As is well known,[1] the character of the number π was a source of disturbance in ancient times because of its connection with the classic problem of the quadrature of the circle. From the Greek period, names of famous mathematicians have been connected with transcendental numbers, but it was not until 1844 that a definite step forward was made in the general investigation of the subject. At this time, Liouville proved the existence of these numbers, thus justifying the classification of algebraic and transcendental. Liouville had already proved that e could not be a root of quadratic equation with rational coefficients. Finally, in 1873, Hermite's proof of the transcendence of e appeared. A few years later (1882), Lindemann, modeling his proof upon that of Hermite, proved the transcendence of π.

The memoir is somewhat over thirty pages long and can be divided roughly into three parts. In the first two parts, two distinct proofs of the transcendence of e are given—but as Hermite says, the second is the more rigorous of the two. In the third part, Hermite obtains, applying the method suggested in the second proof, the following approximations:[2]

$$e = \frac{58291}{21444}, \qquad e^2 = \frac{158452}{21444}$$

The translation here given includes, with indicated omissions, the portion referred to above as the second part of the memoir. Since the time when this paper first appeared, many simplifications have been made, so that now one rarely, if ever, sees more than an acknowledgement of the existence and importance of this proof. The name "Hermite's Theorem" is still, however, given to the statement that e is a transcendental number.

---

[1] *Monographs on Topics of Modern Mathematics*, edited by J. W. A. Young, Monograph IX, "The History and Transcendence of π," by D. E. Smith. Additional references are there given.

[2] Correct to six decimal places, $e = 2.718282$. This fraction gives $e = 2.718289$.
The correction of a numerical mistake pointed out by Picard, in his edition of Hermite's work, increases the accuracy of this approximation.

...But, as a more general case, take

$$F(z) = (z - z_0)^{\mu_0}(z - z_1)^{\mu_1}\ldots(z - z_n)^{\mu_n}$$

for any integral values whatever of the exponents, upon integrating both members of the identity

$$\frac{d[e^{-z}F(z)]}{dz} = e^{-z}[F'(z) - F(z)],$$

one obtains

$$e^{-z}F(z) = \int e^{-z}F'(z)dz - \int e^{-z}F(z)dz,$$

from which it follows that

$$\int_{z_0}^{Z} e^{-z}F(z)dz = \int_{z_0}^{Z} e^{-z}F'(z)dz.^{[1]}$$

Now the formula

$$\frac{F'(z)}{F(z)} = \frac{\mu_0}{z - z_0} + \frac{\mu_1}{z - z_1} + \ldots + \frac{\mu_n}{z - z_n}$$

yields the following decomposition,

$$\int_{z_0}^{Z} e^{-z}F(z)dz = \mu_0 \int_{z_0}^{Z} \frac{e^{-z}F(z)dz}{z - z_0} + \mu_1 \int_{z_0}^{Z} \frac{e^{-z}F(z)dz}{z - z_1}$$
$$+ \ldots + \mu_n \int_{z_0}^{Z} \frac{e^{-z}F(z)dz}{z - z_n}, \ldots$$

...We shall prove that it is always possible to determine two integral polynomials of degree n, $\Theta(z)$ and $\Theta_1(z)$, such that, upon representing one of the roots $z_0, z_1, \ldots z_n$, by ζ, one has the following relation:

$$\int \frac{e^{-z}F(z)f(z)}{z - \zeta} \, dz = \int \frac{e^{-z}F(z)\Theta_1(z)}{f(z)}dz - e^{-z}F(z)\Theta(z).^{[2]}$$

...And further, upon writing $\Theta(z, \zeta)$ in place of $\Theta(z)$, to emphasize the presence of ζ, we have

$$\Theta(z, \zeta) = z^n + \Theta_1(\zeta)z^{n-2} + \Theta_2(\zeta)z^{n-3} + \ldots + \Theta_n(\zeta).^{[3]}$$

---

[1] [Where Z represents any one of the roots $z_0, z_1, \ldots z_n$.]

[2] $f(z) = (Z - z_0)(z - z_1)\ldots(z - z_n)$. The proof of this statement, which is given in detail in the text, is here omitted.

[3] [It is shown in the text that $\Theta_i(\zeta)$ is a polynomial of degree i in ζ, having for coefficients integral functions, with integral coefficients, of the roots z_0, $z_1, \ldots, z_n$.

$\Theta_i(\zeta)$ for $i = 1$, is not to be confused with $\Theta_1(z)$, mentioned above in connection with $\Theta(z)$.]

From this there follows, for the polynomial $\Theta_1(z)$, the formula

$$\frac{\Theta_1(z)}{f(z)} = \frac{\mu_0\Theta(z_0, \zeta)}{z - z_0} + \frac{\mu_1\Theta(z_1, \zeta)}{z - z_1} + \ldots + \frac{\mu_n\Theta(z_n, \zeta)}{z - z_n}.$$

...It is sufficient to take the integrals between the limits z_0 and Z in the relation

$$\int \frac{e^{-z}F(z)f(z)}{z - \zeta}dz = \int \frac{e^{-z}F(z)\Theta_1(z)dz}{f(z)} - e^{-z}F(z)\Theta(z),$$

and thus we obtain the equation

$$\int_{z_0}^{Z} \frac{e^{-z}F(z)f(z)}{z - \zeta}dz = \int_{z_0}^{Z} \frac{e^{-z}F(z)\Theta_1(z)}{f(z)}\,dz$$

$$= \mu_0\Theta(z_0, \zeta)\int_{z_0}^{Z}\frac{e^{-z}F(z)}{z - z_0}\,dz$$

$$+ \mu_1\Theta(z_1, \zeta)\int_{z_0}^{Z}\frac{e^{-z}F(z)}{z - z_1}\,dz$$

$$+\ldots$$

$$+\mu_n\Theta(z_n, \zeta)\int_{z_0}^{Z}\frac{e^{-z}F(z)}{z - z_n}\,dz.$$

We use this equation, in particular, in the case

$$\mu_0 = \mu_1 = \ldots = \mu_n = m;$$

in this case, if one writes

$$m\Theta(z_i, z_k) = (ik)$$

and if one takes ζ successively equal to $z_0, z_1, \ldots, z_n$, the above relations evidently become

$$\int_{z_0}^{Z}\frac{e^{-z}f^{m+1}(z)}{z - z_i}\,dz = (i0)\int_{z_0}^{Z}\frac{e^{-z}f^m(z)}{z - z_0}\,dz$$

$$+(i1)\int_{z_0}^{Z}\frac{e^{-z}f^m(z)}{z - z_1}\,dz$$

$$+\ldots$$

$$+(in)\int_{z_0}^{Z}\frac{e^{-z}f^m(z)}{z - z_n}\,dz,$$

for $i = 0, 1, 2, \ldots, n$. But for the general case, we must still prove the following theorem.

Let Δ and δ be the determinants

$$\begin{vmatrix} \Theta(z_0, z_0) & \Theta(z_1, z_0) \ldots \Theta(z_n, z_0) \\ \Theta(z_0, z_1) & \Theta(z_1, z_1) \ldots \Theta(z_n, z_1) \\ \cdots & \cdots \cdots \cdots \\ \Theta(z_0, z_n) & \Theta(z_1, z_n) \ldots \Theta(z_n, z_n) \end{vmatrix}$$

and

$$\begin{vmatrix} 1 & 1 & \ldots & 1 \\ z_0 & z_1 & \ldots & z_n \\ z_0^2 & z_1^2 & \ldots & z_n^2 \\ \ldots & \ldots & \ldots & \ldots \\ z_0^n & z_1^n & & z_n^n \end{vmatrix} ;$$

then $\Delta = \delta^{2}.$[1] . . . . . .

Now, let

$$\epsilon_m = \frac{1}{1.2 \ldots m} \int_{z_0}^{Z} e^{-z} f^m(z) \, dz,$$

$$\epsilon_m{}^i = \frac{1}{1.2 \ldots m-1} \int_{z_0}^{Z} \frac{e^{-z} f^m(z)}{z - z_i} \, dz,$$

the relation proved above

$$\int_{z_0}^{Z} e^{-z} f^m(z) dz = m \int_{z_0}^{Z} \frac{e^{-z} f^m(z)}{z - z_0} \, dz + m \int_{z_0}^{Z} \frac{e^{-z} f^m(z)}{z - z_1} \, dz$$
$$+ \ldots + m \int_{z_0}^{Z} \frac{e^{-z} f^m(z)}{z - z_n} \, dz$$

becomes simply

$$\epsilon_m = \epsilon_m{}^0 + \epsilon_m{}^1 + \ldots + \epsilon_m{}^n,$$

and the relation

$$\int_{z_0}^{Z} \frac{e^{-z} f^{m+1}(z)}{z - \zeta} \, dz = m\Theta(z_0, \zeta) \int_{z_0}^{Z} \frac{e^{-z} f^m(z)}{z - z_0} \, dz$$
$$+ m\Theta(z_1, \zeta) \int_{z_0}^{Z} \frac{e^{-z} f^m(z)}{z - z_1} \, dz$$
$$+ \ldots$$
$$+ m\Theta(z_n, \zeta) \int_{z_0}^{Z} \frac{e^{-z} f^m(z)}{z - z_n} \, dz,$$

upon taking ζ successively equal to $z_0, z_1, \ldots, z_n$, gives us the following substitution, which we shall represent by S_m, namely

$$\epsilon^0{}_{m+1} = \Theta(z_0, z_0)\epsilon_m{}^0 + \Theta(z_1, z_0)\epsilon_m{}^1 + \ldots + \Theta(z_n, z_0)\epsilon_m{}^n,$$
$$\epsilon^1{}_{m+1} = \Theta(z_0, z_1)\epsilon_m{}^0 + \Theta(z_1, z_1)\epsilon_m{}^1 + \ldots + \Theta(z_n, z_1)\epsilon_m{}^n,$$
$$\ldots \ldots \ldots$$
$$\epsilon^n{}_{m+1} = \Theta(z_0, z_n)\epsilon_m{}^0 + \Theta(z_1, z_n)\epsilon_m{}^1 + \ldots + \Theta(z_n, z_n)\epsilon_m{}^n.$$

If now, one builds up in turn $S_1, S_2, \ldots, S_{m-1}$, one concludes from these, expressions for $\epsilon_m{}^0, \epsilon_m{}^1 \ldots, \epsilon_m{}^n$ in terms of $\epsilon_1{}^0, \epsilon_1{}^1, \ldots, E_1{}^n$, which we shall write as follows.

---

[1] [A short and simple proof for this statement is given in the text.]

$$\epsilon_m{}^0 = A_0\epsilon_1{}^0 + A_1\epsilon_1{}^1 + \ldots + A_n\epsilon_1{}^n,$$
$$\epsilon_m{}^1 = B_0\epsilon_1{}^0 + B_1\epsilon_1{}^1 + \ldots + B_n\epsilon_1{}^n,$$
$$\cdot\cdot\cdot\cdot\cdot\cdot\cdot\cdot\cdot\cdot$$
$$\epsilon_m{}^n = L_0\epsilon_1{}^0 + L_1\epsilon_1{}^1 + \ldots + L_n\epsilon_1{}^n,$$

and the determinant of this new substitution, being equal to the product of the determinants of the partial substitutions, will be $\delta^{2(m-1)}$. It remains for us to replace $\epsilon_1{}^0$, $\epsilon_1{}^1, \ldots, \epsilon_1{}^n$, by their values so that we shall have expressions for the quantities $\epsilon_m{}^i$ in form suitable for our purpose. These values are easily obtained, as will be seen.

For this purpose, we apply the general formula

$$\int e^{-z}F(z)dz = -e^{-z}\gamma(z),$$

taking

$$F(z) = \frac{f(z)}{z - \zeta}$$

that is

$$F(z) = z^n + \begin{vmatrix} \zeta \\ +p_1 \end{vmatrix} z^{n-1} + \begin{vmatrix} \zeta^2 \\ +p_1\zeta \\ +p_2 \end{vmatrix} z^{n-2} + \ldots$$

It is easily seen that $\gamma(z)$ will be an expression integral in z and ζ, entirely similar to $\Theta(z, \zeta)$, such that if one represents it by $\Phi(z, \zeta)$, one has

$$\Phi(z, \zeta) = z^n + \varphi_1(\zeta)z^{n-1} + \varphi_2(\zeta)z^{n-2} + \ldots + \varphi_n(\zeta),$$

where $\varphi_i(\zeta)$ is a polynomial in ζ of degree i, in which the coefficient of ζ^i is unity... and the analogy of the form with $\Theta(z, \zeta)$ shows that the determinant

$$\begin{vmatrix} \Phi(z_0, z_0) & \Phi(z_1, z_0) \ldots \Phi(z_n, z_0) \\ \Phi(z_0, z_1) & \Phi(z_1, z_1) \ldots \Phi(z_n, z_1) \\ \cdots & \cdots \ \cdots \ \cdots \\ \Phi(z_0, z_n) & \Phi(z_1, z_n) \ldots \Phi(z_n, z_n) \end{vmatrix}$$

is also equal to δ^2. Next, we conclude from the relation

$$\int_{z_0}^{Z} \frac{e^{-z}f(z)}{z - \zeta}\, dz = e^{-z_0}\Phi(z_0, \zeta) - e^{-z}\Phi(Z, \zeta),$$

taking $\zeta = z_i$, the desired value

$$\epsilon_1{}^i = e^{-z_0}\Phi(z_0, z_i) - e^{-z}(Z, z_i).$$

Consequently we have the expressions given below for $\epsilon_m{}^i$.

Let

$$\mathfrak{A} = A_o\Phi(Z, z_o) + A_1\Phi(Z, z_1) + \ldots + A_n\Phi(Z, z_n),$$
$$\mathfrak{B} = B_o\Phi(Z, z_o) + B_1\Phi(Z, z_1) + \ldots + B_n\Phi(Z, z_n),$$

$$\cdots\cdots\cdots$$

$$\mathfrak{L} = L_o\Phi(Z, z_o) + L_1\Phi(Z, z_1) + \ldots + L_n\Phi(Z, z_n),$$

and let $\mathfrak{A}_o, \mathfrak{B}_o, \ldots, \mathfrak{L}_o$ be the values obtained for $Z = z_o$; one has

$$\epsilon_m{}^0 = e^{-z_o}\mathfrak{A}_o - e^{-z_o}\mathfrak{A}$$

$$\epsilon_m{}' = e^{-z_o}\mathfrak{B}_o - e^{-z}\mathfrak{B}$$

$$\cdots\cdots\cdots$$

$$\epsilon_m{}^n = e^{-z_o}\mathfrak{L}_o - e^{-z}\mathfrak{L}.$$

In these formulas, Z represents any one whatever of the quantities $z_o, z_1, \ldots, z_n$, now if we wish to state the result for $Z = z_k$, we shall agree at the outset, to represent on the one hand, by $\mathfrak{A}_k, \mathfrak{B}_k, \ldots \mathfrak{L}_k$, and on the other, by $\eta_k{}^0, \eta_k{}', \ldots, \eta_k{}^n$, the values which the coefficients $\mathfrak{A}, \mathfrak{B}, \ldots, \mathfrak{L}$, and the quantities $\epsilon_m{}^0, \epsilon_m{}', \ldots, \epsilon_m{}^n$, take on in this case.   Thus one obtains the equations

$$\eta_k{}^0 = e^{-z_o}\mathfrak{A}_o - e^{-z_k}\mathfrak{A}_k$$

$$\eta_k{}' = e^{-z_o}\mathfrak{B}_o - e^{-z_k}\mathfrak{B}$$

$$\cdots\cdots\cdots$$

$$\eta_k{}^n = e^{-z_o}\mathfrak{L}_o - e^{-z_k}\mathfrak{L}_k,$$

which will lead us to the second proof, we have mentioned, of the impossibility of a relation of the form

$$e^{z_o}N_o + e^{z_1}N_1 + \ldots + e^{z_n}N_n = 0,$$

where the exponents $z_o, z_1, \ldots, z_n$, as also the coefficients $N_o, N_1, \ldots, N_n$, are assumed to be integers.

Note in the first place, that $\epsilon_m{}^i$ can become smaller than any given quantity for a sufficiently large value of m.   For, the exponential e^{-z} being always positive, one has, as is known,

$$\int_{z_o}^{Z} e^{-z}F(z)dz = F(\xi)\int_{z_o}^{Z} e^{-z}dz = F(\xi)(e^{-z_o} - e^{-Z}),$$

$F(z)$ being any function whatever, and ξ a quantity taken between z_o and Z, the limits of the integral.   Now, upon taking

$$F(z) = \frac{f^m(z)}{z - Z_i},$$

one obtains the expression

$$\epsilon_m{}^i = \frac{f^{m-1}(\xi)}{1.2\ldots m - 1} \frac{f(\xi)}{\xi - z_i}(e^{-z_o} - e^{-Z}),$$

which demonstrates the property quoted above. Now, we obtain from the equations

$$\eta_1^{\ o} = e^{-z_0}\mathfrak{A}_o - e^{-z_1}\mathfrak{A}_1,$$
$$\eta_2^{\ o} = e^{-z_0}\mathfrak{A}_o - e^{-z_2}\mathfrak{A}_2,$$
$$\cdots\cdots\cdots$$
$$\eta_n^{\ o} = e^{-z_0}\mathfrak{A}_o - e^{-z_n}\mathfrak{A}_n,$$

the following relation,

$$e^{z_1}\eta_1^{\ o}N_1 + e^{z_2}\eta_2^{\ o}N_2 + \ldots + e^{z_n}\eta_n^{\ o}N_n$$
$$= e^{-z_0}(e^{z_1}N_1 + e^{z_2}N_2 + \ldots + e^{z_n}N_n)$$
$$- (\mathfrak{A}_1 N_1 + \mathfrak{A}_2 N_2 + \ldots + \mathfrak{A}_n N_n).$$

If the condition

$$e^{z_0}N_o + e^{z_1}N_1 + \ldots + e^{z_n}N_n = 0$$

is introduced, this relation becomes

$$e^{z_1}\eta_1^{\ o}N_1 + e^{z_2}\eta_2^{\ o}N_2 + \ldots + e^{z_n}\eta_n^{\ o}N_n$$
$$= -(\mathfrak{A}_o N_o + \mathfrak{A}_1 N_1 + \ldots + \mathfrak{A}_n N_n).$$

However, under the assumption that $z_o, z_1, \ldots, z_n$ are integers, the quantities $\Theta(z_i, z_k)$, $\Phi(z_i, z_k)$ and consequently $\mathfrak{A}_o, \mathfrak{A}_1, \ldots, \mathfrak{A}_n$ are also integers. Then we have a whole number

$$\mathfrak{A}_o N_o + \mathfrak{A}_1 N_1 + \ldots + \mathfrak{A}_n N_n,$$

which decreases indefinitely with $\eta_1^{\ o}, \eta_1^{\ 1}, \ldots, \eta_1^{\ n}$, as m increases; it follows that for a certain value of m and for all larger values,

$$\mathfrak{A}_o N_o + \mathfrak{A}_1 N_1 + \ldots + \mathfrak{A}_n N_n = 0,$$

and, since one obtains similarly the relations

$$\mathfrak{B}_o N_o + \mathfrak{B}_1 N_1 + \ldots + \mathfrak{B}_n N_n = 0,$$
$$\cdots\cdots\cdots\cdots$$
$$\mathfrak{L}_o N_o + \mathfrak{L}_1 N_1 + \ldots + \mathfrak{L}_n N_n = 0.$$

the relation

$$e^{z_0}N_o + e^{z_1}N_1 + \ldots + e^{z_n}N_n = 0$$

demands that the determinant

$$\Delta = \begin{vmatrix} \mathfrak{A}_o & \mathfrak{A}_1 & \ldots & \mathfrak{A}_n \\ \mathfrak{B}_o & \mathfrak{B}_1 & \ldots & \mathfrak{B}_n \\ \cdots & \cdots & \cdots & \cdots \\ \mathfrak{L}_o & \mathfrak{L}_1 & \ldots & \mathfrak{L}_n \end{vmatrix}$$

be equal to zero. But, because of the expressions for $\mathfrak{A}_o, \mathfrak{B}_o, \ldots,$ $\mathfrak{L}_o$, it follows that Δ is the product of these two other determinants

$$\begin{vmatrix} A_o & A_1 & \ldots & A_n \\ B_o & B_1 & \ldots & B_n \\ \cdots & \cdots & \cdots & \cdots \\ L_o & L_1 & \ldots & L_n \end{vmatrix}$$

and

$$\begin{vmatrix} \Phi(z_0, z_0) & \Phi(z_1, z_0)\ldots\Phi(z_n, z_0) \\ \Phi(z_0, z_1) & \Phi(z_1, z_1)\ldots\Phi(z_n, z_1) \\ \cdots & \cdots \quad \cdots \quad \cdots \\ \Phi(z_0, z_n) & \Phi(z, z_n)\ldots\Phi(z_n, z_n) \end{vmatrix}$$

of which the first has for its value $\delta^{2(m-1)}$, and the second δ^2. One has then $\Delta = \delta^{2m}$, and it is easily shown in an entirely rigorous manner, that the assumed relation is impossible,[1] and that therefore, the number e is not among the irrational algebraic numbers.

---

[1] [It can be shown that

$$\delta = \begin{vmatrix} 1 & 1 & \ldots & 1 \\ z_0 & z_1 & \ldots & z_n \\ z_0^2 & z_1^2 & \ldots & z_n^2 \\ \cdots & \cdots & \cdots & \cdots \\ z_0^n & z_1^n & \ldots & z_n^n \end{vmatrix} = \pm (z_n - z_{n-1})(z_n - z_{n-2})\ldots(z_n - z_0) \\ (z_{n-1} - z_{n-2})\ldots(z_1 - z_0)$$

and therefore that δ is not zero, assuming, as is of course assumed throughout, that the exponents, $z_0, z_1, \ldots, z_n$, are distinct.]

GAUSS

On the Congruence of Numbers

(Translated from the Latin by Professor Ralph G. Archibald, Columbia University, New York City.)

Carl Friedrich Gauss (1777–1855), the son of a day laborer, was the founder of the modern school of mathematics in Germany and was, perhaps, equally well known in the fields of physics and astronomy. Kronecker (1823–1891) said of him that "almost everything which the mathematics of our century has brought forth in the way of original scientific ideas is connected with the name of Gauss." His work in the theory of numbers began when he was a student at Göttingen, and much of it appeared in his *Disquisitiones Arithmeticae*, published in 1801, when he was only twenty-four years old. In this is found his treatment of the congruence of numbers, a translation of portions of which is here given. It also appears in the first volume of his *Werke* (Göttingen, 1870).

First Section

Concerning Congruence of Numbers in General

Congruent Numbers, Moduli, Residues, and Non-residues

1

If a number a divides the difference of the numbers b and c, b and c are said to be *congruent with respect to a; but if not, incongruent*. We call a the *modulus*. In the former case, each of the numbers b and c is called a *residue* of the other, but in the latter case, a *non-residue*.

These notions apply to all integral numbers both positive and negative,[1] but not to fractions. For example, -9 and $+16$ are congruent with respect to the modulus 5; -7 is a residue of $+15$ with respect to the modulus 11, but a non-residue with respect to the modulus 3. Now, since every number divides zero, every number must be regarded as congruent to itself with respect to all moduli.

2

If k denotes an indeterminate integral number, all residues of a given number a with respect to the modulus m are contained in

---

[1] Obviously, the modulus is always to be taken *absolutely*,—that is, without any sign.

the formula $a + km$. The easier of the propositions which we shall give can be readily demonstrated from this standpoint; but anyone will just as easily perceive their truth at sight.

We shall denote in future the congruence of two numbers by this sign, $\equiv$, and adjoin the modulus in parentheses when necessary. For example, $-16 \equiv 9 \pmod 5$, $-7 \equiv 15 \pmod{11}$.[1]

3

THEOREM.—*If there be given the m consecutive integral numbers*
$$a, a + 1, a + 2, \ldots, a + m - 1,$$
and another integral number A, then some one of the former will be congruent to this number A with respect to the modulus m; and, in fact, there will be only one such number.

If, for instance, $\dfrac{a - A}{m}$ is an integer, we shall have $a \equiv A$; but if it is fractional, let k be the integer immediately greater (or, when it is negative, immediately *smaller* if no regard is paid to sign). Then $A + km$ will fall between a and $a + m$, and will therefore be the number desired. Now, it is evident that all the quotients $\dfrac{a - A}{m}, \dfrac{a + 1 - A}{m}, \dfrac{a + 2 - A}{m}$, etc., are situated between $k - 1$ and $k + 1$. Therefore not more than one can be integral.

Least Residues

4

Every number, then, will have a residue not only in the sequence $0, 1, 2, \ldots, m - 1$, but also in the sequence $0, -1, -2, \ldots, -(m - 1)$. We shall call these *least residues*. Now, it is evident that, unless 0 is a residue, there will always be two: one *positive*, the other *negative*. If they are of different magnitudes, one of them will be less than $\dfrac{m}{2}$; but if they are of the same magnitude, each will equal $\dfrac{m}{2}$ when no regard is paid to sign. From this it is evident that any number has a residue not exceeding half the modulus. This residue is called the *absolute minimum*.

---

[1] We have adopted this sign on account of the great analogy which exists between an equality and a congruence. For the same reason Legendre, in memoirs which will later be frequently quoted, retained the sign of equality itself for a congruence. We hesitated to follow this notation lest it introduce an ambiguity.

For example, with respect to the modulus 5, −13 has the positive least residue 2, which at the same time is the absolute minimum, and has −3 as the negative least residue. With respect to the modulus 7, +5 is its own positive least residue, −2 is the negative least residue and at the same time the absolute minimum.

Elementary Propositions Concerning Congruences

5

From the notions just established we may derive the following obvious properties of congruent numbers.

The numbers which are congruent with respect to a composite modulus, will certainly be congruent with respect to any one of its divisors.

If several numbers are congruent to the same number with respect to the same modulus, they will be congruent among themselves (with respect to the same modulus).

The same identity of moduli is to be understood in what follows.

Congruent numbers have the same least residues, incongruent numbers different least residues.

6

If the numbers A, B, C, etc. and the numbers a, b, c, etc. are congruent each to each with respect to any modulus, that is, if

$$A \equiv a, B \equiv b, \text{ etc.},$$

then we shall have

$$A + B + C + \text{ etc.} \equiv a + b + c + \text{ etc.}$$

If $A \equiv a$ and $B \equiv b$, we shall have $A - B \equiv a - b$.

7

If $A \equiv a$, we shall also have $kA \equiv ka$.

If k is a positive number, this is merely a particular case of the proposition of the preceding article when we place $A = B = C$ etc. and $a = b = c$ etc. If k is negative, $-k$ will be positive. Then $-kA \equiv -ka$, and consequently $kA \equiv ka$.

If $A \equiv a$ and $B \equiv b$, we shall have $AB \equiv ab$. For, $AB \equiv Ab \equiv ba$.

8

If the numbers A, B, C, etc. and the numbers a, b, c, etc. are congruent each to each, that is, if $A \equiv a$, $B \equiv b$, etc., the products of the numbers of each set will be congruent; that is, ABC etc. $\equiv$ abc etc.

From the preceding article, $AB \equiv ab$, and for the same reason $ABC = abc$; in a like manner we can consider as many factors as desired.

If we take all the numbers A, B, C, etc. equal, and also the corresponding numbers a, b, c, etc., we obtain this theorem:

If $A \equiv a$ and if k is a positive integer, we shall have $A^k \equiv a^k$.

9

Let X be a function of the indeterminate x, of the form
$$Ax^a + Bx^b + Cx^c + \text{etc.},$$
where A, B, C, etc., denote any integral numbers, and a, b, c, etc., non-negative integral numbers. If, now, to the indeterminate x there be assigned values which are congruent with respect to any stated modulus, the resulting values of the function X will then be congruent.

Let f and g be two congruent values of x. Then by the preceding articles $f^a \equiv g^a$ and $Af^a \equiv Ag^a$; in the same way $Bf^b \equiv Bg^b$, etc. Hence
$$Af^a + Bf^b + Cf^c + \text{etc.} \equiv Ag^a + Bg^b + Cg^c + \text{etc.} \quad \text{Q. E. D.}$$

It is easily seen, too, how this theorem can be extended to functions of several indeterminates.

10

If, therefore, all consecutive integral numbers are substituted for x, and if the values of the function X are reduced to least residues, these residues will constitute a sequence in which the same terms repeat after an interval of m terms (m denoting the modulus); or, in other words, this sequence will be formed by a *period of m* terms repeated indefinitely. Let, for example, $X = x^3 - 8x + 6$ and $m = 5$. Then for $x = 0, 1, 2, 3$, etc., the values of X give the positive least residues, 1, 4, 3, 4, 3, 1, 4, etc., where the first five, namely, 1, 4, 3, 4, 3, are repeated without end. And furthermore, if the sequence is continued backwards, that is, if negative values are assigned to x, the same period occurs in the inverse order. It is therefore evident that terms different from those constituting the period cannot occur in the sequence.

11

In this example, then, X can be neither $\equiv 0$ nor $\equiv 2$ (mod 5), and can still less be $= 0$ or $= 2$. Whence it follows that the equations $x^3 - 8x + 6 = 0$ and $x^3 - 8x + 4 = 0$ cannot be solved in integral numbers, and therefore, as we know, cannot be solved in rational numbers. It is obviously true in general that, if it is impossible to satisfy the congruence $X \equiv 0$ with respect to some particular modulus, then the equation $X = 0$ has no rational root when X is a function of the unknown x, of the form
$$x^n + Ax^{n-1} + Bx^{n-2} + \text{etc.} + N,$$

where A, B, C, etc. are integers and n is a positive integer. (It is well known that all algebraic equations can be brought to this form.) This criterion, though presented here in a natural manner, will be treated at greater length in Section VIII. From this brief indication, some idea, no doubt, can be formed regarding the utility of these researches.

Some Applications

12

Many of the theorems commonly taught in arithmetic depend upon theorems given in this section; for example, the rules for testing the divisibility of a given number by 9, 11, or other numbers. *With respect to the modulus* 9, all powers of 10 are congruent to unity. Hence, if the given number is of the form $a + 10b + 100c +$ etc., it will have, with respect to the modulus 9, the same least residue as $a + b + c +$ etc. From this it is evident that, if the individual figures of the number, expressed in the denary scale, are added without regard to their position, this sum and the given number will exhibit the same least residues; and furthermore, the latter can be divided by 9 if the former be divisible by 9, and conversely. The same thing also holds true for the divisor 3. Since *with respect to the modulus* 11, $100 \equiv 1$, we shall have generally $10^{2k} \equiv 1$ and $10^{2k+1} \equiv 10 \equiv -1$. Then a number of the form $a + 10b + 100c +$ etc. will have, with respect to the modulus 11, the same least residue as $a - b + c$ etc.; whence the known rule is immediately derived. On the same principle all similar rules are easily deduced.

The preceding observations also bring out the principle underlying the rules commonly relied upon for the verification of arithmetical operations. These remarks, of course, are applicable when from given numbers we have to deduce others by addition, subtraction, multiplication, or raising to powers: in place of the given numbers, we merely substitute their least residues with respect to an arbitrary modulus (generally 9 or 11; since, as we have just now observed, in our decimal system residues with respect to these moduli can be so very easily found). The numbers thus obtained should be congruent to those which have been deduced from the given numbers. If, on the other hand, this is not the case, we infer that an error has crept into the calculation.

Now as these results and others of a similar nature are so very well known, it would serve no purpose to dwell on them further.

GAUSS

Third Proof of the Law of Quadratic Reciprocity

(Translated from the Latin by D. H. Lehmer, M.Sc., Brown University, Providence, Rhode Island.)

The theorem with which the following pages are concerned and to which Gauss gave the name of Fundamental Theorem is better known today as Legendre's Law of Quadratic Reciprocity. Although a statement of a theorem equivalent to this law is found in the works of Euler[1] without proof, the first enunciation of the law itself is attributed to Legendre,[2] whose proof, however, is invalid. It tacitly assumes that there exist infinitely many primes in certain arithmetical progressions, a fact which was first established by Dirichlet half a century later. The first proof of this theorem was given by Gauss[3] in 1801 and was followed by seven others in an interval of 17 years. The proof given below is the third one published,[4] although it is really his fifth proof. It is considered by Gauss and many others to be the most direct and elegant of his eight demonstrations.

In fact, in the first two paragraphs of the present proof Gauss expresses himself as follows:

§1. The questions of higher arithmetic often present a remarkable characteristic which seldom appears in more general analysis, and increases the beauty of the former subject. While analytic investigations lead to the discovery of new truths only after the fundamental principles of the subject (which to a certain degree open the way to these truths) have been completely mastered; on the contrary in arithmetic the most elegant theorems frequently arise experimentally as the result of a more or less unexpected stroke of good fortune, while their proofs lie so deeply embedded in the darkness that they elude all attempts and defeat the sharpest inquiries. Further, the connection between arithmetical truths which at first glance seem of widely different nature, is so close that one not infrequently has the good fortune to find a proof (in an entirely unexpected way and by means of quite another

---

[1] Euler, *Opuscula*, Vol. 1, p. 64, 1783.

[2] Legendre, *Histoire de l'Académie des Sciences*, pp. 516–517, 1785; *Théorie des Nombres*, Ed. 1, pp. 214–226, 1798; Ed. 2, pp. 198–207, 1808.

[3] Gauss, *Disquisitiones Arithmeticae*, Sect. 4, Leipzig, 1801; *Werke*, Göttingen, 1870, Bd. 1, pp. 73–111.

[4] Gauss, *Commentationes Societatis Regiæ Scientiarum Gottingensis*, Vol. 16, Göttingen, 1808; *Werke*, Göttingen, 1876. Bd. 2, pp. 1–8.

inquiry) of a truth which one greatly desired and sought in vain in spite of much effort. These truths are frequently of such a nature that they may be arrived at by many distinct paths and that the first paths to be discovered are not always the shortest. It is therefore a great pleasure after one has fruitlessly pondered over a truth and has later been able to prove it in a round-about way to find at last the simplest and most natural way to its proof.

§2. The theorem which we have called in sec. 4 of the *Disquisitiones Arithmeticae*, the *Fundamental Theorem* because it contains in itself all the theory of quadratic residues, holds a prominent position among the questions of which we have spoken in the preceding paragraph. We must consider Legendre as the discoverer of this very elegant theorem, although special cases of it had previously been discovered by the celebrated geometers Euler and Lagrange. I will not pause here to enumerate the attempts of these men to furnish a proof; those who are interested may read the above mentioned work. An account of my own trials will suffice to confirm the assertions of the preceeding paragraph. I discovered this theorem independently in 1795 at a time when I was totally ignorant of what had been achieved in higher arithmetic, and consequently had not the slightest aid from the literature on the subject. For a whole year this theorem tormented me and absorbed my greatest efforts until at last I obtained a proof given in the fourth section of the above-mentioned work. Later I ran across three other proofs which were built on entirely different principles. One of these I have already given in the fifth section, the others, which do not compare with it in elegance, I have reserved for future publication. Although these proofs leave nothing to be desired as regards rigor, they are derived from sources much too remote, except perhaps the first, which however proceeds with laborious arguments and is overloaded with extended operations. I do not hesitate to say that till now a *natural* proof has not been produced. I leave it to the authorities to judge whether the following proof which I have recently been fortunate enough to discover deserves this discription.

Inasmuch as Gauss does not give any mathematical background in the introduction to his third proof or even a formal statement of the theorem itself (these having been given in his first proof), we shall attempt to supply in a few sentences the information necessary for the proper understanding of the theorem.

The integer p is said to be a quadratic residue or non-residue of an integer q relatively prime to p according as there exist or not solutions x of the congruence $x^2 \equiv p \pmod{q}$. These two cases may be written symbolically as pRq and pNq, respectively. If p and r are both residues or both non-residues of q, then they are said to have the same *quadratic character* with respect to q. With this understanding, the fundamental theorem may be stated in words as follows: *If p and q are any distinct odd primes, then the quadratic character of p with respect to q is the same as that of q with respect to p except when both p and q are of the form $4n - 1$, in which case the characters are opposite.*

The quadratic character of p with respect to q may be expressed by the symbol of Legendre

$$\left(\frac{p}{q}\right),$$

which has the value $+1$ or -1 according as pRq or pNq. The use of this symbol enables us to state the theorem analytically as follows

$$\left(\frac{p}{q}\right)\left(\frac{q}{p}\right) = (-1)^{\frac{(p-1)(q-1)}{4}}$$

We proceed with the translation of Gauss's proof in full:

§3. THEOREM.[1]—*Let p be a positive prime number and k be any number not divisible by p. Further let A be the set of numbers*

$$1, 2, 3, \ldots, \frac{(p-1)}{2}$$

and B the set

$$\frac{(p+1)}{2}, \frac{(p+3)}{2}, \ldots, p-1.$$

We determine the smallest positive residue modulo p of the product of k by each of the numbers in the set A. These will be distinct and will belong partly to A and partly to B. If we let μ be the number of these residues belonging to B, then k is a quadratic residue of p or a non-residue of p according as μ is odd or even.

Proof.—Let $a, a', a'', \ldots$ be the residues belonging to the class A and $b, b', b'', \ldots$ be those belonging to B. Then it is clear that the complements of these latter: $p - b, p - b', p - b'', \ldots$ are not equal to any of the numbers $a, a', a'', \ldots$, and together with them make up the class A. Consequently we have

$$1.2.3\ldots\frac{p-1}{2} = a.a'.a''\ldots(p-b)(p-b')(p-b'')\ldots$$

The right-hand product evidently becomes, modulo p:

$$\equiv (-1)^{\mu}aa'a''\ldots bb'b''\ldots \equiv (-1)^{\mu}k.2k.3k\ldots k\frac{p-1}{2}$$

$$\equiv (-1)^{\mu}k^{\left(\frac{p-1}{2}\right)}1.2.3\ldots\frac{p-1}{2}$$

---

[1] [This theorem is known to-day as Gauss's Lemma and the number μ is called the characteristic number.]

Hence

$$1 \equiv (-1)^{\mu} k^{\left(\frac{p-1}{2}\right)}$$

that is $k^{\frac{p-1}{2}} \equiv \pm 1$ according as μ is even or odd. Hence our theorem follows at once.[1]

§4. We can shorten the following discussion considerably by introducing certain convenient notations. Let the symbol (k, p)[2] represent the number of products among

$$k, 2k, 3k, \ldots k\frac{p-1}{2}$$

whose smallest positive residues modulo p exceed $p/2$. Further if x is a non-integral quantity we will express by the symbol $[x]$ the greatest integer less than x so that $x - [x]$ is always a positive quantity between 0 and 1. We can readily establish the following relations:

I. $[x] + [-x] = -1$.

II. $[x] + b = [x + b]$, whenever b is an integer.

III. $[x] + [b - x] = b - 1$.

IV. If $x - [x]$ is a fraction less than $\frac{1}{2}$, then $[2x] - 2[x] = 0$. If on the other hand $x - [x]$ is greater than $\frac{1}{2}$, then $[2x] - 2[x] = 1$.

V. If the smallest positive residue of $b(mod\ p)$ is less than $p/2$, then $[2b/p] - 2[b/p] = 0$. If however it is larger than $p/2$, then $[2b/p] - 2[b/p] = 1$.

VI. From this it follows that:

$$(k, p) = \left[\frac{2k}{p}\right] + \left[\frac{4k}{p}\right] + \left[\frac{6k}{p}\right] + \ldots + \left[\frac{(p-1)k}{p}\right]$$
$$- 2\left[\frac{k}{p}\right] - 2\left[\frac{2k}{p}\right] - 2\left[\frac{3k}{p}\right] \ldots - 2\left[\frac{k(p-1)/2}{p}\right]$$

VII. From VI and I we obtain without difficulty:

$$(k, p) + (-k, p) = \frac{p-1}{2}$$

From this it follows that the quadratic character of $-k$ with respect to p is the same as or opposite to the quadratic character

---

[1] [This follows from the famous Euler's criterion: $k^{\frac{p-1}{2}} \equiv \pm 1$ according as k is or is not a quadratic residue of p.]

[2] [The symbol (k, p) replaces the characteristic number μ of the preceding theorem.]

of k with respect to p, according as p is of the form $4n + 1$ or $4n + 3$. It is evident that in the first case -1 is a residue and in the second a non-residue of p.

VIII. We transform the formula given in VI as follows: From III we have

$$\left[\frac{(p-1)k}{p}\right] = k - 1 - \left[\frac{k}{p}\right], \quad \left[\frac{(p-3)k}{p}\right] = k - 1 - \left[\frac{3k}{p}\right],$$

$$\left[\frac{(p-5)k}{p}\right] = k - 1 - \left[\frac{5k}{p}\right]\dots$$

When we apply these substitutions to the last $\frac{p \mp 1}{4}$ terms of the above series we have

first, when p is of the form $4n + 1$,

$$(k, p) = \frac{(k-1)(p-1)}{4}$$
$$- 2\left\{\left[\frac{k}{p}\right] + \left[\frac{3k}{p}\right] + \left[\frac{5k}{p}\right] + \dots + \left[\frac{k(p-3)/2}{p}\right]\right\}$$
$$- \left[\frac{k}{p}\right] + \left[\frac{2k}{p}\right] + \left[\frac{3k}{p}\right] + \dots + \left]\frac{k(p-1)/2}{p}\right]$$

second, when p is of the form $4n + 3$

$$(k, p) = \frac{(k-1)(p+1)}{4}$$
$$- 2\left\{\left[\frac{k}{p}\right] + \left[\frac{3k}{p}\right] + \left[\frac{5k}{p}\right] + \dots + \left[\frac{k(p-1)/2}{p}\right]\right\}$$
$$- \left\{\left[\frac{k}{p}\right] + \left[\frac{2k}{p}\right] + \left[\frac{3k}{p}\right] + \dots + \left[\frac{k(p-1)/2}{p}\right]\right\}.$$

IX. In the special case $k = +2$ it follows from the above formulas[1] that $(2, p) = (p \mp 1)/4$, where we take the upper or lower sign according as p is of the form $4n + 1$ or $4n + 3$. Therefore $(2, p)$ is even and hence $2Rp$ in case p is of the form $8n + 1$ or $8n + 7$; on the other hand $(2, p)$ is odd and hence $2Np$ when p is of the form $8n + 3$ or $8n + 5$.

§5. THEOREM.—*If x is a positive non-integral quantity among whose multiples x, $2x$, $3x$,..., nx there exist no integers; putting*

---

[1] [Each term in the braces is zero in this case, since the quantities in the square brackets are less than unity.]

$[nx] = b$ *we easily conclude that among the multiples of the reciprocal* $\dfrac{1}{x}, \dfrac{2}{x}, \dfrac{3}{x} \ldots \dfrac{b}{x}$ *there appear no integers. Then I say that:*

$$\left.\begin{aligned}&[x] + [2x] + [3x] +\ldots+ [nx]\\ &+\left[\frac{1}{x}\right]+\left[\frac{2}{x}\right]+\left[\frac{3}{x}\right]+\ldots+\left[\frac{b}{x}\right]\end{aligned}\right\} = nb.$$

Proof.—In the series $[x] + [2x] + [3x]+\ldots[nx]$, which we set equal to Ω, all the terms from the first up to and including the $\left[\dfrac{1}{x}\right]^{th}$ are manifestly zero, the following terms up to and including the $\left[\dfrac{2}{x}\right]^{th}$ are equal to 1, and the following up to $\left[\dfrac{3}{x}\right]^{th}$ term are equal to 2 and so on. Hence we have·

$$\left.\begin{aligned}\Omega = \quad & 0 \times \left[\frac{1}{x}\right]\\ &+1 \times \left\{\left[\frac{2}{x}\right] - \left[\frac{1}{x}\right]\right\}\\ &+2 \times \left\{\left[\frac{3}{x}\right] - \left[\frac{2}{x}\right]\right\}\\ &+3 \times \left\{\left[\frac{4}{x}\right] - \left[\frac{3}{x}\right]\right\}\\ &\vdots\\ &+(b-1)\left\{\left[\frac{b}{x}\right] - \left[\frac{b-1}{x}\right]\right\}\\ &+b\left\{n - \left[\frac{b}{x}\right]\right\}\end{aligned}\right\} = bn - \left[\frac{1}{x}\right] - \left[\frac{2}{x}\right] - \left[\frac{3}{x}\right] -\ldots- \left[\frac{b}{x}\right].$$

Q. E. D.

§6. Theorem.—*If k and p are positive odd numbers prime to each other, we have*

$$\left.\begin{aligned}&\left[\frac{k}{p}\right]+\left[\frac{2k}{p}\right]+\left[\frac{3k}{p}\right]+\ldots+\left[\frac{k(p-1)/2}{p}\right]\\ &+\left[\frac{p}{k}\right]+\left[\frac{2p}{k}\right]+\left[\frac{3p}{k}\right]+\ldots+\left[\frac{p(k-1)/2}{k}\right]\end{aligned}\right\} = \frac{(k-1)(p-1)}{4}.$$

Proof.—Supposing that $k < p$ we have $\dfrac{k(p-1)/2}{p} < \dfrac{k}{2}$ but $> \dfrac{k-1}{2}$, and hence

$$\left[\frac{k(p-1)/2}{p}\right] = \frac{k-1}{2}.$$

From this it is clear that the theorem follows at once from the preceding one if we set

$$\frac{k}{p} = x, \quad \frac{p-1}{2} = n, \quad \frac{k-1}{2} = b.$$

It is possible to prove in a similar way that if k is *even* and prime to p, then

$$\left.\begin{array}{l}\left[\dfrac{k}{p}\right] + \left[\dfrac{2k}{p}\right] + \left[\dfrac{3k}{p}\right] + \ldots + \left[\dfrac{k(p-1)/2}{p}\right] \\[2mm] + \left[\dfrac{p}{k}\right] + \left[\dfrac{2p}{k}\right] + \left[\dfrac{3p}{k}\right] + \ldots + \left[\dfrac{kp/2}{k}\right]\end{array}\right\} = k\dfrac{p-1}{4}.$$

However we will not prove this proposition as it is not necessary for our purpose.

§7. Now the main theorem follows from the combination of the last theorem with proposition VIII of paragraph 4. For if we designate by k and p any distinct, positive prime numbers[1] and put

$$(k, p) + \left[\frac{k}{p}\right] + \left[\frac{2k}{p}\right] + \left[\frac{3k}{p}\right] + \ldots + \left[\frac{k(p-1)/2}{p}\right] = L,$$

$$(p, k) + \left[\frac{p}{k}\right] + \left[\frac{2p}{k}\right] + \left[\frac{3p}{k}\right] + \ldots + \left[\frac{p(k-1)/2}{p}\right] = M,$$

then it follows from §4, VIII, that L and M will always be even numbers. It follows from the theorem of §6 that

$$L + M = (k, p) + (p, k) + \frac{(k-1)(p-1)}{4}$$

Therefore, when $(k-1)(p-1)/4$ is even, that is when one or both of the primes k or p is of the form $4n + 1$, then (p, k) and (k, p) are either both even or both odd. On the contrary when $(k-1)(p-1)/4$ is odd, that is when k and p are both of the form $4n + 3$, then necessarily one of the numbers (k, p), (p, k) is even and the other odd. In the first case the relations of k to p, and of p to k (as regards the quadratic character of one with respect to the other) are the same; in the second case they are opposite.

Q. E. D.

---

[1] [In which] k and p should also be different from 2.

KUMMER

On Ideal Numbers

(Translated from the German by Dr. Thomas Freeman Cope, National Research Fellow in Mathematics, Harvard University, Cambridge, Mass.)

Ernst Edward Kummer[1] (1810–1893), who was professor of mathematics in the University of Breslau from 1842 till 1855 and then in the University of Berlin until 1884, made valuable contributions in several branches of mathematics. Among the topics he studied may be mentioned the theory of the hypergeometric (Gaussian) series, the Riccati equation, the question of the convergency of series, the theory of complex numbers, and cubic and biquadratic residues. He was the creator of ideal prime factors of complex numbers and studied intensively surfaces of the fourth order and, in particular, the surfaces which bear his name.

In the following paper which appears in the original in Crelle's *Journal für die reine und angewandte Mathematik* (Vol. 35, pp. 319–326, 1847), Kummer introduces the notion of ideal prime factors of complex numbers, by means of which he was able to restore unique factorization in a field where the fundamental theorem of arithmetic does not hold. Although Kummer's theory has been largely supplanted by the simpler and more general theory of Dedekind, yet the ideas he introduced were of such importance that no less an authority than Professor E. T. Bell is responsible for the statement that[2] "Kummer's introduction of ideals into arithmetic was beyond all dispute one of the greatest mathematical advances of the nineteenth century." For the position of Kummer's theory in the theory of numbers, the reader is referred to the article by Professor Bell from which the above quotation is taken.

On the Theory Of Complex Numbers

(By Professor Kummer of Breslau.)

(Abstract of the *Berichten der Königl. Akad. der Wiss. zu Berlin*, March 1845.)

I have succeeded in completing and in simplifying the theory of those complex numbers which are formed from the higher roots of unity and which, as is well known, play an important rôle in cyclotomy and in the study of power residues and of forms of higher degree; this I have done through the introduction of a peculiar kind of imaginary divisors which I call *ideal complex*

---

[1] For a short biographical sketch, see D. E. Smith, *History of Mathematics*, Vol. I, pp. 507–508, Boston, 1923.
[2] *American Mathematical Monthly*, Vol. 34, pp. 66.

numbers and concerning which I take the liberty of making a few remarks.

If α is an imaginary root of the equation $\alpha^\lambda = 1$, λ a prime number, and a, a_1, a_2, etc. whole numbers, then $f(\alpha) = a + a_1\alpha + a_2\alpha^2 + \ldots + a_{\lambda-1}\alpha^{\lambda-1}$ is a complex whole number. Such a complex number can either be broken up into factors of the same kind or such a decomposition is not possible. In the first case, the number is a composite number; in the second case, it has hitherto been called a complex prime number. I have observed, however, that, even though $f(\alpha)$ cannot in any way be broken up into complex factors, it still does not possess the true nature of a complex prime number, for, quite commonly, it lacks the first and most important property of prime numbers; namely, that the product of two prime numbers is divisible by no other prime numbers. Rather, such numbers $f(\alpha)$, even if they are not capable of decomposition into complex factors, have nevertheless the nature of composite numbers; the factors in this case are, however, not actual but ideal complex numbers. For the introduction of such ideal complex numbers, there is the same, simple, basal motive as for the introduction of imaginary formulas into algebra and analysis; namely, the decomposition of integral rational functions into their simplest factors, the linear. It was, moreover, such a desideratum which prompted Gauss, in his researches on biquadratic residues (for all such prime factors of the form $4m + 1$ exhibit the nature of composite numbers), to introduce for the first time complex numbers of the form $a + b\sqrt{-1}$.

In order to secure a sound definition of the true (usually ideal) prime factors of complex numbers, it was necessary to use the properties of prime factors of complex numbers which hold in every case and which are entirely independent of the contingency of whether or not actual decomposition takes place: just as in geometry, if it is a question of the common chords of two circles even though the circles do not intersect, one seeks an actual definition of these ideal common chords which shall hold for all positions of the circles. There are several such permanent properties of complex numbers which could be used as definitions of ideal prime factors and which would always lead to essentially the same result; of these, I have chosen *one* as the simplest and the most general.

If p is a prime number of the form $m\lambda + 1$, then it can be represented, in many cases, as the product of the following $\lambda - 1$ complex factors: $p = f(\alpha) \cdot f(\alpha^2) \cdot f(\alpha^3) \ldots f(\alpha^{\lambda-1})$; when, however, a decomposition into actual complex prime factors is not possible, let ideals make their appearance in order to bring this about. If $f(\alpha)$ is an actual complex number and a prime factor of p, it has the property that, if instead of the root of the equation $\alpha^\lambda = 1$ a definite root of the congruence $\xi^\lambda \equiv 1$, mod. p, is substituted, then $f(\xi) \equiv 0$, mod. p.   Hence too if the prime factor $f(\alpha)$ is contained in a complex number $\Phi(\alpha)$, it is true that $\Phi(\xi) \equiv 0$, mod. p; and conversely, if $\Phi(\xi) \equiv 0$, mod. p, and p is factorable into $\lambda - 1$ complex prime factors, then $\Phi(\alpha)$ contains the prime factor $f(\alpha)$. Now the property $\Phi(\xi) \equiv 0$, mod. p, is such that it does not depend in any way on the factorability of the number p into prime factors; it can accordingly be used as a definition, since it is agreed that the complex number $\Phi(\alpha)$ shall contain the ideal prime factor of p which belongs to $\alpha = \xi$, if $\Phi(\xi) \equiv 0$, mod. p.   Each of the $\lambda - 1$ complex prime factors of p is thus replaced by a congruence relation.   This suffices to show that complex prime factors, whether they be actual or merely ideal, give to complex numbers the same definite character.   In the process given here, however, we do not use the congruence relations as the definitions of ideal prime factors because they would not be sufficient to represent several equal ideal prime factors of a complex number, and because, being too restrictive, they would yield only ideal prime factors of the real prime numbers of the form $m\lambda - 1$.

Every prime factor of a complex number is also a prime factor of every real prime number q, and the nature of the ideal prime factors is, in particular, dependent on the exponent to which q belongs for the modulus λ.   Let this exponent be f, so that $q^f \equiv 1$, mod. λ, and $\lambda - 1 = e \cdot f$.   Such a prime number q can never be broken up into more than e complex prime factors which, if this decomposition can actually be carried out, are represented as linear functions of the e periods of each set of f terms.   These periods of the roots of the equation $\alpha^\lambda = 1$, I denote by η, η_1, η_2, $\ldots \eta_{e-1}$; and indeed in such an order that each goes over into the following one whenever α is transformed into α^γ, where γ is a primitive root of λ.   As is well known, the periods are the e roots of an equation of the eth degree; and this equation, considered as a congruence for the modulus q, has always e real congruential roots which I denote by u, u_1, u_2, $\ldots u_{e-1}$ and take in an order corresponding to that of the periods, for which, besides the congruence of the eth degree, still other easily found congruences may be used.   If now the complex number $c'\eta + c_1'\eta_1 + c_2'\eta_2 +$

$\ldots + c'_{e-1}\eta_{e-1}$, constructed out of periods, is denoted shortly by $\Phi(\eta)$, then among the prime numbers q which belong to the exponent f, there are always such that can be brought into the form

$$q = \Phi(\eta)\Phi(\eta_1)\Phi(\eta_2)\ldots\Phi(\eta_{e-1}),$$

in which, moreover, the e factors never admit a further decomposition. If one replaces the periods by the congruential roots corresponding to them, where a period can arbitrarily be designated to correspond to a definite congruential root, then one of the e prime factors always becomes congruent to zero for the modulus q. Now if any complex number $f(\alpha)$ contains the prime factor $\Phi(\eta)$, it will always have the property, for $\eta = u_k$, $\eta_1 = u_{k+1}$, $\eta_2 = u_{k+2}$, etc., of becoming congruent to zero for the modulus q. This property (which implies precisely f distinct congruence relations, the development of which would lead too far) is a permanent one even for those prime numbers q which do not admit an actual decomposition into e complex prime factors. It could therefore be used as a definition of complex prime factors; it would, however, have the defect of not being able to express the equal ideal prime factors of a complex number.

The definition of ideal complex prime factors which I have chosen and which is essentially the same as the one described but is simpler and more general, rests on the fact that, as I prove separately, one can always find a complex number $\psi(\eta)$, constructed out of periods, which is of such a nature that $\psi(\eta)\psi(\eta_1)\psi(\eta_2)\ldots$ $\psi(\eta_{e-1})$ (this product being a whole number) is divisible by q but not by q^2. This complex number $\psi(\eta)$ has always the above-mentioned property, namely, that it is congruent to zero, modulo q, if for the periods are substituted the corresponding congruential roots, and therefore $\psi(\eta) \equiv 0$, mod. q, for $\eta = u$, $\eta_1 = u_1$, $\eta_2 = u_2$, etc. I now set $\psi(\eta_1)\psi(\eta_2)\ldots\psi(\eta_{e-1}) = \Psi(\eta)$ and define ideal prime numbers in the following manner:—

If $f(\alpha)$ has the property that the product $f(\alpha).\Psi(\eta_r)$ is divisible by q, this shall be expressed as follows: $f(\alpha)$ contains the ideal prime factor of q which belongs to $u = \eta_r$. Furthermore, if $f(\alpha)$ has the property that $f(\alpha).(\Psi(\eta_r))^\mu$ is divisible by q^μ but $f(\alpha)(\Psi(\eta_r))^{\mu+1}$ is not divisible by $q^{\mu+1}$, this shall be described thus: $f(\alpha)$ contains the ideal prime factor of q which belongs to $u = \eta_r$, exactly μ times.

It would lead too far if I should develop here the connection and the agreement of this definition with those given by congruence relations as described above; I simply remark that the relation: $f(\alpha)\Psi(\eta_r)$ divisible by q, is completely equivalent to f distinct congruence relations, and that the relation: $f(\alpha)(\Psi(\eta_r))^\mu$ divisible by q^μ, can always be entirely replaced by $u \cdot f$ congruence relations. The whole theory of ideal complex numbers which I have already perfected and of which I here announce the principal theorems, is a justification of the definition given as well as of the nomenclature adopted. The principal theorems are the following:

The product of two or more complex numbers has exactly the same ideal prime factors as the factors taken together.

If a complex number (which is a product of factors) contains all the e prime factors of q, it is also divisible by q itself; if, however, it does not contain some one of these e ideal prime factors, it is not divisible by q.

If a complex number (in the form of a product) contains all the e ideal prime factors of q and, indeed, each at least μ times, it is divisible by q^μ.

If $f(\alpha)$ contains exactly m ideal prime factors of q, which may all be different, or partly or wholly alike, then the norm $Nf(\alpha) = f(\alpha)f(\alpha^2)\ldots f(\alpha^{\lambda-1})$ contains exactly the factor q^{mf}.

Every complex number contains only a finite, determinate number of ideal prime factors.

Two complex numbers which have exactly the same ideal prime factors differ only by a complex unit which may enter as a factor.

A complex number is divisible by another if all the ideal prime factors of the divisor are contained in the dividend; and the quotient contains precisely the excess of the ideal prime factors of the dividend over those of the divisor.

From these theorems it follows that computation with complex numbers becomes, by the introduction of ideal prime factors, entirely the same as computation with integers and their real integral prime factors. Consequently, the grounds for the complaint which I voiced in the *Breslauer Programm zur Jubelfeier der Universität Königsberg S.* 18, are removed:—

It seems a great pity that this quality of real numbers, namely, that they can be resolved into prime factors which for the same number are always the same, is not shared by complex numbers; if now this desirable property were part of a complete doctrine, the effecting of which is as yet beset with great difficulties, the matter could easily be resolved and brought to a successful conclusion. Etc. One sees therefore that ideal prime factors disclose the inner nature of complex numbers, make them transparent, as it were, and show their inner crystalline structure. If, in particular, a complex number is given merely in the form $a + a_1\alpha + a_2\alpha^2 + \ldots + a_{\lambda-1}\alpha^{\lambda-1}$, little can be asserted about it until one has determined, by means of its ideal prime factors (which in such a case can always be found by direct methods), its simplest qualitative properties to serve as the basis of all further arithmetical investigations.

Ideal factors of complex numbers arise, as has been shown, as factors of actual complex numbers: hence ideal prime factors multiplied with others suitably chosen must always give actual complex numbers for products. This question of the combination of ideal factors to obtain actual complex numbers is, as I shall show as a consequence of the results which I have already found, of the greatest interest, because it stands in an intimate relationship to the most important sections of number theory. The two most important results relative to this question are the following:

There always exists a finite, determinate number of ideal complex multipliers which are necessary and sufficient to reduce all possible ideal complex numbers to actual complex numbers.[1]

Every ideal complex number has the property that a definite integral power of it will give an actual complex number.

I consider now some more detailed developments from these two theorems. Two ideal complex numbers which, when multiplied by one and the same ideal number, form actual complex numbers, I shall call *equivalent* or of the same class, because this investigation of actual and ideal complex numbers is identical with the classification of a certain set of forms of the $\lambda - 1$st degree and in $\lambda - 1$ variables; the principal results relative to this classification have been found by Dirichlet but not yet published so that I do not know precisely whether or not his principle of classification coincides with that resulting from the theory of complex numbers. For example, the theory of a form of the second degree in two variables with determinant, however, a prime number λ, is closely interwoven with these investigations, and our classification in this case coincides with that of Gauss but not with that of Legendre. The same considerations also throw great light upon Gauss's classification of forms of the second degree and upon the true basis for the differentiation between *Aequivalentia propria et impropria*,[2]

---

[1] A proof of this important theorem, although in far less generality and in an entirely different form, is found in the dissertation: L. Kronecker, *De unitatibus complexis*, Berlin, 1845.

[2] [i. e., proper and improper equivalence.]

which, undeniably, has always an appearance of impropriety when it presents itself in the *Disquisitiones arithmeticae*. If, for example, two forms such as $ax^2 + 2bxy + cy^2$ and $ax^2 - 2bxy + cy^2$, or $ax^2 + 2bxy + cy^2$ and $cx^2 + 2bxy + ay^2$, are considered as belonging to different classes, as is done in the above-mentioned work, while in fact no essential difference between them is to be found; and if on the other hand Gauss's classification must notwithstanding be admitted to be one arising for the most part out of the very nature of the question: then one is forced to consider forms such as $ax^2 + 2bxy + cy^2$ and $ax^2 - 2bxy + cy^2$ which differ from each other in outward appearance only, as merely representative of two new but essentially different concepts of number theory. These however, are in reality nothing more than two different ideal prime factors which belong to one and the same number. The entire theory of forms of the second degree in two variables can be thought of as the theory of complex numbers of the form $x + y\sqrt{D}$ and then leads necessarily to ideal complex numbers of the same sort. The latter, however, classify themselves according to the ideal multipliers which are necessary and sufficient to reduce them to actual complex numbers of the form $x + y\sqrt{D}$. Because of this agreement with the classification of Gauss, ideal complex numbers thus constitute the true basis for it.

The general investigation of ideal complex numbers presents the greatest analogy with the very difficult section by Gauss: *De compositione formarum*, and the principal results which Gauss proved for quadratic forms, pp. 337 and following, hold true also for the combination of general ideal complex numbers. Thus there belongs to every class of ideal numbers another class which, when multiplied by the first class, gives rise to actual complex numbers (here the actual complex numbers are the analogue of the *Classis principalis*).[1] Likewise, there are classes which, when multiplied by themselves, give for the result actual complex numbers (the *Classis principalis*), and these classes are therefore *ancipites*;[2] in particular, the *Classis principalis* itself is always a *Classis anceps*. If one takes an ideal complex number and raises it to powers, then in accordance with the second of the foregoing theorems, one will arrive at a power which is an actual complex number; if h is the smallest number for which $(f(\alpha))^h$ is an actual

---

[1] [Principal class.]
[2] [Dual, or of a double nature.]

complex number, then $f(\alpha)$, $(f(\alpha))^2$, $(f(\alpha))^3$, ... $(f(\alpha))^h$ all belong to different classes. It now may happen that, by a suitable choice of $f(\alpha)$, these exhaust all existing classes: if such is not the case, it is easy to prove that the number of classes is at least always a multiple of h. I have not gone deeper yet into this domain of complex numbers; in particular, I have not undertaken an investigation of the exact number of classes because I have heard that Dirichlet, using principles similar to those employed in his famous treatise on quadratic forms, has already found this number. I shall make only one additional remark about the character of ideal complex numbers, namely, that by the second of the foregoing theorems they can always be considered and represented as definite roots of actual complex numbers, that is, they always take the form $\sqrt[h]{\Phi(\alpha)}$ where $\Phi(\alpha)$ is an actual complex number and h an integer.

Of the different applications which I have already made of this theory of complex number, I shall refer only to the application to cyclotomy to complete the results which I have already announced in the above-mentioned *Programm*. If one sets

$$(\alpha, x) = x + \alpha x^g + \alpha^2 x^{g^2} + \ldots + \alpha^{p-2} x^{g^{p-2}},$$

where $\alpha^\lambda = 1$, $x^p = 1$, $p = m\lambda + 1$, and g is a primitive root of the prime number p, then it is well known that $(\alpha, x)^\lambda$ is a complex number independent of x and formed from the roots of the equation $\alpha^\lambda = 1$. In the *Programm* cited, I have found the following expression for this number, under the assumption that p can be resolved into $\lambda - 1$ actual complex prime factors, one of which is $f(\alpha)$:

$$(\alpha, x)^\lambda = \pm \alpha^h f^{m_1}(\alpha) \cdot f^{m_2}(\alpha^2) \cdot f^{m_3}(\alpha^3) \ldots f^{m_{\lambda-1}}(\alpha^{\lambda-1}),$$

where the power-exponents m_1, m_2, m_3, etc. are so determined that the general m_K, positive, is less than λ and $k \cdot m_k \equiv 1$, mod. λ. Exactly the same simple expression holds in complete generality, as can easily be proved, even when $f(\alpha)$ is not the actual but only the ideal prime factor of p. In order, however, in the latter case, to maintain the expression for $(\alpha, x)^\lambda$ in the form for an actual complex number, one need only represent the ideal $f(\alpha)$ as a root of an actual complex number, or apply one of the methods (although indirect) which serve to represent an actual complex number whose ideal prime factors are given.

CHEBYSHEV (TCHEBYCHEFF)

On the Totality of Primes

(Translated from the French by Professor J. D. Tamarkin, Brown University, Providence, Rhode Island.)

Pafnuty Lvovich Chebyshev (Tchebycheff, Tchebytcheff) was born on May 14, 1821, and died on Nov. 26, 1894. He is one of the most prominent representatives of the Russian mathematical school. He made numerous important contributions to the theory of numbers, algebra, the theory of probabilities, analysis, and applied mathematics. Among the most important of his papers are the two memoirs of which portions are here translated:

1. "Sur la totalité des nombres premiers inférieurs à une limite donnée," *Mémoires presentés à l'Académie Impériale des Sciences de St.-Pétersbourg par divers savants et lus dans ses assemblées*, Vol. 6, pp. 141–157, 1851 (Lu le 24 Mai, 1848); *Journal de Mathématiques pures et appliquées*, (1) Vol. 17, pp. 341–365, 1852; *Oeuvres*, Vol. 1, pp. 29–48, 1899.

2. "Mémoire sur les nombres premiers," *ibid.*, Vol. 7, pp. 15–33, 1854 (lu le 9 Septembre, 1850), *ibid.*, pp. 366–390, *ibid.*, pp. 51–70.

These memoirs represent the first definite progress after Euclid in the investigation of the function $\phi(x)$ which determines the totality of prime numbers less than the given limit x. The problem of finding an asymptotic expression for $\phi(x)$ for large values of x attracted the attention and efforts of some of the most brilliant mathematicians such as Legendre, Gauss, Lejeune-Dirichlet, and Riemann.

Gauss (1791, at the age of fourteen) was the first to suggest, in a purely empirical way, the asymptotic formula $\frac{x}{\log x}$ for $\phi(x)$. (*Werke*, Vol. X₁, p. 11, 1917.) Later on (1792–1793, 1849), he suggested another formula $\int_2^x \frac{dx}{\log x}$, of which $\frac{x}{\log x}$ is the leading term (Gauss's letter to Encke, 1849, *Werke*, Vol. II, pp. 444–447, 1876). Legendre, being, of course, unaware of Gauss's results, suggested another empirical formula $\frac{x}{A \log x + B}$ (*Essai sur la théorie des nombres*, 1st ed., pp. 18–19, 1798) and specified the constants A and B as $A = 1$, $B = -1.08366$ in the second edition of the *Essai* (pp. 394–395, 1808). Legendre's formula, which Abel quoted as "the most marvelous in mathematics" (letter to Holmboe, *Abel Memorial*, 1902, Correspondence, p. 5), is correct up to the leading term only. This fact was recognized by Dirichlet ("Sur l'usage des séries infinies dans la théorie des nombres," *Crelle's Journal*, Vol. 18, p. 272, 1838, in his remark written on the copy presented to Gauss. *Cf.* Dirichlet, *Werke*, Vol. 1, p. 372, 1889). In this note

127

to Gauss, Dirichlet suggested another formula $\sum\limits^{x} \dfrac{1}{\log n}$. The proof of these

results, although announced by Dirichlet, has never been published, so that Chebyshev's (Tchebycheff's) memoirs should be considered as the first attempt at a rigorous investigation of the problem by analytical methods.

Chebyshev did not reach the final goal—to prove that the ratio $\phi(x): \dfrac{x}{\log x}$

tends to 1 as $x \to \infty$. This important theorem was proved some 40 years later by Hadamard ("Sur la distribution des zéros de la fonction $\zeta(s)$ et ses conséquences arithmétiques," *Bulletin de la Société Mathématique de France,* Vol. 24, pp. 199–220, 1896) and by de la Vallée Poussin ("Recherches analytiques sur la théorie des nombres premiers," *Annales de la Société Scientifique de Bruxelles,* Vol. 20, pp. 183–256, 1896), their work being based upon new ideas and suggestions introduced by Riemann ("Über die Anzahl der Primzahlen unter einer gegebenen Grenze," *Monatsberichte der Berliner Akademie,* pp. 671–680, 1859; *Werke,* 2nd ed., pp. 145–153, 1892).

Although Chebyshev did not prove this final theorem, still he succeeded in obtaining important inequalities for the function $\phi(x)$, which enabled him to investigate the possible forms of approximation of $\phi(x)$ by means of expressions containing algebraically x, e^x, $\log x$ (*Memoir* 1, above) with a conclusion concerning the rather limited range of applicability of Legendre's formula. In the *Memoir* 2, Chebyshev obtains rather narrow limits for the ratio $\phi(x): \dfrac{x}{\log x}$, which provide a proof for the famous Bertrand postulate: "If $x \geqq 2$, there is at least one prime number between x and $2x - 2$."

MEMOIR 1: ON THE FUNCTION WHICH DETERMINES THE TOTALITY OF PRIMES LESS THAN A GIVEN LIMIT

§1. Legendre in his *Théorie des nombres*[1] proposes a formula for the number of primes between 1 and any given limit. He begins by comparing his formula with the result of counting the primes in the most extended tables, namely those from 10,000 up to 1,000,000, after which he applies his formula to the solution of many problems. Later the same formula has been the object of investigations of Mr. Lejeune-Dirichlet who announced in one of his memoirs in *Crelle's Journal,* Vol. 18, that he had found a rigorous analytical proof of the formula in question.[2] Despite the authority of the name of Mr. Lejeune-Dirichlet and the pronounced agreement of the formula of Legendre with the tables of primes we permit ourselves to raise certain doubts as to its

---

[1] Volume 2, p. 65 (3rd edition).

[2] [Naturally Chebyshev was unaware of the marginal notation made by Dirichlet in the copy of his paper presented to Gauss, to which we referred above.]

correctness and, consequently, as to the results which have been derived from this formula. We shall base our assertion on a theorem concerning a property of the function which determines the totality of primes less than a given limit,—a theorem from which one might derive numerous curious consequences. We shall first give a proof of the theorem in question; after that we shall indicate some of its applications.

§2. THEOREM 1.—*If* $\phi(x)$ *designates the totality of primes less than* x, n *is any integer, and* ρ *is a quantity* > 0, *the sum*

$$\sum_{x=2}^{x=\infty} \left[\phi(x+1) - \phi(x) - \frac{1}{\log x} \right] \frac{\log^n x}{x^{1+\rho}}$$

will have the property of approaching a finite limit as ρ *converges to zero.*

Proof.—We begin by establishing the property in question for the functions which are obtained by successive differentiations, with respect to ρ, of the three expressions

$$\sum \frac{1}{m^{1+\rho}} - \frac{1}{\rho}, \qquad \log \rho - \sum \log \left(1 - \frac{1}{\mu^{1+\rho}} \right),$$

$$\sum \log \left(1 - \frac{1}{\mu^{1+\rho}} \right) + \sum \frac{1}{\mu^{1+\rho}}.$$

The summation over m is extended, here as well as later, over all integral values from $m = 2$ up to $m = \infty$, while that over μ is taken over primes only, likewise from $\mu = 2$ up to $\mu = \infty$.

Consider the first expression. It is readily seen that[1]

$$\int_0^\infty \frac{e^{-x}}{e^x - 1} x^\rho dx = \sum \frac{1}{m^{1+\rho}} \int_0^\infty e^{-x} x^\rho dx,$$

$$\int_0^\infty e^{-x} x^{-1+\rho} dx = \frac{1}{\rho} \int_0^\infty e^{-x} x^\rho dx,$$

consequently

$$\sum \frac{1}{m^{1+\rho}} - \frac{1}{\rho} = \frac{\int_0^\infty \left(\frac{1}{e^x - 1} - \frac{1}{x} \right) e^{-x} x^\rho dx}{\int_0^\infty e^{-x} x^\rho dx}.$$

---

[1] [The first of these formulas is obtained by expanding $\frac{e^{-x}}{(e^x - 1)}$ in the geometric series Σe^{-mx}, which, being multiplied by x^ρ and integrated termwise, yields the expression

$$\sum \int_0^\infty e^{-mx} x^\rho dx = \sum m^{-1-\rho} \int_0^\infty e^{-x} x^\rho dx.$$

The termwise integration can be readily justified.]

By virtue of this equation the derivative of any order n with respect to ρ of $\sum \dfrac{1}{m^{1+\rho}} - \dfrac{1}{\rho}$ will be equal to a fraction whose denominator is $\left[\displaystyle\int_0^\infty e^{-x}x^\rho dx\right]^{n+1}$ and whose numerator is a polynomial in

$$\int_0^\infty \left(\frac{1}{e^x - 1} - \frac{1}{x}\right)e^{-x}x^\rho dx, \quad \int_0^\infty \left(\frac{1}{e^x - 1} - \frac{1}{x}\right)e^{-x}x^\rho \log x\, dx,$$

$$\int_0^\infty \left(\frac{1}{e^x - 1} - \frac{1}{x}\right)e^{-x}x^\rho \log^2 x\, dx, \ldots \int_0^\infty \left(\frac{1}{e^x - 1} - \frac{1}{x}\right)e^{-x}x^\rho \log^n x\, dx,$$

$$\int_0^\infty e^{-x}x^\rho dx, \int_0^\infty e^{-x}x^\rho \log x\, dx, \int_0^\infty e^{-x}x^\rho \log^2 x\, dx, \ldots \int_0^\infty e^{-x}x^\rho \log^n x\, dx.$$

But a fraction of this type, no matter whether $n = 0$ or $n > 0$, approaches a finite limit at $\rho \to 0$; for, then the limit of the integral $\displaystyle\int_0^\infty e^{-x}x^\rho\, dx$ is 1, and the remaining integrals have finite limiting values.[1]

This proves that the function $\sum \dfrac{1}{m^{1+\rho}} - \dfrac{1}{\rho}$ and its successive derivatives remain finite when $\rho \to 0$.

Consider now the function

$$\log \rho - \sum \log\left(1 - \frac{1}{\mu^{1+\rho}}\right).$$

It is known that

$$\left[\left(1 - \frac{1}{2^{1+\rho}}\right)\left(1 - \frac{1}{3^{1+\rho}}\right)\left(1 - \frac{1}{5^{1+\rho}}\right)\cdots\right]^{-1}$$
$$= 1 + \frac{1}{2^{1+\rho}} + \frac{1}{3^{1+\rho}} + \frac{1}{4^{1+\rho}} + \cdots [2]$$

---

[1] [The reasoning here is justified, since all the integrals in question are uniformly convergent in ρ for $0 \leqq \rho \leqq A$, A being any fixed positive constant.]

[2] [This identity was established by Euler ("Variæ observationes circa series infinitæ," *Commentarii Academiae Scientiarum Petropolitanæ*, 9, pp. 160–188, 1737 (Theorem 8, p. 174); *Leonardi Euleri Opera Omnia*, (1) 14, pp. 216–244 (230). Euler introduces here what is now called Riemann's ζ-function as defined by the series

$$\zeta(\rho) = \sum_{\nu=1}^{\infty} \nu^{-\rho}, \ \rho > 1.$$

The use of this function made by Riemann (*loc. cit.*) gave a most powerful impetus to the modern theory of functions of a complex variable.

The infinite product here is absolutely convergent since $(1 - \mu^{-1-\rho})^{-1} = 1 + \left(\dfrac{1}{\mu^{1+\rho} - 1}\right)$ and the series $\sum\left(\dfrac{1}{\mu^{1+\rho} - 1}\right)$ is absolutely convergent, as well whence, with the notation adopted above,

$$-\sum \log\left(1 - \frac{1}{\mu^{1+\rho}}\right) = \log\left(1 + \sum \frac{1}{m^{1+\rho}}\right).$$

Hence

$$\log \rho - \sum \log\left(1 - \frac{1}{\mu^{1+\rho}}\right) = \log\left(1 + \sum \frac{1}{m^{1+\rho}}\right)\rho,$$

or else

$$\log \rho - \sum \log\left(1 - \frac{1}{\mu^{1+\rho}}\right) = \log\left[1 + \rho + \left(\sum \frac{1}{m^{1+\rho}} - \frac{1}{\rho}\right)\rho\right].$$

This equation shows that all the derivatives with respect to ρ of

$$\log \rho - \sum \log\left(1 - \frac{1}{\mu^{1+\rho}}\right)$$

can be expressed in terms of a finite number of fractions whose denominators are positive integral powers of

$$1 + \rho + \left(\sum \frac{1}{m^{1+\rho}} - \frac{1}{\rho}\right)\rho,$$

and whose numerators are polynomials in ρ and the expression $\sum \frac{1}{m^{1+\rho}} - \frac{1}{\rho}$ and its derivatives with respect to ρ. The fractions of this type tend to finite limits as $\rho \to 0$: the expression $1 + \rho + \left(\sum \frac{1}{m^{1+\rho}} - \frac{1}{\rho}\right)\rho$, which figures in the denominators of these fractions, tends to 1 as $\rho \to 0$, since, as we have proved, the difference $\sum \frac{1}{m^{1+\rho}} - \frac{1}{\rho}$ remains finite; as to the numerators, they are polynomials in $\sum \frac{1}{m^{1+\rho}} - \frac{1}{\rho}$ and its derivatives, and, since all these functions tend to finite limits as $\rho \to 0$, the same will hold true for the numerators in question.

It remains to prove the same property for the derivatives of the function

$$\sum \log\left(1 - \frac{1}{\mu^{1+\rho}}\right) + \sum \frac{1}{\mu^{1+\rho}}.$$

We observe first that its first derivative is

$$\sum \mu^{-2-2\rho} \log \mu.(1 - \mu^{-1-\rho})^{-1}.$$

as the series $\sum \mu^{-1-\rho}$, which is only a part of the absolutely convergent series $\sum m^{-1-\rho}$. All these series and their derived series are also uniformly convergent for $\rho > 0$, which justifies the termwise differentiations in the following work.]

From this it is readily seen that the derivatives of higher order also can be expressed in terms of a finite number of expressions of the form

$$\Sigma \mu^{-2-2\rho-q} \log^p \mu \, (1 - \mu^{-1-\rho})^{-1-r},$$

with p, q, $r \geq 0$. But, each expression of this type has a finite value for $\rho \geq 0$, since the function under the sign Σ is of order higher than 1 in $1/\mu$.

After it has been proved that the derivatives of the three expressions above tend to finite limits as $\rho \to 0$, the same property can be established for the expression

$$\frac{d^n}{d\rho^n}\left[\sum \log \left(1 - \mu^{-1-\rho}\right) + \sum \mu^{-1-\rho} \right] +$$
$$\frac{d^n}{d\rho^n}\left[\log \rho - \sum \log \left(1 - \mu^{-1-\rho}\right) \right] + \frac{d^{n-1}}{d\rho^{n-1}}\left(\sum m^{-1-\rho} - \frac{1}{\rho}\right)$$

which, after the differentiations are performed, reduces to

$$\pm\left(\sum \frac{\log^n \mu}{\mu^{1+\rho}} - \sum \frac{\log^{n-1} m}{m^{1+\rho}} \right).$$

This result implies our theorem above, since it is readily seen that the difference

$$\sum \frac{\log^n \mu}{\mu^{1+\rho}} - \sum \frac{\log^{n-1} m}{m^{1+\rho}}$$

is identical with

$$\sum_{x=2}^{x=\infty}\left[\phi(x + 1) - \phi(x) - \frac{1}{\log x} \right]\frac{\log^n x}{x^{1+\rho}}$$

or, what is the same thing, with

$$\sum_{x=2}^{x=\infty} [\phi(x + 1) - \phi(x)]\frac{\log^n x}{x^{1+\rho}} - \sum_{x=2}^{x=\infty} \frac{\log^{n-1}}{x^{1+\rho}}.$$

To prove this we have only to observe that the first term of the difference above equals $\sum \dfrac{\log^n \mu}{\mu^{1+\rho}}$ since the coefficient $\phi(x + 1) - \phi(x)$ of $\dfrac{\log^n x}{x^{1+\rho}}$, by definition of the function $\phi(x)$ reduces to 1 or to 0 according as x is a prime or a composite number. The second term is transformed into $\sum\dfrac{\log^{n-1} m}{m^{1+\rho}}$ by replacing x by m.[1]

---

[1] [From the modern point of view the essence of Chebyshev's proof above lies in the fact that $\zeta(\rho)$ is analytic for all values of $\rho \neq 1$ while it has a simple pole at $\rho = 1$ with the residue 1, whence $\zeta(\rho) - \dfrac{1}{(\rho - 1)}$ is an entire transcendental function. (Whittaker-Watson, *Modern Analysis*, 3rd edition, 1920, p. 26.)]

This completes the proof of the theorem in question.

§3. The theorem which has been proved above leads to many curious properties of the function which determines the totality of primes less than a given limit. We first observe that the difference

$$\frac{1}{\log x} - \int^{x+} \frac{dx}{\log x}$$

for x very large is an infinitesimal of the first order in $1/x$; consequently the expression

$$\left(\frac{1}{\log x} - \int_x^{x+1} \frac{dx}{\log x} \right) \frac{\log^n x}{x^{1+\rho}}$$

will be of order $2 + \rho$ with respect to $1/x$.[1] Hence the sum

$$\sum_{x=2}^{x=\infty} \left(\frac{1}{\log x} - \int_x^{x+1} \frac{dx}{\log x} \right) \frac{\log^n x}{x^{1+\rho}}$$

remains finite for $\rho \geqq 0$. On adding this sum to the expression

$$\sum_{x=2}^{x=\infty} \left[\phi(x+1) - \phi(x) - \frac{1}{\log x} \right] \frac{\log^n x}{x^{1+\rho}}$$

for which Theorem 1 holds true, we conclude that the expression

$$\sum_{x=2}^{\infty} \left[\phi(x+1) - \phi(x) - \int_x^{x+1} \frac{dx}{\log x} \right] \frac{\log^n x}{x^{1+\rho}}$$

also remains finite as $\rho \to 0$. From this we can derive the following theorem.

THEOREM 2.—*The function $\phi(x)$ which designates the totality of primes less than x, satisfies infinitely many times, between the limits $x = 2$ and $x = \infty$, each of the inequalities*

$$\phi(x) > \int_2^x \frac{dx}{\log x} - \frac{\alpha x}{\log^n x} \text{ and } \phi(x) < \int^x \frac{dx}{\log x} + \frac{\alpha x}{\log^n x},$$

no matter how small is the positive number α and, at the same time, how large is n.

Proof.—We shall restrict ourselves to the proof of one of these two inequalities; the second can be proved exactly in the same fashion. Take for instance the inequality

(1) $$\phi(x) < \int_2^x \frac{dx}{\log x} + \frac{\alpha x}{\log^n x}.$$

To prove that this inequality is satisfied infinitely many times let us assume the contrary and examine the consequences of

---

[1] [By this it is meant that the quotient of the difference in question by any power of $1/x$ less than $(2 + \rho)$ tends to zero as $x \to \infty$.]

this hypothesis. Let a be an integer greater than e^n and, at the same time, greater than the greatest number which satisfies (1). With this assumption we shall have, for $x > a$, the inequality

$$\phi(x) \geqq \int_2^x \frac{dx}{\log x} + \frac{\alpha x}{\log^n x}, \quad \log x > n,$$

whence

(2) $$\phi(x) - \int_2^x \frac{dx}{\log x} \geqq \frac{dx}{\log^n x}, \quad \frac{n}{\log x} < 1.$$

But, if we admit inequalities (2), it will follow, in contradiction with the facts established above, that the expression

$$\sum_{x=2}^{x=\infty} \left[\phi(x+1) - \phi(x) - \int_x^{x+1} \frac{dx}{\log x} \right] \frac{\log^n x}{x^{1+\rho}}$$

will tend to $+\infty$ instead of converging to a finite limit as $\rho \to 0$. Indeed we can consider this expression as the limit of

$$\sum_{x=2}^{x=s} \left[\phi(x+1) - \phi(x) - \int_x^{x+1} \frac{dx}{\log x} \right] \frac{\log^n x}{x^{1+\rho}} \text{ as } s \to \infty.$$

On assuming $s > a$, this can be presented under the form

(3) $$C + \sum_{x=a+1}^{x=s} \left[\phi(x+1) - \phi(x) - \int_x^{x+1} \frac{dx}{\log x} \right] \frac{\log^n x}{x^{1+\rho}},$$

where

$$C = \sum_{x=2}^{x=a} \left[\phi(x+1) - \phi(x) - \int_x^{x+1} \frac{dx}{\log x} \right] \frac{\log^n x}{x^{1+\rho}}$$

remains finite for $\rho \geqq 0$.

On setting

$$u_x = \phi(x) - \int_2^x \frac{dx}{\log x}, \quad u_x = \frac{\log^n x}{x^{1+\rho}}$$

in the known formula

$$\sum_{a+1}^{s} u_x(v_{x+1} - v_x) = u_s v_{s+1} - u_a v_{a+1} - \sum_{a+1}^{s} v_x(u_x - u_{x-1}),$$

we transform expression (3) into

$$C - \left[\phi(a+1) - \int_2^{a+1} \frac{dx}{\log x} \right] \frac{\log^n a}{a^{1+\rho}} + \left[\phi(s+1) \right.$$

$$\left. - \int_2^{s+1} \frac{dx}{\log x} \right] \frac{\log^n s}{s^{1+\rho}} - \sum_{x=a+1}^{x=S} \left[\phi(x) - \int_2^x \frac{dx}{\log x} \right]$$

$$\left[\frac{\log^n x}{x^{1+\rho}} - \frac{\log^n (x-1)}{(x-1)^{1+\rho}} \right]$$

which, in its turn, can be written as

$$C - \left[\phi(a+1) - \int_2^{a+1} \frac{dx}{\log x} \right] \frac{\log^n a}{a^{1+\rho}} + \left[\phi(s+1) \right.$$

$$\left. - \int_2^{s+1} \frac{dx}{\log x} \right] \frac{\log^n s}{s^{1+\rho}} + \sum_{r=a+1}^{s} \left[\phi(x) - \int_a^x \frac{dx}{\log x} \right] \left[1 + \rho - \frac{n}{\log (x-\theta)} \right]$$

$$\frac{\log^n (x-\theta)}{(x-\theta)^{2+\rho}}, \text{ where } 0 < \theta < 1.$$

Let F denote the sum of the two first terms of this expression. Since, by virtue of condition (2), the third term is positive, we conclude that the expression above is greater than

$$F + \sum_{x=a+1}^{x=s} \left[\phi(x) - \int_2^x \frac{dx}{\log x} \right] \left[1 + \rho - \frac{n}{\log (x-\theta)} \right] \frac{\log^n (x-\theta)}{(x-\theta)^{2+\rho}}.$$

The same conditions (2) show that the function under the sign Σ in the last expression remains positive within the limits of summation. Furthermore, we have, within the same limits,

1°. $1 + \rho - \dfrac{n}{\log (x-\theta)} > 1 - \dfrac{n}{\log a}$ since $\rho > 0, x > a+1, \theta < 1$;

2°. $\qquad \phi(x) - \displaystyle\int_2^x \frac{dx}{\log x} > \frac{\alpha(x-\theta)}{\log^n (x-\theta)}$

since, by the first of inequalities (2),

$$\phi(x) - \int_2^x \frac{dx}{\log x} \geqq \frac{\alpha x}{\log^n x},$$

while, by the second one, the derivative of $\dfrac{\alpha x}{\log^n x}$, which equals

$\dfrac{\alpha}{\log^n x} \left(1 - \dfrac{n}{\log x} \right)$, is positive, whence,

$$\frac{\alpha x}{\log^n x} > \frac{\alpha (x-\theta)}{\log^n (x-\theta)}.$$

Hence our expression is greater than the sum

$$F + \sum_{x=a+1}^{x=s} \frac{\alpha (x-\theta)}{\log^n (x-\theta)} \left(1 - \frac{n}{\log a} \right) \frac{\log^n (x-\theta)}{(x-\theta)^{2+\rho}} =$$

$$F + \alpha \left(1 - \frac{n}{\log a} \right) \sum_{x=a+1}^{s} \frac{1}{(x-\theta)^{1+\rho}}.$$

But this is obviously greater than

$$F + \alpha \left(1 - \frac{n}{\log a} \right) \sum_{x=a+1}^{s} \frac{1}{x^{1+\rho}}.$$

which, for $s \to \infty$, reduces to

$$F + \alpha\left(1 - \frac{n}{\log a}\right) \sum_{x=a+1}^{x=\infty} \frac{1}{x^{1+\rho}} = F + \alpha\left(1 - \frac{n}{\log a}\right)\frac{\int_0^\infty \frac{e^{-ax}}{e^x-1} x^\rho dx}{\int_0^\infty e^{-x}x^\rho dx}.$$

It is readily seen that the expression at which we have arrived tends to $+\infty$ as $\rho \to 0$. For, we have

$$\int_0^\infty \frac{e^{-ax}}{e^x-1}\, dx + \infty, \qquad \int_0^\infty e^{-x}dx = 1.$$

while both α and $1 - \dfrac{n}{\log a}$ are positive, the former by hypothesis and the latter by the second of inequalities (2).

Thus, with the assumption made, it is assured that not only the sum

$$\sum_{x=a}^{x=\infty}\left[\phi(x + 1) - \phi(x) - \int_x^{x+1}\frac{dx}{\log x}\right]\frac{\log^n x}{x^{1+\rho}},$$

but even a quantity which is less than this sum, tends to $+\infty$, whence we conclude that the assumption in question is not admissible; this immediately proves Theorem 2.

§4. On the basis of the preceding proposition it will be easy now to prove the following theorem.

THEOREM 3.—*The expression* $\dfrac{x}{\phi(x)} - \log x$ *can not have a limit distinct from* -1 *as* $x \to \infty$.

Proof.—Let L be the limit as $x \to \infty$ of the difference $\dfrac{x}{\phi(x)} - \log x$. Under this assumption we always can find a number N so large that for $x > N$ the value of $\dfrac{x}{\phi(x)} - \log x$ will be within the limits $L - \epsilon$ and $L + \epsilon$, $\epsilon > 0$ being as small as we please. For such values of x and ϵ

$$(4) \qquad \frac{x}{\phi(x)} - \log x > L - \epsilon, \qquad \frac{x}{\phi(x)} - \log x < L + \epsilon.$$

But, by the preceding theorem, the inequalities

$$\phi(x) > \int_2^x \frac{dx}{\log x} - \frac{\alpha x}{\log^n x}, \qquad \phi(x) < \int_2^x \frac{dx}{\log x} + \frac{\alpha x}{\log^n x}$$

are satisfied for infinitely many values of x, consequently also for values of x greater than N, for which inequalities (4) hold true. The inequalities (4), combined with those written above, imply

$$\frac{x}{\displaystyle\int_2^x \frac{dx}{\log x} - \frac{\alpha x}{\log^n x}} - \log x > L - \epsilon,$$

$$\frac{x}{\displaystyle\int_2^x \frac{dx}{\log x} + \frac{\alpha x}{\log^n x}} - \log x < L + \epsilon,$$

whence

$$L + 1 < \frac{x - (\log x - 1)\left(\displaystyle\int_2^x \frac{dx}{\log x} - \frac{\alpha x}{\log^n x}\right)}{\displaystyle\int \frac{dx}{\log x} - \frac{\alpha x}{\log^n x}} + \epsilon.$$

$$L + 1 > \frac{x - (\log x - 1)\left(\displaystyle\int_2^x \frac{dx}{\log x} + \frac{\alpha x}{\log^n x}\right)}{\displaystyle\int \frac{dx}{\log x} + \frac{\alpha x}{\log^n x}} - \epsilon.$$

Thus the absolute value of $L + 1$ does not exceed that of each of the expressions which figure in the right-hand members of the preceding inequalities. Furthermore, ϵ can be made as small as we please by taking N sufficiently large, and the same will be true also of each of the quantities

$$\frac{x - (\log x - 1)\left(\displaystyle\int_2^x \frac{dx}{\log x} \mp \frac{\alpha x}{\log^n x}\right)}{\displaystyle\int_2^x \frac{dx}{\log x} \mp \frac{\alpha x}{\log^n x}},$$

for, it can be found by the principles of the differential calculus that their common limit for $x = \infty$ is zero.

Thus it is shown that the limits between which the absolute value of $L + 1$ is included can be made arbitrarily small; hence $L + 1 = 0$ or $L = -1$, which was to be proved.

The fact established above concerning the limit of $\frac{x}{\phi(x)} - \log x$ for $x = \infty$ does not agree with a formula given by Legendre for approximate computation of the totality of primes less than a given limit. According to Legendre the function $\phi(x)$ for x large is expressed with a sufficient degree of approximation by the formula

$$\phi(x) = \frac{x}{\log x - 1.08366},$$

which gives for the limit of $\frac{x}{\phi(x)} - \log x$ the number -1.08366 instead of -1.

§5. Starting from Theorem 2 it is possible to estimate the degree of approximation of the function $\phi(x)$ by any other given function $f(x)$. In what follows we shall compare the difference $(x) - \phi(x)$ with the expressions

$$\frac{x}{\log x}, \frac{x}{\log^2 x}, \frac{x}{\log^3 x}, \cdots$$

To simplify the discussion we shall say that a quantity A is of order $\frac{x}{\log^m x}$ if, as $x \to \infty$, the ratio of A to $\frac{x}{\log^m x}$ is infinite for $m > n$ and zero for $m < n$. We proceed now to prove the following theorem.

THEOREM 4.—*If the expression*

$$\frac{\log^n x}{x}\left(f(x) - \int_2^x \frac{dx}{\log x}\right)$$

has a finite [$\neq 0$] or infinite limit as $x \to \infty$, the function $f(x)$ can not represent $\phi(x)$ up to terms of order $\frac{x}{\log^n x}$ inclusive.[1]

Proof.—Let L be the limit of the expression

$$\frac{\log^n x}{x}\left(f(x) - \int_2^x \frac{dx}{\log x}\right)$$

as $x \to \infty$. Since, by hypothesis, L is distinct from zero, it is either positive or negative. Assume L to be positive; our reasoning is readily applied to the case of $L < 0$.

If $L > 0$ we can find a number N so large that for $x > N$ the expression

$$\frac{\log^n x}{x}\left(f(x) - \int_2^x \frac{dx}{\log x}\right)$$

remains always greater than a positive number l.

---

[1] [This means to imply that the difference $f(x) - \phi(x)$ can not be of order $\frac{x}{\log^m x}$ with $m > n$.]

Hence, for $x > N$,

(5)
$$\frac{\log^n x}{x}\left(f(x) - \int_2^x \frac{dx}{\log x}\right) > l.$$

But, by Theorem 2, no matter how small $\alpha = l/2$ may be, the inequality

(6)
$$\phi(x) < \int_2^x \frac{dx}{\log x} + \frac{\alpha x}{\log^n x}$$

will be satisfied for infinitely many values of x, which gives

$$f(x) - \int_2^x \frac{dx}{\log x} < f(x) - \phi(x) + \frac{\alpha x}{\log^n x};$$

on multiplying this by $\dfrac{\log^n x}{x}$ and observing that $\alpha = l/2$ we find

$$\frac{\log^n x}{x}\left[f(x) - \int_2^x \frac{dx}{\log x}\right] < \frac{\log^n x}{x}\left[f(x) - \phi(x)\right] + \frac{l}{2}$$

or, in view of (5)

$$\frac{\log^n x}{x}\left[f(x) - \phi(x)\right] > \frac{l}{2}.$$

Since $l/2 > 0$ and the preceding inequality, as well as inequalities (5) and (6), are satisfied for infinitely many values of x, the limit of

$$\frac{\log^n x}{x}\left[f(x) - \phi(x)\right]$$

as $x \to \infty$ can not be equal to zero. Then the difference $f(x) - \phi(x)$, according to the agreement above, is either of order $\dfrac{x}{\log^n x}$ or of a lower order, which was to be proved.

On the basis of this theorem we can show that the formula of Legendre, $\dfrac{x}{\log x - 1.08366}$, for which the limit as $x \to \infty$ of the expression

$$\frac{\log^2 x}{x}\left(\frac{x}{\log x - 1.08366} - \int_2^x \frac{dx}{\log x}\right)$$

equals 0.08366, can not represent $\phi(x)$ up to terms of order $\dfrac{x}{\log^2 x}$ inclusive.

It is also easy to determine the constants A and B so that the function $\dfrac{x}{A \log x + B}$ will represent $\phi(x)$ up to terms of order $\dfrac{x}{\log^2 x}$

inclusive.   By the preceding theorem the constants A and B must satisfy the equation

$$\lim \left[\frac{\log^2 x}{x} \left(\frac{x}{A \log x + B} - \int_2^x \frac{dx}{\log x} \right) \right] = 0.$$

On expanding we have

$$\frac{x}{A \log x + B} = \frac{1}{A} \frac{x}{\log x} - \frac{B}{A^2} \frac{x}{\log^2 x} + \frac{B^2}{A^3} \frac{x}{\log^3 x} - \cdots,$$

while an integration by parts yields

$$\int_2^x \frac{dx}{\log x} = \frac{x}{\log x} + \frac{x}{\log^2 x} + 2 \int_2^x \frac{dx}{\log^3 x} + C.$$

The equation above then reduces to

$$\lim_{x \to \infty} \left\{ \begin{array}{c} \dfrac{\log^2 x}{x} \left(\dfrac{1}{A} \dfrac{x}{\log x} - \dfrac{B}{A^2} \dfrac{x}{\log^2 x} + \dfrac{B^2}{A^3} \dfrac{x}{\log^3 x} - \cdots \right) \\[2mm] \cdots - \dfrac{x}{\log x} - \dfrac{x}{\log^2 x} - 2 \int_2^x \dfrac{dx}{\log^3 x} + C \end{array} \right\} = 0$$

or else to

$$\lim_{x \to \infty} \left\{ \begin{array}{c} \left(\dfrac{1}{A} - 1 \right) \log x - \left(\dfrac{B}{A^2} + 1 \right) + \dfrac{B^2}{A^3} \cdot \dfrac{1}{\log x} - \cdots \\[2mm] \cdots - 2 \dfrac{\log^2 x}{x} \int_2^x \dfrac{dx}{\log^3 x} - C \dfrac{\log^2 x}{x} \end{array} \right\} = 0.$$

On observing that all the terms beginning with the third converge to zero when x increases indefinitely, it is seen at once that the preceding equation can not be satisfied unless $\frac{1}{A} - 1 = 0$. $\frac{B}{A^2} + 1 = 0$.   Hence $A = 1$, $B = -1$.

Thus among all the functions of the form $\dfrac{x}{A \log x + B}$ only $\dfrac{x}{\log x - 1}$ can represent $\phi(x)$ up to terms of order $\dfrac{x}{\log^2 x}$ inclusive.[1]

---

[1] [We omit §§6 and 7 of this memoir.   In §6 Chebyshev proves by a method analogous to that used above that if $\phi(x)$ can be represented up to terms of order $\dfrac{x}{\log^n x}$ inclusive by an expression algebraic in x, $\log x$, e^z, then $\phi(x)$ can be represented also with the same degree of approximation by the expression

$$\frac{x}{\log x} + \frac{1.x}{\log^2 x} + \cdots + \frac{1.2 \ldots (n-1)x}{\log^n x}$$

which is obtained from $\int \dfrac{dx}{\log x}$ by repeated integration by parts.   §7 contains

MEMOIR 2: MEMOIR ON PRIME NUMBERS[1]

§2. Let us designate by $\theta(z)$ the sum of logarithms of all the primes which do not exceed z. This function equals zero when x is less than the smallest prime, viz. 2. It is not difficult to show that this function satisfies the following equation[2]

$$
\left.
\begin{aligned}
&\theta(x) + \theta(x)^{\frac{1}{2}} + \theta(x)^{\frac{1}{3}} + \ldots \\
&+ \theta\!\left(\frac{x}{2}\right) + \theta\!\left(\frac{x}{2}\right)^{\frac{1}{2}} + \theta\!\left(\frac{x}{3}\right)^{\frac{1}{3}} + \ldots \\
&+ \theta\!\left(\frac{x}{3}\right) + \theta\!\left(\frac{x}{3}\right)^{\frac{1}{2}} + \theta\!\left(\frac{x}{3}\right)^{\frac{1}{3}} + \ldots \\
&\cdots\cdots\cdots\cdots\cdots\cdots\cdots\cdots \\
&\cdots\cdots\cdots\cdots\cdots\cdots\cdots\cdots
\end{aligned}
\right\} = \log 1.2.3\ldots[x],
$$

where the symbol $[x]$ is used to designate the greatest integer contained in x.

To verify this equation we note that both its members are made up of terms of the form $K \log a$, where a is a prime and K is an integer. In the left-hand member K is equal to the number of terms in the sequence

$$
\begin{aligned}
&x, \quad \frac{x}{2}, \quad \frac{x}{3}, \; \ldots \\
(1) \quad &(x)^{\frac{1}{2}}, \quad \left(\frac{x}{2}\right)^{\frac{1}{2}}, \quad \left(\frac{x}{3}\right)^{\frac{1}{2}}, \ldots \\
&(x)^{\frac{1}{3}}, \quad \left(\frac{x}{3}\right)^{\frac{1}{3}}, \quad \left(\frac{x}{3}\right)^{\frac{1}{3}}, \ldots
\end{aligned}
$$

which are not less than a, since the expression for $\theta(z)$ will contain the term $\log a$ only in the case where $z \geqq a$. As to the coefficient of $\log a$ in the right-hand member, it is equal to the highest power of a which divides $1.2.3\ldots[x]$. It is found however, that this power is also equal to the number of terms in the sequence (1) which are not less than a; for, the number of terms of the sequence

$$
x, \quad \frac{x}{2}, \quad \frac{x}{3}, \ldots
$$

an attempt (not rigorous) to prove the remarkable asymptotic relations

$$
\sum_{\mu \leqq P} \frac{1}{\mu} \sim \log \log P + C_1, \quad \prod_{\mu \leqq P}\left(1 - \frac{1}{\mu}\right) \sim \frac{C_2}{\log P}
$$

where C_1, C_2 are fixed constants, P is any prime number, and the summation and product are extended over all primes $\mu \leqq P$.]

[1] [We omit the introductory §1 of this memoir.]

[2] To abbreviate we write $\theta(x/n)^m$ instead of $\theta\{(x/n)^m\}$.

which are not less than a, is equal to that of the terms of the sequence

$$1, 2, 3, \ldots, [x]$$

which are divisible by a.

The same relationship exists between the number of terms of this sequence, which are divisible by a^2, a^3, a^4, ... and the number of terms of the sequence

$$(x)^{1/2}, \quad \left(\frac{x}{2}\right)^{1/2}, \quad \left(\frac{x}{3}\right)^{1/2}, \ldots$$

$$(x)^{1/3}, \quad \left(\frac{x}{2}\right)^{1/3}, \quad \left(\frac{x}{3}\right)^{1/3}, \ldots$$

. . . .

which are not less than a.

Hence both members of our equation are composed of the same terms, which proves that they are identical.

The equation just established can be presented as

$$(2) \qquad \psi(x) + \psi\left(\frac{x}{2}\right) + \psi\left(\frac{x}{3}\right) + \ldots = T(x)$$

on setting for abbreviation

$$(3) \qquad \begin{cases} \theta(z) + \theta(z)^{1/2} + \theta(z)^{1/3} + \ldots = \psi(z), \\ \log 1.2.3 \ldots [x] = T(x). \end{cases}$$

In applications of these formulas we shall observe that, in view of what has been said about the value of $\theta(z)$ when $z < 2$ the function $\psi(z)$ vanishes when $z < 2$, and consequently, equation (2) will be valid in the limiting cases $x = 0$, $x = 2$ if we agree to take zero as the value of $T(x)$ when $x < 2$.

§3. By means of this equation it is not difficult to find numerous inequalities which are satisfied by the function $\psi(x)$; those we shall use in this memoir are the following:

$$\psi(x) > T(x) + T\left(\frac{x}{30}\right) - T\left(\frac{x}{2}\right) - T\left(\frac{x}{3}\right) - T\left(\frac{x}{5}\right),$$

$$\psi(x) - \psi\left(\frac{x}{6}\right) > T(x) + T\left(\frac{x}{30}\right) - T\left(\frac{x}{2}\right) - T\left(\frac{x}{3}\right) - T\left(\frac{x}{5}\right).$$

To prove these inequalities we shall compute the value of

$$T(x) + T\left(\frac{x}{30}\right) - T\left(\frac{x}{2}\right) - T\left(\frac{x}{3}\right) - T\left(\frac{x}{5}\right)$$

by means of (2), which leads to the equation

$$(4) \quad
\begin{vmatrix}
\psi(x) + \psi\left(\dfrac{x}{2}\right) + \psi\left(\dfrac{x}{3}\right) + \ldots \\[1ex]
+\psi\left(\dfrac{x}{30}\right) + \psi\left(\dfrac{x}{2.30}\right) + \psi\left(\dfrac{x}{3.30}\right) + \ldots \\[1ex]
-\psi\left(\dfrac{x}{2}\right) - \psi\left(\dfrac{x}{2.2}\right) - \psi\left(\dfrac{x}{3.2}\right) - \ldots \\[1ex]
-\psi\left(\dfrac{x}{3}\right) - \psi\left(\dfrac{x}{2.3}\right) - \psi\left(\dfrac{x}{3.3}\right) - \ldots \\[1ex]
-\psi\left(\dfrac{x}{5}\right) - \psi\left(\dfrac{x}{2.5}\right) - \psi\left(\dfrac{x}{3.5}\right) - \ldots
\end{vmatrix}
\begin{aligned}
&= T(x) + T\left(\dfrac{x}{30}\right) \\[1ex]
&\quad - T\left(\dfrac{x}{2}\right) - T\left(\dfrac{x}{3}\right) \\[1ex]
&\qquad\qquad - T\left(\dfrac{x}{5}\right).
\end{aligned}$$

whose left-hand member reduces to

$$A_1\psi(x) + A_2\psi\left(\frac{x}{2}\right) + \ldots + A_n\psi\left(\frac{x}{n}\right) + \ldots$$

$A_1, A_2, \ldots A_n \ldots$ being numerical coefficients. Upon examining their values it is not difficult to establish that

$A_n = 1$ if $n = 30m + 1, 7, 11, 13, 17, 19, 23, 29,$

$A_n = 0$ if $n = 30m + 2, 3, 4, 5, 8, 9, 14, 16, 21, 22, 25, 26, 27,$
$$28,$$

$A_n = -1$ if $n = 30m + 6, 10, 12, 15, 18, 20, 24,$

$A_n = -1$ if $n = 30m + 30.$

Indeed, in the first case n is not divisible by any of the numbers 2, 3, 5, hence the term $\psi(x/n)$ figures only in the first line in equation (4). In the second case n is divisible by one of the numbers 2, 3, 5, hence, besides the term $\psi(x/n)$ in the first line, the term $-\psi(x/n)$ will be found in one of the last three lines, and, after reduction, the coefficient of $\psi(x/n)$ will become 0. In the third case n is divisible by two of the numbers 2, 3, 5. Hence the last three lines will contain two terms equal to $-\psi(x/n)$, while the first line contains $\psi(x/n)$ with the plus sign, so that the result will be $-\psi(x/n)$. In the last case where n is divisible by 30 we arrive at the same conclusion, since the term $\pm\psi(x/n)$ will figure in all the five lines, twice with the plus and three times with the minus sign.

Hence for

$n = 30m + 1, 2, 3, 4, 5, 6, 7, 8, 9, 10, 11, 12, 13, 14, 15$
$\qquad\qquad 16, 17, 18, 19, 20, 21, 22, 23, 24, 25, 26, 27, 28, 29, 30$

we find respectively

$A_n = 1, 0, 0, 0, 0, -1, 1, 0, 0, -1, 1, -1, 1, 0, -1,$
$\qquad\quad 0, 1, -1, 1, -1, 0, 0, 1, -1, 0, 0, 0, 0, 1, -1,$

which shows that equation (4) reduces to

$$\psi(x) - \psi\!\left(\frac{x}{6}\right) + \psi\!\left(\frac{x}{7}\right) - \psi\!\left(\frac{x}{10}\right) + \psi\!\left(\frac{x}{11}\right) - \psi\!\left(\frac{x}{12}\right) + \cdots$$

$$= T(x) + T\!\left(\frac{x}{30}\right) - T\!\left(\frac{x}{2}\right) - T\!\left(\frac{x}{3}\right) - T\!\left(\frac{x}{5}\right)$$

where the terms of the left-hand member have the coefficient 1 alternately with the plus and minus signs. Furthermore, since by nature of the function $\psi(x)$ the series of the left-hand member is decreasing, its value will be included within the limits $\psi(x)$ and $\psi(x) - \psi(x/6)$. Hence, by the preceding equation, we shall have necessarily

$$\psi(x) \geqq T(x) + T\!\left(\frac{x}{30}\right) - T\!\left(\frac{x}{2}\right) - T\!\left(\frac{x}{3}\right) - T\!\left(\frac{x}{5}\right),$$

$$\psi(x) - \psi\!\left(\frac{x}{6}\right) \leqq T(x) + T\!\left(\frac{x}{30}\right) - T\!\left(\frac{x}{2}\right) - T\!\left(\frac{x}{3}\right) - T\!\left(\frac{x}{5}\right).$$

§4. Let us examine now the function $T(x)$ which figures in these formulas. On denoting by a the greatest integer contained in x, which we shall assume to be $\geqq 1$, we have from (3)

$$T(x) = \log 1.2.3 \ldots a$$

or, what amounts to the same thing,

$$T(x) = \log 1.2.3 \ldots a(a + 1) - \log (a + 1).$$

But it is known that

$$\log 1.2.3 \ldots a < \log \sqrt{2\pi} + a \log a - a + \tfrac{1}{2} \log a +$$
$$\log 1.2.3 \ldots a(a + 1) > \log \sqrt{2\pi} + (a + 1) \log (a + 1) -$$
$$\tfrac{1}{12a}(a + 1) + \tfrac{1}{2} \log (a + 1);$$

hence

$$T(x) < \log \sqrt{2\pi} + a \log a - a + \tfrac{1}{2} \log a + \tfrac{1}{12a},$$
$$T(x) > \log \sqrt{2\pi} + (a + 1) \log (a + 1) - (a + 1) - \tfrac{1}{2} \log (a + 1)$$

and consequently

$$T(x) < \log \sqrt{2\pi} + x \log x - x + \tfrac{1}{2} \log x + \tfrac{1}{12},$$
$$T(x) > \log \sqrt{2\pi} + x \log x - x - \tfrac{1}{2} \log x,$$

since the inequalities

$$a \leqq x < a + 1, \quad a \geqq 1$$

obviously imply the conditions

$$x \log x - x + \frac{1}{2} \log x + \frac{1}{12} \geqq a \log a - a + \frac{1}{2} \log a + \frac{1}{12a},$$
$$x \log x - x - \tfrac{1}{2} \log x \leqq (a + 1) \log (a + 1) - (a + 1)$$
$$- \tfrac{1}{2} \log (a + 1).$$

The inequalities above concerning $T(x)$ give

$$T(x) + T\left(\frac{x}{30}\right) < 2 \log \sqrt{2\pi} + \frac{2}{12} + \frac{31}{30}x \log x - x \log 30^{\frac{1}{30}}$$

$$- \frac{31}{30}x + \log x - \frac{1}{2} \log 30,$$

$$T(x) + T\left(\frac{1}{30}\right) > 2 \log \sqrt{2\pi} + \frac{31}{30}x \log x - x \log 30^{\frac{1}{30}} - \frac{31}{30}x$$

$$- \log x + \frac{1}{2} \log 30,$$

$$T\left(\frac{x}{2}\right) + T\left(\frac{x}{3}\right) + T\left(\frac{x}{5}\right) < 3 \log \sqrt{2\pi} + \frac{3}{12} + \frac{31}{30}x \log x$$

$$- x \log 2^{\frac{1}{2}}3^{\frac{1}{3}}5^{\frac{1}{5}} - \frac{31}{30}x + \frac{3}{2} \log x - \frac{1}{2} \log 30,$$

$$T\left(\frac{x}{2}\right) + T\left(\frac{x}{3}\right) + T\left(\frac{x}{5}\right) > 3 \log \sqrt{2\pi} + \frac{31}{30}x \log x$$

$$- x \log 2^{\frac{1}{2}}3^{\frac{1}{3}}5^{\frac{1}{5}} - \frac{31}{30}x - \frac{3}{2} \log x + \frac{1}{2} \log 30.$$

On subtracting the last of these inequalities from the first and the third from the second we find

$$T(x) + T\left(\frac{x}{30}\right) - T\left(\frac{x}{2}\right) - T\left(\frac{x}{3}\right) - T\left(\frac{x}{5}\right) < Ax + \frac{5}{2} \log x$$

$$- \frac{1}{2} \log 1800\pi + \frac{2}{12}$$

$$T(x) + T\left(\frac{x}{30}\right) - T\left(\frac{x}{2}\right) - T\left(\frac{x}{3}\right) - T\left(\frac{x}{5}\right) > Ax - \frac{5}{2} \log x$$

$$+ \frac{1}{2} \log \frac{450}{\pi} - \frac{3}{12},$$

where to abbreviate we have set

(5) $\qquad A = \log 2^{\frac{1}{2}}3^{\frac{1}{3}}5^{\frac{1}{5}}30^{-\frac{1}{30}} = 0.92129202\ldots$

The analysis used in proving these inequalities assumes that $x \geqq 30$, since, in discussing $T(x)$ we have assumed $x \geqq 1$ and after that we have replaced x successively by $x/2$, $x/3$, $x/5$ and $x/30$. It is not difficult, however, to obtain formulas which can be used for all values of $x > 1$, if we replace the preceding inequalities by simpler ones

$$T(x) + T\left(\frac{x}{30}\right) - T\left(\frac{x}{2}\right) - T\left(\frac{x}{3}\right) - T\left(\frac{x}{5}\right) < Ax + \frac{5}{2} \log x,$$

$$T(x) + T\left(\frac{x}{30}\right) - T\left(\frac{x}{2}\right) - T\left(\frac{x}{3}\right) - T\left(\frac{x}{5}\right) > Ax - \frac{5}{2} \log x - 1;$$

an examination readily shows that these inequalities are valid for values of x between 1 and 30.

§5. On combining these inequalities with those derived above for the function $\psi(x)$ (§3) we arrive at two formulas

$$\psi(x) > Ax - \frac{5}{2} \log x - 1, \quad \psi(x) - \psi\left(\frac{x}{6}\right) < Ax + \frac{5}{2} \log x,$$

of which the first gives a lower limit for $\psi(x)$.

As to the second formula, it will be used in assigning another limit for $\psi(x)$. For this purpose we observe that the function

$$f(x) = \frac{6}{5} Ax + \frac{5}{4 \log 6} \log^2 x + \frac{5}{4} \log x$$

satisfies the equation

$$f(x) - f\left(\frac{x}{6}\right) = Ax + \frac{5}{2} \log x,$$

which, being subtracted from the inequality

$$\psi(x) - \psi\left(\frac{x}{6}\right) < Ax + \frac{5}{2} \log x$$

gives

$$\psi(x) - \psi\left(\frac{x}{6}\right) - f(x) + f\left(\frac{x}{6}\right) < 0$$

or else

$$\psi(x) - f(x) < \psi\left(\frac{x}{6}\right) - f\left(\frac{x}{6}\right).$$

On replacing x successively by $x/6, x/6^2, \ldots x/6^m$ in this formula we find

$$\psi(x) - f(x) < \psi\left(\frac{x}{6}\right) - f\left(\frac{x}{6}\right) < \ldots < \psi\left(\frac{x}{6^{m+1}}\right) - f\left(\frac{x}{6^{m+1}}\right).$$

Assume now that m is the greatest integer which satisfies the condition $\frac{x}{6^m} \geqq 1$. Then $x/6^{m+1}$ will be between 1 and $\frac{1}{6}$, while $\psi(z) = 0$ and $-f(z)$ remains greater than 1 within the limits $z = 1$, $z = \frac{1}{6}$. Hence $\psi(x/6^{m+1}) - f(x/6^{m+1}) < 1$, and by the preceding inequalities

$$\psi(x) - f(x) < 1.$$

Finally, on substituting the value of $f(x)$ we have

$$\psi(x) < \frac{6}{5} Ax + \frac{5}{4 \log 6} \log^2 x + \frac{5}{4} \log x + 1.$$

On the basis of the formulas just found it is not difficult to assign two limits including the value of $\theta(x)$.

Indeed, we find from (3)

$$\psi(x) - \psi(x)^{1/2} = \theta(x) + \theta(x)^{1/3} + \theta(x)^{1/5} + \dots,$$
$$\psi(x) - 2\psi(x)^{1/2} = \theta(x) - [\theta(x)^{1/2} - \theta(x)^{1/3}] - \dots$$

which shows that

(6) $\theta(x) \leqq \psi(x) - \psi(x)^{1/2}, \quad \theta(x) \leqq \psi(x) - 2\psi(x)^{1/2},$

since the terms

$$\theta(x)^{1/3}, \quad \theta(x)^{1/5}, \dots, \quad \theta(x)^{1/2} - \theta(x)^{1/3}, \dots$$

obviously are positive or zero.

But we have found

$$\psi(x) < \frac{6}{5}Ax + \frac{5}{4 \log 6} \log^2 x + \frac{5}{4} \log x + 1,$$
$$\psi(x) > Ax - \tfrac{5}{2} \log x - 1,$$

which gives

$$\psi(x)^{1/2} < \frac{6}{5}Ax^{1/2} + \frac{5}{16 \log 6} \log^2 x + \frac{5}{8} \log x + 1,$$
$$\psi(x)^{1/2} > Ax^{1/2} - \frac{5}{4} \log x - 1,$$

and consequently

$$\psi(x) - \psi(x)^{1/2} < \frac{6}{5}Ax - Ax^{1/2} + \frac{5}{4 \log 6} \log^2 x + \frac{5}{2} \log x + 2,$$

$$\psi(x) - 2\psi(x)^{1/2} > Ax - \frac{12}{5}Ax^{1/2} - \frac{5}{8 \log 6} \log^2 x - \frac{15}{4} \log x - 3$$

Hence, by (6),

(7) $\begin{cases} \theta(x) < \dfrac{6}{5}Ax - Ax^{1/2} + \dfrac{5}{4 \log 6} \log^2 x + \dfrac{5}{2} \log x + 2 \\[2mm] \theta(x) > Ax - \dfrac{12}{5}Ax^{1/2} - \dfrac{5}{8 \log 6} \log^2 x - \dfrac{15}{4} \log x - 3.^{[1]} \end{cases}$

---

[1] [We omit the concluding §§6–9 of the memoir. In §6 Chebyshev gives the proof of the Bertrand postulate, taking as the point of departure the obvious inequalities

$$\theta(L) - \theta(l) > m \log l, \quad \theta(L) - \theta(l) < m \log L$$

where m is the number of primes between l and L, and using the inequalities obtained above for $\theta(x)$. §7 contains a proof of the following remarkable theorem: If for x sufficiently large $F(x)$ is positive and $\dfrac{F(x)}{\log x}$ is not increasing, then the convergence of the series $\sum \dfrac{F(m)}{\log m}$ is a necessary and sufficient condition for the convergence of the series $\Sigma F(\mu)$. The proof is based upon the simple transformation formula

$$\sum_{l,L} F(\mu) = \sum_{m=l}^{L} F(m) \left\{ \frac{\theta(m) - \theta(m-1)}{\log m} \right\}$$

where the summation over μ is extended over all primes, while that over m over all integers between the two given limits l and L.   Thus the series

$$\frac{1}{2 \log 2} + \frac{1}{3 \log 3} + \frac{1}{5 \log 5} + \cdots ; \frac{1}{2 \log^2 (\log 2)} + \frac{1}{3 \log^2 (\log 3)} + \frac{1}{5 \log^2 (\log 5)} + \cdots$$

are convergent while the series

$$\frac{1}{2} + \frac{1}{3} + \frac{1}{5} + \cdots ; \quad \frac{1}{2 \log 2} + \frac{1}{3 \log 3} + \frac{1}{5 \log 5} + \cdots$$

are divergent.   §§8 and 9 contain some applications of the above results to the approximate computation of sums of the form $\Sigma F(\mu)$ and, in the special case where $F(x) = 1$, to the computation of the totality of primes.]

NAPIER

On the Table of Logarithms

(Selections Made by Professor W. D. Cairns, Oberlin College, Oberlin, Ohio.)

John Napier (1550–1617), Baron of Merchiston, Scotland, has been given undisputed priority with regard to the publication of a table of logarithms and an account of their meaning and use. His work is the more important since, through improvements by himself, Henry Briggs, and others, it quickly became a system practical for purposes of calculation and nearly in the modern form. He published his system in 1614 in *Mirifici logarithmorum canonis descriptio* and gave therein a description of the nature of logarithms and a table of his logarithms of the sines of angles for successive minutes. The present account is, however, taken from his *Mirifici logarithmorum canonis constructio*, which appeared posthumously in 1619 but which was written several years earlier than the *Descriptio*. Sufficient extracts are given, with the original numbers of the articles, to show his method of construction of the table, his definition of logarithms, and the rules for combining these.

The *Descriptio* was translated into English by Edward Wright under the title *A Description of the Admirable Table of Logarithmes* and was published posthumously at London in 1616. The *Constructio* was translated into English by W. R. Macdonald (Edinburgh, Wm. Blackwood & Sons, Ltd., 1889). The following selections are taken from the latter work with the kind permission of the publishers, the numbers of the paragraphs being as in the original. Only the more important parts of the numbered paragraphs have been selected, there being sufficient to show Napier's method of constructing a logarithmic table. Upon the question of the invention of logarithms, see the articles on prosthaphæresis (pp. 455 and 459).

1.   A logarithmic table is a small table by the use of which we can obtain a knowledge of all geometrical dimensions and motions in space, by a very easy calculation...It is picked out from numbers progressing in continuous proportion.

2.   Of continuous progressions, an arithmetical is one which proceeds by equal intervals; a geometrical, one which advances by unequal and proportionally increasing or decreasing intervals.

16.   If from the radius with seven ciphers added you subtract its 10000000th part, and from the number thence arising its 10000000th part, and so on, a hundred numbers may very easily be continued geometrically in the proportion subsisting

149

between the radius and the sine less than it by unity, namely between 10000000 and 9999999; and this series of proportionals we name the First table.

Thus from the radius, with seven ciphers added for greater accuracy, namely, 10000000.0000000, subtract 1.0000000, you get 9999999.0000000; from this subtract .9999999, you get 9999998.0000001; and proceed in this way until you create a hundred proportionals, the last of which, if you have computed rightly, will be 9999900.0004950.

17. The Second table proceeds from the radius with six ciphers added, through fifty other numbers decreasing proportionally in the proportion which is easiest, and as near as possible to that subsisting between the first and last numbers of the First table

Thus the first and last numbers of the First table are 10000000.0000000 and 9999900.0004950, in which proportion it is difficult to form fifty proportional numbers. A near and at the same time an easy proportion is 100000 to 99999, which may be continued with sufficient exactness by adding six ciphers to the radius and continually subtracting from each number its own 100000th part; and this table contains, besides the radius which is the first, fifty other proportional numbers, the last of which, if you have not erred, you will find to be 9995001.222927.[1]

18. The Third table consists of sixty-nine columns, and in each column are placed twenty-one numbers, proceeding in the proportion which is easiest, and as near as possible to that subsisting between the first and last numbers of the Second table.

Whence its first column is very easily obtained from the radius with five ciphers added, by subtracting its 2000th part, and so from the other numbers as they arise.

In forming this progression, as the proportion between 10000000.000000, the first of the Second table, and 9995001.222927, the last of the same, is troublesome; therefore compute the twenty-one numbers in the easy proportion of 10000 to 9995, which is sufficiently near to it; the last of these, if you have not erred, will be 9900473.57808.

From these numbers, when computed, the last figure of each may be rejected without sensible error, so that others may hereafter be more easily computed from them.

---

[1] [This should be 9995001.224804.]

19. The first numbers of all the columns must proceed from the radius with four ciphers added, in the proportion easiest and nearest to that subsisting between the first and the last numbers of the first column.

As the first and the last numbers of the first column are 10000000.0000 and 9900473.5780, the easiest proportion very near to this is 100 to 99. Accordingly sixty-eight numbers are to be continued from the radius in the ratio of 100 to 99 by subtracting from each one of them its hundredth part.

20. In the same proportion a progression is to be made from the second number of the first column through the second numbers in all the columns, and from the third through the third, and from the fourth through the fourth, and from the others respectively through the others.

Thus from any number in one column, by subtracting its hundredth part, the number of the same rank in the following column is made, and the numbers should be placed in order.

Remark: The last number in the Sixty-ninth column is 4998609.4034, roughly half the original number.

21. Thus, in the Third table, between the radius and half the radius, you have sixty-eight numbers interpolated, in the proportion of 100 to 99, and between each two of these you have twenty numbers interpolated in the proportion of 10000 to 9995; and again, in the Second table, between the first two of these, namely between 10000000 and 9995000, you have fifty numbers interpolated in the proportion of 100000 to 99999; and finally, in the First table, between the latter, you have a hundred numbers interpolated in the proportion of the radius or 10000000 to 9999999; and since the difference of these is never more than unity, there is no need to divide it more minutely by interpolating means, whence these three tables, after they have been completed, will suffice for computing a Logarithmic table.

Hitherto we have explained how we may most easily place in tables sines or natural numbers progressing in geometrical proportion.

22. It remains, in the Third table at least, to place beside the sines or natural numbers decreasing geometrically their logarithms or artificial numbers increasing arithmetically.

26.   The logarithm of a given sine is that number which has
increased arithmetically with the same velocity throughout
as that with which the radius began to decrease geometrically,
and in the same time as the radius has decreased to the given
sine.[1]

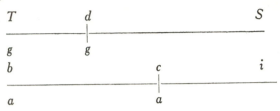

Let the line *TS* be the radius, and *dS* a given sine in the
same line; let *g* move geometrically from *T* to *d* in certain
determinate moments of time.   Again, let *bi* be another
line, infinite towards *i*, along which, from *b*, let *a* move
arithmetically with the same velocity as *g* had at first
when at *T*; and from the fixed point *b* in the direction of *i*
let *a* advance in just the same moments of time up to the
point *c*.   The number measuring the line *bc* is called the
logarithm of the given sine *dS*.

27.   Zero is the logarithm of the radius.

28.   Whence also it follows that the logarithm of any given sine
is greater than the difference between the radius and the
given sine, and less than the difference between the radius
and the quantity which exceeds it in the ratio of the radius
to the given sine.   And these differences are therefore called
the limits of the logarithm.

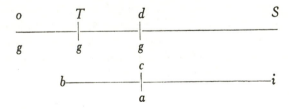

Thus, the preceding figure being repeated, and *ST* being
produced beyond *T* to *o*, so that *oS* is to *TS* as *TS* to *dS*.

---

[1] [To Napier the sine was a line, or the number measuring the line, as in the
present-day line representation of functions of angles.]

I say that *bc*, the logarithm of the sine *dS*, is greater than *Td* and less than *oT*. For in the same time that *g* is borne from *o* to *T*, *g* is borne from *T* to *d*, because (by 24) *oT* is such a part of *oS* as *Td* is of *TS*, and in the same time (by the definition of a logarithm) is *a* borne from *b* to *c*; so that *oT*, *Td*, and *bc* are distances traversed in equal times. But since *g* when moving between *T* and *o* is swifter than at *T*, and between *T* and *d* slower, but at *T* is equally swift with *a* (by 26); it follows that *oT* the distance traversed by *g* moving swiftly is greater, and *Td* the distance traversed by *g* moving slowly is less, than *bc* the distance traversed by the point *a* with its medium motion, in just the same moments of time; the latter is, consequently, a certain mean between the two former.

Therefore *oT* is called the greater limit, and *Td* the less limit of the logarithm which *bc* represents.

29. To find the limits of the logarithm of a given sine.

By the preceding it is proved that the given sine being subtracted from the radius, the less limit remains, and that the radius being multiplied into the less limit and the product divided by the given sine, the greater limit is produced.

30. Whence the first proportional of the First table, which is 9999999, has its logarithm between the limits 1.0000001 and 1.0000000.

31. The limits themselves differing insensibly, they or anything between them may be taken as the true logarithm.

32. There being any number of sines decreasing from the radius in geometrical proportions, of one of which the logarithm or its limits is given, to find those of the others.

This necessarily follows from the definitions of arithmetical increase, of geometrical decrease, and of a logarithm...So that, if the first logarithm corresponding to the first sine after the radius be given, the second logarithm will be double of it, the third triple, and so of the others; until the logarithms of all the sines are known.

36. The logarithms of similarly proportioned sines differ equally.

This necessarily follows from the definitions of a logarithm and of the two motions. Also there is the same ratio of equality between the differences of the respective limits of the logarithms, namely as the differences of the less among themselves, so also of the greater among themselves, of which logarithms the sines are similarly proportioned.

38.   Of four geometrical proportionals, as the product of the means is equal to the product of the extremes; so of their logarithms, the sum of the means is equal to the sum of the extremes.   Whence any three of these logarithms being given, the fourth becomes known.[1]

39.   The difference of the logarithms of two sines lies between two limits; the greater limit being to the radius as the difference of the sines to the less sine, and the less limit being to the radius as the difference of the sines to the greater sine.[2]

47.   In the Third table, beside the natural numbers, are to be written their logarithms; so that the Third table, which after this we shall always call the Radical table, may be made complete and perfect.

48.   The Radical table being now completed, we take the numbers for the logarithmic table from it alone.

For as the first two tables were of service in the formation of the third, so this Radical table serves for the construction of the principal Logarithmic table, with great ease and no sensible error.

51.   All sines in the proportion of two to one have 6931469.22 for the difference of their logarithms.[3]

52.   All sines in the proportion of ten to one have 23025842.34 for the difference of their logarithms.

---

[1] [The modern theorem for the logarithm of a product does not hold here, since the logarithm of unity is not zero.]

[2] [This is proved by the principle of proportion and of Article 36.   This rule is used first in Articles 40–41 as an illustration to find the logarithm of 9999975.5 from that of the nearest sine in the First table, 9999975.0000300, noting that the limits of the logarithms of the latter number are 25.0000025 and 25.0000000, that the difference of the logarithms of the two numbers by the rule just given is .4999712 and that the limits for the logarithm of 9999975.5 are therefore 24.5000313 and 24.5000288, whence he lists the logarithm as 24.5000300.

In Articles 41–45 he illustrates the fact that one may now calculate the logarithms of all the "proportionals" in the First, Second, and Third tables, as well as of the sines or natural numbers not proportionals in these tables but near or between them.]

[3] [Napier obtains this result by first calculating the logarithm of 7071068, which is to the nearest unit, the square root of 50 × 10^{12} and which is, to his "radius," the sine of 45°.   By Article 39 its logarithm is 3465734.5, whence the result in Article 51.]

55.   As the half radius is to the sine of half a given arc, so is the sine of the complement of the half arc to the sine of the whole arc.[1]

56.   Double the logarithm of an arc of 45 degrees is the logarithm of half the radius.

57.   The sum of the logarithms of half the radius and any given arc is equal to the sum of the logarithms of half the arc and the complement of the half arc. Whence the logarithm of the half arc may be found if the logarithms of the other three are given.

59. To form a logarithmic table.[2]

---

[1] [Only here does Napier begin to introduce angles into the construction of his tables. Napier proves Articles 55–57 by geometric principles and the preceding theorems concerning logarithms.]

[2] [Napier's table is constructed in quite the same form as used at present, except that the second (sixth) column gives sines for the number of degrees indicated at the top (bottom) and of minutes in the first (seventh) column, the third (fifth) column gives the corresponding logarithm and the fourth column gives the "differentiæ" between the logarithms in the third and fifth columns, these being therefore essentially logarithmic tangents or cotangents. A reproduction of one page may be seen in Macdonald's translation, page 138.

DELAMAIN

On The Slide Rule

(Edited by Professor Florian Cajori, University of California, Berkeley, California.)

The earliest publication describing a slide rule (an instrument differing from Gunter's scale, which had no sliding parts) was brought out in the year 1630 by Richard Delamain, a teacher of mathematics in London. It was a pamphlet of 30 pages, entitled *Grammelogia*[1] and describing a circular slide rule. There is a copy in the Cambridge University Library in England. This tract has no drawing of the slide rule. During the next 2 or 3 years there were issued at least four undated new editions, or impressions, of the *Grammelogia*, with new parts added. The Cambridge University Library has a copy which is the same as the 1630 publication but with an appendix[2] of 17 pages added. In the British Museum at London and in the Bodleian Library at Oxford, there are copies of another edition of 113 pages, which was published in 1632 or 1633, as is shown by its reference to Oughtred's book, the *Circles of Proportion* of 1632. It has two title pages[3] which we reproduce in facsimile.

---

[1] The full title of the *Grammelogia* of 1630 is as follows:
Gramelogia|or,|The Mathematicall Ring.|Shewing (any reasonable Capacity that hath| not Arithmeticke) how to resolve and worke|all ordinary operations of Arithmeticke.|And those which are most difficult with greatest|facilitie: The extraction of Roots, the valuation of| Leases, &c. The measuring of Plaines|and Solids.|With the resolution of Plaine and Sphericall|Triangles.|And that onely by an Ocular Inspection,|and a Circular Motion.| Naturae secreta tempus aperit.|London printed by John Haviland, 1630.

[2] The appendix is entitled:
De la Mains|Appendix|Vpon his|Mathematicall|Ring. Attribuit nullo (praescripto tempore) vitae|vsuram nobis ingeniique Deus.|London,|
...The next line or two of this title page which probably contained the date of publication, were cut off by the binder in trimming the edges of this and several other pamphlets for binding into one volume.

[3] The first title page (engraved) is as follows:
Mirifica Logarithmoru' Projectio Circularis. There follows a diagram of a circular slide rule, with the inscription within the innermost ring: *Nil Finis, Motvs, Circvlvs vllvs Habet.*
The second title page is as follows:
Grammelogia|Or, the Mathematicall Ring.|Extracted from the Logarythmes, and projected Circular: Now published in the|inlargement thereof unto any magnitude fit for use; shewing any reason-|able capacity that hath not Arithmeticke how to resolve and worke,|all ordinary operations of Arithmeticke:|And those that are most difficult with greatest facilitie, the extracti-|on of Rootes, the valuation of Leases, &c. the measuring of Plaines and Solids,|with the resolution of Plaine and Sphericall Triangles applied to the|Practicall parts of Geometrie, Horologographie, Geographie|Fortification, Navigation, Astronomie, &c.|And that onely by an ocular inspection, and a Circular motion, Invented and first published, by R. Delamain, Teacher, and Student of the Mathematicks.|Naturae secreta tempus aperit.|
There is no date. There follows the diagram of a second circular slide rule, with the inscription within the innermost ring: *Typus proiectionis Annuli adaucti vt in Conslusione Lybri praelo commissi, Anno 1630 promisi.* There are numerous drawings in the *Grammelogia*, all

156

In the *Grammelogia* of 1630, Delamain, in an address to King Charles I, emphasizes the ease of operating with his slide rule by stating that it is "fit for use...as well on Horse backe as on Foot." Speaking "To the Reader," he states that he has "for many yeares taught the Mathematicks in this Towne" and made efforts to improve Gunter's scale "by some Motion, so that the whole body of Logarithmes might move proportionally the one to the

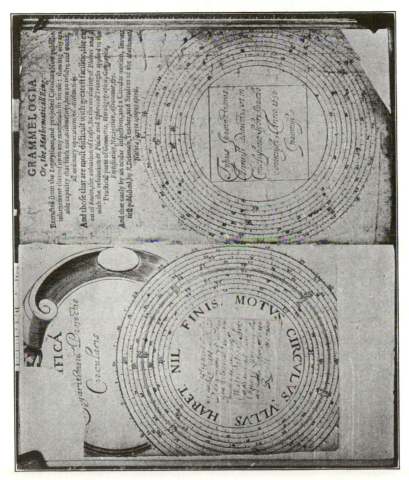

other, as occasion required. This conceit in February last [1629] I struke upon, and so composed my *Grammelogia* or *Mathematicall Ring;* by which only with an *ocular inspection*, there is had at one instant all proportionalls through

of which, excepting the drawings of slide rules on the engraved title pages were printed upon separate pieces of paper and then inserted by hand into the vacant spaces on the printed pages reserved for them. Some drawings are missing, so that the Bodleian *Grammelogia* differs in this respect slightly from the two copies in the British Museum.

the said body of Numbers." He dates his preface "first of January, 1630."
The term *Grammelogia* is applied to the instrument, as well as to the book.
Delamain's description of this *Grammelogia* is as follows:

The parts of the Instrument are two Circles, the one moveable,
and the other fixed: The moveable is that unto which is fastened
a small pin to move it by; the other Circle may be conceived to be
fixed; The circumference of the moveable Circle is divided into
unequall parts, charactered with figures thus, 1. 2. 3. 4. 5. 6. 7. 8. 9.
these figures doe represent themselves, or such numbers unto
which a Cipher or Ciphers are added, and are varied as the occasion
falls out in *the speech of Numbers*, so 1. stands for 1. or 10. or 100.,
&c. the 2. stands for 2. or 20. or 200. or 2000., &c. the 3. stands for
30. or 300. or 3000.; &c.

"How to perform the Golden Rule" (the rule of proportion), is explained
thus:

Seeke the first number in the moveable, and bring it to the
second number in the fixed, so right against the third number in
the moveable, is the answer in the fixed.
If the Interest of 100. li. be 8. li. in the yeare, what is the Interest
of 65. li. for the same time.
Bring 100. in the moveable to 8. in the fixed, so right against 65.
in the moveable is 5.2. in the fixed, and so much is the Interest of
65. li. for the yeare at 8. li. for 100. li. *per annum.*
The *Instrument* not removed, you may at one instant right
against any summe of money in the moveable, see the Interest
thereof in the fixed: the reason of this is from the *Definition of
Logarithmes.*

Relating to the "resolution of Plaine and Sphericall Triangles," Delamain
says:

If there be composed three Circles of equal thicknesse, A.B.C.
so that the inner edge of D [should be B] and the outward edge of
A bee answerably graduated with *Logarithmall signes* [sines], and
the outward edge of B and the inner edge of A with *Logarithmes;*
and then on the backside be graduated the *Logarithmall Tangents,*
and againe the *Logarithmall signes* oppositly to the former graduations, it shall be fitted for the resolution of *Plaine* and *Sphericall
Triangles.*

After twelve lines of further remarks on this point, he adds:

Hence from the forme, I have called it a *Ring*, and *Grammelogia* by annoligie of a *Lineary speech;* which *Ring*, if it were projected in the *convex* unto two yards *Diameter*, or thereabouts, and the line *Decupled*, it would worke *Trigonometrie* unto seconds, and give *proportionall numbers* unto six places only by an *ocular inspection*, which would compendiate *Astronomicall calculations*, and be sufficient for the *Prostbapbaeresis* of the Motions: But of this as God shall give life and ability to health and time.

The patent and copyright on the instrument and book are as follows:

Whereas Richard Delamain, Teacher of Mathematicks, hath presented vnto Vs an Instrument called Grammelogia, or The Mathematicall Ring, together with a Booke so instituled, expressing the use thereof, being his owne Invention; we of our Gracious and Princely favour have granted unto the said Richard Delamain and his Assignes, Privilege, Licence, and Authority, for the sole Making, Printing and Selling of the said Instrument and Booke: straightly forbidding any other to Make, Imprint, or Sell, or cause to be Made, or Imprinted, or Sold, the said Instrument or Booke within any our Dominions, during the space of ten years next ensuing the date hereof, upon paine of Our high displeasure, Given under our hand and Signet at our Palace of Westminster. the fourth day of January, in the sixth yeare of our Raigne.

OUGHTRED

ON THE SLIDE RULE

(Edited by Professor Florian Cajori, University of California, Berkeley, California.)

William Oughtred (1574–1660) was a clergyman living near London and intensely interested in mathematics. He taught mathematics at his residence, without compensation, to promising pupils. At one time, Oughtred had assisted Delamain in his mathematical studies. His *Circles of Proportion*[1] appeared in 1632, translated into English from his Latin manuscript by one of his pupils, William Forster. Forster wrote a preface in which he makes the charge (without naming Delamain) that "another...went about to pre-ocupate" the new invention. This led to verbal disputes and to the publication by Delamain of the several new editions of the *Grammelogia*, describing further designs of circular slide rules and also stating his side of the controversy. Oughtred prepared an *Epistle*, in reply, which was published in the 1633 edition of his *Circles of Proportion*. Each combatant accuses the other of stealing the invention of the circular slide rule. After reading both sides of the controversy, we conclude that Oughtred invented the circular slide rule before the time when Delamain claimed to have made his invention, but it is not shown conclusively that the latter was dishonest; we incline to the opinion that he was an independent inventor. In 1633, Oughtred published the description of a rectilinear slide rule, in the invention of which he has no rival.

Extracts from the *Circles of Proportion*

1 There are two sides of this Instrument. On the one side, as it were in the *plaine of the Horizon*, is delineated the *proiection of the Sphere*. On the other side there are divers kindes of Circles, divided after many severall Waies; together with an *Index* to be opened after the manner of a paire of Compasses. And of this side we will speake in the first place.

---

[1] There are two title pages. The first is engraved and reads thus:

The|Circles|of|Proportion|and|The Horizontall|Instrument.|Both invented, and|the vses of both|Written in Latine by|Mr. W. O.|Translated into English: and set forth|for the publique benefit by|William Forster.|London|Printed for Elias Allen maker|of these and all other Mathe:|matical Instruments, and are to|be sold at his shop ouer against| St Clements church with out Temple-barr.|1632. T. Cecill Sculp|

The second title page is:

The|Circle|of|Proportion,|and |The Horizontall|Instrvment.|Both invented, and the vses of both|written in Latine by that learned Mathe-|matician Mr W. O.|Bvt|Translated into English: and set forth for|the publique benefit by William Forster, louer|and prac-tizer of the Mathematicall Sciences.|London|Printed by Avg. Mathevves,|dwelling in the Parsonage Court, neere|St Brides. 1632.|

160

2 The *First*, or outermost circle is of *Sines*, from 5 degrees 45 minuts almost, vntill 90. Every degree till 30 is divided into 12 parts, each part being 5 min : from thence vntill 50 deg : into sixe parts which are 10 min : a peece : from thence vntill 75 degrees into two parts which are 30 minutes a peece. After that vnto 85 deg : they are not divided.

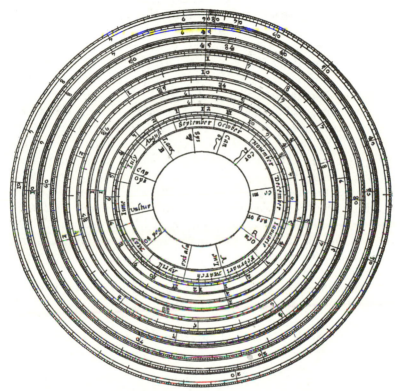

Oughtred's Circular Slide Rule, from his *Circles of Proportions*, 1632.

3 The *Second circle* is of *Tangents*, from 5 degrees 45. min : almost, untill 45 degrees. Every degree being divided into 12 parts which are 5 min : a peece.

4 The *Third circle* is of *Tangents*, from 45 degrees untill 84 degrees 15 minutes. Each degree being divided into 12 parts, which are 5 min : a peece.

5 The *Sixt circle* is of *Tangents* from 84 degrees till about 89 degrees 25 minutes.

The *Seventh circle* is of *Tangents* from about 35.Min : till 6 degrees.

The *Eight circle* is of *Sines*, from about 35 minutes til 6 degrees.

6 The *Fourth circle* is of *Vnæquall Numbers*, which are noted with the Figures 2, 3, 4, 5, 6, 7, 8, 9, 1.   Whether you vnderstand them to bee single Numbers, or Tenns. or Hundreds, or Thousands, etc.   And every space of the numbers till 5, is divided into 100 parts, but after 5 till 1, into 50 parts.

The *Fourth circle* also sheweth the *true* or *naturall Sines*, and *Tangents*.   For if the *Index* bee applyed to any Sine or Tangent, it will cut the *true Sine* or *Tangent* in the fourth circle.   And wee are to knowe that if the *Sine* or *Tangent* be in the *First*, or *Second circle*, the figures of the *Fourth circle* doe signifie so many thousands. But if the *Sine* or *Tangent* be in the *Seventh* or *Eight circle*, the figures in the *Fourth circle* signifie so many hundreds.   And if the *Tangent* bee in the *Sixt circle*, the figures of the *Fourth circle*, signifie so many times tenne thousand, or whole *Radij*.

And by this meanes the Sine of 23°, 30' will bee found 3987: and the Sine of it's complement 9171.   And the Tangent of 23°, 30 will be found 4348: and the Tangent of it's complement, 22998. And the Radius is 10000, that is the figure 1 with foure cyphers, or circles.   And hereby you may finde out both the summe, and also the difference of Sines, and Tangents.

7 The *Fift circle* is of *Aequall numbers*, which are noted with the figures 1, 2, 3, 4, 5, 6, 7, 8, 9, 0; and every space is divided into 100 aequall parts.

This *Fift circle* is scarse of any use, but onely that by helpe thereof the given distance of numbers may be multiplied, or divided, as neede shall require.

As for example, if the space between 1|00 and 1|0833+ bee to bee septupled.   Apply the Index vnto 1|0833+ in the Fourth circle, and it will cut in the Fift circle 03476+; which multiplyed by 7 makes 24333: then againe, apply the Index vnto this number 24333 in the Fift circle, and it will cut in the Fourth circle 1|7512+. And this is the space betweene 1|00 and 1|0833+ septupled, or the Ratio betweene 100 and 108⅓ seven times multiplied into it selfe.

And contrarily, if 1|7512 bee to bee divided by 7: Apply the Index vnto 1|7568 in the fourth circle, and it will cut in the fit circle 24333: which divided by 7 giveth 03476+ Then againe vnto

this Number in the Fift circle apply the Index, and in the Fourth
circle it wil cut vpon 1|0833+ for the Septupartion sought for.

The reason of which Operation is, because this *Fift circle* doth
shew the *Logarithmes* of Numbers. *For if the Index be applyed
unto any number* in the Fourth circle, *it will in the* Fift circle *cut
vpon the* Logarithme *of the same number, so that to the* Logarithme
found you praefixe a Caracteristicall (*as Master* Brigs *termes it*)
one lesse then in the number of the places of the integers proposed
(Which *you may rather call the* Graduall Number).   So the Logarithme of the number 2 will be 0.30103.   And the Logarithme of
the Number 43|6 will bee found 1.63949.

Numbers are multiplied by Addition of their Logarithmes: *and
they are Divided by Substraction of their* Logarithmes.

8 In the middest among the Circles, is a *double Nocturnall*
instrument, to shew the hower of the night.

9 The right line passing through the Center, through 90, and
45 I call the *Line of Vnitie*, or of the *Radius*.

10 That *Arme of the Index* which in euery Operation is placed
at the Antecedent, or first terme, I call the *Antecedent arme:* and that
which is placed at the consequent terme, I call the *Consequent Arme*.

Oughtred's Rectilinear Slide Rule

*An Addition Vnto the Vse of the Instrvment called the Circles of Proportion,
For the Working of Nauticall Questions... London*, 1633, contains the following
description of the rectilinear slide rule, consisting of "two Rvlers:"

I call the *longer* of the two *Rulers* the *Staffe*, and the *Shorter* the
Transversarie.   And are in length one to the other almost as 3 to 2.

The *Rulers* are just foure square, with right angles: and equall
in bignesse: they are thus divided.

The *Transversarie* at the upper end noted with the letters S, T,
N, E, on the severall sides, hath a *pinnicide* or *sight:* at the lower
edge of which sight is the *line of the Radius*, or *Vnite line*, where the
divisions beginne.

After explaining the different lines on the two rulers, Oughtred continues:

Thus have you on the *two Rulers* the very same lines which are
in the *Circles of Proportion:* and whatsoever can be done by those
Circles, may also as well be performed by the two Rulers: and the
Rules which have bin here formerly set downe for that Instrument,
may also be practised upon these: so that you bee carefull to observe
in both the different propriety in working.   It will not therfore
be needfull, to make any new and long discourse, concerning these

Rulers, but onely to shew the manner, how they are to be used, for the calculation of any proportion given.

In working a Proportion by the Rulers, *hold the Transversary in your left hand, with the end at which the line of the Radius or Vnite line is, from you ward: turning that side of the Ruler upward, on which the line of the kind of the first terme is, whether it be Number, Sine, or Tangent: and therein seeke both the first terme, and the other which is homogene to it.   Then take the Staffe in your right hand with that side upward, in which the line of the kind of the fourth terme sought for is: and seeke in it the terme homogene to the fourth. Apply this to the first terme in the Transversarie: and the other homogene terme shall in the Staffe shew the fourth terme.*

As if you would multiply 355 by 48: Say

$$1 . 355 :: 48 . 17040.$$

For if in the line of Numbers on the Staffe you reckon 355, and apply the same to 1 in the line of Numbers on the Transversarie: then shall 48 on the Transversarie shew 17040 on the Staffe.[1]

---

[1] For additional details relating to Delamain's and Oughtred's slide rules, consult an article entitled "On the History of Gunter's Scale and the Slide Rule during the Seventeenth Century," in the *University of California Publications in Mathematics*, Vol. I, No. 9, pp. 187–209, Feb. 17, 1920.

PASCAL

On His Calculating Machine

(Translated from the French by L. Leland Locke, A. M.,
Brooklyn, New York.)

Blaise Pascal, philosopher, mathematician, physicist, inventor, was born
at Clermont, June 19, 1623, and died at Paris, Aug. 19, 1662.

To Pascal (see p. 67) must be given the credit of having conceived and
constructed the first machine for performing the four fundamental operations
of arithmetic, of which a complete description and authentic models have been
preserved. His first machine was completed in 1642, and the Privilege was
granted by Chancellor Seguier on May 22, 1649. Pascal's account of his
invention, here reproduced, was written subsequent to the Privilege. It is
found in his collected works, the most recent edition being the *Oeuvres de
Blaise Pascal*," by Brunschvieg and Boutroux, Paris, Vol. I, pp. 303–314, 1908.
The perfected machine, made at the direction of and dedicated to Seguier, is
one of the most interesting relics of Pascal as well as being the first significant
model in the development of machine calculation.

Pascal's "Advis," the first discussion of the problem of machine calculation,
may be somewhat clarified, if his accomplishment is viewed in the light of
present-day design. The instrument may be classed as a machine on two
grounds: First, the transfer of tens, usually designated as the "carry," is
performed automatically; second, an initial installation is transmitted through
the medium of intermediate parts and finally registered on the result dials.
The primary conception due to Pascal is that of building a machine which
would automatically produce the carry. A simple ratchet or latch is introduced between successive orders which has the property of moving the dial of
higher order one unit forward as the dial of lower order passes from 9 to 0.
The Pascal machine may be classed as a modern counting machine, with
provision for the entry of numbers in all orders, provided such entries are made
separately. The common disk and cylinder dials of today originated in this
model. Pascal used crown wheels with pin teeth, a device which resulted in a
minimum of friction. The chief problem in the design of a machine of this
kind is so to adjust the load of the carry that a minimum of effort expended
on the initial installation will produce the carry as far as desired. Such a
condition arises when the dials all register 9 and when 1 is then added in the
lowest order. Pascal's solution would do credit to a designer of much later
date. A weighted ratchet is gradually raised as the number being installed
approaches 9. As it passes from 9 to 0, the ratchet is released and in falling
transfers 1 to the next higher order. If all dials stand at 9, only a slight lifting
remains for each ratchet. This cumulative load would become excessive over
many orders, thus definitely limiting the capacity of the machine. Pascal's

statement that a thousand dials may be turned as easily as one would, of course, fail in practice.

The one feature of his design which more than any other reveals his genius is the use of the complements of numbers in subtraction, enabling him to perform all four operations with a single direction of operation, a modification of this device being still in use in many key-driven machines.

ADVERTISEMENT

Necessary to those who have curiosity to see the Arithmetic Machine, and to operate it

Dear reader, this notice will serve to inform you that I submit to the public a small machine of my invention, by means of which you alone may, without any effort, perform all the operations of arithmetic, and may be relieved of the work which has often times fatigued your spirit, when you have worked with the counters or

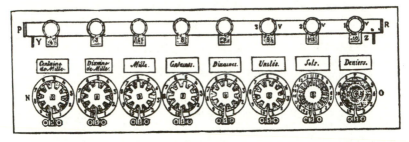

with the pen. I can, without presumption, hope that it will not be displeasing to you, since Monseigneur le Chancelier has honored it by his favorable opinion, and since, in Paris, those who are most versed in mathematics have judged it not unworthy of their approbation. However in order not to appear negligent in making it acquire yours also, I have felt obliged to make clear all the difficulties that I have judged capable of confronting your understanding when you take the trouble to consider it.

I have no doubt that after having seen it, there will come at once to your thought that I should have explained by writing both its construction, and its use, and that, to render the discourse intelligible, I should myself be obliged, following the method of the geometers,[1] to represent by figures the dimensions, the disposition and the relation of all of the parts and how each should be placed to compose the instrument, and to place its movement in

---

[1] [A word then used to denote mathematicians in general as well as those concerned with the study of geometric forms.]

its perfection; but do not believe that, after having spared neither time nor labor nor expense to put it in a state of being useful, I have failed to do everything necessary to satisfy you on that point, —which would indicate failure of accomplishment if I have not been prevented from doing it by a consideration so powerful that I myself hope that it will compel you to excuse me. I trust that you will approve my refraining from this phase of the work, when you reflect on the one hand on the ease with which by a brief conference the construction and the use of this machine can be explained, and on the other hand on the embarassment and the difficulty there would be in trying to express by writing the dimensions, the forms, the proportions, the position, and the rest of the properties of all the different pieces. Furthermore, you may well consider what learning means to the number of those who can be taught by word of mouth, and how an explanation in writing will be as useless as a written description of all the parts of a watch, although the verbal explanation is so easy. It is therefore apparent that such a written discourse would produce no other effect than a distaste on the part of many persons, making them conceive of a thousand difficulties at every point where there is none at all.

Now, dear reader, I deem it necessary to say that I foresee two things capable of forming some clouds in your spirit. I know that there are a number of persons who make profession of finding fault everywhere, and that among them will be found those who will say that the machine may be simplified, and this is the first mist that I feel it necessary to dispel. Such a proposition can only be made by certain persons who have indeed some knowledge of mechanics or of geometry, but who, not knowing how to join the one to the other, and both of these to physics, flatter themselves or are deceived by their imaginary conceptions, and persuade themselves that many things are possible which are not. Having only an imperfect knowledge in general, this is not sufficient to make them forsee the inconveniences arising, either from the nature of the case or as to the places which the pieces of the machine should occupy. The movements of these parts are different, and they must be so free as not to interfere with one another. When, therefore, those whose knowledge is so imperfect propose that this machine be simplified, I ask you to say to them that I will reply for myself, if they will simply ask me; and to assure them on my part, that I will let them see, whenever they wish, many other models, together with a perfect instrument which is much less complex, and which I have publicly operated for six months.   It will thus appear that I am quite aware that the machine may be simplified.   In particular I could, if I had wished, institute the movement of the operation on the front face,[1] but this could only be substituted with much inconvenience for that which it is now done on the top face with all the convenience that one should wish and with pleasure.   You may also say to them that my design does not always have in view the reducing in controlled movement all the operations of arithmetic.   [Hence] I am persuaded that it will be successful only if the movement is simple, easy, convenient and quick of execution, and if the machine is durable, solid, and capable of undergoing without alteration the strain of transportation.   Finally you may say that if they had thought as much as I on this matter and had considered all the means which I have taken to reach my goal, experience would have shown them that a more simple instrument could not have all of the qualities that I have successfully given to this little machine.

As for the simplicity of movement of the operations, I have so devised it that, although the operations of arithmetic are in a way opposed the one to the other,—as addition to subtraction, and multiplication to division,—nevertheless they are all performed on this machine by a single unique movement.

The facility of this movement of operation is very evident since it is just as easy to move one thousand or ten thousand dials, all at one time, if one desires as to make a single dial move, although all accomplish the movement perfectly.   (I do not know if there remains another principle in nature such as the one upon which I have based this ease of operation.)   In addition to the facility of movement in the operation, if you wish to appreciate it, you may compare it with the methods of counters and with the pen.   You know that, in operating with counters, the calculator (especially when he lacks practice) is often obliged, for fear of making an error, to make a long series and extension of counters, being afterward compelled to gather up and retake those which are found to be extended unnecessarily, in which you see these two useless tasks, with the double loss of time.   This machine facili-

---

[1] [The ratchet design was based on a horizontal position of the intermediate shaft.   The stylus-operated dial, being on the top face with a vertical shaft, required a pair of crown wheels at right angles to transmit the motion to the horizontal shaft.   Pascal here proposes placing the dial on the front face.]

tates the work and eliminates all unnecessary features. The most ignorant find as many advantages as the most experienced. The instrument makes up for ignorance and for lack of practice, and even without any effort of the operator, it makes possible shortcuts by itself, whenever the numbers are set down. In the same way you know that in operating with the pen one is obliged to retain or to borrow the necessary numbers, and that errors slip in, in these retentions and borrowings, except through very long practice, and in spite of a profound attention which soon fatigues the mind. This machine frees the operator from that vexation; it suffices that he have judgment; he is relieved from the failing of memory; and without any retaining or borrowing, it does by itself what he wishes, without any thinking on his part. There are a hundred other advantages which practice will reveal, the details of which it would be wearisome to mention.

As to the amount[1] of movement, it is sufficient to say that it is imperceptible, going from left to right and following our method of common writing except that it proceeds in a circle.

And, finally, its speed is evident at once in comparing it with the other two methods of the counters and the pen. If you still wish a more particular explanation of its rapidity, I shall tell you that it is equal to the agility of the hand of the operator. This speed is based, not only on the facility of the movements which have no resistance, but also on the smallness of the dials which are moved with the hand. The result is that, the key board being very short, the movement can be performed in a short time. Thus, the machine is small and hence is easily handled and carried.

And as to the lasting and wearing qualities of the instrument, the durability of the metal of which it is made should be a sufficient warrant: I have been able to give entire assurance to others only after having had the experience of carrying the instrument over more than two hundred and fifty leagues of road, without its showing any damage.

Therefore, dear reader I ask you again not to consider it an imperfection for this machine to be composed of so many parts, because without these I could not give to it all the qualities which I have explained, and which are absolutely necessary. In this you may notice a kind of paradox, that to render the movement of operation more simple, it is necessary that the machine should be constructed of a movement more complex.

---

[1] This has reference to the movement of the stylus in setting down a number.

The second possibility for distrust, dear reader, might be the imperfect reproductions of this machine which have been produced by the presumption of certain artisans. In these cases I beg of you to carefully consider [the product], to guard yourself from surprise, to distinguish between "la lepre et la lepre,"[1] and not to judge the true original by the imperfect productions of the ignorance and the temerity of the mechanics. The more excellent they are in their art, the more we should fear that vanity forces them to consider themselves capable of undertaking and of producing new instruments, of which the principles and the rules of which they are ignorant. Intoxicated by that false persuasion, they grope aimlessly about, without precise measurements carefully determined and without propositions. The result is that after much time and labor, either they do not produce anything equal to what they have attempted, or, at most, they produce a small monstrosity of which the principal members are lacking, the others being formless and without any proportion. These imperfections, rendering it ridiculous, never fail to attract the contempt of all those who see it, and most of them blame without reason, the inventor instead of inquiring about it from him, and then censuring the presumption of these artisans, who by their unjustified daring undertake more than they are equal to, producing these useless abortions. It is important to the public to recognize their weakness and to learn from them that, for new inventions, it is necessary that art should be aided by theory until usage has made the rules of theory so common that it has finally reduced them to an art and until continued practice has given the artisans the habit of following and practicing these rules with certainty. It was not in my power, with all the theory imaginable, to execute alone my own design without the aid of a mechanic who knew perfectly the practice of the lathe, of the file, and of the hammer to reduce the parts of the machine in the measures and proportions that I prescribed to him. Likewise it is impossible for simple artisans, skillful as they may be in their art, to make perfectly a new instrument which consists, like this one, of complicated movements, without the aid of a person who, by the rules of theory, gives him the measures and proportions of all of the pieces of which it shall be composed.

Dear reader, I have good reason to give you this last advice, after having seen with my own eyes a wrong production of my idea

---

[1] [Bossut reads, "between the copy and the copy."]

by a workman of the city of Rouen, a clockmaker by profession, who, from a simple description which had been given him of my first model, which I had made some months previously, had the presumption, to undertake to make another; and what is more, by another type of movement. Since the good man has no other talent than that of handling his tools skillfully, and has no knowledge of geometry and mechanics (although he is very skillful in his art and also very industrious in many things which are not related to it), he made only a useless piece, apparently true, polished and well filed on the outside, but so wholly imperfect on the inside that it was of no use. Because of its novelty alone, it was not without value to those who did not understand about it; and, notwithstanding all these essential defects which trial shows, it found in the same city a place in a collector's cabinet which is filled with many other rare and curious pieces. The appearance of that small abortion displeased me to the last degree and so cooled the ardor with which I had worked to the accomplishment of my model, that I at once discharged all my workmen, resolved to give up entirely my enterprise because of the just apprehension that many others would feel a similar boldness and that the false copies which they would produce of this new idea would only ruin its value at its beginning and its usefulness to the public. But, some time afterward, Monseigneur le Chancelier, having deigned to examine my first model and to give testimony of the regard which he held for that invention, commanded me to perfect it. In order to eliminate the fear which held me back for some time, it pleased him to check the evil at its root, and to prevent the course it could take in prejudicing my reputation and inconveniencing the public. This was shown in the kindness that he did in granting me an unusual privilege, and which stamped out with their birth all those illegitimate abortions which might be produced by others than by the legitimate alliance of the theory with art.

For the rest, if at any time you have thought of the invention of machines, I can readily persuade you that the form of the instrument, in the state in which it is at present, is not the first attempt that I have made on that subject. I began my project by a machine very different from this both in material and in form, which (although it would have pleased many) did not give me entire satisfaction. The result was that in altering it gradually I unknowingly made a second type, in which I still found inconveniences to which I would not agree. In order to find a remedy, I have devised a third, which works by springs and which is very simple in construction. It is that one which, as I have just said, I have operated many times, at the request of many persons, and which is still in perfect condition. Nevertheless, in constantly perfecting it, I have found reasons to change it, and finally recognizing in all these reasons, whether of difficulty of operation, or in the roughness of its movements, or in the disposition to get out of order too easily by weather or by transportation, I have had the patience to make as many as fifty models, wholly different, some of wood, some of ivory and ebony, and others of copper, before having arrived at the accomplishment of this machine which I now make known. Although it is composed of many different small parts, as you can see, at the same time it is so solid that, after the experience of which I have spoken before, I assure you that all the jarring that it receives in transportation, however far, will not disarrange it.

Finally, dear reader, now that I deem it ready to be seen, and in order that you yourself can see and operate it, if you are interested, I pray you to grant me the liberty of hoping that this same idea of finding a third method of performing all the operations of arithmetic, totally new, and which has nothing in common with the two ordinary methods of the pen and of the counters, will receive from you some esteem. I hope that in approving my aim of pleasing and assisting you, you will be grateful to me for the care that I have taken to make all of the operations, which by the preceding methods are painful, complex, long, and uncertain, hereafter easy, simple, prompt, and assured.

LEIBNIZ

On His Calculating Machine

(Translated from the Latin by Dr. Mark Kormes, New York City.)

Pascal's calculating machine, described in the preceding article, was intended to add numbers mechanically. The first one made for the purpose of multiplying was constructed by Gottfried Wilhelm, Freiherr von Leibniz (1646-1716) about the year 1671. One of these machines is still to be seen in the Kästner Museum in Hannover, the city in which Leibniz spent his later years. An article by Jordan, published in *Die Zeitschrift für Vermessungswesen*, in 1897, brought to light a manuscript by Leibniz describing his machine and is now in the Royal Library of the same city. This manuscript was written in 1685, some years after the machine was invented, and it bears the title: "Machina arithmetica in qua non additio tantum et subtractio sed et multiplicatio nullo, divisio vero pæne nullo animi labore peragantur."

When, several years ago, I saw for the first time an instrument which, when carried, automatically records the numbers of steps

taken by a pedestrian, it occurred to me at once that the entire arithmetic could be subjected to a similar kind of machinery so that not only counting but also addition and subtraction, multiplication and division could be accomplished by a suitably arranged machine easily, promptly, and with sure results.

The calculating box of Pascal was not known to me at that time. I believe it has not gained sufficient publicity. When I noticed, however, the mere name of a calculating machine in the preface of his "posthumous thoughts" (his arithmetical triangle I saw first in Paris) I immediately inquired about it in a letter to a Parisian friend. When I learned from him that such a machine exists I requested the most distinguished Carcavius by letter to give me an explanation of the work which it is capable of performing. He replied that addition and subtraction are accomplished by it directly, the other [operations] in a round-about way by repeating additions and subtractions and performing still another calculation. I wrote back that I venture to promise something more, namely, that multiplication could be performed by the machine as well as addition, and with greatest speed and accuracy.

He replied that this would be desirable and encouraged me to present my plans before the illustrious King's Academy of that place.

In the first place it should be understood that there are two parts of the machine, one designed for addition (subtraction) the other for multiplication (division) and that they should fit together.

The adding (subtracting) machine coincides completely with the calculating box of Pascal. Something, however, must be added for the sake of multiplication so that several and even all the wheels of addition could rotate without disturbing each other, and nevertheless anyone of them should precede the other in such a manner that after a single complete turn unity would be transferred into the next following. If this is not performed by the calculating box of Pascal it may be added to it without difficulty.

The multiplying machine will consist of two rows of wheels, equal ones and unequal ones. Hence the whole machine will have three kinds of wheels: the wheels of addition, the wheels of the multiplicand and the wheels of the multiplier. The wheels of addition or the decadic wheels are now visible in Pascal's adding box and are designated in the accompanying figure by the numbers 1, 10, 100, etc. Everyone of these wheels has ten fixed teeth.

The wheels which represent the multiplicand are all of the same size, equal to that of the wheels of addition, and are also provided with ten teeth which, however, are movable so that at one time there should protrude 5, at another 6 teeth, etc., according to whether the multiplicand is to be represented five times or six times, etc. For example, the multiplicand 365 consists of three digits 3, 6 and 5. Hence the same number of wheels is to be used. On these wheels the multiplicand will be set, if from the right wheel there protrude 5 teeth, from the middle wheel 6, and from the left wheel 3 teeth.

In order that this could be performed quickly and easily a peculiar arrangement would be needed, the exposition of which would lead too far into details. The wheels of the multiplicand should now be adjoined to the wheels of addition in such a manner that the last corresponds to the last, the last but one to the last but one, and that before the last but one to that before the last but one, or 5 should correspond to 1, 6 to 10, and 3 to 100. In the addition box itself there should show through small openings the number set as 0, 0, 0, etc. or zero. If after making such an arrangement we suppose that 365 be multiplied by one, the wheels 3, 6, and 5 must make one complete turn (but while one is being rotated all are being rotated because they are equal and are connected by cords as it will be made apparent subsequently) and their teeth

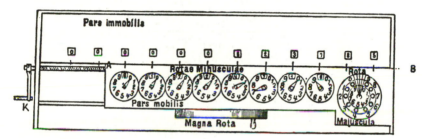

now protruding will turn the same number of fixed teeth of the wheels 100, 10, 1 and thus the number 365 will be transferred to the addition box.

Assuming, however, that the number 365 is to be multiplied by an arbitrary multiplier (124) there arises the need of a third kind of wheels, or the wheels of the multiplier. Let there be nine such wheels and while the wheels of the multiplicand are variable so that the same wheel can at one time represent 1 and at another time 9 according to whether there protrude less or more teeth, the wheels of the multiplier shall on the contrary be designated by fixed numbers, one for 9, one for 1, etc.

This is accomplished in the following manner: Everyone of the wheels of the multiplier is connected by means of a cord or a chain to a little pulley which is affixed to the corresponding wheel of the multiplicand: Thus the wheel of the multiplier will represent a number of units equal to the number of times the diameter of the multiplier-wheel contains the diameter of the corresponding pulley. The pulley will turn namely this number of times while the wheel turns but once. Hence if the diameter of the wheel contains the diameter of the pulley four times the wheel will represent 4.

Thus at a single turn of the multiplier-wheel to which there corresponds a pulley having a quarter of its diameter the pulley will turn four times and with it also the multiplicand-wheel to which it [the pulley] is affixed. When, however, the multiplicand-wheel is turned four times its teeth will meet the corresponding wheel of addition four times and hence the number of its units will be repeated as many times in the box of addition.

An example will clarify the matter best: Let 365 be multiplied by 124. In the first place the entire number 365 must be multi-

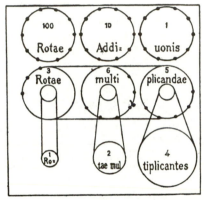

plied by four. Turn the multiplier-wheel 4 by hand once; at the same time the corresponding pulley will turn four times (being as many times smaller) and with it the wheel of the multiplicand 5, to which it is attached, will also turn four times. Since the wheel 5 has five teeth protruding at every turn 5 teeth of the corresponding wheel of addition will turn once and hence in the addition box there will be produced four times 5 or 20 units.

The multiplicand-wheel 6 is connected with the multiplicand-wheel 5 by another cord or chain and the multiplicand-wheel 3 is connected with wheel 6. As they are equal, whenever wheel 5 turns four times, at the same time wheel 6 by turning four times will give 24 tens (it namely catches the decadic addition-wheel 10) and wheel 3 catching the addition-wheel 100 will give twelve hundred so that the sum of 1460 will be produced.

In this way 365 is multiplied by 4, which is the first operation. In order that we may also multiply by 2 (or rather by 20) it is necessary to move the entire adding machine by one step so to say, so that the multiplicand-wheel 5 and the multiplier-wheel 4 are under addition-wheel 10, while they were previously under 1, and in the same manner 6 and 2 under 100 and also 3 and 1 under 1000. After this is done let the multiplier-wheel 2 be turned once: at the same time 5 and 6 and 3 will turn twice and 5 catching twice [the addition-wheel] 10 will give 10 tens, 6 catching 100 will give twelve hundred and 3 catching 1000 will give six thousand, together 7300. This number is being added at the very same turn to the previous result of 1460.

In order to perform as the third operation, the multiplication by 1 (or rather by 100), let the multiplication machine be moved again (of course the multiplicand-wheels together with the multiplier-wheels while the addition-wheels remain in their position) so that the wheels 5 and 4 be placed under 100 and in the same way 6 and 2 under 1000 and 3 and 1 under 10,000. If wheel 1 be turned once at the same time the wheels 3, 6, and 5 will turn once and thus add in the addition box that many units, namely, 36,500. As a product we obtain, therefore:

$$
\begin{array}{r}
1,460 \\
7,300 \\
36,500 \\
\hline
45,260
\end{array}
$$

It should be noted here that for the sake of greater convenience the pulleys should be affixed to the multiplicand-wheels in such a manner that the wheels must move when the pulleys move but that the pulleys do not need to move while the wheels are turned. Otherwise when one multiplier-wheel (*e. g.*, 1) be turned and thus all the multiplicand-wheels moved, all the other multiplier wheels (*e. g.*, 2 and 4) would necessarily move, which would increase the difficulty and perturb the motion.

It should be also noted that it does not make any difference in what order the multiplier-wheels 1, 2, 4, etc. be arranged but they could very well be placed in numerical order 1, 2, 3, 4, 5. For even then one is at liberty to decide which one to turn first and which afterwards.

In order that the multiplier-wheel, *e. g.*, the one representing 9 or whose diameter is nine times as great as the diameter of the corresponding pulley, should not be too large we can make the pulley so much smaller preserving the same proportion between the pulley and the wheel.

In order that no irregularity should follow the tension of the cords and the motion of pulleys tiny iron chains could be used in place of the cords and on the circumference of the wheels and pulleys where the chains would rest there should be put little brass teeth corresponding always to the individual links of the chain; or in place of cords there could be teeth affixed to both the pulleys and the wheels so that the teeth of the multiplier-wheel would immediately catch the teeth of the pulley.

If we wanted to produce a more admirable machine it could be so arranged that it would not be necessary for the human hand to turn the wheels or to move the multiplication machine from operation to operation: Things could be arranged in the beginning so that everything should be done by the machine itself. This, however, would render the machine more costly and complicated and perhaps in no way better for practical use.

It remains for me to describe the method of dividing on the machine, which [task] I think no one has accomplished by a machine alone and without any mental labor whatever, especially where great numbers are concerned.

But whatever labor remains to be done in [the case of] our machine it could not be compared with that intricate labyrinth of the common division which is in the case of large numbers the most tedious [procedure] and [the one] most abundant in errors that can be conceived. Behold our method of division! Let the number 45,260 be divided by 124. Begin as usual and ask for the first simple quotient or how many times 452 contains 124.

It is but very easy for anyone with mediocre ability to estimate the correct quotient at first sight. Hence let 452 contain 124 thrice. Multiply the entire divisor by this simple quotient which can be easily accomplished by one simple turn of the wheel. The product will be 372. Subtract this from 452. Combine the remainder 80 with the rest of the dividend 60. This gives 8060. (But that will be effected by itself in the machine during the multiplication if we arrange in it the dividend in such a manner that whatever shall be produced by multiplication will be automatically deducted. The subtraction also takes place in the machine if we arrange in it the dividend in the beginning; the performed multiplications are then deducted from it and a new dividend is given by the machine itself without any mental labor whatever.)

Again divide this [8060] by 124 and ask how many times 806 contains 124. It will be clear to every beginner at first sight that it is contained six times. Multiply 124 by 6. (One turn of the multiplier wheel) gives 744. Subtract this result from 806, there remains 62. Combine this with the rest of the dividend, giving 620. Divide this third result again by 124. It is clear immediately that it is contained 5 times. Multiply 124 by 5; [this] gives 620. Deduct this from 620 and nothing remains; hence the quotient is 365.

The advantage of this division over the common division consists mostly in the fact (apart from infallibility) that in our method there are but few multiplications, namely as many as there are digits in the entire quotient or as many as there are simple quotients. In the common multiplication a far greater number is needed, namely, as many as [are given by] the product of the number of digits of the quotient by the number of the digits of the divisor. Thus in the preceding example our method required three multiplications because the entire divisor, 124, had to be multiplied by the single digits of the quotient 365,—that is, three. In the common method, however, single digits of the divisor are multiplied by single digits of the quotient and hence there are nine multiplications in the given example.

It also does not make any difference whether the few multiplications are large, but in the common method there are more and smaller ones; similarly one could say that also in the common method few multiplications but large ones could be done if the entire divisor be multiplied by an arbitrary number of the quotient. But the answer is obvious, our single large multiplication being so easy, even easier than any of the other kind no matter how small. It is effected instantly by a simple turn of a single wheel and at that without any fear of error. On the other hand in the common method the larger the multiplication the more difficult it is and the more subject to errors. For that reason it seemed to the teachers of arithmetic that in division there should be used many and small multiplications rather than one large one. It should be added that the largest part of the work already so trifling consists in the setting of the number to be multiplied, or to change according to the circumstances the number of the variable teeth on the multiplicand-wheels. In dividing, however, the multiplicand (namely the divisor) remains always the same, and only the multiplier (namely the simple quotient) changes without the necessity of moving the machine. Finally, it is to be added that our method does not require any work of subtraction; for while multiplying in the machine the subtraction is done automatically. From the above it is apparent that the advantage of the machine becomes the more conspicuous the larger the divisor.

It is sufficiently clear how many applications will be found for this machine, as the elimination of all errors and of almost all work from the calculations with numbers is of great utility to the government and science.   It is well known with what enthusiasm the calculating rods [baculi] of Napier,[1] were accepted, the use of which, however, in division is neither much quicker nor surer than the common calculation.   For in his [Napier's] multiplication there is need of continual additions, but division is in no way faster than by the ordinary [method].   Hence the calculating rods [baculi] soon fell into disuse.   But in our [machine] there is no work when multiplying and very little when dividing.

Pascal's machine is an example of the most fortunate genius but while it facilitates only additions and subtractions, the difficulty of which is not very great in themselves, it commits the multiplication and division to a previous calculation so that it commended itself rather by refinement to the curious than as of practical use to people engaged in business affairs.

And now that we may give final praise to the machine we may say that it will be desirable to all who are engaged in computations which, it is well known, are the managers of financial affairs, the administrators of others' estates, merchants, surveyors, geographers, navigators, astronomers, and [those connected with] any of the crafts that use mathematics.

But limiting ourselves to scientific uses, the old geometric and astronomic tables could be corrected and new ones constructed by the help of which we could measure all kinds of curves and figures, whether composed or decomposed and unnamed, with no less certainty than we are now able to treat the angles according to the work of Regiomontanus and the circle according to that of Ludolphus of Cologne, in the same manner as straight lines.   If this could take place at least for the curves and figures that are most important and used most often, then after the establishment of tables not only for lines and polygons but also for ellipses, parabolas, hyperbolas, and other figures of major importance, whether described by motion or by points, it could be assumed that geometry would then be perfect for practical use.

---

[1] [See p. 182.]

Furthermore, although optical demonstration or astronomical observation or the composition of motions will bring us new figures, it will be easy for anyone to construct tables for himself so that he may conduct his investigations with little toil and with great accuracy; for it is known from the failures [of those] who attempted the quadrature of the circle that arithmetic is the surest custodian of geometrical exactness. Hence it will pay to undertake the work of extending as far as possible the major Pythagorean tables; the table of squares, cubes, and other powers; and the tables of combinations, variations, and progressions of all kinds, so as to facilitate the labor.

Also the astronomers surely will not have to continue to exercise the patience which is required for computation. It is this that deters them from computing or correcting tables, from the construction of Ephemerides, from working on hypotheses, and from discussions of observations with each other. For it is unworthy of excellent men to lose hours like slaves in the labor of calculation, which could be safely relegated to anyone else if the machine were used.

What I have said about the construction and future use [of the machine], should be sufficient, and I believe will become absolutely clear to the observers [when completed].

NAPIER

The Napier Rods

(Translated from the Latin by Professor Jekuthiel Ginsburg, Yeshiva College, New York City.)

John Napier (1550–1617), Laird of Merchiston, Edinburgh (see p. 149), was known in his time quite as widely for his computing rods as for his invention of logarithms. While these rods are almost unknown at the present time, their advent was a distinct step in advance in mechanical computation. As is well known, they consist merely in putting on rods a scheme of multiplication which had long been in use among the Arabs and then using the rods for other operations as well. The following translation is from certain particularly significant parts of the *Rabdologiæ, sev nvmerationis per virgulas libri duo*, Edinburgh, 1617 (posthumously published). The word *rabdologia* is from the Greek ράβδος (*rhab′dos*. "rod") and λογία (*logi′a*, "collection"). When Leybourn published his English translation (*The Art of Numbring By Speaking-Rods: Vulgarly termed Nepeir′s Bones*, London, 1667), he used a false etymology, not recognizing Napier's use of λογία.

Napier gives (p. 2) the number of the rods. Ten rods will suffice for calculations with numbers less than 11,111; twenty for numbers less than 111, 111, 111; and thirty rods for 13-place numbers less than 111, 111, 111, 111, 1.

Each rod is divided lengthwise into ten equal parts in the following way: nine parts in the middle, one half of a part above and another half below. Horizontal lines joining the points of division will divide the surface into nine squares plus two half squares. The diagonals are then drawn as here shown.

To mark the faces of the rods: the face turned toward the eye during the marking is called the "first;" the one to the right side of the observer, the "second;" the one toward the left, the "fourth."

The nine little areas on each face serve for entering the multiples of one of the nine digits by 1, 2, 3, 4, 5, 6, 7, 8, 9. If the products are expressed by one digit, the lower half of the square is used; if by two, then one of the digits namely, the digits of the tens) is put in the upper area and the other in the lower one.

The first four rods are marked as follows: In the squares of the first face of each (that is, the face turned to the eye of the observer) we put zeros. This uses up four of the sixteen available faces.

Then, turning around each rod lengthwise so that the third face will now be turned toward the eye, but upside down, we write in the nine squares the

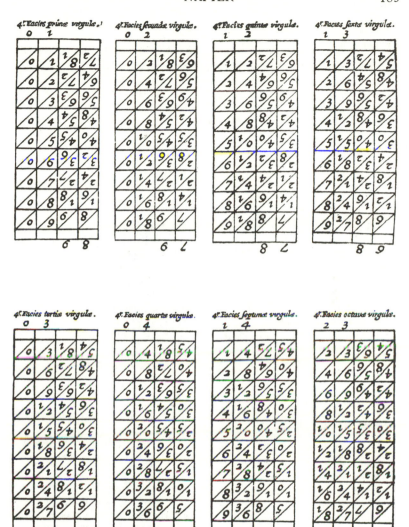

products of 9 by 1, 2, 3, 4, 5, 6, 7, 8, 9—namely, the numbers 9, 18, 27, 36, 45, 54, 63, 72, 81.

This takes care of four more of the available sixteen faces, leaving eight faces for the remaining eight digits. These will be filled in the following way: The second face of the first rod will be given to the multiples of 1 by the first nine digits (that is, the numbers 1, 2, 3, 4, 5, 6, 7, 8, 9), while the opposite face will be given to similar multiples of its complement to the number 9—namely, the multiples of 8 (8, 16, 24, 32, 40, 48, 56, 64, 72).

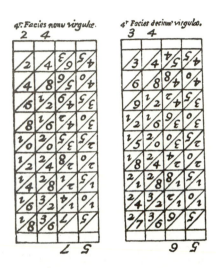

Similarly, the second and fourth faces of rod 2 will be marked, respectively, by the multiples of 2 and 7 by the first nine digits.

On the second and fourth faces of the third rod will be inscribed the products of the numbers 3 and 6, while the same faces of the rod will serve the numbers 4 and 5.

Hence, the first four rods will contain all the products of the first nine numbers by each other, besides four zero columns and three 9 columns, making in all four 9 columns. The twenty-four faces of the remaining six rods will have to contain the columns of other digits repeated three times each.

On the first face of each of the following three rods (namely, the fifth, sixth, and seventh), we enter the unit column (*i. e.*, 1, 2, 3, 4, 5, 6, 7, 8, 9) on the opposite face (the third) of each of these columns—the corresponding product of its complement to 9, namely, 8 (the numbers 8, 16, 24, 32, 40, 48, 56, 64, 72). Hence, each of the columns 0, 1, 8, 9 has been repeated four times.

Of the remaining faces, we enter in the second place of each rod the columns headed by 2, 3, 4 and the columns headed by the complements 7, 6, 5 on the fourth place. Hence, we used up seven rods and we entered the columns 0, 1, 8, 9, . . . four times each and the columns 2, 3, 4, 5, 6 twice. We still have three rods (the eighth, ninth, and tenth). On the first face of the eighth and ninth, we enter the column headed by 2, and on the third face of each the column headed by 7, its complement. The remaining two second faces we give to 3 and 4, and the opposite faces to their complements 6 and 5.

If an inventory be taken now, we still find that each of the columns 0, 1, 2, 7, 8, 9, has occurred just four times, while each of the columns 3, 4, 5, 6, only three times, and we still have one unused rod.

Hence, we enter on this the columns 3, 4, 5, 6 in the way indicated. So that now each column is found to have occurred just four times.

The translation of Napier's rule for multiplying numbers by the use of rods (pp. 16–17) follows.

Set up one of the numbers given for multiplication (preferably the larger) by means of the rods. Write the other on paper with a line under it. Then under each written figure put that multiple found in the rods of which the figure is so to speak an index. It makes no difference whether the first figures on the right side of each multiple follow each other obliquely in the same order as the numbers signifying their indexes, or as the first figures to the left. The multiples thus arranged are to be added arithmetically and this will give the product of the multiplication.

Thus let it be required to multiply the year of the Lord 1615 by 365.

The first number is to be formed by the rods, the second written on paper as here shown. The triple, sextuple, and quintuple of the tabulated numbers are taken, corresponding to the figures in the numbers on paper (3, 6, 5) which are the indices.

365	365
4845	8075
9690	9690
8075	4845
589475	589475

Thus the triple of the number 1615 which is to be transcribed from the rods is 4845. The sextuple which is 9690 and the quintuple 8075 are written obliquely under their indices 3, 6, 5, either beginning under them as in the first scheme, or terminating under them as in the second...The multiples arranged in this way are to be added arithmetically, and the desired number 589,475 will thus be obtained, which is the product of the multiplication.

GALILEO GALILEI

On the Proportional or Sector Compasses

(Translated from the Italian by Professor David Eugene Smith, Teachers
College, Columbia University, New York City.)

Galileo Galilei (1564–1642), the greatest physicist, astronomer, and mathematician of Italy in his time, and one of the greatest in the world, was interested
not only in the higher branches of his chosen subjects but also in the improvement of methods of computation and of measuring.   Before the slide rule was
invented (see p. 156) or logarithms were known (see p. 149), he devised the
simple but ingenious proportional compasses, or, as he called them, the
geometric and military compasses (*compasso*).   They were first described in
Le Operazioni del Compasso Geometrico et Militare (Padua, 1606).   The following extract from this work will suffice to make the general purpose of the
instrument known.

The Operations of the Geometric and Military Compasses.[1]
On Arithmetic Lines.   Division of the Line.   First Operation

Coming to the special explanation of the methods of using the
new geometric and military compasses (Fig. 1), we will first consider the side in which are shown four pairs of lines, with their
divisions and numbers.   Of these we shall first speak of the innermost ones.   These are called the arithmetic lines because their
divisions are in arithmetic proportion; that is, they proceed by
equal increments up to 250.   We shall find several ways of using
these lines.   First, we shall by their help show how to divide a
proposed straight line into as many equal parts as we wish, using
any one of several methods mentioned below.   When the proposed line is of medium length, not exceeding the spread of the
instrument, we open an ordinary pair of compasses[2] the full
length of the line and transfer this length to any number on these
arithmetic lines, taking care that there is a smaller number that
is contained in this one as often as the part of the proposed line is
contained in the whole...[For example], to divide the line into
five equal parts, let us take two numbers, one being five times the

---

[1] The Italian usage is *compasso*, the singular form.

[2] He speaks of the "geometric and military compasses" as "the instrument,"
and of ordinary compasses used in transferring lengths as "compasses."

other,—say 100 and 20. Now open the instrument so that the given line as transferred by the compasses shall reach from 100

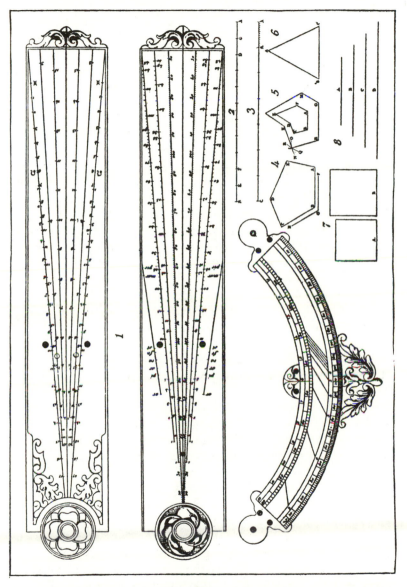

[on one leg] to 100 [on the other]. Now, without moving the instrument, let us take the distance between the points marked 20 and 20, and this will manifestly be the fifth part of the proposed line.   In the same way we can find every other division, taking care that we do not use numbers beyond 250...

The same result will be obtained by solving the problem another way, like this: If we wish to divide the line AB (Fig. 2) into 11 parts, take a number that is eleven times another,—say 110 and 10.   Then transfer the whole line AB by the compasses so that it reaches from 110 [on one leg] to 110 [on the other].   It is impossible in this figure to get the distances between the points 10 and 10, because each is covered by the nut.   Instead of this, we take the distances between 100 and 100, closing the compasses a little so that one point [100] lies on B and the other on C.   Then the remaining distance AC will be $\frac{1}{11}$ of AB.   In the same way we may place one point of the compasses on A and let the other lie on E, leaving EB equal to CA.   Then close up the compasses and take the distance between the points 90 and 90, transferring it from B to D and from A to F, after which CD and EF will each be $\frac{1}{11}$ of the whole line.   In the same way, transferring the distances from 80 to 80, 70 to 70, etc., we shall find the other divisions, as can be seen in the line AB.

If, however, we have a very short line to divide into many parts, such as AB (Fig. 3) to be divided into thirteen parts, we proceed by another rule, as follows: Produce AB to any point C, laying off on it as many lines as you wish, say six, so that AC shall be seven times AB.   It is then evident that if AB contains 13 equal parts, AC will contain 91 [of the same length].   We therefore transfer the distance between 90 and 90 to the line CA from C toward A, thus leaving the 91st part of CA, or the 13th part of AB, toward A. If we wish, we may now close up, point by point, the [transferring] compasses at 89, 88, 87, etc., transferring the distances from C toward A, and we shall find the other parts of the proposed line AB.

Finally, if the line to be divided is very long, so as to greatly exceed the maximum opening of the instrument, we can nevertheless divide it, say into seven equal parts.   First, take two numbers, one seven times the other,—say 140 and 20.   Now open the instrument as far as you please, taking with the compasses the distance from 140 to 140.   Then as many times as this distance is contained in the length of the given line, that number of times the distance from 20 to 20 will be the seventh part of it...

Galileus Galileus   Florentinus

28

Superior   licentia

16   24

Eques Octauius Leonis   Roman' pictor fecit

(Facing page 188.)

How from a Proposed Line we can take any stated Parts.
Second Operation

This operation is much more useful and necessary [than the first], since without our instrument it would be very difficult, while with it the solution is found at once. Suppose, for example, we are required to take from the 197 parts of a given line 113 parts. We open the instrument until the given line can be transferred by the compasses so that it reaches from 197 [on one leg] to 197 [on the other]. Without moving it, the distance from 113 to 113 will then be $113/197$ of the given line...

How the Same Lines furnish two or even an infinite number of Scales for increasing or decreasing the Scale of a Drawing.
Third Operation

If we wish to reduce a drawing to another scale, it is evidently necessary to use two scales, one for the given drawing and the other for the new one. Such scales are at once given by the instrument. One will be the line as already divided into equal parts, and will be used in measuring the given figure. The other will be used for the new drawing, and this has to be adjustable; that is, it must be constructed so that we can lengthen or shorten it according as the new drawing is to be larger or smaller. Such an adjustable scale is the one that we get from the same lines by adjusting the instrument. That you may understand more clearly the process, consider this example:

Suppose that we have the figure *ABCDE* (Fig. 4) and wish to draw a similar figure with side *FG* corresponding to side *AB*. We must evidently use two scales, one to measure the lines of *ABCDE* and the other to measure those of the new drawing, these being longer or shorter than the former according to the ratio which *FG* has to *AB*. Take therefore the length of *AB* with a pair of compasses and then place one of the points at the vertex of the instrument, noting where the other falls on one of the lines,—say at 60. Then transfer *FG* with the compasses so that one point rests on this 60 and the other on the corresponding 60 [on the other arm of the instrument]. If the instrument be now allowed to remain fixed, all the lines in the given figure can be measured on the straight scale, and the corresponding lines of the new figure can be measured transversely. For example, if we wish the length of *CH* corresponding to the given *BC*, we simply lay off *BC* from the vertex,—say to 66,—and then turn the other [leg of the measuring compasses] until the point rests on the 66 [of the other arm of the instrument]. This will then have to *BC* the same ratio as *FG* to *AB*.

If you wish to greatly enlarge a figure, you will need to use two scales in the opposite way [from that shown above]; that is, you will have to use the straight scale [on the arm] for the required drawing and the transverse measurement [from one arm to the other] for the given one. For example, suppose that we have the figure *ABCDEF* (Fig. 5) which we wish to enlarge so that *GH* corresponds to *AB*. We measure *GH*, supposing it to be, say, 60 points on one of the arms. We then open the instrument so that the distance from 60 to 60 is *AB*. Leaving the instrument fixed, we then find *HI* corresponding to *BC* by seeing what two corresponding points, say 46 and 46, determine the ends of *BC*. Then the length from the vertex to 46 will be *HI*[1]. . .

The Rule of Three. Solved by Means of the Compasses and the same Arithmetic Lines. Operation IV

The lines [of the proportional compasses] serve not so much for solving geometric linear problems as for certain arithmetic rules, among which we place one corresponding to one of Euclid's problem, thus: Given three numbers, find their fourth proportional. This is merely the Golden Rule, which experts call the Rule of Three,—to find the fourth number proportional to three that are proposed. To illustrate by examples for the purpose of a clearer understanding,—if 80 gives us 120, what will 100 give? We now have three numbers in this order: 80, 120, 100, and to find the fourth number sought [we proceed as follows:] Find on one arm 120; connect this with 80 on the other arm; find 100 [on the same arm as 80] and draw a parallel to the connecting line[2] and what you find will be 150, the fourth number sought. Observe also that the same thing would result if instead of taking the second number [120] you had taken the third [100], and instead of the third you had taken the second [120][3]. . .

---

[1] [Galileo then proceeds to show how the vertices are found, but this is obvious.]

[2] [In the original:. . .prendi sopra lo strumento rettamente il secondo numero de' proposti, cioè 120, ed applicato trasversalmente al primo, cioè all' 80; dipoi prendi trasversalmente il terzo numero, cioè 100, e mesuralo rettamente sopra la scala, equello che troverai, cioè 150, sarà il quarto numero cercato.]

[3] [Galileo then proceeds to discuss the question when the numbers are such as to require other adjustments, as in the "First Operation" already explained.]

Inverse Rule of Three, solved by means of the Same Lines.
Operation V

In the same way we can solve problems involving the inverse Rule of Three, as in this example: If there is food sufficient for 100 soldiers for 60 days, how many would it feed for 75 days. The numbers may be arranged as 60, 100, 75. Find 60 on one arm of the instrument. Connect it with the third number, 75, on the other arm. Without moving it, take 100 on the same arm as 60 and draw a parallel to the connecting line and what you find will be 80, the number sought...

Rule of Exchange. Operation VI

By means of these same arithmetic lines we can change money by finding the equal values. This is done very easily and quickly as follows: Adjust the instrument by finding on one of the lines the value of the piece of money to be exchanged. Connect it with the value of the other piece which we wish to exchange; but in order that you may understand the matter more clearly, we shall illustrate it by an example. Suppose that we wish to exchange gold scudi into Venetian ducats, and that the value of the gold scudo is 8 lire and the value of the Venetian ducat is 6 lire 4 soldi. Since the ducat is not precicely measured by the lire, there being 4 soldi to be considered, it is best to reduce both to soldi, the value of the scudo being 160 soldi and that of the ducat 124 [soldi]. To adjust the instrument for translating scudi into ducats lay off the value of the scudo, or 160, and then open the instrument and connect the 160 to 124, the value of the ducat. Now leave the instrument unchanged. Then any proposed number of scudi can be changed into ducats by laying off the number of scudi on the arm [of ducats] and drawing a parallel to the line already drawn from 160 to 124. For example. 186 scudi will then be found equal to 240 ducats.[1]

---

[1] [This section closes with the *Rule of Compound Interest,...Operation VII.* The next section discusses geometric lines; the third, stereometric lines, including cube root; the fourth, "metallic lines," finding the size of bodies with respect to weight; and the rest dealing with mensuration, closing with an extended discussion of operations with the quadrant.]

D'OCAGNE

On Nomography

(Translated from the French by Nevin C. Fisk, M.S., University of Michigan, Ann Arbor, Michigan.)

Philbert Maurice d'Ocagne was born in Paris, March 25, 1862. His education was received at the Collège Chaptal, the Lycée Fontanges, and the Ecole Polytechnique.

For many years he has been professor of geometry at the Ecole Polytechnique and professor of topometry and applied geometry at the Ecole des Ponts et Chaussées. He is a member of the Académie des Sciences and an officer of the Légion d'Honneur.

D'Ocagne has published numerous books and articles on nomography, graphical and mechanical calculus, and geometry. The selections following are taken from his *Traité de Nomographie*, published by Gauthier-Villars, Paris, in 1899. A second edition of this book appeared in 1921.

D'Ocagone's *Traité de Nomographie* presents a collection and correlation of important developments in graphical proceedure during the latter part of the nineteenth century. Outstanding among these developments is the alignment chart, the principle of which is due to d'Ocagne himself. D'Ocagne may also be credited with the application of the alignment chart to many engineering formulas.

The subject of nomography received much of its impetus from the problems arising in connection with the construction of railroads in France. Most of the men contributing to its growth during the nineteenth century were engineers. Nomography has thus been essentially a branch of applied mathematics finding use in engineering, military science, and industry. At present it is one of the most useful mathematical tools of the technical man.

The[1] purpose of Nomography is to reduce to simple readings on graphical charts, constructed once for all, the computations which necessarily intervene in the practice of various technical arts. If one makes a system of geometric elements (points or lines) correspond to each of the variable connected by a certain equation, the elements of each system being numbered[2] in terms of the values of the corresponding variable, and if the relationship between the variables established by the equation may be translated geometri-

---

[1] [Introduction, page v. Pages given in the footnotes refer to the *Traité de Nomographie*.]

[2] [French; *côtés*.]

192

cally into terms of a certain relation of position easy to set up between the corresponding geometric elements, then the set of elements constitutes a chart[1] of the equation considered. This is the theory of charts, that is to say the graphical representation of mathematical laws defined by equations in any number of variables, which is understood today under the name of Nomography.

1. *Normal Scale of a Function.*[2]—Let $f(\alpha)$ be a function of the variable α, taken in an interval where it is uniform, that is, where it has for each value of α only a single determinate value. Let us lay off on an axis Ox, starting from the origin O, the lengths

$$x_1 = lf(\alpha_1), \quad x_2 = lf(\alpha_2), \quad x_3 = lf(\alpha_3), \ldots$$

l being an arbitrarily chosen length, and let us inscribe beside the points which limit these segments, points which are marked by a fine stroke per-pendicular to the axis, the corresponding values of the variables $\alpha_1, \alpha_2, \alpha_3, \ldots$ The set of points thus obtained constitutes the scale of the function $f(\alpha)$. The length l is called the modulus of this scale.

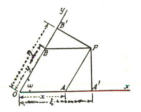

· · · · · · · · · ·

3. *Geometrical Construction of a Scale.*[3]—In order to construct the scale of the function $f(\alpha)$, we may have recourse to the curve C whose equation is

$$x = lf(y)$$

It is sufficient to take on the curve the point whose ordinate is α in order that the extremity of its abcissa may give on Ox the point numbered α for the desired scale. If the curve C may be obtained point by point by means of a simple geometric construction, all calculations can be dispensed with.

· · · · · · · · · ·

II. CHARTS OF EQUATIONS WITH TWO VARIABLES.

9. *Charts with adjacent*[4] *scales.*[5] Let us first take the equation to be represented under the form

$$\alpha_2 = f_1(\alpha_1),$$

[1] [French; *abaque.*]
[2] [Page 1 et seq.]
[3] [Page 7.]
[4] [French; *accolées.*]
[5] [Page 17.]

a form frequently occurring in practice.   Let suppose the scales

$$x = l\alpha_2 \text{ and } x = lf_1(\alpha_1)$$

to be constructed on the same axis, starting from the same origin. Then two values of α_1 and α_2, corresponding by virtue of the preceding equation, are inscribed at the same point of the axis bearing the two scales.   If a value of one of the variables, say α_1 is given, the corresponding value of α_2 is read from the second scale at the point graduated α_1 on the first.

. . . .

12. *Cartesian Charts for Two Variables.*[1]—We may apply a uniform scale to each of the two variables by establishing the linkage between corresponding points through the medium of a curve.   Let us imagine that the scales

$$x = l_1\alpha_1 \text{ and } y = l_2\alpha_2$$

are carried on two rectangular axes Ox and Oy, and let us suppose that perpendiculars are erected at the points of division marked on each axis.   If the values α_1 and α_2 together satisfy the equation

(1) $F(\alpha_1, \alpha_2) = 0$

the perpendiculars to the axes at the points graduated α_1 and α_2 intersect at a certain point.   The points corresponding to various couples of values of α_1 and α_2 satisfying the equation (1) are distributed along a curve C whose equation referred to the axes Ox and Oy is

$$F\left(\frac{x}{l_1}, \frac{y}{l_2}\right) = 0$$

The various points of the curve C determined individually are easily marked on the plane, thanks to the cross-section[2] system defined above, and one observes that the curve C obtained by connecting all these points constitutes a chart of the equation (1). Such a chart derived by the use of cartesian coordinates is called cartesian.

In order to find one of the variables, say α_2, when α_1 is given, it is sufficient to note the point (P) where the curve C is met by the perpendicular to Ox passing through the point on that axis numbered α_1, and to read the graduation α_2 at the foot of the perpendicular dropped from the point (P) upon the axis Oy.

. . . .

---

[1] [Page 24 et seq.]

[2] [French; *quadrillage.*]

16. *Cartesian Charts for Three Variables.*[1]—Suppose it be desired to construct a chart for the equation

(1) $$F(\alpha_1, \alpha_2, \alpha_3) = 0$$

The first plan to present itself is this. Let us give a determined value to one of the variables, preferably that one which will usually be calculated as a function of the other two, say α_3. We shall then have an equation in the two variables α_1 and α_2 which we may represent as indicated in section 12 by mean of a curve traced on the cross-section network[2] defined by the equations

(α_1) $$x = l_1\alpha_1$$
(α_2) $$y = l_2\alpha_2$$

l_1 and l_2 being moduli chosen to give the most satisfactory chart. The equation of this curve will be

(α_3) $$F\left(\frac{x}{l_1}, \frac{y}{l_2}, \alpha_3\right) = 0$$

This curve along which the element α_3 conserves a constant value has been called by Lalanne a curve "d'égal élément," and by the German author Vogler an isopleth curve. The latter term has since been adopted by Lalanne. We shall call it simply a numbered curve.[3] In the same manner as above let us construct curves corresponding to a series of values of α_3 increasing by regular steps, taking care to label each curve with the corresponding value of α_3. Let us remark further that it is necessary only to trace the portion of each curve contained within a rectangle bounded by the perpendiculars erected to Ox and Oy respectively at the points corresponding to the limiting values a_1 and b_1 for α_1, and a_2 and b_2 for α_2, values which are given in the problem since we have assumed that α_1 and α_2 are the independent variables. We thus obtain within a ruled[4] rectangle a system of numbered curves which furnishes the representation desired within the limits admitted for the independent variables. This ruled rectangle, resembling a sort of checkerboard, has given the name "abaque" to diagrams of this kind, and, by extension, to every sort of numbered chart. Making the convention once for all of designating by the terms horizontal and vertical the lines parallel to Ox and Oy respectively, we may state that the method of using such a

---

[1] [Page 32 et seq.]
[2] [French; *quadrillage.*]
[3] [French; *courbe cotée.*]
[4] [French; *quadrillé.*]

chart in order to obtain the value of α_3 when α_1 and α_2 are given
is to read the graduation α_3 of the curve passing through the point
of intersection of the vertical numbered α_1 with the horizontal
numbered α_2.

. . . .

24. *Principle*.[1]—We have seen in section 15 that in substituting
other functional scales for uniform scales which at first sight one
would be inclined to use along the axes Ox and Oy, we may always
transform into a straight line the curve representative of an equation linking the variables to which the two scales correspond. In
what case may such a modification applied to the scales Ox and
Oy of a cartesian chart for three variables transform simultaneously all the curves of the chart into straight lines? The answer
to this question is easy to obtain. In order that the curves (α_3)
constitute a straight line diagram with graduations

(α_1) $\qquad\qquad\qquad x = l_1 f_1(\alpha_1)$

(α_2) $\qquad\qquad\qquad y = l_2 f_2(\alpha_2)$

it is necessary and sufficient that their equation be of the form

(α_3) $\qquad\qquad \dfrac{x}{l_1} f_3(\alpha_3) + \dfrac{y}{l_2} \varphi_3(\alpha_3) + \psi_3(\alpha_3) = 0$

This will be the case if the proposed equation is of the form

$$f_1(\alpha_1) f_3(\alpha_3) + f_2(\alpha_2) \varphi_3(\alpha_3) + \psi_3(\alpha_3) = 0$$

We thus obtain at the same time the form of the equations to
which this artifice is applicable, and an indication of the way in
which it may be put into play.

In the case in which a single curve constitutes the chart for an
equation in two variables as in section 15, such a transformation
offers no appreciable advantage, the work required for the establishment of a functional scale being practically equivalent to that
required for the determination of points for the corresponding
curve; the sole difference is in the tracing of the curve joining the
points individually obtained. This is not the case here and the
advantage becomes appreciable. If the labor demanded by
the change in graduation is equivalent to that involved in the construction of a curve, one sees that when there are n curves to be
drawn the economy achieved may be represented approximately
by the work required to draw $n - 1$ curves. In fact, once the
new scale system is established, it is necessary to locate only two

---

[1] [Page 50 et seq.]

points to determine each of the straight lines intended to replace the curves which had to be constructed point by point in the primitive system, straight lines which have the further advantage of being easy to draw with accuracy.

The principle of such a transformation was indicated for the first time by Léon Lalanne who gave it the name "geometrical anamorphosis."[1] It was under the form indicated in section 28[2] that the idea first occurred to him in connection with applications treated in sections 29[3] and 108.[4]

. . . .

56. *Principle of Aligned Points.*[5]—We have already explained in section 30 the reasons for which it is desirable whenever possible to have only points appear as numbered elements in the representation of an equation. We have seen furthermore how the use of a "transparent"[6] with three indices allows the realization of this end for equations representable by three systems of parallel straight lines, (section 26) that is to say, equations of the form[7]

$$f_1 + f_2 = f_3$$

We shall now expound another method which allows the attainment of the same end for the much more general category of equations representable by three systems of any straight lines whatsoever, comprising as a consequence the foregoing as a particular case; that is to say, those equations which are of the form

$$\begin{vmatrix} f_1 & \varphi_1 & \psi_1 \\ f_2 & \varphi_2 & \psi_2 \\ f_3 & \varphi_3 & \psi_3 \end{vmatrix} = 0$$

The idea which allows us a priori to take account of the possibility of such a result is blended with that of the principle of duality

---

[1] "Mémoire sur les tables graphiques et sur la Géométrie anamorphique," *Annales des Ponts et Chaussées ponts chaussées*, -erer semestre, 1846.

[2] [Section 28 of the present work treats of logarithmic anamorphosis. It appears that Lalanne is to be credited with the invention of logarithmic paper.]

[3] [Section 29 of the *Traité* presents a straight-line chart on logarithmic paper for the multiplication of two variables.]

[4] [Section 108 treats of superposed charts with logarithmic graduations.]

[5] [Page 123 *et seq.* This principle was first announced by d'Ocagne in the *Annales des Ponts et Chaussées* for November, 1884, p. 531, under the title "Procédé nouveau de calcul graphique."]

[6] [A sheet of celluloid or similar transparent material on which straight lines are engraved. In the case referred to, three intersecting lines were used.]

[7] [The author has made the convention of writing f_1 for $f(\alpha_1)$, f_2 for $f(\alpha_2)$, etc.]

which is fundamental in the field of pure geometry today. We know that it is possible in an infinity of ways to construct a figure composed of points corresponding to a given figure composed of straight lines, so that to any three concurrent lines of the given figure there correspond three collinear points on the other. Every transformation possessing such a property, of which the typical case is transformation by polar reciprocals, is said to be dualistic.

Suppose then that we have applied such a transformation to a chart made up of three systems of straight lines, retaining, let it be well understood, the graduation of each element in the passage

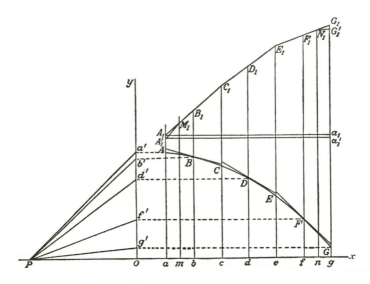

from one figure to the other. We thus obtain a new diagram on which to each of the variables α_1, α_2, and α_3 there corresponds a system of numbered points distributed along a curve called their support. In the transformation effected, this curve will be the correlative of the envelope of the corresponding system of straight lines on the first chart. These three systems of numbered points constitute curvilinear scales. Just as on the first chart the three straight lines, numbered in terms of a system of values of α_1, α_2, α_3 satisfying the equation represented, are concurrent, so here the three corresponding points are collinear. The method of using the chart follows from this fact. The straight line joining the points numbered α_1 and α_2 on the first two curvilinear scales intersects the third scale at the point graduated α_3.

To avoid drawing this line, one may make use of a transparent with one index line or a fine thread which is stretched between the points α_1 and α_2.

. . . .

88. *General Principle.*[1]—Suppose that the variables α, α_1 and α_2 on the one side and α, α_3 and α_4 on the other are linked by equations such as

$$(E) \quad \begin{vmatrix} f(\alpha) & \varphi(\alpha) & \psi(\alpha) \\ f_1(\alpha_1) & \varphi_1(\alpha_1) & \psi_1(\alpha_1) \\ f_2(\alpha_2) & \varphi_2(\alpha_2) & \psi_2(\alpha_2) \end{vmatrix} = 0$$

$$(E') \quad \begin{vmatrix} f(\alpha) & \varphi(\alpha) & \psi(\alpha) \\ f_3(\alpha_3) & \varphi_3(\alpha_3) & \psi_3(\alpha_3) \\ f_4(\alpha_4) & \varphi_4(\alpha_4) & \psi_4(\alpha_4) \end{vmatrix} = 0$$

Each of these will be representable by an alignment chart, and since the functions f, φ, ψ are the same in the two equations, we observe that their curvilinear scale (α) will be the same in the two

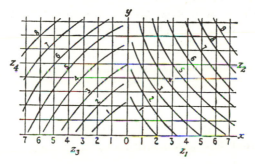

charts. Consequently the two charts can be constructed with this scale in common. If values are given for three of the four variables α_1, α_2, α_3, α_4, the equations (E) and (E') allow the calculation of the fourth as well as the value of α. On the chart the straight line passing through the points numbered α_1 and α_2 and the straight line passing through the points numbered α_3 and α_4 intersect in a point numbered α.

Now it may happen that an equation in four variables α_1, α_2, α_3, α_4 results from the elimination of an auxiliary variable α between two equations such as (E) and (E'). In this case the chart which has just been constructed furnishes a representation of the equation considered. The method of proceedure to obtain say α_4 when α_1, α_2 and α_3 are known is reduced to the following.

---

[1] [Page 213, *et seq.*]

Make the index pass through the points numbered α_1 and α_2, then pivot the index bout the point where it intersects the scale (α) until it passes through the point numbered α_3.   It then cuts the last scale in the point numbered α_4.   Since in general it is not necessary to know the corresponding value of the auxiliary variable α, we may dispense with graduating the scale of this variable; it is sufficient to draw its support which will be called the pivot line. If however it is desired in any case to note the position of the pivot, the line may be graduated in any manner whatsoever.   The chart thus constructed by combining two alignment charts having a scale in common is called a double alignment chart in order to recall the way in which it is used.[1]

. . . .

In practice it is rarely necessary to apply this method except in the case in which the auxiliary scale is uniform, that is to say, when the function $f(\alpha)$ reduces to α, the function φ reduces to 1, and the function ψ reduces to 0 in the equations (E) and (E'). These equations may then be written

$$\begin{vmatrix} \alpha & 1 & 0 \\ f_1 & \varphi_1 & \psi_1 \\ f_2 & \varphi_2 & \psi_2 \end{vmatrix} = 0$$

$$\begin{vmatrix} \alpha & 1 & 0 \\ f_3 & \varphi_3 & \psi_3 \\ f_4 & \varphi_4 & \psi_4 \end{vmatrix} = 0$$

The elimination of α immediately carried out gives

(2) $\begin{vmatrix} \psi_1 & f_1 \\ \psi_2 & f_2 \end{vmatrix} \cdot \begin{vmatrix} \varphi_3 & \psi_3 \\ \varphi_4 & \psi_4 \end{vmatrix} = \begin{vmatrix} \psi_3 & f_3 \\ \psi_4 & f_4 \end{vmatrix} \cdot \begin{vmatrix} \varphi_1 & \psi_1 \\ \varphi_2 & \psi_2 \end{vmatrix}.$

---

[1] [The portion of Sec. 88 omitted at this point deals with a generalization of the principle under discussion.]

II. FIELD OF ALGEBRA

CARDAN'S TREATMENT OF IMAGINARY ROOTS

(Translated from the Latin by Professor Vera Sanford, Western Reserve University, Cleveland, Ohio.)

For a biographical note on Cardan see page 203. Although Cardan (1501–1576) spoke of the complex roots of a certain equation as "impossible," he seems to have been the first to use such numbers in computation, and he even devoted a full page of his *Ars Magna* (1545) to showing the solution of the problem in which this question occurred.   The translation which follows was made from the first edition of the *Ars Magna*, ff. 65*v.* and 66*r.*

A second type of false position[1] makes use of roots of negative numbers.[2] I will give as an example: If some one says to you, divide 10 into two parts, one of which multiplied into the other shall produce 30 or 40, it is evident that this case or question is impossible.   Nevertheless, we shall solve it in this fashion.   Let us divide 10 into equal parts and 5 will be its half.   Multiplied by itself, this yields 25.   From 25 subtract the product itself, that is 40, which, as I taught you in the chapter on operations in the sixth book[3] leaves a remainder m: 15.   The root[4] of this added and then subtracted from 5 gives the parts which multiplied together will produce 40.   These, therefore, are 5 p: ℞ m: 15 and 5 m: ℞m: 15.[5]

Proof

That the true significance of this rule may be made clear, let the line *AB* which is called 10, be the line which is to be divided

---

[1] [The preceding section of this chapter discusses the solution of equations of the type $x^2 = 4x + 32$, which Cardan wrote in the form

$$\overline{\text{qdratu}} \text{ ae}\overline{\text{q}}\text{tur } 4 \text{ rebus p: } 32.]$$

[2] ["...est per radicem m."]

[3] [The *Ars Magna* begins with Book X, the preceding nine being Cardan's arithmetic, the *Practica aritbmetice*, Milan, 1539.]

[4] [For "root," Cardan uses the symbol ℞.]

[5] [Although the symbols + and − appeared in print in Widman's arithmetic of 1489, the signs were not generally adopted for some time, and the use of the letters p and m continued in Italy until the beginning of the seventeenth century.

It should be noted that Cardan made use of the letter ℞ to represent both an unknown quantity (*res*) and a root (*radix*).]

into two parts whose rectangle is to be 40. Now since 40 is the quadruple of 10, we wish four times the whole of *AB*. Therefore, make *AD* the square on *AC*, the half of *AB*. From *AD* subtract four times *AB*. If there is a remainder, its root should be added to and subtracted from *AC* thus showing the parts (into which *AB* was to be divided). Even when such a residue is minus, you will nevertheless imagine ℞ m:15 to be the difference between

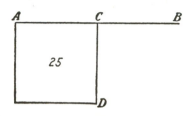

AD and the quadruple of *AB* which you should add and subtract from *AC* to find what was sought. That is 5 p:℞v:25 m:40[1] and 5 m:℞v:25 m:40 or 5 p:℞ − 15 and 5 m:℞ − 15. Multiplying 5 p:℞m:15 by 5 m:℞m:15, the imaginary parts being lost,[2] gives 25 m:m:15 which is p. 15. Therefore the product is 40.

| 5 p:℞m:15 |
| 5 m:℞m:15 |
| 25 m:m:15 qd. est 40 |

However, the nature of *AD* is not the same as that of 40 or *AB* because a surface is far from a number or a line. This, however, is closest to this quantity which is truly imaginary[3] since operations may not be performed with it as with a pure negative number, nor as in other numbers. Nor can we find it by adding the square of half the number in producing the number and take away from the root of the sum and add half of the dividend. For example, in the case of dividing 10 into two parts whose product is 40, you add 25, the square of one half of 10, to 40 making 65. From the root of this subtract 5 and then add 5 and according to similar reasoning you will have ℞ 65 p:5 and ℞ 65 m:5. But these numbers differ by 10, and do not make 10 jointly. This subtility results from arithmetic of which this final point is as I have said as subtile as it is useless.

---

[1] [The v acts as a sign of aggregation and might be considered an abbreviation for the *radix universalis*, or *vniversalis*.]

[2] ["...dimissis incruciationibus."]

[3] [...uere est sophistica.]

CARDAN

Solution of the Cubic Equation

(Translated from the Latin by Professor R. B. McClenon, Grinnell College, Grinnell, Iowa.)

In his *Ars Magna* (Nürnberg, 1545) Girolamo Cardano (Hieronymus Cardanus, 1501–1576) states that Scipio del Ferro discovered the method of solving an equation of the type $x^3 + px = q$ about the year 1515. Nicolo Tartaglia (in the Latin texts, Tartalea) agrees to this but claims for himself the method of solving the type $x^3 + px^2 = q$ and also the independent discovery already made by Scipio del Ferro. Cardan secured the solution from Tartaglia and published it in his work above mentioned. The merits of the discoveries and the ethics involved in the publication may be found discussed in any of the histories of mathematics.

The selection here made is from Chapter XI of the *Ars Magna*, "De cubo & rebus æqualibus numero," the first edition, the type considered being $x^3 + px = q$, the particular equation being cub⁹ p; 6 reb⁹ æqlis 20; that is, $x^3 + 6x = 20$. The edition of 1570 differs considerably in the text. A facsimile of the two pages is given in Smith's *History of Mathematics*, vol. II, pp. 462, 463.

The translation can be more easily followed by considering the general plan as set forth in modern symbols.

Given

$$x^3 + 6x = 20.$$

Let

$$u^3 - v^3 = 20 \text{ and } u^3 v^3 = (\tfrac{1}{3} \times 6)^3 = 8.$$

Then

$$(u - v)^3 + 6(u - v) = u^3 - v^3,$$

for

$$u^3 - 3u^2 v + 3uv^2 - v^3 + 6u - 6v = u^3 - v^3,$$

whence

$$3uv(v - u) = 6(v - u)$$

and

$$uv = 2.$$

Hence

$$x = u - v.$$

But

$$u^3 = 20 + v^3 = 20 + \frac{8}{u^3}.$$

whence

$$u^6 = 20u^3 + 8,$$

which is a quadratic in u^3. Hence u^3 can be found, and therefore v^3, and therefore $u - v$.

Concerning a Cube and "Things"[1] Equal to a Number

Chapter XI

Scipio del Ferro of Bologna about thirty years ago invented [the method set forth in] this chapter, [and] communicated it to Antonio Maria Florido of Venice, who when he once engaged in a contest with Nicolo Tartalea of Brescia announced that Nicolo also invented it; and he [Nicolo] communicated it to us when we asked for it, but suppressed the demonstration. With this aid we sought the demonstration, and found it, though with great difficulty, in the manner which we set out in the following.

Demonstration

For example, let the cube of *GH* and six times the side *GH* be equal[2] to 20. I take two cubes *AE* and *CL* whose difference shall

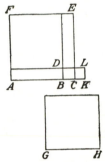

be 20, so that the product of the side *AC* by the side *CK* shall be 2,—i.e., a third of the number of "things;"[3] and I lay off *CB* equal to *CK*, then I say that if it is done thus, the remaining line *AB* is equal to *GH* and therefore to the value of the "thing," for it was supposed of *GH* that it was so [i. e., equal to *x*], therefore I complete, after the manner of the first theorem of the 6th chapter of this book, the solids *DA, DC, DE, DF*, so that we understand by *DC* the cube of *BC*, by *DF* the cube of *AB*, by *DA* three times *CB* times the square of *AB*, by *DE* three times AB[4] times the square of *BC*. Since therefore from *AC* times *CK* the result is 2, from 3 times *AC* times *CK* will result 6, the number of "things;" and

---

[1] [We shall render by "thing" Cardan's *res* or *positio*, the two words he employs to designate the unknown quantity in an equation.]

[2] [This is, $x^3 + 6x = 20$.]

[3] [Here $AC = u$, $CK = v$, $uv = 2 = \frac{1}{3}$ of the coefficient of x.]

[4] [In modern form, we have $DC = v^3$, $DF = (u - v)^3 = x^3$, $DA = 3(u - v)^2v$, and $DE = 3(u - v)v^2$.]

Cardan

(Facing page 204.)

therefore from *AB* times 3 *AC* times *CK* there results 6 "things" *AB*, or 6 times *AB*, so that 3 times the product of *AB*, *BC*, and *AC* is 6 times *AB*. But the difference of the cube *AC* from the cube *CK*, and likewise from the cube *BC*, equal to it by hypothesis, is 20;[1] and from the first theorem of the 6th chapter, this is the sum of the solids *DA*, *DE*, and *DF*, so that these three solids make 20.[2] But taking *BC minus*, the cube of *AB* is equal to the cube of *AC* and 3 times *AC* into the square of *CB* and minus the cube of *BC* and minus 3 times *BC* into the square of *AC*.[3] By the demonstration, the difference between 3 times *CB* times the square of *AC*, and 3 times *AC* times the square of *BC*, is [3 times][4] the product of *AB*, *BC*, and *AC*.[5] Therefore since this, as has been shown, is equal to 6 times *AB*, adding 6 times *AB* to that which results from *AC* into 3 times the square of *BC* there results 3 times *BC* times the square of *AC*, since *BC* is minus.[6] Now it has been shown that the product of *CB* into 3 times the square of *AC* is minus; and the remainder which is equal to that is plus, hence 3 times *CB* into the square of *AC*[7] and 3 times *AC* into the square of *CB* and 6 times *AB* make nothing.[8] Accordingly, by common sense, the difference between the cubes *AC* and *BC* is as much as the totality of the cube of *AC*, and 3 times *AC* into the square of *CB*, and 3 times *CB* into the square of *AC* (minus), and the cube of *BC* (minus), and 6 times *AB*.[9] This therefore is 20, since the difference of the cubes *AC* and *CB* was 20.[10] Moreover, by the second theorem of the 6th chapter, putting *BC* minus, the cube of *AB* will be equal to the cube of *AC* and 3 times *AC* into the square of *BC* minus the cube of *BC* and minus 3 times *BC* into the square of *AC*.[11] Therefore the cube of *AB*, with 6 times *AB*, by common sense, since it is equal to the cube of *AC* and 3 times *AC* into the square of *CB*, and minus 3 times *CB* into the square of *AC*,[12] and

---

[1] [That is, $u^3 - v^3 = 20$.]
[2] [That is, $(u - v)^3 + 3(u - v)^2v + 3(u - v)v^2 = 20$.]
[3] [That is, $(u - v)^3 = u^3 + 3uv^2 - v^3 - 3vu^2$.]
[4] [The original omits "triplum" here.]
[5] [That is, $3vu^2 - 3uv^2 = 3(u - v)uv$.]
[6] [That is, $6(u - v) + 3uv^2 = 3u^2v$.]
[7] [In the text this is *AB*.]
[8] [That is $-3vu^2 + 3uv^2 + 6(u - v) = 0$.]
[9] [That is, $u^3 - v^3 = u^3 + 3uv^2 - 3vu^2 - v^3 + 6(u - v) = 20$.]
[10] [That is, $u^3 - v^3 = 20$.]
[11] [That is, $(u - v)^3 = u^3 + 3uv^2 - v^3 - 3vu^2$.]
[12] [The text has *AB*.]

minus the cube of *CB* and 6 times *AB*, which is now equal to 20, as has been shown, will also be equal to 20.[1]   Since therefore the cube of *AB* and 6 times *AB* will equal 20, and the cube of *GH*, together with 6 times *GH*, will equal 20, by common sense and from what has been said in the 35th and 31st of the 11th Book of the *Elements*,[2] *GH* will be equal to *AB*, therefore *GH* is the difference of *AC* and *CB*.   But *AC* and *CB*, or *AC* and *CK*, are numbers or lines containing an area equal to a third part of the number of "things" whose cubes differ by the number in the equation, wherefore we have the

RULE

Cube the third part of the number of "things," to which you add the square of half the number of the equation,[3] and take the root of the whole, that is, the square root, which you will use, in the one case adding the half of the number which you just multiplied by itself,[4] in the other case subtracting the same half, and you will have a "binomial" and "apotome" respectively; then subtract the cube root of the apotome from the cube root of the binomial, and the remainder from this is the value of the "thing."[5]   In the example, the cube and 6 "things"[6] equals 20; raise 2, the 3rd part of 6, to the cube, that makes 8; multiply 10, half the number, by itself, that makes 100; add 100 and 8, that makes 108; take the root, which is $\sqrt{108}$, and use this, in the first place adding 10, half the number, and in the second place subtracting the same amount, and you will have the binomial $\sqrt{108} + 10$, and the apotome $\sqrt{108} - 10$; take the cube root of these and subtract that of the apotome from that of the binomial, and you will have the value of the "thing," $\sqrt[3]{\sqrt{108} + 10} - \sqrt[3]{\sqrt{108} - 10}$.

---

[1] [That is, $x^3 + 6x = u^3 + 3uv^2 - 3vu^2 - v^3 + 6(u - v) = 20.$]

[2] [Evidently an incorrect reference to Euclid. It does not appear in the edition of 1570.]

[3] [That is, if the equation is $x^3 + px = q$, take $(\tfrac{1}{3}p)^3 + (\tfrac{1}{2}q)^2$.]

[4] [That is, adding $\tfrac{1}{2}q$.]

[5] $\left[\text{That is, } \sqrt[3]{\sqrt{(\tfrac{1}{3}p)^3 + (\tfrac{1}{2}q)^2} + \tfrac{1}{2}q} - \sqrt[3]{\sqrt{(\tfrac{1}{3}p)^3 + (\tfrac{1}{2}q)^2} - \tfrac{1}{2}q.} \right]$

[6] $[x^3 + 6x = 20.]$

FERRARI-CARDAN

Solution of the Biquadratic Equation

(Translated from the Latin by Professor R. B. McClenon, Grinnell College, Grinnell, Iowa, with notes by Professor Jekuthiel Ginsburg, Yeshiva College, New York City.)

Luigi (Ludovico) Ferrari (1522–c. 1560), a man of humble birth, was taken into Cardan's household as a servant at the age of fifteen. He showed such unusual ability that the latter made him his secretary. After three years of service Ferrari left and took up the work of teaching. Such was his success that he became professor of mathematics at Bologna but died in the first year of his service there. Zuanne de Tonini da Coi, a teacher at Brescia had proposed a problem which involved the equation

$$x^4 + 6x^2 + 36 = 60x.$$

Cardan, being unable to solve it, gave it to Ferrari. The latter succeeded in finding a solution and this was published, with due credit, by Cardan in his *Ars Magna* (Nürnberg, 1545). For an outline of the solution in modern symbolism see Smith, *History of Mathematics* (Boston, Mass., 1925), vol. II, p. 468.

Rule II

Another rule...is due to Luigi Ferrari, who invented it at my request. By it we have the solutions of absolutely all types of fourth powers, squares, and numbers; or fourth powers, cubes, squares, and numbers [1]...

Demonstration

Let the square *AF* be divided into two squares *AD* and *DF*, and two supplementary parts *DC* and *DE*; and I wish to add the gnomon *KFG* around this so that the whole *AH* may remain a square. I say that such a gnomon consists of twice the product of *GC*, the added line, by *CA*, with the square of *GC*; for *FG* is con-

---

[1] [Cardan uses "square-square" for fourth power. He now proceeds to state all types of biquadratics, beginning with the equivalents of

$$(1) \quad x^4 = ax^2 + bx + c,$$
$$(2) \quad x^4 = ax^2 + bx^3 + c,$$
$$(3) \quad x^4 = ax^3 + b.$$

His list includes twenty types.
He then considers one of these types, as shown in the translation.]

tained by the lines *GC* and *CF*, from the definition given at the beginning of the 2nd [book] of the *Elements*[1]; and *CF* is equal to *CA*, from the definition of a square; and by the 44th [proposition] of the 1st [book] of the *Elements*,[2] *KF* is equal to *FG*. Therefore the two areas *GF* and *FK* consist of *GC* into twice *CA*. Also the square of *GC* is *FH*, in consequence of the 4th [proposition] of the 2nd [book] of the *Elements*.[3] Therefore the proposition is evident.[4]

If therefore *AD* is made 1 square-square, and *CD* and *DE* [are made] 3 "squares," and *DF* [is made] 9,[5] *BA* will necessarily be a

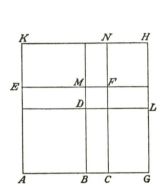
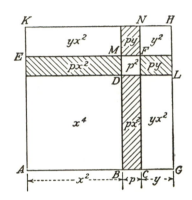

square and *BC* will necessarily be 3. Since we wish to add some "squares"[6] to *DC*[7] and *DE*, let these [additions] be [the

---

[1] [See Heath's *Euclid*, vol. I, p. 370. The general plan of attack will be better understood from the diagram here shown, which should be compared with Cardan's as given in the text. What he and Ferrari meant to do was to geometrize

$$(x^2 + p + y)^2$$

With this diagram before the reader, the proof will be more clearly understood. It shows the components of

$$(x^2 + p + y)^2 = x^4 + p^2 + y^2 + 2x^2 p + 2x^2 y + 2py$$

[2] [The 43d as usually numbered in later Editions.]

[3] [See Heath's *Euclid*, vol. I, p. 380.]

[4] [$(AC + CG)^2 = \overline{AC^2} + 2AC.CG + \overline{CG.^2}$]

[5] [He assumes the side of the square to be itself a square, say x^2. Then the increase *BC* will be 3. Then the area of the square *AD* is x^2, the area of rectangle *DE* will be $3x^2$, that of *DC* will be the same as that of *DE*, or $3x^2$, and that of *DF* will be 9.]

[6] [$AB = x^2$, $BC = 3$, $CG = y$. Each of *CL* and *KM* equals x^2y.]

[7] [Which, as stated above, is $3s^2$, as is also *DE*.]

rectangles] *CL* and *KM*.[1] Then in order to complete the square it will be necessary to add the area *LNM*. This has been shown to consist of the square on *GC*, which is half the number of [added] squares, since *CL* is the area [made] from [the product of] *GC* times *AB*, where *AB* is a square, *AD* having been assumed to be a fourth power.[2] But *FL* and *MN* are each equal to *GC* times *CB*, by Euclid, I, 42, and hence the area *LMN*, which is the number to be added, is a sum composed of the product of *GC* into twice *CB*, that is, into the number of squares which was 6, and *GC* into itself, which is the number of squares to be added. This is our proof.

This having been completed, you will always reduce the part containing the square-square to a root, *viz.*, by adding enough to each side so that the square-square with the square and number may have a root.[3] This is easy when you take half the number of the squares as the root of the number; and you will at the same time make the extreme terms on both sides plus, for otherwise the trinomial or binomial changed to a trinomial will necessarily fail to have a root. Having done this, you will add enough squares and a number to the one side, by the 3rd rule,[4] so that the same being added to the other side (in which the unknowns were) will make a trinomial having a square root by assumption; and you will have a number of squares and a number to be added to each side, after which you will extract the square root of each side, which will be, on the one side, 1 square plus a number (or minus a number)

---

[1] [Such an addition will convert the original figure, *AF* into the following.]

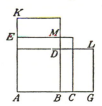

[2] [Cardan uses *quadratum* to mean both the second power of a number and a square figure.]

[3] [Let the equation $x^4 + px^2 + qx + r = 0$ be given. This may be written

$$x^4 + px^2 = -(qx + r).$$

Adding $px^2 + p^2$ to each side,

$$x^4 + 2px^2 + p^2 = p^2 + px^2 - qx - r,$$

or

$$(x^2 + p)^2 = p^2 + px^2 - qx - r.]$$

[4] [Given elsewhere in the *Ars Magna*.]

and **on the** other side 1 unknown or more, plus a number (or minus a number; or a number minus unknowns) wherefore by the 5th chapter of this book you will have what has been proposed.

Question V

Example.—Divide 10 into 3 parts in continued proportion such that the 1st multiplied by the 2nd gives 6 as product. This problem was proposed by Johannes Colla, who said he could not solve it. I nevertheless said I could solve it, but did not know how until Ferrari found this solution. Put then 1 unknown as the middle number, then the 1st will be $\dfrac{6}{1\ \text{unknown}}$, and the 3rd will be $\frac{1}{6}$ of a cube. Hence these together will be equal to 10. Multiplying all by 6 unknowns we shall have 60 unknowns equal to one square-square plus 6 squares plus 36. Add, according to the 5th rule, 6 squares to each side, and you will have 1 square-square plus 12 squares plus 36, equal to 6 squares plus 60 unknowns; for if equals are added to equals, the totals are equal. But 1 square-square plus 12 squares plus 36 has a root, which is 1 square plus 6.

1q̄dq̄d. *p*: 6 q̄d. *p*:36 æqualia 60 pos.[1]
6 q̄d.   6 q̄d.
1q̄dq̄d. *p*:12 q̄d. *p*:36 æq̄lia 6 q̄d. *p*:60 pos.
2 pos.   1 q̄d. *p*:12 pos.

If 6 squares plus 60 unknowns also had a root, we should have the job done; but they do not have; hence we must add so many squares and a number to each side, that on the one side there may remain a trinomial having a root, while on the other side it should be made so. Let therefore a number of squares be 1 unknown[2] and since, as you see in the figure of the 3rd rule, *CL* and *MK* are formed from twice *GC* into *AB*, and *GC* is 1 unknown,[3] I

---

[1] [That is, $x^4 + 6x^2 + 36 = 60x$, hence $x^4 + 12x^2 + 36 = 6x^2 + 60x$.]

[2] [Having reduced the equation to the form
$$(x^2 + p)^2 = p^2 + px^2 - qx - r,$$
he makes use of another unknown for the purpose of converting the left side into $(x^2 + p + y)^2$. This is done by adding $2y(x^2 + p) + y^2$ to each side. The equation then becomes
$$(x^2 + p + y)^2 = p^2 + px^2 - qx - r + 2y(x^2 + p) + y^2,$$
an equation in the form
$$x^2 + a = bx + c.$$
The problem now reduces to one of finding such a value of y as shall make the right side a square.]

[3] [That is, $GC = y$, and y is half the coefficient of x^2 in the part to be added.]

will always take the number of squares to be added as 2 unknowns, that is, twice *GC*; and since the number to be added to 36 is *LNM* it therefore is the square of *GC* together with the product of twice *GC* into *CB* or of *GC* into twice *CB*, which is 12, the number of the squares in the original equation. I will therefore always multiply 1 unknown, half the number of squares to be added, into the number of squares in the original equation and into itself and this will make 1 square plus 12 unknowns to be added on each side, and also 2 unknowns for the number of the squares.[1] We shall therefore have again, by common sense, the quantities written below equal to each other; and each side will have a root, the first, by the 3rd rule, but the 2nd

| 1 q̄dq̄d. p:2 pos. p:12. q̄d ℞ p:1 q̄d. p:12 pos. additi numeri p:36 æqualia. |
| 2 pos. 6 q̄dratorū, p:60 pos. p:1 q̄d. p:12 pos. numeri additi.[2] |

quantity by an assumption as to *y*. Therefore the first part of the trinomial multiplied by the third makes the square of half the 2nd part of the trinomial. Thus from half the 2nd part multiplied by itself there results 900, a square, and from the 1st [multiplied] into the 3rd there results 2 cubes plus 30 squares plus 72 unknowns. Likewise, this may be reduced ... since equals divided by equals produce equals, as 2 cubes plus 30 squares plus 72 unknowns equals 900,[3] therefore 1 cube plus 15 squares plus 36 unknowns equals 450.[4]

---

[1] [The problem has been reduced to
$$x^4 + 12x^2 + 36 = 6x^2 + 60x,$$
or
$$(x^2 + 6)^2 = 6x^2 + 60x.$$
To convert the left-hand side into $(x^2 + 6 + y)^2$, it is necessary to add $2y(x^2 + 6) + y^2$ to both sides, which converts the equation into
$$(x^2 + 6 + y)^2 = 6x^2 + 60x + y^2 + 12y + 2yx^2.]$$

[2] [That is,
$$x^4 + (2y + 12)x^2 + y^2 + 12y + 36$$
$$= (2y + 6)x^2 + 60x + y^2 + 12y$$
in which the first member reduces to $(x^2 + 6 + y)^2$.]

[3] [To find the value of *y* that will make the second member a square, Cardan had to consider the trinomial $ax^2 + 2bx + c$, as we should write it. This is a square when $b^2 = ac$, for then $b^2 - ac = 0$. But here $b = 30$, and so b^2 ("the square of half the second part") is 900, and *a* ("the first") is $2y + 6$, and $c = y^2 + 12y$. Then "the first into the third" is *ac*, or
$$(2y + 6)(y^2 + 12y) = 2y^3 + 30y^2 + 72y,$$
which Cardan describes as "2 cubes plus 30 squares plus 72 unknowns."]

[4] [Since $2y^3 + 30y^2 + 72y = 900$, $y^3 + 15y^2 + 36y = 450$.]

It is therefore sufficient for reducing to the rule, to have always
1 cube plus the number of the former squares, with a 4th of it
added to it plus such a multiple of the assumed quantity as the
first number of the equation indicates;[1] so that if we had 1 square-
square plus 12 squares plus 36 equals 6 squares plus 60 unknowns
we should have 1 cube plus 15 squares plus 36 unknowns equal to
450, half the square of half the number of unknowns. And if we
had 1 square-square plus 16 squares plus 64 equal to 80 unknowns
we should have 1 cube plus 20 squares plus 64 unknowns equal to
800.[2] And if we had 1 square-square plus 20 squares plus 100
equal to 80 unknowns we should have 1 cube plus 25 squares plus
100 unknowns equal to 800.[3] This being understood, in the
former example we had 1 cube plus 15 squares plus 36 unknowns
equal to 450; therefore the value of the unknown, by the 17th
chapter, is

$$\sqrt[3]{287\tfrac{1}{2} + \sqrt{+\,80449\tfrac{1}{4}}} + \sqrt[3]{287\tfrac{1}{2} - \sqrt{80449\tfrac{1}{4}}} - 5$$

This then is the number of squares which is to be doubled and
added to each side (since we assumed 2 unknowns to be added)
and the number to be added to each side, by the demonstration,
is the square of this, with the product of this by 12, the number
of squares.[4]

---

[1] $[y^3 + (12 + 1\tfrac{3}{4})y^2 + 36y = \tfrac{1}{2}(6\tfrac{9}{2})^2$. The "number of former squares"
means the coefficient of x^2, 12; and the "first number of the equation" means
the constant term, 36.]

[2] $[y^3 + (16 + 1\tfrac{9}{4})y^2 + 64y = \tfrac{1}{2}(8\tfrac{9}{2})^2.]$

[3] $[y^3 + (20 + 2\tfrac{9}{4})y^2 + 100y = \tfrac{1}{2}(8\tfrac{9}{2})^2.]$

[4] $[2yx^2 + 12y + y^2$, 12 being the coefficient of x^2 in the original equation,
$x^4 + 12x^2 + 36 = 60x + 6x^2.]$

FERMAT

Note on the Equation $x^n + y^n = z^n$

(Translated from the French by Professor Vera Sanford, Western Reserve University, Cleveland Ohio.)

Pierre de Fermat (c. 1608–1665), a member of the provincial parliament of Toulouse, became interested in the theory of numbers through Bachet's translation of Diophantus. Fermat's many discoveries in this field were given in letters to other mathematicians or were noted on the pages of the books which he read. The theorem which follows appears beside the eighth proposition of the second book of Diophantus:—"To divide a square number into two other square numbers." Fermat's note[1] reads:

To divide a cube into two other cubes, a fourth power, or in general any power whatever into two powers of the same denomination above the second is impossible, and I have assuredly found an admirable proof of this, but the margin is too narrow to contain it.

---

[1] *Précis des Oeuvres Mathématiques de P. Fermat et de l'Arithmétique de Diophante*, E. Brassinne, Paris, 1853, pp. 53–54. It should be noted that no one as yet has proved this theorem except for special cases.

FERMAT

The So-called Pell Equation

(Translated from the Latin by Professor Edward E. Whitford, College of the City of New York.)

Pierre de Fermat (c. 1608–1665) was the first to assert that the equation

$$x^2 - Ay^2 = 1$$

where A is any non-square integer, always has an unlimited number of solutions in integers. This equation may have been suggested to him by the study of some of the double equations of Diophantus; for he says in a note on the works of the latter (IV, 39), "Suppose if you will, that the double equation to be solved is

$$2m + 5 = \text{square},$$
$$6m + 3 = \text{square}.$$

The first square must be made equal to 16 and the second to 36; and others will be found ad infinitum satisfying the equation. Nor is it difficult to propound a general rule for the solution of this kind of equation."

Fermat was a profound scholar in all branches of learning and a mathematician of exceptional power. He has left the impression of his genius upon all branches of mathematics known in his time.

Fermat[1] first proposed the general problem of the Pell equation as a challenge problem to the English mathematicians Lord Brouncker and John Wallis (see p. 46). This was written in Latin in the form of a letter. In these contests of wits the Englishmen did not use French and the Frenchmen did not use English and the letters passed through intermediaries. The name Pell equation originated in a mistaken notion of Léonard Euler[2] (see p. 91) that John Pell was the author of the solution which was really the work of Lord Brouncker. Euler in his cursory reading of Wallis's algebra must have confused the contributions of Pell and Brouncker.[3] Nevertheless it seems not improbable that Pell solved the equation, for we find it discussed in Rahn's algebra[4] under the form

$$x = 12yy - 33.$$

This shows that Pell had some acquaintance with the general equation, and that Euler was not so far out of the way when he attributed to him some

---

[1] *Oeuvres de Fermat*, publiées par les soins de MM. Paul Tannery et Charles Henry, Paris, 1894, vol. II, p. 333-5.

[2] P. H. Fuss, editor, *Correspondance mathématique et physique de quelques célèbres géometres du XVIII ième siècle*, letter IX of L. Euler to C. Goldbach, Aug. 10, 1732, St. Petersburg, 1843, p. 37.

[3] G. Wertheim, "Über den Ursprung des Ausdruckes 'Pellsche Gleichung,'" *Bibliotheca Mathematica*, vol. II (3), p. 360, Leipzig, 1901.

[4] J. H. Rahn, *An introduction to algebra, translated out of the High Dutch into English by Thomas Brancker, M. A. Much altered and augmented by D. P.*, London, 1668, p. 143.

work upon it. Pell was an extensive contributor to Rahn's algebra and is referred to in the title of this book by the initials D. P.

The Pell equation affords the simplest case of Dirichlet's elegant and very general theorem on the existence of units in any algebraic field or domain. It is of great importance in the theory of binary quadratic forms. The problem to find all the rational solutions of the most general equation of the second degree in two unknowns reduces readily to that for $x^2 - Ay^2 = B$, all of whose solutions follow from the solution of $x^2 - Ay^2 = 1$. The honor of having first recognized the deep importance of the Pell equation for the general solution of the indeterminate equation of the second degree belongs to Euler.[1] The first admissible proof of the solvability of the equation $x^2 - Ay^2 = 1$ was given by Lagrange.[2]

Useful tables of solutions have been given by Euler, Legendre, Degen, Tenner, Koenig, Arndt, Cayley, Stern, Seeling, Roberts, Bickmore, Cunningham, and Whitford.

The letter of Fermat, dated February, 1657, which is called the second challenge of Fermat to the mathematicians, runs as follows:

There is scarcely any one that sets forth purely arithmetical questions, and scarcely any one that understands them. Is it not because arithmetic has heretofore been treated geometrically rather than arithmetically? This is certainly intimated by many works of ancient and modern writers, including Diophantus himself. Although he got away from geometry a little more than the rest, while limiting his analysis to rational numbers only, yet the "Zetetica" of Vieta, in which the method of Diophantus is extended to continuous quantity and therefore to geometry, sufficiently proves that this branch is not wholly separated from geometry.

Therefore arithmetic claims for itself the theory of whole numbers as its own estate. Arithmeticians ("children of arithmetic") should strive either to advance or restore it, which was only imperfectly represented by Euclid in his *Elements*, and moreover not sufficiently perfected by those who followed him. Perhaps it lies concealed in those books of Diophantus which the damages done by time have destroyed.

To these, in order to show them the light which may lead the way, I propose the following theorem or problem to be either proved or solved. Moreover, if they discover this, they will admit that questions of this sort are not inferior to the more celebrated ones from geometry, either in subtlety or in difficulty or in method of proof.

---

[1] L. Euler. "De solutione problematum Diophantaeorum per numeros integros," *Commentarii Academiae scientiarum imperialis Petropolitanae*, 1732, vol. VI, p. 175, St. Petersburg, 1738.

[2] J. L. Lagrange, "Solution d'un problème d'arithmétique," *Miscellanea Taurinensis*, vol. IV, p. 41, Turin, 1766. *Oeuvres de Lagrange*, Paris, 1867, vol. I, p. 671

Given any number not a square, then there are an infinite number of squares which, when multiplied into the given number, make a square when unity is added.

Example.—Given 3, a non-square number; this number multiplied into the square number 1, and 1 being added, produces 4, which is a square.

Moreover, the same 3 multiplied into the square 16, with 1 added makes 49, which is a square.

And instead of 1 and 16, an infinite number of squares may be found showing the same property; I demand, however, a general rule, any number being given which is not a square.

It is sought, for example, to find a square which when multiplied into 149, 109, 433, etc., becomes a square when unity is added.

In the same month (February, 1657) Fermat, in a letter to Frénicle, suggests the same problem, and expressly states the important condition implied in the foregoing that the solution be in integers:

Every non-square is of such a nature that one can find an infinite number of squares by which if you multiply the number given and if you add unity to the product, it becomes a square.

Example: 3 is a non-square number, which multiplied by 1, which is a square, makes 3, and by adding unity makes 4, which is a square.

The same 3, multiplied by 16, which is a square, makes 48, and with unity added makes 49, which is a square.

There is an infinity of such squares which when multiplied by 3 with unity added likewise make a square number.

I demand a general rule,—given a non-square number, find squares which multiplied by the given number, and with unity added, make squares.

What is for example the smallest square which multiplied by 61 with unity added, makes a square?

Moreover, what is the smallest square which, when multiplied by 109 and with unity added, makes a square?

If you do not give me the general solution, then give the particular solution for these two numbers, which I have chosen small in order not to give too much difficulty.

After I have received your reply, I will propose another matter. It appears without saying that my proposition is to find integers which satisfy the question, for in the case of fractions the lowest type of arithmetician could find the solution.

WALLIS

On General Exponents

(Translated from the Latin by Professor Eva M. Sanford, College for Women, Western Reserve University, Cleveland, Ohio.)

John Wallis (1616–1703), Savilian professor of geometry at Oxford (1649–1703), contemporary of Newton, was the first writer to set forth with any completeness the meaning of negative and fractional exponents. Steps in this direction had already been taken by Nicole Oresme (c. 1360), Chuquet (1484), Stifel (1544), and Girard (1629), but it remained for Wallis (1655) and Newton (1669) to generalize the subject for rational exponents. The following extract is from Wallis's *Arithmetica Infinitorum* as published in his *Opera Mathematica*, Oxford, 1695, vol. I, pp. 410, 411. The *Arithmetica Infinitorum* first appeared in 1655, and the use of the generalized exponent occurs in connection with the study of series, Proposition CVI.

Prop. CVI

If any series of reciprocals be multiplied or divided by another series (whether reciprocal or direct) or even if the series multiplies or divides another; the same laws are to be observed as in direct series (see propositions 73 and 81).

Example.—If the series of the reciprocals of squares ($\frac{1}{1}$, $\frac{1}{4}$, $\frac{1}{9}$, &c.) whose index is -2 be multiplied term by term into the series of the reciprocals of cubes ($\frac{1}{1}$, $\frac{1}{8}$, $\frac{1}{27}$, &c.) whose index is -3, the product will be a series of the reciprocals of the fifth powers[1] ($\frac{1}{1}$, $\frac{1}{32}$, $\frac{1}{243}$, &c.) whose index $-5 = -2 - 3$ as is evident.

Furthermore, if a series of the reciprocals of cubes ($\frac{1}{1}$, $\frac{1}{8}$, $\frac{1}{27}$, &c.) whose index is -3 be multiplied term by term by a series of squares (1, 4, 9, &c.) whose index is 2, the result is the series $\frac{1}{1}$, $\frac{4}{8}$, $\frac{9}{27}$, &c. This is 1, $\frac{1}{2}$, $\frac{1}{3}$, &c., a series of the reciprocals of first powers whose index $-1 = -3 + 2$.

Likewise, if the series of the reciprocals of the square roots[2]

$$\frac{1}{\sqrt{1}}, \frac{1}{\sqrt{2}}, \frac{1}{\sqrt{3}}, \text{&c.}$$

[1] [*Subquintanis.*]
[2] [*Subsecundans.*]

whose index is $-\frac{1}{2}$ be multiplied term by term by the series of squares (1, 4, 9, &c.) whose index is 2, the product will be the series

$$\frac{1}{\sqrt{1}}, \frac{4}{\sqrt{2}}, \frac{9}{\sqrt{3}}, \&c.,$$

or

$$\tfrac{1}{1}\sqrt{1}, \tfrac{4}{2}\sqrt{2}, \tfrac{9}{3}\sqrt{3}, \&c.,$$

or

$$\sqrt{1}, \sqrt{8}, \sqrt{27}, \&c.,$$

the square roots of the cubes or third powers, whose index $\frac{3}{2} = -\frac{1}{2} + 2$.

Furthermore, if a series of the reciprocals of squares whose index is -2 divides a series of the reciprocals of integers whose index is -1, the product will be the series of first powers whose index $1 = -1 + 2$, or -1 minus -2.

Likewise if a series of the reciprocals of integers whose index is -1 divides a series of the reciprocals of squares whose index is -2, then the product will be a series of the reciprocals of first powers whose index $-1 = -2 + 1$, that is, -2 minus -1.

Likewise if a series of reciprocals of first powers whose index is -1 divides a series of squares whose index is 2, the product will be a series of third powers whose index $3 = 2 + 1$, that is 2 minus -1.

Likewise if a series of the reciprocals of first powers whose index is -1 be divided by a series of squares whose index is 2, the product will be a series of the reciprocals of third powers whose index $-3 = -1 - 2$, that is -1 minus 2.

And the same thing will happen in any other cases whatsoever of this sort, and hence the proposition is proved.[1]

---

[1] [In the wide page of the original there are arranged alongside the above paragraphs, beginning with the fourth one, the following series.

$\frac{1}{1}$)	$\frac{1}{1}$	$(\frac{1}{1} = 1$
$\frac{1}{4}$)	$\frac{1}{2}$	$(\frac{4}{2} = 2$
$\frac{1}{9}$)	$\frac{1}{3}$	$(\frac{9}{3} = 3$
	&c.	
$\frac{1}{1}$)	$\frac{1}{1}$	$(\frac{1}{1} = \frac{1}{1}$
$\frac{1}{2}$)	$\frac{1}{4}$	$(\frac{2}{4} = \frac{1}{2}$
$\frac{1}{3}$)	$\frac{1}{9}$	$(\frac{3}{9} = \frac{1}{3}$
	&c.	
1)	$\frac{1}{1}$	$(\frac{1}{1}$
4)	$\frac{1}{2}$	$(\frac{1}{8}$
9)	$\frac{1}{3}$	$(\frac{1}{27}$
	&c.]	

WALLIS AND NEWTON

On the Binomial Theorem for Fractional and Negative Exponents

(Selected from the English version by Professor David Eugene Smith, Teachers College, Columbia University, New York City.)

John Wallis (1616–1703), in his work *De Algebra Tractatus, Historicus & Practicus* was instrumental in making known several discoveries made by Newton. Among them is Newton's generalization of the Binomial Theorem to include fractional and negative exponents. This was first published in Latin, and was later translated by Wallis into English. The following extract is from this translation. In it Wallis assigns credit to Newton and sets forth his results, as yet unpublished. These results appear in the article which follows this one.

CHAP. XCI.[1]

The Doctrine of INFINITE SERIES, *further prosecuted by* Mr. Newton.

Now (to return where we left off:) Those Approximations (in the Arithmetick of Infinites) above mentioned, (for the Circle or Ellipse, and the Hyperbola;) have given occasion to others (as is before intimated,) to make further inquiry into that subject; and seek out other the like Approximations, (or continual approaches) in other cases. Which are now wont to be called by the name of *Infinite Series*, or *Converging Series*, or other names of a like import. (Thereby intimating, the designation of some particular quantity, by a regular Progression or rank of quantities, continually approaching to it; and which, if infinitely continued, must be equal to it.) Though it be but little of this nature which hath yet been made publick in print.

Of all that I have seen in this kind; I do not find any that hath better prosecuted that notion, nor with better success, than Mr. *Isaac Newton*, the worthy Professor of Mathematicks in *Cambridge*: Who about the Year 1664, or1665, (though he did afterwards for divers years intermit those thoughts, diverting to other Studies,) did with great sagacity apply himself to that Speculation. This I find by Two Letters of his (which I have seen,) written to Mr.

---

[1] Page 330.

Oldenburg, on that Subject, (dated *June* 13, and *Octob.* 24. 1676,) full of very ingenious discoveries, and well deserving to be made more publick.   In the latter of which Letters, he says, that by the Plague (which happened in the Year 1665), he was driven from *Cambridge*; and gave over the prosecution of it for divers years. And when he did again resume it, about the Year 1671, with intention then to make it publick; (together with his new discoveries concerning the Refractions of Light,) he was then by other accidents diverted.

He doth therein, not only give us many such Approximations fitted to particular cases; but lays down general Rules and Methods, easily applicable to cases innumerable; from whence such Infinite Series or Progressions may be deduced at pleasure; and those in great varieties for the same particular case.   And gives instances, how those Infinite or Interminate Progressions may be accommodated, to the Rectifying of Curve Lines (Geometrick or Mechanick;) Squaring of Curve-lined Figures; finding the length of Archs, by their given Chords, Sines, or Versed Sines; and of these by those; fitting Logarithms to Numbers, and Numbers to Logarithms given; with many other of the most perplexed Inquiries in Mathematicks.

In order hereunto, he applies not only Division in Species; (such as we have before described;) but Extraction of Roots in Species, (Quadratick, Cubick, and of other consequent, and intermediate Powers;) as well in Single, as in Affected Equations.

How this was by him made use of in the way of Interpolation, we have shewed before; upon a discovery that the *Vnciæ* or Numbers prefixed to the members of Powers, created from a Binomial Root, (the Exponent of which Powers respectively he calls *m*,) doth arise from such continual Multiplication as this,

$$1 \times \frac{m-0}{1} \times \frac{m-1}{2} \times \frac{m-2}{3} \times \frac{m-3}{4} \times \frac{m-4}{5}, \times \&c.[1]$$

Which Process, if *m* (the Exponent of the Power) be an Integer will (after a certain number of places, such as the nature of each Power requires) terminate again at 1, as it did begin: But if *m* be a Fraction, it will (passing it) run on to Negative numbers infinitely.

According to this notion; having found the numbers answering the Power commonly expressed by $\sqrt{q}$, (which is the intermediate

---
[1] Page 331.

between an Unite and the Lateral,) whose Exponent is $\frac{1}{2}=m$; to be these

$$1 \cdot \tfrac{1}{2} \cdot - \tfrac{1}{8} \cdot + \cdot \tfrac{1}{16} \cdot - \tfrac{5}{128} \cdot + \tfrac{7}{256} \cdot - \&c.$$

He applys this (for instance) to that of mine, (accommodated as is before shewed, to the Quadrature of the Circle, or a Quadrant thereof, ($\sqrt{}:RR-cc$; or (putting $R=1$,) $\sqrt{}:1-cc$. And finds $\sqrt{}:1-cc:=1-\tfrac{1}{2}cc-\tfrac{1}{8}c^4-\tfrac{1}{16}c^6$, &c. (Which multiplied into itself, restores $1-cc$.) The Process thus.

$$1-cc(1,-\tfrac{1}{2}cc,-\tfrac{1}{8}c^4,-\tfrac{1}{16}c^6, \qquad \&c.$$
$$1$$
$$\overline{0-cc}$$
$$\underline{-cc+\tfrac{1}{4}c^4}$$
$$\overline{-\tfrac{1}{4}c^4}$$
$$-\tfrac{1}{4}c^4+\tfrac{1}{8}c^6+\tfrac{1}{64}c^3$$
$$\overline{-\tfrac{1}{8}c^6-\tfrac{1}{64}c^3}$$
$$\&c. \quad \&c.$$

$$into \quad \begin{array}{l} 1-\tfrac{1}{2}cc-\tfrac{1}{8}c^4-\tfrac{1}{16}c^6 \quad \&c. \\ 1-\tfrac{1}{2}cc-\tfrac{1}{8}c^4-\tfrac{1}{16}c^6 \quad \&c. \\ \hline 1-cc-\tfrac{1}{4}c^4-\tfrac{1}{8}c^6 \quad \&c. \\ +\tfrac{1}{4}c^4+\tfrac{1}{8}c^6 \quad \&c. \\ \hline \qquad\qquad\qquad \&c. \\ \hline 1-cc. \end{array}$$

From whence (and from others of the like nature) he derives this Theorem for such Extractions,

$$\overline{P+PQ}\Big|\tfrac{m}{n}=P\tfrac{m}{n}+\tfrac{m}{n}AQ+\tfrac{m-n}{2n}BQ+\tfrac{m-2n}{3n}CQ+\tfrac{m-3n}{4n}DQ+\&c.$$

Where $P+PQ$ is the Quantity, whose Root is to be extracted, or any Power formed from it, or the Root of any such Power extracted. P is the first Term of such Quantity; Q, the rest (of such proposed Quantity) divided by that first Term. And $\frac{m}{n}$ the Exponent of such Root or Dimension sought. That is, in the present case, (for a Quadratick Root,) $\frac{1}{2}$.

(Note here, for preventing mistakes, that whereas it is usual to express the Exponent of a Power, or the number of its Dimensions, by a small Figure, at the head of the letter, as a^3 for aaa; the same is here done by a Fraction, when such Exponent is not an Integer Number, as $a\ \tfrac{3}{2}$ for $\sqrt{aaa}$; which Fraction is so to be understood, as if the whole of it were above the letter; and signifies the Exponent of the Power; not as at other times, a Fraction adjoined, as if it were $a+\tfrac{3}{2}$: And the same is to be understood afterwards in many places; where the like happens, by reason that there is not room to set the whole Fraction above the Letter, but equal with it.)

And according to this Method; if of any such Quantity proposed, we seek a Square, Cube, or Higher Power, whose Exponent is an

Integer; we shall find for it, a Series terminated, consisting of so many members as the nature of each Power requires; (the Side of 2, the Square of 3; the Cube of 4; &c.) But if a Root or Intermediate Power be sought, whose Exponent is a Fraction, or an Integer[1] with a Fraction annexed, (as $\frac{1}{2}$, $1\frac{1}{2}$, $2\frac{1}{2}$, &c; that is, $\frac{1}{2}$, $\frac{3}{2}$, $\frac{5}{2}$, &c: Or $\frac{1}{3}$, $\frac{2}{3}$, $1\frac{1}{3}$, $1\frac{2}{3}$, that is, $\frac{1}{3}$, $\frac{2}{3}$, $\frac{4}{3}$, $\frac{5}{3}$, &c:) We shall have (for its value) an Interminate or Infinite Series; to be continued as far as we please. And the farther it is continued, the more exactly doth it represent the quantity sought.

Of this Process, he giveth divers Examples; which (because they are not yet Extant in Print,) I have thought fit here to transcribe.

Example I. $\sqrt{}:cc+xx$, or $\overline{cc+xx}|\frac{1}{2}$, $=c+\dfrac{xx}{2c}-\dfrac{x^4}{8c^3}+\dfrac{x^6}{16c^6}-\dfrac{5x^8}{128c^7}$ $+\dfrac{7x^{10}}{256c^9}$&c. For in this case, is $P=cc$. $Q=\dfrac{xx}{cc}$. $m=1$. $n=2$.

$A(=\overline{P}|\dfrac{m}{n}=\overline{cc}|\frac{1}{2},)=c$. $B(=\dfrac{m}{n}AQ)=\dfrac{xx}{2c}$. $C(=\dfrac{m-n}{2n}BQ)=\dfrac{-x^4}{8c^3}$.&c.

Examp. II. $\sqrt{}^5:c^5+c^4x-x^5:$ or $\overline{c^5+c^4x-x^5}|\frac{1}{5}$.

$=c+\dfrac{c^4x-x^5}{5c^4},\dfrac{-2c^8xx+4c^4x^6-2x^{10}}{25c^9}+$&c. As will be evident by substituting $1=m$. $5=n$. $c^5=P$. and $c^5)c^4x-x^5(Q$.

Or we might in like manner substitute $-x^5=P$, and $-x^5)c^4x+c^5(Q$.

And then $\sqrt{}^5:c^5+c^4x-x^5:=-x+\dfrac{c^4x+c^5}{5x^4}+\dfrac{2c^8xx+4c^9+c^{10}}{25x^9}+$&c.

The former way is most eligible, if x be very small; the latter if x be very great.

Examp. III. $\dfrac{N}{\sqrt{}^3:y^3-aay:}$ That is, $N\times\overline{y^3-aay}|-\frac{1}{3}$. $=N$ into $\dfrac{1}{y}+\dfrac{aa}{3y^3}+\dfrac{a^4}{9y^5}+\dfrac{7a^6}{81y^7}+$&c. For here, $P=y^3$. $Q=\dfrac{-aa}{yy}$. $m=-1$.

$n=3$. $A(=\overline{P}\dfrac{m}{n}=y^3\times-\frac{1}{3})=y^{-1}$, that is, $\dfrac{1}{y}$. $B\left(=\dfrac{m}{n}AQ=-\frac{1}{3}\times\dfrac{1}{y}\times\right.$ $\left.\dfrac{-aa}{yy}\right)=\dfrac{aa}{3y^3}$. &c.

Examp. IV. The Cubick Root of the Biquadrate of $d+e$; that is, $\overline{d+e}|\frac{4}{3}$. is $d^{\frac{4}{3}}+\dfrac{4ed^{\frac{1}{3}}}{3}+\dfrac{2ee}{9d^{\frac{2}{3}}}-\dfrac{e^3}{9d^{\frac{2}{3}}}+$&c. For $P=d$. $d)e(Q$. $m=4$.

$n=3$. $A\left(=P\dfrac{m}{n}\right)=d^{\frac{4}{3}}$. &c.

---

[1] Page 332.

Examp. V. After the same manner may Single Powers be formed; as the Sursolid, or Fifth Power of $d+e$: That is, $\overline{d+e}|^5$, or $\overline{d+e}|^{\frac{5}{1}}$. For then $P=d.$ $d)e(Q.$ $m=5.$ $n=1.$ $A\left(=P\dfrac{m}{n}\right)=$ $d^5.$ $B\left(=\dfrac{m}{n}AQ\right)=5d^4e.$ $C=10d^3ee.$ $D=10dde^3.$ $E=5dc^4.$ $F=e.^5$ $G\left(=\dfrac{m-5n}{6n}FQ\right)=0.$ That is, $\overline{d+e}|^5=d^5+5d^4e+10d^3ee+10dde^3$ $+5de^4+e^5.$

Examp. VI. And even bare Divison, (whether single, or repeated,) may be performed by the same Rule. As $\dfrac{1}{d+e}$, that is $\overline{d+e}|^{-1}$, or $\overline{d+e}|^{-\frac{1}{1}}$. For then $P=d.$ $d)e(Q.$ $m=-1.$ $n=1.$ $A\left(=P\dfrac{m}{n}=d^{-\frac{1}{1}}\right)=d^{-1},$ or $\dfrac{1}{d}.$ $B\left(=\dfrac{m}{n}AQ=-1\times\dfrac{1}{d}\times\dfrac{e}{d}\right)=\dfrac{-e}{dd}.$ And in like manner, $C=\dfrac{ee}{d^3}.$ $D=\dfrac{-e^3}{d^4}.$ &c.

That[1] is, $\dfrac{1}{d+e}=\dfrac{1}{d}-\dfrac{e}{dd}+\dfrac{ee}{d^3}-\dfrac{e^3}{d^4}+$&c.

Examp. VII. In like manner $\overline{d+e}|^{-3}$: That is, an Unite Three times divided by $d+e$, or divided by the Cube of $d+e$: Is $\dfrac{1}{d^3}-\dfrac{3e}{d^4}$ $+\dfrac{6ee}{d^5}-\dfrac{10e^3}{d^6}+$&c.

Examp. VIII. And $N\times\overline{d+e}|^{-\frac{1}{3}}$; That is, N divided by the Cubick Root of $d+e$: is $N:\times\dfrac{1}{d^{\frac{1}{3}}}-\dfrac{e}{3d^{\frac{4}{3}}}+\dfrac{2ee}{9d^{\frac{7}{3}}}-\dfrac{14e^3}{81d^{\frac{10}{3}}}+$&c.

Examp. IX. And $N\times\overline{d+e}|^{-\frac{3}{5}}$; That is, N divided by the Sursolidal Root of the Cube of $d+e$: Or $\dfrac{N}{\sqrt{5}:d^3+3d^2e+3dee+e^3}:$. Is N into: $\dfrac{1}{d^{\frac{3}{5}}}-\dfrac{3e}{5d^{\frac{8}{5}}}+\dfrac{12ee}{25d^{\frac{13}{5}}}-\dfrac{52e^3}{125d^{\frac{18}{5}}}+$&c.

And by the same Rule, we may in Numbers (as well as Species,) perform the Generation of Powers; Division by Powers, or by Radical Quantities; and the Extraction of Roots of higher Powers; and the like.

---

[1] Page 333.

NEWTON

On the Binomial Theorem for Fractional and Negative Exponents

(Translated from the Latin by Professor Eva M. Sanford, College for Women, Western Reserve University, Cleveland, Ohio.)

Isaac Newton (1642–1727) was the first to state the binomial theorem for negative and fractional exponents. The formula appears in a letter written on June 13, 1676, to Oldenburg, the Secretary of the Royal Society, for transmission to Leibniz who had asked for information regarding Newton's work with infinite series On the receipt of this communication, Leibniz requested further details and Newton replied on October 24th of the same year. Both letters were printed in the *Commercium Epistolicum* (1712) with other papers that bore upon the Newton-Leibniz controversy. For a biographical note on Newton, see page 613.

Letter of June 13, 1676[1]

Although the modesty of Dr. Leibniz in the Excerpts which you recently sent me from his Letter, attributes much to my work in certain Speculations regarding *Infinite Series*,[2] rumor of which is already beginning to spread, I have no doubt that he has found not only a method of reducing any Quantities whatsoever into Series of this type, *as he himself asserts*, but also that he has found various Compendia, similar to ours if not even better.

Since, however, he may wish to know the discoveries that have been made in this direction by the English (I myself fell into this Speculation some years ago) and in order to satisfy his wishes to some degree at least, I have sent you certain of the points which have occurred to me.

Fractions may be reduced to Infinite Series by Division, and Radical Quantities may be so reduced by the Extraction of Roots. These Operations may be extended to Species[3] in the same way as

---

[1] [*Commercium Epistolicum* (1712; 1725 edition, pp. 131–132).]

[2] [Probably as early as 1666, Newton had told Barrow and others of his work in infinite series in connection with the problem of finding the area under a curve, but this work was not published until 1704 when it appeared as an appendix to Newton's *Opticks*.]

[3] [That is "to algebraic numbers." In his *Arithmetica Universalis* (1707; 1728 edition) Newton says, "Computation is either perform'd by *Numbers*, as in Vulgar Arithmetick, or by *Species*, as usual among Algebraists..."]

that in which they apply to Decimal Numbers. These are the Foundations of the Reductions.

The Extractions of Roots are much shortened by the Theorem

$$\overline{P + PQ}\Big|\frac{m}{n} = P\frac{m}{n} + \frac{m}{n}AQ + \frac{m-n}{2n}BQ + \frac{m-2n}{4n^1}CQ$$

$$+ \frac{m-3n}{4n}DQ + \&c.$$

where $P + PQ$ stands for a Quantity whose Root or Power or whose Root of a Power is to be found, P being the first Term of that quantity, Q being the remaining terms divided by the first term, and $\frac{m}{n}$ the numerical Index of the powers of $P + PQ$. This may be a Whole Number or (so to speak) a Broken Number; a positive number or a negative one. For, as the Analysts write a^2 and a^3 &c. for aa and aaa, so for $\sqrt{a}$, $\sqrt{a^3}$, $\sqrt{c \cdot a^5}$, &c. I write $a^{\frac{1}{2}}$, $a^{\frac{3}{2}}$, $a^{\frac{5}{2}}$, &c.; for $\frac{1}{a}$, $\frac{1}{aa}$, $\frac{1}{aaa}$, a^{-1}, a^{-2}, a^{-3}; for $\dfrac{aa}{\sqrt{c \cdot a^3 + bbx}}$, $aa \times \overline{a^3 + bbx}\big|^{-\frac{1}{2}}$; and for $\dfrac{aab}{\sqrt{c : a^3 + bbx \times a^3 + bbx :}}$, I write $aab \times \overline{a^3 + bbx}\big|^{-\frac{3}{2}}$. In this last case, if $\overline{a^3 + bbx}\big|^{-\frac{3}{2}}$ be taken to mean $P + PQ$ in the Formula, then will $P = a^3$, $Q = bbx/a^3$, $m = -2$, $n = 3$. Finally, in place of the terms that occur in the course of the work in the Quotient, I shall use A, B, C, D, &c. Thus A stands for the first term $P^{\frac{m}{n}}$; B for the second term $\frac{m}{n}AQ$; and so on. The use of this Formula will become clear through Examples."[2]

. . . .

Letter of October 24, 1676[3]

One of my own [methods of deriving infinite series] I described before; and now I shall add another, namely the way in which I discovered these Series, for I found them before I knew the Divisions and Extractions of Roots which I now use. The explanation of this method will give the basis of the Theorem given at the beginning of my former Letter which Dr. Leibniz desires of me.

---

[1] [Evidently a misprint for $3n$.]

[2] [The examples show the application of the formula in cases in which the exponents are $\frac{1}{2}$, $\frac{1}{6}$, $-\frac{1}{3}$, $\frac{5}{3}$, 5, -1, $-\frac{3}{5}$.]

[3] [*Commercium Epistolicum* (1712; 1725 edition, pp. 142–145). This letter begins with a note of appreciation of the work in series done by Leibniz.]

"Towards the beginning of my study of Mathematics, I happened on the works of our most Celebrated Wallis[1] and his considerations of the Series by whose intercalation he himself shows the values of the Area of a Circle and Hyperbola, and of that series of curves that have a common Base or Axis x and whose Ordinates are in the Form $\overline{1-xx}|^{\frac{1}{2}} \cdot \overline{1-xx}|^{\frac{1}{2}} \cdot \overline{1-xx}|^{\frac{1}{2}} \cdot \overline{1-xx}|^{\frac{1}{2}}$ $\overline{1-xx}|^{\frac{1}{2}} \cdot \overline{1-xx}|^{\frac{1}{2}}$. &c. Then if the Areas of the alternate ones which are x, $x - \frac{1}{3}x^3$, $x - \frac{2}{3}x^3 + \frac{1}{5}x^5$, $x - \frac{3}{3}x^3 + \frac{3}{5}x^5 - \frac{1}{7}x^7$, &c. could have values interpolated between these terms, we should have the Areas of the intermediates, the first of which $\overline{1-xx}|^{\frac{1}{2}}$ is the Circle. For these interpolations, I noticed that the first term in each is x and that the second term $\frac{0}{3}x^3$, $\frac{1}{3}x^3$, $\frac{2}{3}x^3$, $\frac{3}{3}x^3$, &c., are in Arithmetic progression. Thus the two first terms of the Series to be intercalated should be $x - \frac{\frac{1}{2}x^3}{3}$, $x - \frac{\frac{3}{2}x^3}{3}$, $x - \frac{\frac{5}{2}x^3}{5}$. &c.

For intercalcating the rest, I considered that the Denominators 1, 3, 5, 7, &c. were in Arithmetic progression and so only the Numerical Coefficients of the Numerators would require investigation. Moreover, in the alternate Areas given, these were the figures of the powers of the eleventh number, namely, 11^0, 11^1, 11^2, 11^3, 11^4. That is, first 1, then 1, 1, thirdly 1, 2, 1, fourth, 1, 3, 3, 1, fifth 1, 4, 6, 4, 1, &c. Therefore, I sought a method of deriving the remaining elements in these Series, having given the two first figures. I found that when the second figure m was supplied, the rest would be produced by continuous multiplication of the terms of this Series:

$$\frac{m-0}{1} \times \frac{m-1}{2} \times \frac{m-2}{3} \times \frac{m-3}{4} \times \frac{m-4}{5} \text{ &c.}$$

For Example: Let (second term) $m = 4$, then the third term will be $4 \times \frac{m-1}{2}$, that is 6; and $6 \times \frac{m-2}{3}$, that is 4, the fourth; and $4 \times \frac{m-3}{4}$ that is 1, the fifth; and $1 \times \frac{m-4}{5}$, that is 0, the sixth at which the series ended in this case.

I therefore, applied this Rule to the Series to be inserted. Thus

---

[1] [John Wallis (1616–1703) Savilian professor at Oxford, whose *Arithmetica Infinitorum* appeared in 1655. At a later date, Newton wrote an appendix to Wallis's *Algebra*.]

for a Circle, the second term would be $\frac{\frac{1}{2}x^3}{3}$, I then let $m = \frac{1}{2}$, and the terms which resulted were $\frac{1}{2} \times \frac{\frac{1}{2} - 1}{2}$ or $-\frac{1}{8}$, $-\frac{1}{8} \times \frac{\frac{1}{2} - 2}{3}$ or $+\frac{1}{16}$, $+\frac{1}{16} \times \frac{\frac{1}{2} - 3}{4}$ or $-\frac{5}{128}$, and so infinity. From this I learned that the desired Area of a segment of a Circle is

$$x - \frac{\frac{1}{2}x^3}{3} - \frac{\frac{1}{8}x^5}{5} - \frac{\frac{1}{16}x^7}{7} - \frac{\frac{5}{128}x^9}{9} \text{ \&c.}$$

By the same process the areas of the remaining Curves to be inserted were found, as the area of a Hyperbola, and of the other alternates in this Series $\overline{1 + xx}|^{\frac{3}{2}}$, $\overline{1 + xx}|^{\frac{5}{2}}$, $\overline{1 + xx}|^{\frac{7}{2}}$, $\overline{1 + xx}|^{\frac{9}{2}}$, &c.

The same method may be used for intercalating other Series, even with intervals of two or more terms lacking at once.

This was my first entry into these studies; which would surely have slipped from my memory had I not referred to certain notes a few weeks ago.

But when I had learned this, I soon considered that the terms $\overline{1 - xx}|^{\frac{2}{2}}$, $\overline{1 - xx}|^{\frac{4}{2}}$, $\overline{1 - xx}|^{\frac{6}{2}}$, $\overline{1 - xx}|^{\frac{8}{2}}$, &c. that is 1, $1 - xx$, $1 - 2xx + x^4$, $1 - 3xx + 3x^4 - x^6$, &c. could be interpolated in the same way and areas could be derived from them; and that for this nothing more is required than the omission of the denominators 1, 3, 5, 7, &c. in the terms expressing the areas, that is, the coefficients of the terms of the quantity to be intercalated $\overline{1 - xx}|^{\frac{1}{2}}$, or $\overline{1 - xx}|^{\frac{3}{2}}$, or more generally $\overline{1 - xx}|^m$ could be produced by continuous multiplication of the terms of this Series $m \times \frac{m - 1}{2} \times \frac{m - 2}{3} \times \frac{m - 3}{4}$ &c.

Thus, (for example), $\overline{1 - xx}|^{\frac{1}{2}}$ would amount to $1 - \frac{1}{2}x^2 - \frac{1}{8}x^4 - \frac{1}{16}x^6$ &c. And $\overline{1 - xx}|^{\frac{3}{2}}$ would come to $1 - \frac{3}{2}x^2 + \frac{3}{8}x^4 + \frac{1}{16}x^6$ &c. And $\overline{1 - xx}|^{\frac{1}{3}}$ would be $1 - \frac{1}{3}xx - \frac{1}{9}x^4 - \frac{1}{81}x^6$ &c.

Thus the general Reduction of Radicals into infinite Series became known to me through the Rule which I set at the beginning of the former Letter, before I knew the Extractions of Roots.

But, having learned this, the other could not long remain hidden from me. To prove these operations, I multiplied $1 - \frac{1}{2}x^2 -$

$\frac{1}{8}x^4 - \frac{1}{16}x^6$ &c. by itself, and $1 - xx$ resulted, the remaining terms vanishing into infinity by the continuance of the series. Similarly $1 - \frac{1}{2}xx - \frac{1}{8}x^4 - \frac{5}{81}x^6$ &c. twice multiplied by itself produced $1 - xx$. Which, that these might be a Demonstration of these conclusions, led me naturally to try the converse, to see whether these Series which it was certain were Roots of the quantity $1 - xx$ could not be extracted by Arithmetical means. The attempt succeeded well...[1]

Having discovered this, I gave up entirely the interpolation of Series, and used these operations alone as a more genuine basis, nor did I fail to discover Reduction by Division, a method certainly easier.

---

[1] [The algebraic work is given here.]

LEIBNIZ AND THE BERNOULLIS

On the Polynomial Theorem

(Translated from the Latin by Professor Jekuthiel Ginsburg, Yeshiva College, New York City.)

In a letter to Jean (I) Bernoulli, dated May 16, 1695, Leibniz speaks of a rule invented by him for finding the coefficients of a polynomial raised to any power whatsoever. This led to some correspondence between them, which appears in the *Commercium Philosophicum et Mathematicum* (Lausanne and Geneva, 1745). Extracts from this correspondence are given here in translation, the footnotes explaining their origin. Jacques (I) Bernoulli discussed the matter in his *Ars Conjectandi* (posthumously printed at Basel in 1713), and his method is given after the extracts from the correspondence of Leibniz with Jean I, his brother.

Jacques (1) Bernoulli returned to the same problem in a note published posthumously in his *Opera* (Vol. 2, Geneva 1744, pp. 995–6). It represents an attempt to apply the polynomial theorem to the solution of the related but more complicated problem of finding the power of an infinite series of terms arranged according to the ascending power of x. The attempt was inspired by one of the two articles on the subject of the "infinitonome" published by De Moivre in the *Philosophical Transactions* for the years 1697 and 1698. Bernoulli's plan apparently was to apply the previously-found theorem, first to the case of an infinite number of terms and then to an infinite series arranged according to ascending powers of x. This plan was not carried beyond a few theoretical remarks which are of little practical applicability.

The special problem of finding any power of an infinite power series has been the subject of a large number of ingenious memoirs, first among which are those by De Moivre.

[Leibniz to Jean Bernoulli, May 16, 1695]

I have conceived then of a wonderful rule for the coefficients of powers not only of the binomial $x + y$, but also of the trinomial $x + y + z$, in fact, of any polynomial; so that when given the power of any degree say the tenth, and any term contained in it, as $x^5 y^3 z^2$,[1] it should be possible to assign the coefficient (numerum coefficientem) which it must have...

---

[1] [*Commercium Philosophicum...*, I, p. 47. The term $x^5 y^1 z^1$ there given is evidently a misprint.]

[Jean Bernoulli to Leibniz, June 8, 1695[1]]

Let it be required to raise any polynomial $s + x + y + z$ etc. to an arbitrary power r, and let it be required to find the coefficient of the term $s^a x^b y^c z^e$ etc.   I say that that coefficient will be

$$\frac{r \cdot r - 1 \cdot r - 2 \cdot r - 3 \cdot r - 4 \ldots a + 1}{1 \cdot 2 \cdot 3 \ldots b \times 1 \cdot 2 \cdot 3 \ldots c \times 1 \cdot 2 \cdot 3 \ldots e \text{ etc.}};$$

that is, the required coefficient will be given by the product of all terms of the arithmetic progression which begins with the number of the power of the multinomial and decreases by 1 until the number is reached which is greater by one than the power of the first character, this product to be divided by the product of all terms of all the arithmetic progressions ascending from 1 up to the respective numbers of powers of all the letters except the first.   Note that the tedious division and a considerable part of the multiplication could be eliminated by cancelling before the operation those multiplicative parts [Bernoulli's terms for factors] that the numerator has in common with the denominator.   As an example we will take what you proposed: It is required to find the coefficient of the term $s^5 x^3 y^2$ comprehended in the value of the trinomial $s + x + y$ raised to the tenth power.   Substituting in the general formula the values $r = 10$, $b = 3$, $c = 2$.   We will have for the required coefficient $\dfrac{10 \cdot 9 \cdot 8 \cdot 7 \cdot 6}{1 \cdot 2 \cdot 3 \times 1 \cdot 2} = 10 \cdot 9 \cdot 4 \cdot 7 = 2520$.   If the coefficient of $s^8 x^6 y^4 z^2$ in the expansion of the quadrinomial $s + x + y + z$ to the 20th power be required, it will be =

$$\frac{20 \cdot 19 \cdot 18 \cdot 17 \cdot 16 \cdot 15 \cdot 15 \cdot 13 \cdot 12 \cdot 11 \cdot 10 \cdot 9}{1 \cdot 2 \cdot 3 \cdot 4 \cdot 5 \cdot 6 \times 1 \cdot 2 \cdot 3 \cdot 4 \times 1 \cdot 2} =$$
$$19 \cdot 17 \cdot 5 \cdot 7 \cdot 13 \cdot 12 \cdot 11 \cdot 10 \cdot 9 = 1745944200.$$

It would be a pleasure to see your rule and it would be well to test whether they agree [experiri liceret, ut inter se consentiant]; yours is possibly simpler:

[Jacques(I) Bernoulli, in the *Ars Conjectandi*[2]]

It is proper here to note the peculiar συμπάθειαν between combinations and powers of multinomials.   At the beginning of the

---

[1] [*Ibid.*, p. 55.   In this Jean Bernoulli derives the *regula mirabilis* referred to above by Leibniz.]

[2] [The formula is found at the end of Chapter VIII of the second part of his classical work.   Having disposed of the problem to find the number of arrangements of many various things when each one may also be combined with itself according to one exponent, he proceeds as in the above translation.]

chapter it has been shown that in order to find the "binions" of all the letters a, b, c, d, each of the letters must be put before every other one (including itself); and in order to get the "ternions," every one of the "binions" must be written before every one of the letters. But the same must take place when the literal quantity [*quantitas literalis*] $a + b + c + d$ is raised to the second, third, and higher powers. It follows from this that when the same symbols are regarded as parts of any multinomial, the "binions" will represent all the terms of the square; the "ternions," the terms of the cube; and the "quaternions," the terms of the fourth power. The terms of the power will be expressed by the addition of the combinations of the parts of the base to the order indicated by the index of the power. Since, however, all the terms containing the same letters, arranged in different ways, represent the same quantity, they should be combined in one term for the sake of brevity. To this terms should be prefixed the number of such equivalent terms, which number is called the coefficient of the term. It is evident that the coefficient of any term is equal to the number of permutations of its characters. The total number of terms is equal to the number of combinations of the order of the index of the power that can be formed of the elements of the base (when the order of terms is disregarded). This number can be found by the rule explained in Chapter V.

The great value of this observation may be seen from the fact that by its means, it is possible to promptly determine both the number of terms in a power and the coefficient of any term. Thus, for example, the tenth power of the trinomial $(a + b + c)$, consists of $\dfrac{11 \cdot 12}{1 \cdot 2} \infty 66$ terms, by the rule of chapter V; and the coefficient of the term $a^5 b^3 cc$ [1] by the second rule of chapter I, is

$$\frac{1 \cdot 2 \cdot 3 \cdot 4 \cdot 5 \cdot 6 \cdot 7 \cdot 8 \cdot 9 \cdot 10}{1 \cdot 2 \cdot 3 \cdot 4 \text{ in } 1 \cdot 2 \cdot 3 \text{ in } 1 \cdot 2} \infty 2520.$$

Similarly the cube of the quadrinomial $a + b + c + d$ will be composed of $\dfrac{4 \cdot 5 \cdot 6}{1 \cdot 2 \cdot 3} \infty 20$ terms, and the terms aab and abc will have as coefficients the numbers 3 and 6.

---

[1] [It is interesting that Jacques (I) Bernoulli used the example discussed in the correspondence between Leibniz and Jean (I) Bernoulli.]

HORNER'S METHOD

(Selected and Edited by Margaret McGuire, A.M., Teachers College, Columbia University, New York City.)

William George Horner (1786–1837) was educated at Kingswood School, near Bristol, but had no university training and was not a noted mathematician. In 1809, he established a school at Bath, where he remained until his death. It was there that he discovered the method of approximating roots of higher numerical equations which is his sole claim for fame. His method closely resembles that which seems to have been developed during the thirteenth century by the Chinese and perfected by Chin Kiu-shao about 1250.[1] It is also very similar to the approximation process effected in 1804 by Paolo Ruffini (1765–1822).[2] The probability is, however, that neither Horner nor Ruffini knew of the work of the other and that neither was aware of the ancient Chinese method. Apparently Horner knew very little of any previous work in approximation, as he did not mention in his article the contributions of Vieta, Harriot, Oughtred, or Wallis.

The paper here reproduced, the first written by Horner on approximations, was read before the Royal Society July 1, 1819 and was published in the *Philosophical Transactions of the Royal Society of London*, 1819, pp. 308–335. The modern student of mathematics will notice at once the length and difficulty of Horner's treatment when comparing it with the simple, elementary explanation in modern texts. In speaking of the publication of the article in the *Transactions*, T. S. Davies said: "The elementary character of the subject was the professed objection; his recondite mode of treating it was the professed passport for its admission." The paper was reprinted in the *Ladies' Diary* of 1838 and two revisions were published,—the first in Leybourn's *Repository*, (1830) and the second (posthumously) in the first volume (1845) of *The Mathematician*. In his original article, Horner made use of Taylor's Theorem, obtaining his transformations by methods of the calculus, but in his revisions he used ordinary algebra and gave a more simple explanation of the process.

XXI. A new method of solving numerical equations of all orders, by continuous approximation.[3] By W. G. Horner, Esq. Communicated by Davies Gilbert, Esq. F. R. S.

Read July 1, 1819.

1. THE process which it is the object of this Essay to establish, being nothing else than the leading theorem in the Calculus of

---

[1] Smith, D. E., *History of Mathematics* (New York, 1925) Vol. II, p. 381.

[2] Cajori, F. "Horner's Method of Approximation Anticipated by Ruffini" *Bulletin of American Mathematical Society*, XVII (1911), pp. 409–414.

[3] The only object proposed by the author in offering this Essay to the acceptance of the Royal Society, for admission into the Philosophical Transactions, is to secure beyond the hazard of controversy, an Englishman's property

Derivations, presented under a new aspect, may be regarded as a
universal instrument of calculation, extending to the composition
as well as analysis of functions of every kind.   But it comes into
most useful application in the numerical solution of equations.

2. ARBOGAST's developement of

$$\varphi(\alpha+\beta x+\gamma x^2+\delta x^3+\epsilon x^4+\ldots.)$$

(See *Calc. des Der.* § 33) supposes all the coefficients within the
parenthesis to be known previously to the operation of φ.   To the
important cases in which the discovery of γ, δ, &c. depends on the
previous developement of the partial functions

$$\varphi(\alpha+\beta x), \ \varphi(\alpha+\beta x+\gamma x^2), \ \&c.$$

it is totally inapplicable.   A theorem which should meet this
deficiency, without sacrificing the great facilitating principle
of attaching the functional symbols to α alone, does not appear
to have engaged the attention of mathematicians, in any degree
proportionate to the utility of the research.   This desideratum
it has been my object to supply.   The train of considerations
pursued is sufficiently simple; and as they have been regulated
by a particular regard to the genius of arithmetic, and have been
carried to the utmost extent, the result seems to possess all the
harmony and simplicity that can be desired; and to unite to continuity and perfect accuracy, a degree of facility superior even to
that of the best popular methods.

Investigation of the Method.

3. In the general equation

$$\varphi x = 0$$

I assume $x = R+r+r'+r''+\ldots\ldots$
and preserve the binomial and continuous character of the operations, by making successively

$$x = R \ +z \ = R \ +r \ +z'$$
$$= R' \ +z' \ = R'+r'+z''$$
$$= R''+z'' = \&c.$$

in a useful discovery.   Useful he may certainly be allowed to call it, though the
produce of a purely mathematical speculation; for of all the investigations of
pure mathematics, the subject of *approximation* is that which comes most
directly, and most frequently into contact with the practical wants of the
calculator.

How far the manner in which he has been so fortunate as to contemplate it
has conduced, by the result, to satisfy those wants, it is not for him to determine; but his belief is, that both Arithmetic and Algebra have received some
degree of improvement, and a more intimate union.   The abruptness of
transition has been broken down into a gentle and uniform acclivity.

Where R^n represents the whole portion of x which has already been subjected to φ, and $z_x = r^x + z^{x'}$ the portion still excluded; but of which the part r^x is immediately ready for use, and is to be transferred from the side of z to that of R, so as to change φR^x to $\varphi R^{x'}$ without suspending the corrective process.

4. By TAYLOR's theorem, expressed in the more convenient manner of ARBOGAST, we have

$$\varphi x = \varphi(R+z) =$$
$$\varphi R + D\varphi R \cdot z + D^2\varphi R \cdot z^2 + D^3\varphi R \cdot z^3 + \ldots\ldots$$

Where by $D^n\varphi R$ is to be understood $\dfrac{d^n\varphi R}{1.2\ldots\ldots n.dR^n}$, viz. the nth derivee with its proper denominator; or, that function which ARBOGAST calls the *derivée divisée*, and distinguishes by a c subscribed. Having no occasion to refer to any other form of the derivative functions, I drop the distinctive symbol for the sake of convenience. Occasionally these derivees will be represented by a, b, c, &c.

5. Supposing φR and its derivees to be known, the mode of valuing $\varphi R'$ or $\varphi(R+r)$ is obvious. We have only to say in the manner of LAGRANGE, when preparing to develope his Theory of Functions,

$$\varphi R' = \varphi R + Ar$$
$$A = D\varphi R + Br$$
$$B = D^2\varphi R + Cr$$
$$C = D^3\varphi R + Dr$$

$$\cdot \; \cdot \; \cdot \; \cdot \; \cdot \; \cdot \; \cdot \; \cdot$$

$$V = D^{n-2}\varphi R + Ur$$
$$U = D^{n-1}\varphi R + r \; \cdot \; \cdot \; \cdot \; \cdot \; \cdot \; [I.]$$

Taking these operations in reverse order, we ascend with rapidity to the value of $\varphi(R+r)$ or $\varphi R'$.

6. The next point is, to apply a similar principle to discover the value of $\varphi(R+r+r') = \varphi(R'+r') = \varphi R''$. We here have

$$\varphi R'' = \varphi R' + A'r'$$
$$A' = D\varphi R' + B'r'$$
$$B' = D^2\varphi R' + C'r'$$
$$C' = D^3\varphi R' + D'r'$$

$$\cdot \; \cdot \; \cdot \; \cdot \; \cdot \; \cdot \; \cdot \; \cdot$$

$$V' = D^{n-2}\varphi R' + U'r'$$
$$U' = D^{n-1}\varphi R' + r'$$

But the former operation determined $\varphi R'$ only, without giving the value of any of the derived functions. The very simple scale of known quantities, therefore, by which we advance so rapidly in the first process, fails in those which follow.

7. Still we can reduce these formulæ to known terms; for since we have in general

$$D^r D^s \varphi\alpha = \frac{r+1 \cdot r+2 \ldots\ldots r+s}{1 \quad 2 \qquad s} D^{r+s}\varphi\alpha$$

(See ARBOGAST, § 137); by applying a similar reduction to the successive terms in the developement of $D^m\varphi R' = D^m\varphi(R+r)$, we obtain [1]

$$D^m\varphi R' = D^m\varphi R + \frac{m+1}{1}D^{m+1}\varphi R \cdot r + \frac{m+1}{1} \cdot \frac{m+2}{2}D^{m+2}\varphi R \cdot r^2$$
$$+ \frac{m+1\ldots\ldots m+3}{1 \qquad 3}D^{m+3}\varphi R \cdot r^3 + \&c.$$

And it is manifest that this expression may be reduced to a form somewhat more simple, and at the same time be accommodated to our principle of successive derivation, by introducing the letters A, B, C, &c. instead of the functional expressions.

8. As a general example, let

$$M = D^m \quad \varphi R + Nr$$
$$N = D^{m+1}\varphi R + Pr$$
$$P = D^{m+2}\varphi R + Qr$$
$$\cdots\cdots\cdots\cdots$$

represent any successive steps in the series in Art. 5; then are

$$D^m \quad \varphi R = M - Nr$$
$$D^{m+1}\varphi R = N - Pr$$
$$D^{m+2}\varphi R = P - Qr$$
$$\cdots\cdots\cdots\cdots$$

---

[1] This theorem, of which that in Art. 4 is a particular case [$m = o$], has been long in use under a more or less restricted enunciation, in aid of the transformation of equations. HALLEY's *Speculum Analyticum*, NEWTON's limiting equations, and the formulæ in SIMPSON's Algebra (ed. 5, p. 166, *circa fin.*) are instances. In a form still more circumscribed [$r = 1$, $R = o$, 1, 2, &c.] it constitutes the *Nouvelle Methode* of BUDAN; which has been deservedly characterized by LAGRANGE as simple and elegant. To a purpose which will be noticed hereafter, it applies very happily; but regarded as an instrument of approximation, its extremely slow operation renders it perfectly nugatory: and as LEGENDRE justly reported, and these remarks prove, it has not the merit of originality.

And by substituting these equivalents in the developement just enounced, it becomes

$$D^m_\varphi R' = M + mNr + \frac{m \cdot m+1}{1 \quad 2}Pr^2 + \frac{m...m+2}{1....3}Qr^3 + \&c.$$

9. With this advantage, we may now return to the process of Art. 6, which becomes

$$\varphi R'' = \varphi R' + A'r'$$
$$A' = (A + Br + Cr^2 + Dr^3 + Er^4 + \&c.) + B'r'$$
$$B' = (B + 2Cr + 3Dr^2 + 4Er^3 + \&c.) + C'r'$$
$$C' = (C + 3Dr + 6Er^2 + \&c.) + D'r'$$

$$V' = (V + \frac{n-2}{1}Ur + \frac{n-2 \cdot n-1}{1 \cdot 2}r^2) + U'r'$$

$$U' = (U + \frac{n-1}{1}r) + r' \qquad \cdot \qquad \cdot \qquad \cdot \qquad \cdot \qquad [\text{II.}]$$

Taking these operations in reverse order as before, by determining U', V'....C', B', A', we ascend to the value of $\varphi R''$.

10. In this theorem, the principle of successive derivation already discovers all its efficacy; for it is obvious that the next functions U'', V''.....C'', B'', A'', $\varphi R'''$, flow from the substitution of A', B', C',....V', U', $\varphi R''$, r', r'', for A, B, C....V, U, $\varphi R'$, r, r', in these formulæ; and from these U''', V''', &c.; and so on to any desirable extent. In this respect, Theorem II, algebraically considered, perfectly answers the end proposed in Art. 2.

11. We perceive also, that some advance has been made toward arithmetical facility; for all the figurate coefficients here employed are lower by one order than those which naturally occur in transforming an equation of the nth degree. But it is much to be wished, that these coefficients could be entirely dispensed with. Were this object effected, no multipliers would remain, except the successive corrections of the root, and the operations would thus arrange themselves, in point of simplicity, in the same class as those of division and the square root.

12. Nor will this end appear unattainable, if we recur to the known properties of figurate numbers; which present to our view, as *equivalent to the nth term of the mth series:*

1. *The difference of the nth and $n-1^{th}$ term of the $m+1^{th}$ series.*

2. *The sum of the first n terms of the $m-1^{th}$ series.*

3. *The sum of the nth term of the $m-1^{th}$, and the $n-1^{th}$ term of the mth series.*

The depression already attained has resulted from the first of these properties, and a slight effort of reflection will convince us that the second may immediately be called to our aid.

13. For this purpose, let the results of Art. 9 be expressed by the following notation:

$$\varphi R'' = \varphi R' + A'r'$$
$$A' = A_1 + B'r'$$
$$B' = B_2 + C'r'$$
$$C' = C_3 + D'r'$$
$$\cdot \quad \cdot \quad \cdot \quad \cdot$$
$$V' = V_{n-2} + U'r'$$
$$U' = U_{n-1} + r'$$

the exponents subjoined to any letter indicating the degree of the figurate coefficients in that formula of the theorem, of which such letter is the first term.

14. Although this statement appears only to have returned to us the conditions of Art. 6, with all their disadvantages, and to have merely substituted

$$A_1 \text{ for } D\varphi R' \text{ or } a'$$
$$B_2 \text{ for } D^2\varphi R' \text{ or } b'$$
$$C_3 \text{ for } D^3\varphi R' \text{ or } c'$$

&c. yet, by means of the property just alluded to, the essential data A, B, C, &c. which have disappeared, will again be extricated. For the developement of $D^m\varphi R'$, found in Art. 8, undergoes thereby the following analysis:

$$M + mNr + \frac{m \cdot m+1}{1 \cdot 2}Pr^2 + \frac{m \cdot m+1 \cdot m+2}{1 \cdot 2 \cdot 3}Qr^3 + \dots$$
$$= M + Nr + Pr^2 + Qr^3 + \dots$$
$$+ Nr + 2Pr^2 + 3Qr^3 + \dots$$
$$+ Nr + 3Pr^2 + 6Qr^3 + \dots$$
$$\cdot \quad \cdot \quad \cdot \quad \cdot \quad \cdot \quad \cdot \quad \cdot$$
$$+ Nr + mPr^3 + \frac{m \cdot m+1}{1 \cdot 2}Qr^3 + \dots$$

which equivalence will be thus expressed:

$$M_m = M + N_1r + N_2r + N_3r + \dots + N_mr$$

Returning therefore once more to our theorem, we now have

$$\varphi R'' = \varphi R' + A'r'$$
$$A = (A + B_1 r) + B'r'$$
$$B' = (B + C_1 r + C_2 r) + C'r'$$
$$C' = (C + D_1 r + D_2 r + D_3 r) + D'r'$$

$$\cdot \quad \cdot \quad \cdot \quad \cdot \quad \cdot \quad \cdot \quad \cdot$$

$$V' = (V + U_1 r + U_2 r + U_3 r + \ldots\ldots U_{n-1} r) + U'r'$$
$$U' = (U + \overline{n-1} \cdot r) + r'$$

15. This theorem employs exactly the same total number of addends as Theorem II, but with the important improvement, that the number of addends to each derivee is inversely as their magnitude, contrary to what happened before. Figurate multipliers are also excluded. And it is easy to convince ourselves that no embarrassment will arise from the newly introduced functions. For if we expand any of the addends $N_k r$ in the general formula equivalent to M_m, and analyze it by means of the *third property* of figurate series, we shall find

$$M_k r = N_{k-1} r + P_k rr.$$

And since we take the scale in our Theorem in a reverse or ascending order, this formula merely instructs us to multiply an addend already determined by r, and to add the product to another known addend; and if we trace its effect through all the descending scale, to the first operations, we observe that the addends to the last derivee, from which the work begins, are simply r repeated $n-1$ times.

16. Because $N_o = N$, the addend exterior to the parenthesis, might for the sake of uniformity be written $N_o'r'$. The harmony of the whole scheme would then be more completely displayed. To render the simplicity of it equally perfect, we may reflect that as the factors r, r', &c. are engaged in no other manner than has just been stated, viz. in effecting the subordinate derivations, their appearance among the principal ones is superfluous, and tends to create embarrassment. Assume therefore

$$_kN = N_k r,$$

and we have

$$\varphi R'' = \varphi R' + {}_0 A'$$
$$A' = (A + {}_1 B) + {}_0 B'$$

$$B' = (B +_1 C +_2 C) +_0 C'$$
$$C' = (C +_1 D +_2 D +_3 D) +_0 D'$$

.   .   .   .

$$V' = (V +_1 U +_2 U +_3 U + \ldots\ldots_{n-2} U) +_0 U'$$
$$U' = (U + \overline{n-1} \cdot r) + r' \qquad . \qquad . \qquad . \qquad \text{[III]}$$

the subordinate derivations being understood.

17. The Theorems hitherto give only the synthesis of φx, when $x = R + r + r' + \&c.$ is known. To adapt them to the inverse or analytical process, we have only to subtract each side of the first equation from the value of φx; then assuming $\varphi x - \varphi R^x = \Delta^x$, we have

$$\Delta' = \Delta -_0 A$$
$$A = a +_0 B$$

&c. as in Theorem I.

$$\Delta'' = \Delta' -_0 A'$$
$$A' = (A +_1 B) +_0 B'$$

&c. as in Theorem II. or III.

The successive invention of R, r, r', &c. will be explained among the numerical details. In the mean time, let it be observed that these results equally apply to the popular formula $\varphi x = $ constant, as to $\varphi x = 0$.

18. I shall close this investigation, by exhibiting the whole chain of derivation in a tabular form. The calculator will then perceive, that the algebraic composition of the addends no longer requires his attention. He is at liberty to regard the characters by which they are represented, in the light of mere corresponding symbols, whose origin is fully explained at their first occurrence in the table, and their ultimate application at the second. The operations included in the parentheses may be mentally effected, whenever r is a simple digit. And lastly, the vertical arrangement of the addends adapts them at once to the purposes of arithmetic, on every scale of notation.

General Synopsis.

	n—1st Der. / n—2th Derivee.	n—3th Derivee.	3rd. Derivee.	2nd. Derivee.	1st. Derivee.	Synthesis.	Anal. Root.
	u v	t	c	b	a	φR	$\triangle$ R+r.....
r	r $(Ur=)$ $_0U$	$(_rV=)$ $_0V$	$(Dr=)$ $_0D$	$(Cr=)$ $_0C$	$(Br=)$ $_0B$	$(Ar=)$ $_0A$	$\triangle$ $-_0A$
$\overline{n-1}.r$	U V $(_0U+r^2=)\,_1U$ $(_1U+r^2=)\,_2U$ &c. to $_{n-2}U$	$(_0V+_1Ur=)\,_1V$ $(_1V+_2Ur=)\,_2V$ &c. to $_{n-3}V$	$(_0D+_1Er=)\,_1D$ $(_1D+_2Er=)\,_2D$ $(_2D+_3Er=)\,_3D$	$(_0C+_1Dr=)\,_1C$ $(_1C+_2Dr=)\,_2C$	$(_0B+_1Cr=)\,_1B'$	φR	$\triangle'$
					$(B'r'=)$ $_0B'$	$(A'r'=)$ $_0A'$	$-_0A'$
r'	r' $(U'r'=)$ $_0U'$	$(V'r'=)$ $_0V'$	$(D'r'=)$ $_0D'$	$(C'r'=)$ $_0C'$	A'	$\varphi R''$	$\triangle''$
$\overline{n-1}.r'$	U' V' $(_0U'+r'^2=)\,_1U$	$(_0V+_1U'r'=)\,_1V'$	$(_0D'+_1E'r'=)\,_1D'$	$(_0C'+_1D'r'=)\,_1C'$	$(_0B'+_1C'r'=)\,_1B'$	$(A''r''=)$ $_0A''$	$-_0A''$
&c.	&c.	&c.	&c.	&c.	&c.	&c.	&c.

Illustrations.

19. The remarks which are yet to be adduced will bear almost exclusively on the Analytic portion of the Theorem, from which the Synthetic differs only in the less intricate management of the first derivee; this function having no concern with the discovery of the root, and its multiple being additive like all the rest, instead of subtractive.

From the unrestricted nature of the notation employed, it is evident that no class of equations, whether finite, irrational or transcendental, is excluded from our design. In this respect indeed, the new method agrees with the established popular methods of approximation; a circumstance in favour of the latter, which is overlooked by many algebraists, both in employing those methods, and in comparing them with processes pretending to superior accuracy. The radical feature which distinguishes them from ours is this: they forego the influence of all the derivees, excepting the first and perhaps the second; ours provides for the effectual action of all.

20. Concerning these *derivées* little need be said, as their nature and properties are well known. It is sufficient to state that they may be contemplated either as differential coefficients, as the limiting equations of NEWTON, or as the numerical coefficients of the transformed equation in R+z. This last elementary view will suffice for determining them, in most of the cases to which the popular solutions are adequate; viz. in finite equations where R, an unambiguous limiting value of x, is readily to be conjectured. When perplexity arises in consequence of some roots being imaginary, or differing by small quantities,[1] the second notation must be called in aid. The first, in general, when φx is irrational or transcendental.

21. The fact just stated, namely, that our theorem contains within itself the requisite conditions for investigating the limits, or presumptive impossibility, of the roots, demonstrates its sufficiency for effecting the developement of the real roots, independently of any previous knowledge of R. For this purpose, we might assume R=o; r, r, &c.=1 or .1 &c. and adopt, as most suitable to these conditions, the algorithm of Theorem II, until we had arrived at R^z, an unambiguous limiting value of x. But

---

[1] [Horner did not show how to separate two nearly equal roots. He elaborated this discussion in his second paper published in Leybourn's *Repository*. Vol. V, part II, London, 1830, pp. 21–75.]

since these initiatory researches seem more naturally to depend on the simple derivees, *a*, *b*, &c. than on A, B, &c. their aggregates; and since, in fact, as long as *r* is assumptive or independent of R, our system of derivation offers no peculiar advantage; I should prefer applying the limiting formulæ in the usual way; passing however from column to column (Wood, § 318.) of the results, at first by means of the neat algorithm suggested in the note on Art. 7, and afterwards by differencing, &c. as recommended by Lagrange, (*Res. des Eq. Num.* § 13), when the number of columns has exceeded the dimensions of the equation. (Vide Addendum.)

If, during this process the observation of De Gua be kept in view, that whenever all the roots of φx are real, $D^{m-1}\varphi x$ and $D^{m+1}\varphi x$ will have contrary signs when $D^m \phi x$ is made to vanish, we shall seldom be under the necessity of resorting to more recondite criteria of impossibility. Every column in which *o* appears between results affected with like signs, will apprize us of a distinct pair of imaginary roots; and even a horizontal change of signs, occurring between two horizontal permanences of an identical sign, will induce a suspicion, which it will in general be easy, in regard of the existing case, either to confirm or to overthrow.

22. The facilities here brought into a focus, constitute, I believe, a perfectly novel combination; and which, on that account, as well as on account of its natural affinity to our own principles, and still more on account of the extreme degree of simplicity it confers on the practical investigation of limits, appears to merit the illustration of one or two familiar examples.

Ex. 1. Has the equation $x^4 - 4x^3 + 8x^2 - 16x + 20 = 0$ any real root?—See Euler, C. D. p. 678.

$x =$	0	1	2
	20	9	4
	−16	−8	0
	8	2	8
	−4	0	4
	.1	1	1

Here the first column consists of the given co-efficients taken in reverse order. In the second, 9 is = the sum of the first column, -8 is $= -16 + 2(8) + 3(-4) + 4(1)$, 2 is $= 8 + 3(-4) + 6(1)$, &c. The third column is formed from the second, by the same easy process. We need proceed no farther; for the sequences 2, 0, 1 in the second column, and 4, 0, 8 in the third, show that the equation has two pairs of imaginary roots. Consequently it has no real root.

Ex. 2. To determine the nearest distinct limits of the positive roots of $x^3 - 7x + 7 = 0$. See LAGRANGE, *Res. des E. N.* § 27, and note 4. § 8.

Operating as in the former example, we have

$x =$	0	1	2
	7	1	1
	-7	-4	5
	0	3	6
	1	1	1

Since all the signs are now positive, 2 is greater than any of the positive roots. Again, between -4 and $+5$, it is manifest, that o will occur as a value of the first derivee, and that the simultaneous value of the second derivee will be affirmative. But as the principal result has evidently converged and subsequently diverged again in this interval, no conclusion relative to the simultaneous sign of that result can be immediately drawn. We will return to complete the transformations.

For $x =$	1.0	1.1	1.2	1.3	1.4	1.5	1.6	1.7
	1000	631	328	97	-56	-125	-104	13
	-400	-337	-268					
	30	33	36					
	1	1	1					

Here the first column was formed from that under $x = 1$, by annexing ciphers according to the dimensions of the functions; the 2nd and 3rd columns and the number 97 were found as in the former Example; the remaining numbers by differencing and extending the series 1000, 631, 328, 97. We have no need to continue the work, since the changes of signs in the principal results indicate the first digits of the roots in question to be 1.3 and 1.6. But if we proceed by farther differencing to complete all the lines, the columns standing under these numbers will give the co-efficients of $\varphi(1.3 + z)$ and $\varphi(1.6 + z)$ without farther trouble.

23. Assuming, then, that R has been determined, and R+z substituted for x in the proposed equation, thereby transforming it to

$$\Delta = az + bz^2 + cz^3 + dz^4 + \ldots\ldots\ldots$$

it is to this latter equation that the analytical part of our theorem is more immediately adapted. Now the slightest degree of reflection will evince, that our method is absolutely identical for all equations of the same order, whether they be binomial or adfected, as soon as the transformation in R has been accomplished. The following description, therefore, of a familiar process in arithmetic, will convey an accurate general idea of our more extensive calculus, and obviate the necessity of any formal precepts.

In EVOLUTION, the first step is unique, and if not assisted by an effort of memory, could only be tentative. The whole subsequent process may be defined, *division by a variable divisor.* ·For an accurate illustration of this idea, as discoverable in the existing practice of arithmeticians, we cannot however refer to the mode of extracting any root, except that of the square; and to this, only in its most recently improved state. Here, in passing from one divisor to another, *two additive corrections* are introduced; the first depending on the last correction of the root, the second on the correction actually making. And this *new quotient correction* of the root, since it must exist previously to the completion of the divisor by which it is to be verified, *is required to be found by means of the incomplete divisor*; and may be taken out, either to one digit only, as is most usual, or to a number of digits equal to that which the complete and incomplete divisors possess in common. And farther, as these *divisors* may not, *in the first* instance, agree accurately even in a single digit, it is necessary at that stage of the operation, mentally to anticipate the effect of the new quotient, so as to obtain a sufficiently correct idea of the magnitude of the new divisor.

24. This is an accurate statement of the relation which the column headed by the first derivee bears to the analysis. The remaining columns contribute their aid, as successively subsidiary to each other; the contributions commencing with the last or $n-1^{th}$ derivee, and being conveyed to the first through a regular system of *preparatory addends* dependent on the last quotient-correction, and of *closing addends* dependent on the new one. The *overt and registered* manner of conducting the whole calculation, enables us to derive important advantage from *anticipated corrections* of the divisors, not only at the first step, but, if requisite, through the whole performance, and also, without the necessity of a minute's bye-calculation, communicates, with the result, its *verification*.

25. Let us trace the operation of the theorem as far as may be requisite, through the ascending scale of equations.

1. In *Simple equations*, the reduced equation may be represented by $\Delta = az$; whence $z = \dfrac{\Delta}{a}$. Now the theorem directs us to proceed thus:

$$
a \qquad\qquad \Delta(r + r' + \ldots
$$
$$
\frac{-ar}{\Delta'}
$$
$$
\frac{-ar'}{\Delta''}
$$
$$
\frac{-ar'}{\Delta'''}
$$
$$
\&c.
$$

precisely the common arithmetical process of division.

2. In *Quadratics*, we have $\Delta = az + z^2$, and proceed in this manner:

$$
1 \qquad a \qquad\qquad \Delta(r + r' + \ldots
$$
$$
\frac{r}{A} \qquad\qquad \frac{-Ar}{\Delta'}
$$
$$
r \qquad\qquad \frac{-A'r'}{\Delta''}
$$
$$
\frac{r'}{A'} \qquad\qquad \&c.
$$
$$
\&c.
$$

the known arithmetical process for extracting the square root.

3. At *Cubic equations*, the aberration of the old practice of evolution commences, and our theorem places us at once on new ground. We have here

$$
\Delta = az + bz^2 + z^3
$$

and must proceed thus:

$$
1 \qquad b \qquad\qquad a \qquad\qquad \Delta(r + r + \ldots
$$
$$
\frac{r}{B} \qquad Br = {}_0B \qquad \frac{-Ar}{\Delta'}
$$
$$
2r \qquad\quad \frac{{}_0B + r^2 = {}_1B}{A} \qquad \frac{-A'r'}{\Delta''}
$$
$$
\frac{r'}{B'} \qquad\qquad \frac{B'r'}{A'} \qquad\qquad \&c.
$$
$$
\&c. \qquad\qquad \&c.
$$

This *ought to be* the arithmetical practice of the cube root, as an example will prove.

. . .

Ex. I. *Extract the cube root of* 48228544.

Having distributed the number into tridigital periods as usual, we immediately perceive that the first figure of the root is $3 = R$. Consequently, the first subtrahend is $R^3 = 27$, the first derivee $3R^2 = 27$, the second $3R = 9$; the third ($=1$,) need not be written. Hence   . . .

```
                                           48228544(364
   9.                    27..              27
   6  ··················· 576              ——
  ——                     ——               21228
  96····                 3276··············19656
  12.                    612..             ——
   4  ·················· 4336              1572544
  ——                     ——    ···········1572544
 1084····                393136····
```

In this example the reader will perceive that no supplementary operations are concealed. The work before him is complete, and may be verified mentally. I need not intimate *how much more concise* it is *than even the abbreviated statement of the old process.* (See HUTTON's *Course.*)

The station of 1, 2, &c. numeral places respectively, which the closing addends occupy in advance of the preparatory ones, is an obvious consequence of combining the numeral relation of the successive root-figures with the potential relation of the successive derivees. In fact, as is usual in arithmetic, we tacitly regard the last root-figure as units, and the new one as a decimal fraction; then the common rules of decimal addition and multiplication regulate the vertical alineation of the addends.

26. The advantage of mental verification is common to the solution of equations of every order, provided the successive corrections of the root be simple digits: for the parenthetic derivations will, in that case, consist of multiplying a given number by a digit, and adding the successive digital products to the corresponding digits of another given number; all which may readily be done without writing a figure intermediate to these given numbers and the combined result. For this reason the procedure by single digits appears generally preferable.

Nevertheless, to assist the reader in forming his own option, and at the same time to institute a comparison with known methods on their own grounds, I introduce one example illustrative of the advantage which arises from the anticipatory correction of the divisors spoken of in Art. 24, when the object is to secure a high degree of convergency by as few approximations as possible. The example is that by which NEWTON elucidates his method. I premise as the depreciators of NEWTON do, that it is an extremely easy problem; and I say this to invite comparison, not so much with his mode of treating it, as with theirs.

Ex. II. What is the value of x in the equation $x^3 - 2x = 5$.[1]

The root is manifestly a very little greater than 2. Make it $x = 2 + z$, and the equation becomes

$$1 = 10z + 6z^2 + z^3.$$

Hence, arranging the derivees,

$$6. \qquad\qquad 10.. \qquad\qquad \overset{\text{.}\quad\text{.}}{1.000(}$$
$$6$$

The first digit will obviously be so nearly 1, that by anticipating its effect on the divisor, we are sure this will be very nearly 106. Hence

$$10.6)1.000(.094 \text{ first correction}$$

The square is $94^2 = 8836$.
Hence we have

$$
\begin{array}{lll}
6... & 10...... & \overset{\text{.}\quad\text{.}\quad\text{.}\quad\text{.}}{1.000000000(.094} \\
094 & \quad\quad 572836 & \quad\quad 993846584 \\
\overline{6094} \times 94 = & \overline{10572836} & \overline{\quad 6153416} \\
188 & 581672 & \\
& 3 &
\end{array}
$$

The first digit of the next correction will evidently be 5; the effect of which we have as before anticipated as far as one digit. The divisor will therefore be 11158 correct to the last figure. Hence

$$11158)6153416(55148, \text{ second correction.}$$

The square is 30413, &c. to 10 digits.

---

[1] [The equation $x^3 - 2x - 5 = 0$ is Newton's classic example, also used by Ruffini.]

Hence,

6094	10572836	
18855148	581672	
$\overline{628255148}\times 5$ &c. =	·············34647014901904	6153416
110296.	111579727014901204 ·····	615339878541781019
	34650056	1721458218981
	1	

Consequently,

$$1116143772)1721458218979(1542326590,22$$

This third correction is carried two places beyond the extent of the divisor, for the sake of ascertaining rigidly the degree of accuracy now attained. For this purpose, we proceed thus:

628 &c. $\times$ 154 &c. = ,968, &c. is the true correction of the last divisor. Our anticipated correction was 1,000. For which if we substitute 968 &c. it will appear that our divisor should have ended in 1,678, &c. instead of 2. The error is, ,322 &c. which induces an ultimate error of (111 &c. : 154 &c. ::, 322 &c. &c. :),44 &c.

Consequently, our third correction should be.... 1542326590,66, &c. agreeing to 10 figures with the value previously determined. And the root is

$$x = 2.094551481542326590, \text{ \&c.}$$

correct in the 18th decimal place at three approximations.

So rapid an advance is to be expected only under very favorable data. Yet this example clearly affixes to the new method, a character of unusual boldness and certainty; advantages derived from the overt manner of conducting the work, which thus contains its own proof.

The abbreviations used in the close of this example, are of a description sufficiently obvious and inartificial; but in order to perfect the algorithm of our method in its application to higher equations, and to the progress by simple digits, attention must be given to the following general principles of

Compendious Operation.[1]

28. From these principles we form the following conclusions, demonstrative of the facilities introduced by this improvement on the original process:

1. Whatever be the dimensions (n) of the proposed equation, whose root is to be determined to a certain number of places, only

---

[1] [These principles are stated in Article 27.]

$\frac{1}{n}$th part of that number (reckoning from the point at which the highest place of the closing addend begins to advance to the right of that of the first derivee) needs to be found by means of the process peculiar to the complete order of the equation; after which, $\frac{1}{n \cdot n-1}$ may be found by the process of the $n-1^{th}$ order, $\frac{1}{n-1 \cdot n-2}$ by that of the $n-2^{th}$ order, &c.

2. Several of these inferior processes will often be passed over *per saltum;* and when this advantage ceases, or does not occur, the higher the order of the process, the fewer will be the places determinable by it. And in every case, the latter half of the root will be found by division simply. Meantime, the number of figures employed in verification of each successive root-digit, instead of increasing, is rapidly diminishing.

3. The process with which we commence, need not be of a higher order than is indicated by the number of places to which we would extend the root; and may be even reduced to an order as much lower as we please, by means of an introductory approximation.

Ex. III. Let the root of the equation in Ex. II. be determined to the tenth place of decimals.

Arranging the derivees as before, we proceeded thus: [1]

---

[1] [Horner's arrangement differs from that of Ruffini in that the coefficients of the transformed equation appear in a diagonal line, while Ruffini arranges them in the extreme right hand column. The modern arrangement would begin as follows:

```
1    0   -2   -5  |2
    +2   +4   +4
1   +2   +2  |-1
    +2   +8       |
1   +4  |+10
    +2       |
1   +6       +10        -1          |.09
    + .09   +   .5481  + .949329
1   +6.09  +10.5481   |-  .050671
    + .09  +   .5562        |
1   +6.18 |+11.1043
    + .09      |
1   +6.27
```

etc.]

```
  6..              10....            1.000000( 0945514815
  ──          ──────────5481        ──────────949329
609······                ────
  184           105481····             50671000
  ──              5562..          ────────44517584
62|74·················25096
   |.8.              ────              6153416
  ──             11129396·····         5578825
|..|62|82          2511|2              ────
                   314|12              574591
                  ──────               558055
                 1115764|92            ────
                    31|4|1       1|1|1|6|1)16536(14815
                     3|1|4             11161
                    ──────             ────
                  111611|0|4            5375
                      |3|1              4465
                    ──────             ────
                  11161|4|1             910
                                        893
                                        ───
                                         17
                                         11
                                        ──
                                         6
                                         6
```

Consequently the root is 2.0945514815, correct to the proposed extent, as appears on comparing it with the more enlarged value already found. The work occupied a very few minutes, and may be verified by mere perusal, as not a figure was written besides those which appear. By a similar operation, in less than half an hour, I have verified the root to the whole extent found in Ex. II.[1]

. . . .

Ex. VI. If it were proposed to obtain a very accurate solution of an equation of very high dimensions, or of the irrational or transcendental kind, a plan similar to the following might be adopted. Suppose, for example, the root of

$$x^x = 100, \text{ or } x \log x = 2$$

were required correct to 60 decimal places. By an easy experiment we find $x = 3.6$ nearly; and thence, by a process of the *third* order, $x = 3.597285$ more accurately.

---

[1] [Examples IV and V, which have been omitted, show the extraction of roots of the equations $x^3 - 7x = -7$ and $x^5 + 2x^4 + 3x^3 + 4x^2 - 5x = 321$.]

Now, $3597286 = 98 \times 71 \times 47 \times 11$, whose logarithms, found to 61 decimals in Sharpe's Tables, give R log $R = 2.00000096658$, &c. correct to 7 figures; whence the subsequent functions need be taken out to 55 figures only. They are

$$a = \text{Mod} + \log R = .990269449408, \&c.$$
$$b = \text{Mod} \div 2R \quad = \ldots \ldots .0^7 60364, \&c.$$
$$c = -b \quad \div 3R \quad = \ldots \ldots -.0^{14}55, \&c.$$

&c. The significant part disappears after the 8th derivee; consequently, the process will at first be of the *eighth* order. If the root is now made to advance by single digits, the first of these will reduce the process to the *seventh* order; one more reduces it to the *sixth* order; two more, to the *fifth*, &c. The last 27 figures will be found by division alone.

But if the first additional correction is taken to 8 figures, and the second to 16, on the principle of Example II, we pass from the 8th order to the 4th at once, and thence to the 1st or mere division, which will give the remaining 29 figures. This mode appears in description to possess the greater simplicity, but is perhaps the more laborious.

It cannot fail to be observed, that in all these examples a great proportion of the whole labour of solution is expended on the comparatively small portion of the root, which is connected with the leading process. The toil attending this part of the solution, in examples similar in kind to the last, is very considerable; since every derivee is at this stage to receive its utmost digital extent. To obviate an unjust prejudice, I must therefore invite the reader's candid attention to the following particulars:

In all other methods the difficulty increases with the extent of the root, nearly through the whole work; in ours, it is in a great measure surmounted at the first step: in most others, there is a periodical recurrence to first conditions, under circumstances of accumulating inconvenience; in the new method, the given conditions affect the first derivees alone, and the remaining process is *arithmetically direct*, and increasingly easy to the end.

The question of practical facility may be decided by a very simple criterion; by comparing the *times* of calculation which I have specified, with a similar datum by Dr. Halley in favor of his own favorite method of approximation. (Philosophical Transactions for 1694.)

Addendum I. (*Vide* Art. 21.) *Note.* But in this case, it will be more elegant to find the differences at once by the theorem

$$\Delta^{t+1}D^m\varphi R' = \frac{m+1}{1}\Delta^t D^{m+1}\varphi R.r + \frac{m+1}{1} \cdot \frac{m+2}{2}\Delta^t D^{m+2}\varphi R.r + \quad \&c.$$

which, supposing r to be constant, is a sufficiently obvious corollary to the theorem in Art. 7. All the results may then be derived from the first column by addition. Thus, for the latter transformations in Ex. II. Art. 22, the preparatory operation would be

1st. Terms.	Diff. 1st.	2nd.	3rd.
1000	−369	66	6
−400	63	6	
30	3		
1			

and the succeeding terms would be found by adding these differences in the usual way to the respective first terms.

Addendum II. It is with pleasure that I refer to the Imperial Encyclopædia (Art. Arithmetic) for an improved method of extracting the cube root, which should have been noticed in the proper place, had I been aware of its existence; but it was pointed out to me, for the first time, by the discoverer, Mr. EXLEY, of Bristol, after this Essay was completed. It agrees in substance with the method deduced in Art. 25, from my general principle, and affords an additional illustration of the affinity between that principle and the most improved processes of common arithmetic.

ROLLE'S THEOREM

(Translated from the French by Professor Florian Cajori, University of California, Berkeley, Calif.)

Writers on the history of mathematics of the early part of the present century did not know where in the writings of Michel Rolle the theorem named after him could be found—the theorem according to which $f'(x) = 0$ has at least one real root lying between two consecutive real roots of $f(x) = 0$. One historian went so far as to express the opinion that the theorem is wrongly attributed to Rolle. Finally, in 1910, the theorem was found in a little-known book of Rolle, entitled, *Démonstration d'une Methode pour résoudre les Egalitez de tous les degrez; suivie de deux autres Méthodes, dont la première donne les moyens de resoudre ces mêmes égalitez par la Geometrie, et la seconde, pour resoudre plusieurs questions de Diophante qui n'ont pas encore esté resoluës.* A Paris, Chez Jean Cusson, ruë Saint Jacques, à l'Image de Saint Jean Baptiste. M.DC.XCI. (pp. 128).[1] Copies of this book are in the "Bibliothèque Nationale," in the "Bibliothèque de L'Arsenal," and in the "Bibliothèque de l'Institut de France," in Paris. In this treatise the theorem in question is established only incidentally, in Rolle's demonstration of the "method of cascades" for the approximation to the roots of numerical equations.

Nowhere in his *Démonstration*, nor in his *Traité d'algebre*, a widely read book published at Paris a year earlier (1690), is there given a formal definition of a "cascade." But it is implied in what Rolle states that, if in an equation $f(x) = 0$, $f(x)$ is "multiplied by a progression," the result when simplified and equated to zero is a "cascade." He prefers to use the progression 0, 1, 2, 3, ...Then, after multiplying each term of an equation by the corresponding term of the progression, he divides the resulting expression by x and equates the quotient to zero. Thus, multiplying the terms of $a + bz + cz^2 + ...$ by the respective terms of 0, 1, 2,..., he obtains $bz + 2cz^2 + ...$; dividing this by z and equating to zero, he arrives at the first or proximate "cascade," $b + 2cz + ... = 0$. It will be seen that this result is the first derivative of the initial expression, equated to zero.

Rolle's "method of cascades" is given in his *Traité d'algebre*, without sufficient proof. To meet this criticism leveled against it, Rolle wrote the *Démonstration*. In both treatises Rolle used certain technical terms which we must explain. Complex roots of an equation, as well as all but one root of each multiple root, are called "racines défaillantes." We shall translate this phrase by "imaginary roots and multiple roots." Roots which are not "défaillantes" he calls "racines effectives;" we shall translate this by "real and distinct roots." Another term used by Rolle is "hypotheses" or "limits" of the roots. If two numbers a and b are substituted for z in $f(z)$, and $f(a)$ and

[1] See an historical article on Rolle's theorem in *Bibliotheca Mathematica*, 3rd. S., Vol. 11. pp. 300–313.

$f(b)$ have opposite signs, then between a and b there is a root of $f(z) = 0$, and a and b are called "limits" (hypotheses) of the roots.

In the extracts given below, it will be seen that Rolle's theorem is proved in Article IX for the case when the roots of the equation are all positive, and in Artile XI when the roots may be any real or complex numbers.[1]

Before applying to a given equation his "method of cascades," Rolle transforms the equation so that the coefficient of the highest power of the unknown is unity and all real roots are positive. When this is achieved he calls the equation "prepared." In the reasoning which follows the equations are assumed to be "prepared."

Before making quotations from Rolle's *Démonstration*, it is desirable to give an example of his "method of cascades." To find an upper limit of the real roots, he takes the numerically largest negative coefficient $-g$, divides g by the coefficient of the highest power of the unknown, and then adds 1 to the quotient and enough more to get a positive integer; this result is his upper limit. Given the limits 0, 6, 13 of an equation $f(v) \equiv 6v^2 - 72v + 198 = 0$, Rolle approximates to the root between, say, 6 and 13 in this manner: The mean of 6 and 13 is $9\frac{1}{2}$. By substitution of 6 and 9 in $f(v)$, opposite signs are obtained. Hence 6 and 9 are closer limits. Repeating this process yields the limits 6 and 8, and finally 7 and 8. Take 7 as the approximate root.

The "method of cascades" is illustrated by the following quotation from Rolle's *Traité d'algebre*, 1790, p. 133:

Take the equation $v^4 - 24v^3 + 198vv - 648v + 473 \backsim \theta$,[2] and the first rule [rule for finding cascades] gives

$$4v - 24 \backsim \theta$$
$$6vv - 72v + 198 \backsim \theta$$
$$4v^3 - 72vv + 396v - 648 \backsim \theta$$
$$v^4 - 24v^3 + 198vv - 648v + 473 \backsim \theta$$

In the first cascade one finds $v \backsim 6$; then the second has $\theta . 6 . 13$ for limits; and by the means of these limits one finds 4 and 7 as approximate roots of the second cascade. If one regards these approximate roots as veritable roots, they may be taken as intermediate limits of the next cascade. Accordingly the limits of the third cascade are $\theta . 4 . 7 . 163$, by which one discovers that $3 . 6 . 9$ are three roots of this third cascade. Consequently, the fourth cascade has as limits $\theta . 3 . 6 . 9 . 649$. With the aid of these one finds that unity is an exact root of the proposed equation and that $6 . 8 . 10$. are approximate roots."

---

[1] The first occurence of the name "Rolle's Theorem" appears to be in the writings of the Italian Mathematician Giusto Bellavitis. He used the expression "teorema del Rolle" in 1846 in the *Memorie dell' I. R. Istituto Veneto di Scienze, Lettere ed Arte*, Vol. III (reprint), p. 46, and again in 1860 in Vol. 9, § 14, p. 187.

[2] [Rolle uses the small Greek letter θ as the symbol for zero. See F. Cajori, *History of Mathematical Notations* (1928), Vol. I, §82. Rolle expresses equality by the sign $\backsim$ used by Descartes. See F. Cajori, *op. cit.*, §191.]

The first five articles in the *Démonstration* refer to elementary matters which it is not necessary to reproduce here. In our quotation we begin with article VI.

Article VI.—Take in order [of magnitude] any number of roots which are positive and different from one another, such as 3, 7, 12, 20, and form equations containing them, such as[1]

$$z - 3 \cdot z - 7 \cdot z - 12 \cdot z - 20 \cdot \text{etc.}$$

This done, in the order shown here, it is evident, that if one substitutes θ in place of z, or else a number smaller than the first root, the results [the resulting factors] are all negative; that if one substitutes a number greater than the first root and less than the others, the results are all negative except one; if one substitutes a number greater than the first two roots and less than the others, the results are all negative except two; and so on. But if one limits the number of roots, the substitution of a number which surpasses the greatest root will give + everywhere. This is clear. Therefrom it follows that if one multiplies together all the results obtained from the substitution of each number, so that there are as many respective products as there are numbers, these products will be alternatively positive and negative, or negative and positive...

Coroll. I.—It is clear that these numbers thus chosen, give by their substitution, a regular sequence of signs and thereby serve the purpose of limits of the roots.

Coroll. II.—It is evident also that if all the limits, except the first and last, are not placed singly between the roots, so to speak, the regular sequence of the signs will be broken.

Coroll. III.—It is likewise clear that the roots [all positive and distinct] are numbers placed singly between these limits, and consequently, if the roots are substituted in an equation whose roots are these limits, this substitution will yield results alternately positive and negative, or negative and positive. One sees this in the example,

$$y - 6 \cdot y - 21 \cdot y - 30. \quad \text{Roots of the equation,}[2]$$
$$y - \theta \cdot y - 12 \cdot y - 26. \quad \text{Roots of the cascade,}$$

---

[1] [The omission of parentheses as seen in Rolle is not infrequent in books of the seventeenth and eighteenth centuries. Rolle's notation is equivalent to $(z - 3)(z - 7)(z - 12)(z - 20)$ etc. See F. Cajori, *op. cit.*, Vol. I, §354.]

[2] [Rolle uses the term "equation" (égalité) even when the polynomial is not equated to zero, or the equality is not indicated symbolically.]

where it appears that 6 substituted for y at the cascade...gives the factors whose product is positive; that 21 gives factors whose product is negative, and that 30 gives factors whose product is positive; and consequently the roots 6, 21, 30, being each substituted for y in the cascade, in turn will give alternately + and − ...Thus the roots are limits[1] of their own limits [taken as roots]...

Coroll. VI.—If one is able to prove that the *Methode* gives necessarily the limits of all [distinct, positive] roots, it follows that there are imaginary and multiple roots when it does not yield limits. But to establish this truth, other principles are necessary.

Article VII.—If one takes each of the letters y and v to represent any number, all the arithmetical progressions which have only three terms are comprised in the following:

$$y \cdot y + v \cdot y + 2v.$$

This is unquestionable.

If one has any arithmetical progression and if one takes in that progression several successive terms, it is evident that these terms are in arithmetical progression. For example, if one has the progression $\theta . 1 . 2 . 3 . 4 . 5$. etc., and if one takes θ, 1, 2 or 1, 2, 3 or also 2, 3, 4, etc., it is clear that the terms in each are in arithmetical progression.

When I say that an equation is multiplied by a progression, it must be understood that the first term of the equation is multiplied by the first term of the progression; that the second term of the equation is multiplied by the second term of the progression, and so on. When the sum of these products is taken to be equal to θ, one says that this equation is generated by the progression.

Article VIII.—When the product of the two quantities $z - a$, $z - b$, is multiplied by the progression $y + 2v, y + v, y$, and b is substituted in place of the unknown in the product of the progression, the result of the substitution is measured by [*i. e.* will have the factor] $b - a$. Here is the proof:

$$\left. \begin{array}{l} ab - az + zz \\ - bz \end{array} \right\} \text{ Product of } z - a \text{ and } z - b.$$

$y \cdot y + v \cdot y + 2v$ The progression.

---

[1] [This is the first reference to what we now call "Rolle's theorem," restricted as yet to the case of equations all of whose roots are real and positive.]

Multiplying,

$$
\left.\begin{array}{l}
aby - ayz + yzz \\
\quad - byz + 2vzz \\
\quad - avz \\
\quad - bvz
\end{array}\right\} \text{Product which the progression gives.}
$$

Upon substituting b in place of z in the last product, one obtains $bbv - abv$, having the factor $b - a$, which was to be proved...

Coroll.—Having as above the quantity

$$
\begin{array}{l}
ab - az + zz: \\
\quad - bz
\end{array}
$$

If one multiplies it by z raised to any arbitrary power and if the product is multiplied by an arithmetical progression, it is clear that on substituting b for z in the product of the progression the result has the factor $b - a$...

Article IX.—Having as above, the given quantity,

$$
\begin{array}{l}
ab - az + zz, \\
\quad - bz
\end{array}
$$

if one multiplies it by $f + gz + bzz + rz^3 + nz^4$, and so on, so that the unknown z attains any given degree, I say that the partial products may always be disposed as follows:

$$
A\ldots \left.\begin{array}{l}
abf - afz + fzz \\
\quad - bfz
\end{array}\right\} \text{First product}
$$

$$
B\ldots \left.\begin{array}{l}
+ gabz - agzz + gz^3 \\
\quad - bgzz
\end{array}\right\} \text{Second product}
$$

$$
C\ldots \left.\begin{array}{l}
+ babzz - baz^3 + bz^4 \\
\quad - bbz^3
\end{array}\right\} \text{Third product}
$$

Prog. θ . 1   . 2   . 3   . 4 etc.

And so on to infinity, where one sees that each of the partial products which are marked by $A . B . C$. etc. is always measured by [i. e., has as a factor] the given quantity, since that quantity is one of the generators.

Coroll. I.—If the sum of the partial products is multiplied by the arithmetical progression θ . 1 . 2 . 3 . 4 . etc., each of the products $A . B . C$. etc. is also multiplied by the progression: that is to say, the product A by θ . 1 . 2, the product B by 1, 2, 3, and so on.

Observe for the understanding of what follows, that the product A when altered by the progression by which it is multiplied, is designated D; that the product B thus altered is designated E; that the product C after a similar change is marked F, etc.

Coroll. II.—From this first corollary and Articles VII and VIII, one may conclude that on substituting b for z in each of the products $D \cdot E \cdot F$. etc., each of the results is divisible by $b - a$ without a remainder. But substituting b for z in each of these products amounts to making the substitution in the total product. From this it is evident that after this substitution is made, the total product is measured by $b - a$.

Coroll. III.—If in place of the quantity $f + gz + bzz + rz^3 +$ etc. one takes the product of $z - c$, $z - d$, $z - e$, etc., one arrives at all the conclusions which have been reached; that is to say, after substituting b for z in the total product which the progression brings forth, the result is divisible by $b - a$ without a remainder. This is evident, since, as we see, f, g, b, r, etc. stand for any given quantities.

Coroll. IV.—It is also clear from the formation of the total product that all the letters a, b, c, d, etc. are on the same footing and all that has been established for b with respect to a, may be concluded for any of the letters with regard to any of the others. From this it follows that on substituting separately, in the total product of the progression, any of the letters a, b, c, d, e, etc. in place of the unknown z, the result will be divisible by the letter substituted less any of the others that we may wish. So that the result which the substitution of a gives, is divisible by $a - b$, by $a - c$, by $a - d$, etc.

In the same way, the substitution of c for z must yield a result, divisible without a remainder, by c minus any one of the other roots taken separately. Similarly for the others.

Coroll. V.—If one supposes...that the root a is greater than the root b, that b is greater than c, that c is greater than d, etc., it follows from Article V[1] and the preceeding corollary that the results [products] which give [*i. e.*, which limit] the roots of the proximate cascade, are alternately positive and negative or negative and positive.

Coroll. VI.—If an equation is formed as in Article I [all the roots being positive and distinct], its roots are the limits of the roots of the proximate cascade, for this cascade is derived by multiplying by the progression $\theta \cdot 1 \cdot 2$. etc. as in this Article IX. Moreover, these roots have the limitations imposed upon them in the preced-

---

[1] [Article V in the *Démonstration*, which we omitted, states that if the positive roots a, b, c, d,...are so related tht $a > b > c > d > ...$, then the products $(b - a)(b - c)(b - d)$.....and $(c - a)(c - b)(c - d)$...have opposite signs.]

ing Coroll. V. Hence it follows from this same corollary and Coroll. III of Article VI, that the roots are limits of the roots of its cascade.[1]

Coroll. VII.—Since the roots of equations thus formed are limits of the roots of their cascade, it follows from Coroll. VI. . . . that the roots of the proximate cascade are limits of the roots of the equation of which it is the cascade.

Since the progression [in Article IX] gives a cascade which is divisible by the unknown z, one sees that θ is one of the roots [of the cascade] and it is evident from this that θ is the lower limit [of the roots of the given equation], according to our suppositions.

If one substitutes the roots [of an equation] in its cascade before dividing it by z, the results are divisible by the letter substituted. But as this letter represents only some positive number, according to our assumptions, it does not bring about any change in the sequence of signs. . .

Coroll. VIII.—If the roots are irrational, the limits will give the regular sequence signs, on the supposition that the roots satisfy the conditions specified in Article I [*i. e.*, are positive and distinct], for these roots are determined by the equation which contains them, and the proximate cascade is formed from that equation.

Coroll. IX.—The roots being all positive and distinct, there are as many of them as the number indicating the degree of the equation which contains them. . .

Article X.—Let all the signs of an equation be alternating as the result of the "preparation" of equations, then it always transpires that the real and distinct roots are all positive; and one may prove this truth as shown in what follows.

Let all the powers of an unknown, such as x, which are arranged in order, have alternately the signs $-$ and $+$, as seen in $-x + xx - x^3 + x^4$ etc. From this it is clear that, if one substitutes a negative unknown in an equation the terms of which are alternating [in sign], such as $+q - pz + nzz - pz^3 +$ etc., it comes about that the signs of the resulting equation are all positive. And if the proposed one should be $-q + pz - nzz + rz^3$ etc., a transposition [of terms] after the substitution, produces the same effect. And reciprocally, a [complete] equation all the signs of which are positive, is changed into another in which all terms are alternating, when one substitutes a negative unknown in place of the unknown

---

[1] [This is another passage containing "Rolle's theorem." See also Coroll. VIII.]

of the equation. This is clear. It is clear also that an equation of which all the terms are positive can not have positive roots, for when such a root is substituted in the equation, the sum of the positive terms should destroy that of the negative ones, when as we suppose, all the terms are in the same member; which is impossible...

Article XI.—If some of the roots are real and distinct, the others imaginary, these imaginary ones do not prevent the limits from giving suitable signs to the real and distinct roots. For, the proposed equation may always be conceived to be formed by the multiplication of two simpler equations, the one having all roots real and distinct, the other having all roots imaginary, and by Article IX the cascade involves limits which agree with the real and distinct roots [of the given equation]. And one can see that the imaginary roots do not give rise to the sequence [of signs] which one finds in real and distinct roots...

Article XII.—There are at least as many imaginary roots in an equation as there are in its proximate cascade. For, if the roots of the equation which correspond to these imaginaries were real, it would follow that upon substituting them in the cascade they would give the regular sequence referred in Coroll. V of Article IX, while according to the definition of imaginary roots, they give when substituted always +. This is contrary to supposition.

If one does not take zero as one of the terms of a progression, and if this θ is not placed beneath the last term or beneath the first term of the equation, the proximate cascade will have the same degree as the equation itself. Thus the Method would suppose what is in question.[1] But taking zero for one of the extremes of the progression and marking this progression in general terms, the letter which serves in this general expression is found only of the first degree in each term of the cascade, and disappears in the ordinary cancellation. From this it follows that this progression produces no other effect on the limits than does θ . 1 . 2. This happens also when θ is placed under the last term, for the reasons just stated.

---

[1] [That is, the solution of the cascade equation presents the same problem as does the original equation.]

ABEL

ON THE QUINTIC EQUATION

(Translated from the French by Dr. W. H. Langdon, with Notes by Professor
Oystein Ore, Yale University, New Haven, Conn.)

The Norwegian mathematician, Niels Henrik Abel (1802–1829) very early
showed an unusual mathematical ability, and in spite of the fact that his short
life was a constant struggle against poverty and illness, he wrote a series of
scientific papers that secures him a position among the greatest mathematicians
of all time. In his "Mémoire sur les équations algébriques ou l'on démontre
l'impossibilité de la resolution de l'équation générale du cinquième degré *Œuvres
complètes*, (Vol. I, Christiania (Oslo) 1881, p. 28–33), Abel proves the impossibility of solving general equations of the fifth and higher degrees by means
of radicals. The paper was published as a pamphlet at Oslo in 1824 at Abel's
own expense. In order to save printing costs, he had to give the paper in a very
summary form, which in a few places affects the lucidity of his reasoning.

After the solutions of the third and fourth degrees had been found by
Cardano and Ferrari, the problem of solving the equation of the fifth degree
had been the object of innumerable futile attempts by the mathematicians of
the 17th and 18th centuries. Abel's paper shows clearly why these attempts
must fail, and opens the road to the modern theory of equations, including
group theory and the solution of equations by means of transcendental functions.

Abel proposed himself the problem of finding all equations solvable by
radicals, and succeeded in solving all equations with communtative groups,
now called Abelian equations. Among Abel's numerous other achievements
are his discovery of the elliptic functions and their fundamental properties,
his famous theorem on the integration of algebraic functions, theorems on
power series (see p. 286), where further biographical notes appear, etc.

A Memoir on Algebraic Equations, Proving the Impossibility of a Solution of the General Equation of the Fifth Degree

The mathematicians have been very much absorbed with finding
the general solution of algebraic equations, and several of them
have tried to prove the impossibility of it. However, if I am not
mistaken, they have not as yet succeeded. I therefore dare hope
that the mathematicians will receive this memoir with good will,
for its purpose is to fill this gap in the theory of algebraic equations.
Let

$$y^5 - ay^4 + by^3 - cy^2 + dy - e = 0$$

be the general equation of fifth degree and suppose that it can be
solved algebraically,—i. e., that y can be expressed as a function

of the quantities a, b, c, d, and e, composed of radicals. In this case, it is clear that y can be written in the form

$$y = p + p_1R^{\frac{1}{m}} + p_2R^{\frac{2}{m}} + \ldots + p_{m-1}R^{\frac{m-1}{m}},$$

m being a prime number, and R, p, p_1, p_2, etc. being functions of the same form as y. We can continue in this way until we reach rational functions of a, b, c, d, and e. We may also assume that $R^{\frac{1}{m}}$ cannot be expressed as a rational function of a, b, etc., p, p_1, p_2, etc., and substituting $\dfrac{R}{p_1^m}$ for R, it is obvious that we can make $p_1 = 1$.
Then

$$y = p + R^{\frac{1}{m}} + P_2R^{\frac{2}{m}} + \ldots + p_{m-1}R^{\frac{m-1}{m}}.$$

Substituting this value of y in the proposed equation, we obtain, on reducing, a result in the form

$$P = q + q_1R^{\frac{1}{m}} + q_2R^{\frac{2}{m}} + \ldots + q_{m-1}R^{\frac{m-1}{m}} = 0,$$

q, q_1, q_2, etc. being integral rational functions of a, b, c, d, e, p, p_2, etc. and R.

For this equation to be satisfied, it is necessary that $q = 0$, $q_1 = 0$, $q_2 = 0, \ldots q_{m-1} = 0$. In fact, letting $z = R^{\frac{1}{m}}$, we have the two equations

$$z^m - R = 0, \text{ and } q + q_1z + \ldots + q_{m-1}z^{m-1} = 0.$$

If now the quantities q, q_1, etc. are not equal to zero, these equations must necessarily have one or more common roots. If k is the number of these roots, we know that we can find an equation of degree k, whose roots are the k roots mentioned, and whose coefficients are rational functions of R, q, q_1, and q_{m-1}. Let this equation be

$$r + r_1z + r_2z^2 + \ldots + r_kz^k = 0.$$

It has all its roots in common with the equation $z^m - R = 0$; now all the roots of this equation are of the form $\alpha_\mu z$, α_μ being one of the roots of the equation $\alpha_\mu^m - 1 = 0$. On substituting, we obtain the following equations

$$r + r_1z + r_2z^2 + \ldots + r_kz^k = 0,$$
$$r + \alpha r_1z + \alpha^2 r_2z^2 + \ldots + \alpha^k r_kz^k = 0,$$
$$\cdots\cdots\cdots\cdots$$
$$r + \alpha_{k-2}r_1z + \alpha_{k-2}^2 r_2z^2 + \ldots + \alpha_{k-2}^k r_kz^k = 0.$$

From these k equations we can always find the value of z, expressed as a rational function of the quantities r, r_1, . . . r_k; and as these quantities are themselves rational functions of a, b, c, d, e, R, p, p_2, . . ., it follows that z is also a rational function of these latter quantities; but that is contrary to the hypotheses. Thus it is necessary that

$$q = 0, \; q_1 = 0, \ldots q_{m-1} = 0.$$

If now these equations are satisfied, it is clear that the proposed equation is satisfied by all those values which y assumes when $R^{\frac{1}{m}}$ is assigned the values

$$R^{\frac{1}{m}}, \; \alpha R^{\frac{1}{m}}, \; \alpha^2 R^{\frac{1}{m}}, \ldots, \; \alpha^{m-1} R^{\frac{1}{m}},$$

α being a root of the equation

$$\alpha^{m-1} + \alpha^{m-2} + \ldots + \alpha + 1 = 0.$$

We also note that all the values of y are different; for otherwise we should have an equation of the same form as the equation $P = 0$, and we have just seen that such an equation leads to a contradictory result. The number m cannot exceed 5. Letting y_1, y_2, y_3, y_4, and y_5 be the roots of the proposed equation, we have

$$y_1 = p + R^{\frac{1}{m}} + p_2 R^{\frac{2}{m}} + \ldots + p_{m-1} R^{\frac{m-1}{m}},$$

$$y_2 = p + \alpha R^{\frac{1}{m}} + \alpha^2 p R^{\frac{2}{m}} + \ldots + \alpha^{m-1} p_{m-1} R^{\frac{m-1}{m}},$$

$$\ldots \ldots \ldots$$

$$y_m = p + \alpha^{m-1} R^{\frac{1}{m}} + \alpha^{m-2} p_2 R^{\frac{2}{m}} + \ldots + \alpha p_{m-1} R^{\frac{m-1}{m}}.$$

Whence it is easily seen that

$$p = \frac{1}{m}(y_1 + y_2 + \ldots + y_m),$$

$$R^{\frac{1}{m}} = \frac{1}{m}(y_1 + \alpha^{m-1} y_2 + \ldots + \alpha y_m),$$

$$p_2 R^{\frac{2}{m}} = \frac{1}{m}(y_1 + \alpha^{m-2} y_2 + \ldots + \alpha^2 y_m),$$

$$\ldots \ldots \ldots$$

$$p_{m-1} R^{\frac{1}{m}} = \frac{1}{m}(y_1 + \alpha y_2 + \ldots + \alpha^{m-1} y_m).$$

Thus p, p_2, . . . p_{m-1}, R, and $R^{\frac{1}{m}}$ are rational functions of the roots of the proposed equation.

Let us now consider any one of these quantities, say R.   Let

$$R = S + v^{\frac{1}{n}} + S_2 v^{\frac{2}{n}} + \ldots + S_{n-1} v^{\frac{n-1}{n}}.$$

Treating this quantity as we have just treated y, we obtain the similar result that the quantities S, S_2,..., S_{n-1}, v, and $v^{\frac{1}{n}}$ are rational functions of the different values of R; and since these are rational functions of y_1, y_2, etc., the functions $v^{\frac{1}{n}}$, v, S, S_2 etc. have the same property.   Reasoning in this way, we conclude that all the irrational functions contained in the expression for y, are rational functions of the roots of the proposed equation.

This being established, it is not difficult to complete the demonstration.   Let us first consider irrational functions of the form $R^{\frac{1}{m}}$, R being a rational function of a, b, c, d, and e.   Let $R^{\frac{1}{m}} = r$.   Then r is a rational function of y_1, y_2, y_3, y_4, and y_5, and R is a symmetric function of these quantities.   Now as we are interested in the solution of the general equation of the fifth degree, it is clear that we can consider y_1, y_2, y_3, y_4, and y_5 as independent variables; thus the equation $R^{\frac{1}{m}} = r$ must be satisfied under this supposition.   Consequently we can interchange the quantities y_1, y_2, y_3, y_4, and y_5 in the equation $R^{\frac{1}{m}} = r$; and, remarking that R is a symmetric function, $R^{\frac{1}{m}}$ takes on m different values by this interchange.   Thus the function r must have the property of assuming m values, when the five variables which it contains are permuted in all possible ways.   Thus either $m = 5$, or $m = 2$, since m is a prime number, (see the memoir by M. Cauchy in the *Journal de l'école polytechnique*, vol. 17).[1]   Suppose that $m = 5$. Then the function r has five different values, and hence can be put in the form

$$R^{\frac{1}{5}} = r = p + p_1 y_1 + p_2 y_1^2 + p_3 y_1^3 + p_4 y_1^4,$$

---

[1] ["Mémoire sur le nombre des valeurs qu'une fonction peut acquérir," etc.
Let p be the greatest prime dividing n.   Cauchy then proves (p. 9) that a function of n variables, taking less than p values, either is symmetric or takes only two values.   In the latter case the function can be written in the form $A + B\Delta$ where A and B are symmetric, and Δ is the special two-valued function

$$\Delta = (y_1 - y_2)(y_1 - y_3)\ldots(y_{n-1} - y_n).]$$

p, p_1, p_2,...being symmetric functions of y_1, y_2, etc. This equation gives, on interchanging y_1 and y_2,

$$p + p_1 y_1 + p_2 y_1{}^2 + p_3 y_1{}^3 + p_4 y_1{}^4 = \alpha p + \alpha p_1 y_2 + \alpha p_2 y_2{}^2$$
$$+ \alpha p_3 y_2{}^3 + \alpha p_4 y_2{}^4,$$

where

$$\alpha^4 + \alpha^3 + \alpha^2 + \alpha + 1 = 0.$$

But this equation (is impossible);[1] hence m must equal two. Then

$$R^{1/2} = r,$$

and so r must have two different values, of opposite sign. We then have,[2] (see the memoir of M. Cauchy),

$$R^{1/2} = r = v(y_1 - y_2)(y_1 - y_3)\ldots(y_2 - y_3)\ldots(y_4 - y_5) = vS^{1/2}.$$

v being a symmetric function.

Let us now consider irrational functions of the form

$$(p + p_1 R^\nu + p_2 R_1{}^\mu + \ldots)^{\frac{1}{m}},$$

p, p_1, p_2, etc., R, E_1, etc., being rational functions of a, b, c, d, and e, and consequently symmetric functions of y_1, y_2, y_3, y_4, and y_5. We have seen that it is necessary that $\nu = \mu = \ldots = 2$, $R = v^2 S$, $R_1 = v_1{}^2 S$, etc. The preceeding function can thus be written in the form

$$(p + p_1 S^{\frac{1}{2}})^{\frac{1}{m}},$$

Let

$$r = (p + p_1 S^{\frac{1}{2}})^{\frac{1}{m}},$$

$$r_1 = (p^2 - p_1 S^{\frac{1}{2}})^{\frac{1}{m}}.$$

Multiplying, we have

$$r r_1 = (p^2 - p_1{}^2 S)^{\frac{1}{m}}.$$

---

[1] [In a later paper (*Journal für die reine und angewandte Mathematik* Vol. 1, 1826) Abel gives a more detailed proof of the main theorem, based on the same principles. At the corresponding point he gives the following more elaborate proof. By considering y_1 as a common root of the given equation, the relation defining R, y_1 can be expressed in the form

$$y_1 = s_0 + s_1 R^{1/5} + s_2 R^{2/5} + s_3 R^{3/5} + s_4 R^{4/5}.$$

Substituting $\alpha^i R^{1/5}$ for R we obtain the other roots of the equation, and solving the corresponding system of five linear equations gives

$$s_1 R^{1/5} = \tfrac{1}{5}(y_1 + \alpha^4 y_2 + \alpha^3 y_3 + \alpha^2 y_4 + \alpha y_5).$$

This identity is impossible, however, since the right-hand side has 120 values, and the left-hand side has only 5.]

[2] [Compare 1.]

If now rr_1 is not a symmetric function, m must equal two; but then r would have four different values, which is impossible; hence rr_1 must be a symmetric function. Let v be this function, then

$$r + r_1 = (p + p_1 S^2)^{\frac{1}{m}} + v(p + p_1 S^2)^{-\frac{1}{m}} = z.$$

This function having m different values, m must equal five, since m is a prime number. We thus have

$$z = q + q_1 y + q_2 y^2 + q_3 y^3 + q_4 y^4 = (p + p_1 S^{\frac{1}{2}})^{\frac{1}{5}} + v(p + p_1 S^{\frac{1}{2}})^{-\frac{1}{5}},$$

q, q_1, q_2, etc. being symmetric functions of y_1, y_2, y_3, etc., and consequently rational functions of a, b, c, d, and e. Combining this equation with the proposed equation, we can find y expressed as a rational function of z, a, b, c, d, and e. Now such a function can always be reduced to the form

$$y = P + R^{\frac{1}{5}} + P_2 R^{\frac{2}{5}} + P_3 R^{\frac{3}{5}} + P_4 R^{\frac{4}{5}},$$

where P, R, P_2, P_3, and P_4 are functions of the form $p + p_1 S^{\frac{1}{2}}$, where p, p_1, and S are rational functions of a, b, c, d, and e. From this value of y we obtain

$$R^{\frac{1}{5}} = \tfrac{1}{5}(y_1 + \alpha^4 y_2 + \alpha^3 y_3 + \alpha^2 y_4 + \alpha y_5) = (p + p_1 S^{\frac{1}{2}})^{\frac{1}{5}},$$

where

$$\alpha^4 + \alpha^3 + \alpha^2 + \alpha + 1 = 0.$$

Now the first member has 120 different values, while the second member has only 10; hence y can not have the form that we have found: but we have proved that y must necessarily have this form, if the proposed equation can be solved: hence we conclude that

It is impossible to solve the general equation of the fifth degree in terms of radicals.

It follows immediately from this theorem, that it is also impossible to solve the general equations of degrees higher than the fifth, in terms of radicals.

LEIBNIZ

ON DETERMINANTS

(Translated from the French and Latin by Dr. Thomas Freeman Cope, National Research Fellow in Mathematics, Harvard University, Cambridge, Mass.)

The work on determinants of Gottfried Wilhelm Leibniz (1646–1716), who was almost equally distinguished as a philosopher, mathematician, and man-of-affairs, is far less widely known than his work on the calculus. In fact, his contributions to this domain of algebra were entirely overlooked until the publication, in 1850, of the correspondence between him and the Marquis de l'Hospital. The letters to L'Hospital disclose the remarkable fact that, more than fifty years before the time of Cramer, who was the real moving spirit in the development of the theory, the fundamental idea of determinants had been clear to Leibniz and had been expounded by him in considerable detail in one of these letters. His work, however, had little or no influence on succeeding investigators.

A study of the following extracts from the writings of Leibniz shows that his contributions to this phase of algebra are at least two in number: (1) a new notation, numerical in character and appearance; (2) a rule for writing out the resultant of a set of linear equations.

The first of the extracts here given is from a letter of Leibniz to L'Hospital, which was dated April 28, 1693, and published for the first time (1850) at Berlin in *Leibnizens Mathematische Schriften*, herausg. von C. I. Gerhardt, Ie. Abth., Band II, pp. 238–240. The second extract is from a manuscript which was published for the first time (1863) at Halle in a subsequent volume of the above-mentioned work, namely, in the 2e. Abth., Band III, pp. 5–6. The original manuscript bears no date, but it was probably written before 1693 and possibly goes back to 1678. Each of the articles was published in Muir's well-known *Theory of Determinants*, the second edition (Macmillan & Co.) of which appeared in 1906. For an excellent account of Leibniz's life and work, the reader is referred to the *Encyclopaedia Britannica*, 12th ed., and, for an analysis of his contributions to the theory of determinants, to the scholarly treatise of Muir mentioned above. For further biographical notes relating particularly to his work on the calculating machine and on the calculus, see pages 173 and 619.

Leibniz on Determinants

I

Since you say that you have difficulty in believing that it is as general and as convenient to use numbers instead of letters, I must not have explained myself very well. There can be no doubt

about the generality if one considers that it is permissible to use
2, 3, etc., like *a* or *b*, provided that it is understood that these are
not really numbers.   Thus 2 . 3 does *not* denote 6 but rather *ab*.
As regards convenience, it is so considerable that I myself often
use them,[1] especially in long and difficult computations where it is
easy to make mistakes.   For besides the convenience of checking
by numbers and even by the casting out of nines, I find their use
a very great advantage even in the analysis itself.   As this is
quite an extraordinary discovery, I have not yet spoken to any
others about it, but here is what it is.   When one has need of
many letters, is it not true that these letters do not at all express
the relationship among the magnitudes they represent, while by
the use of numbers I am able to express this relationship.   For
example, consider three simple equations in two unknowns, the
object being to eliminate the two unknowns and indeed by a
general law.   I suppose that

$$10 + 11x + 12y = 0 \qquad (1),$$

and

$$20 + 21x + 22y = 0 \qquad (2),$$

and

$$30 + 31x + 32y = 0 \qquad (3),$$

where, in the pseudo number of two digits, the first tells me the
equation in which it is found, the second, the letter to which it
belongs.   Thus on carrying out the computation, we find throughout a harmony which not only serves as a check but even makes us
suspect at first glance some rules or theorems.   For example,
eliminating *y* first from the first and second equations, we shall
have:

$$\frac{10 . 22 + 11 . 22x}{-12 . 20 - 12 . 21..} = 0 \qquad (4)[2];$$

and from the first and third:

$$\frac{10 . 32 + 11 . 32x}{-12 . 30 - 12 . 31..} = 0 \qquad (5),$$

where it is easy to recognize that these two equations differ only
in that the anterior character 2 is changed to the anterior character
3.   Moreover, in similar terms of an equation, the anterior

---

[1] [*I. e.* numbers in place of letters.]

[2] [This is an abbreviated form, as Muir points out, for

$$\left. \begin{array}{l} +10.22 + 11.22x = 0 \\ -12.20 - 12.21x = 0 \end{array} \right\} \Bigg].$$

characters are the same and the posterior characters have the same sum. It remains now to eliminate the letter x from the fourth and fifth equations, and, as the result, we shall have:[1]

$$1_0 . 2_1 . 3_2 \quad 1_0 . 2_2 . 3_1$$
$$1_1 . 2_2 . 3_0 = 1_1 . 2_0 . 3_2$$
$$1_2 . 2_0 . 3_1 \quad 1_2 . 2_1 . 3_0,$$

which is the final equation freed from the two unknowns that we wished to eliminate, which carries its own proof along with itself from the harmony observable throughout, and which we should find very troublesome to discover using the letters a, b, c, especially when the number of letters and equations is large. A part of the secret of analysis is the characteristic, rather the art, of using notation well, and you see, Sir, by this little example, that Vieta and Descartes did not even know all of its mysteries. Continuing the calculation in this fashion, one will come to a *general theorem* for any desired numbers of letters and simple equations. Here is what I have found it to be on other occasions:—

Given any number of equations which is sufficient for eliminating the unknown quantities which do not exceed the first degree:—for the final equation are to be taken, first, all possible combinations of coefficients, in which one coefficient only from each equation is to enter; secondly, those combinations, after they are placed on the same side of the final equation, have different signs if they have as many factors alike as is indicated by the number which is less by one than the number of unknown quantities: the rest have the same sign.

II

I have found a rule for eliminating the unknowns in any number of equations of the first degree, provided that the number of equations exceeds by one the number of unknowns. It is as follows:—

Make all possible combinations of the coefficients of the letters, in such a way that more than one coefficient of the same unknown and of the same equation never appear together.[2] These combinations, which are to be given signs in accordance with the law which will soon be stated, are placed together, and the result set equal to zero will give an equation lacking all the unknowns.

---

[1] [The notation here has been slightly changed. What is clearly meant is, as Muir notes,
$$10.21.32 + 11.22.30 + 12.20.31 = 10.22.31 + 11.20.32 + 12.21.30.]$$
[2] [*I. e.*, in the same combination.]

The law of signs is this:—To one of the combinations a sign will be arbitrarily assigned, and the other combinations which differ from this one with respect to two, four, six, etc. factors will take the opposite sign: those which differ from it with respect to three, five, seven, etc. factors will of course take its own sign. For example, let

$$10 + 11x + 12y = 0, \quad 20 + 21x + 22y = 0, \quad 30 + 31x + 32y = 0;$$
there will result

$$\frac{+10 . 21 . 32 - 10 . 22 . 31 - 11 . 20 . 32}{+11 . 22 . 30 + 12 . 20 . 31 - 12 . 21 . 30} = 0.$$

I consider also as coefficients those characters which do not belong to any of the unknowns, as 10, 20, 30.

THE VERSES OF JACQUES BERNOULLI

On Infinite Series

(Translated from the Latin by Professor Helen M. Walker, Teachers College, Columbia University, New York City.)

Jacques (Jakob, Jacobus, James) Bernoulli (1654–1705), the first of the Bernoulli family of mathematicians, a native of Basel, wrote one of the earliest treatises on probability,—the *Ars Conjectandi*. This was published posthumously in 1713. At the close of a section entitled "Tractatus de Seriebus Infinitis Earumque Summa Finita et Usu in Quadraturis Spatiorum & Rectificationibus Curvarum," following Pars Quarta, these six verses appear. Because they represent one of the clearest of the early statements relating to the limit of an infinite series, they are given place in this symposium. Their brevity permits of inserting both the Latin form and the translation.

Ut non-finitam Seriem finita cöercet,
   Summula, & in nullo limite limes adest:
Sic modico immensi vestigia Numinis haerent
   Corpore, & angusto limite limes abest.
Cernere in immenso parvum, dic, quanta voluptas!
   In parvo immensum cernere, quanta, Deum!

Even as the finite encloses an infinite series
   And in the unlimited limits appear,
So the soul of immensity dwells in minutia
   And in narrowest limits no limits inhere.
What joy to discern the minute in infinity!
   The vast to perceive in the small, what divinity!

JACQUES BERNOULLI

ON THE THEORY OF COMBINATIONS

(Translated from the Latin by Mary M. Taylor, M. A., University of Pittsburgh, Pittsburgh, Penn.)

The following translation is taken from Part 2 of Jacques (Jakob, James) Bernoulli's *Ars Conjectandi*. Although Bernoulli (1654–1705) was also interested and active in other branches of science, it is for his mathematical works that he is particularly known. The *Ars Conjectandi* was published eight years after his death, and contains, in addition to the work on combinations, a treatise on Infinite Series. The first part of the book is attributed to Huygens, but Part 2 is Bernoulli's own. This selection is part of Chapter V and is from the first edition, pages 112 to 118, inclusive.

While this is by no means the earliest material published on the subject, it si among the earliest scientific treatments and is so authoritative as to deserve a place in a source book of this nature. The subject matter chosen presents for solution a situation which occurs in various problems of higher mathematics. The method of solution is typical of the rest of the work.

Chapter V. Part 2.

To find the number of combinations, when each of the objects to be combined, whatever they are, is different from the others, but may be used more than once in each combination.

In the combinations of the preceding chapters we have assumed that an object could not be joined with itself, and could not even be accepted more than once in the same combination; but now we shall add this condition—that each object can be placed next to itself, and further that it can occur repeatedly in the same combination.

Thus let the letters to be combined by this plan be *a, b, c, d*, etc. Let there be made as many series as there are letters, and let the individual letters, just as so many units, occupy the first place in each, as was done in chapter two.

In finding the combinations of two, or binary terms of each series, the letter which heads that sequence is to be combined not only with each of the letters preceding it, but also with itself. Thus we shall have in the first series one binary *aa*, in the second two binaries *ab, bb*, in the third three, *ac, bc, cc*, in the fourth four, *ad, bd, cd, dd*, etc.

So also in forming the ternaries,—each letter must be joined not only with the binaries of all the preceding series, but also with those of its own series. In this way we shall have in the first series one ternary, *aaa*; in the second series three, *aab*, *abb*, *bbb*; in the third series six, *aac*, *abc*, *bbc*, *acc*, *bcc*, *ccc*; and so on.

This same plan is to be followed in combinations of every other degree, by which plan it is clear that none of the possible selections among the given objects can be overlooked. In tabular form;

a. aa. aaa.

b. ab. bb. aab. abb. bbb.

c. ac. bc. cc. aac. abc. bbc. acc. bcc. ccc.

d. ad. bd. cd. dd. aad. abd. bbd acd. bcd. ccd. add bdd. cdd. ddd.

From this, with no great difficulty, we infer that the single terms of all the series form a group of ones; the binaries, a series of positive integers (or natural numbers); the ternaries, a series of three-sided figures; and the other combinations of higher degree likewise constitute series of other figures of higher order, just as did the combinations of the preceding chapters, with this one difference, that there the series began with zeros, and here they start directly from the ones. Thence if the series are collected into tabular form, they present this arrangement:

TABULA COMBINATORIA

Exponentes Combinationum

I	II	III	IV	V	VI	VII	VIII	IX	X	XI	XII
1	1	1	1	1	1	1	1	1	1	1	1
2	1	2	3	4	5	6	7	8	9	10	11	12
3	1	3	6	10	15	21	28	36	45	55	66	78
4	1	4	10	20	35	56	84	120	165	220	286	364
5	1	5	15	35	70	126	210	330	495	715	1001	1365
6	1	6	21	56	126	252	462	792	1 287	2002	3003	4368
7	1	7	28	84	210	462	924	1716	3 003	5005	8008	12376
8	1	8	36	120	330	792	1716	3432	6435	11440	19448	31824
9	1	9	45	165	495	1287	3003	6435	12870	24310	43758	75582
10	1	10	55	220	715	2002	5005	11440	24310	48620	92378	167960

Moreover it is worth while to note especially two properties of the table thus formed: 1. That the transverse columns are congruent to the vertical, the first to the first, the second to the second, etc. 2. That if two contiguous columns are chosen, whether vertical or horizontal, with an equal number of terms, the sum of the terms of the first column is equal to the last term of the second column.

From these properties it is easy to find the sum of the terms of any series whatever, and so the number of combinations according to the degree thereof. For if the number of terms, that is of things to be combined, is called n, the sum of the ones, or terms of the first series, will be the last term of the second series, likewise n.

We may suppose the second series to have a zero prefixed, so that the number of terms becomes $n + 1$; if the last term n is multiplied by half of this $n + 1$, the product $\frac{n.n + 1}{1.2}$ will be the sum of the twos or terms of the second series (according to property 12,[1] chapter 3), and the last term of the third, (by property 2 of this chapter).

We may suppose two ciphers to be prefixed to the third series, and the number of terms will become $n + 2$. If the last term just found, $\frac{n.n + 1}{1.2}$, be multiplied by one-third of this, it will become $\frac{n.n + 1.n + 2}{1.2.3}$, the sum of the ternaries or terms of the third series, and at the same time, by the same properties, the last term of the fourth series.

In the same way the sum of the terms of the fourth series (quaternaries) is found to be $\frac{n.n + 1.n + 2.n + 3}{1.2.3.4}$, of the fifth series $\frac{n.n + 1.n + 2.n + 3.n + 4}{1.2.3.4.5}$; and in general the sum of the terms of the c series, or combinations of degree c, is found to be

$$\frac{n.n + 1.n + 2.n + 3.n + 4 \ldots (n + c - 1)}{1.2.3.4.5 \ldots c}.$$

Here it should be noted that if $c > n$ the factors of the fraction can be diminished by dividing numerator and denominator by $n.n + 1$. $\ldots c$, so that we have $\frac{c + 1.c + 2.c + 3.c + n - 1}{1.2.3.4 \ldots n - 1}$, and since this fraction, worked out according to the formula, should at the

---

[1] [Property 12, Chapter 3. The sum of any number of terms of any vertical column beginning with the proper number of ciphers has the same ratio to the sum of as many terms equal to the last, as unity has to the number of that series; that is, the sum of any number of the natural numbers, beginning the series with one cipher, is to the sum of as many terms, each equal to the greatest of these, or the last, as 1:2; of the third order series beginning with two ciphers as 1:3, etc. This same is also true of the ratio which the sum of the terms of any series beginning with unity has to the sum of as many terms, equal to the term following the last.]

same time indicate the sum of $c + 1$ terms in the series $n - 1$, it follows that the sum of n terms in the series c is always equal to the sum of $c + 1$ terms in the series $n - 1$, which is another by no means negligible property of this table. Thence results the following

Rule

for finding the number of combinations according to a given degree, when the same objects can enter into the same combination more than once.

Let two increasing arithmetic progressions be formed, the first starting from the number of things to be combined, the other from unity, of both of which the common difference is unity, and let each have as many terms as the degree of the combination has units. Then let the product of the terms of the first progression be divided by the product of the terms of the second progression, and the quotient will be the desired number of the combinations according to the given degree. With this understanding, the number of combinations by four among ten different things is

$$\frac{10 . 11 . 12 . 13}{1 . 2 . 3 . 4} \quad \infty \quad \frac{17160}{24} \quad \infty \quad 715.$$

Note.—If the degree of the combination is greater than the number of objects, (as is clearly possible under the present hypothesis) it will be shorter to begin the first progression with that degree increased by one, and to make each series of one fewer terms than there are objects. Thus the number of combinations of degree 10 among four objects is

$$\frac{11 . 12 . 13}{1 . 2 . 3} \quad \infty \quad \frac{1716}{6} \quad \infty \quad 286.$$

But also we can find with no more difficulty the number of combinations according to several degrees following each other successively from unity up, that is, the sum of as many vertical series. For since, for example, the first 10 terms of the first 4 vertical columns are the same as the first 4 terms of the first 10 transverse columns, and moreover the sums of these terms are equal to eleven terms of the first vertical column, decreased by the first or unity (of course the sums are equal one by one to these terms, as is evident from the second property of the table), it is clear, also, that the 10 first terms of the first four vertical columns, i. e., the sum of all the ones, twos, threes, and fours selected from ten things, is less by one than the eleven first terms of the fourth column, *i. e.*, than the number of quaternaries formed of eleven things, or than number of combinations formed from one more than the given number of objects and of degree equal to the greatest of the given degrees. This same fact may also be shown in this way: Obviously, the eleventh object either does not occur in a particular combination of four from the given eleven objects, or it occurs once, twice, thrice, or four times; but it is evident that those quaternaries in which the eleventh object does not appear are just the ones which the ten remaining objects can form among themselves. And it is no less evident that the number of those into which the eleventh enters only once should equal the number of ternaries to be formed from the remaining ten; so also the number of those in which it occurs twice (should equal) the number of binaries, and of those in which it occurs three times the number of ones, since when joined once to the ternaries, twice to the binaries, and three times to the units, it forms quaternaries; besides it is known that there is one quaternion which is formed by the eleventh object, repeated four times.

From this, we conclude that the number of combinations of four included in eleven objects, that is in one more than the given number of objects, exceeds by one all the combinations by one, two, three, and four, of the given ten objects, unless we wish to add to the latter the zero combination, in which case the two are equal.

Wherefore, since, when the number of objects given is n, and the greatest degree c, the number of combinations of that degree in $n + 1$ things is found by Rule of chapter 4 to be

$$\frac{n + 1 \cdot n + 2 \cdot n + 3 \cdot n + 4 \ldots n + c}{1 \cdot 2 \cdot 3 \cdot 4 \ldots c},$$

the number of combinations of n things according to all degrees from one to c becomes (as it is one less than this)

$$\frac{n + 1 \cdot n + 2 \cdot n + 3 \cdot n + 4 \ldots n + c}{1 \cdot 2 \cdot 3 \cdot 4 \ldots c} - 1.$$

But if c is greater than n itself, *i. e.*, if the greatest of the degrees is higher than the number of objects, the terms of the fraction can in this case be divided by $n + 1 \cdot n + 2 \cdot n + 3 \ldots c$, and hence the quantity can be expressed more briefly as

$$\frac{c + 1 \cdot c + 2 \cdot c + 3 \ldots c + n}{1 \cdot 2 \cdot 3 \cdot 4 \ldots n} - 1.$$

From this comes the

Rule

for finding the number of combinations according to several degrees following successively from unity.

Let two increasing arithmetic progressions be formed, the first starting from one more than the number of objects to be combined, the other from unity, of which the common difference is one, and let each have as many terms as the highest degree has units. (But if the greatest of the degrees is larger than the number of objects, it is easier to begin the first progression with that degree increased by one, and to make each of as many terms as there are given objects.) Then the product of the terms of the first progression is to be divided by the product of the terms of the second progression; and the quotient will be the required number of combinations if, of course, we wish the zero combination included; but if not, the quotient diminished by one will indicate the desired quantity. Thus the number of units, binaries, ternaries, and quaternaries, together with the zero combination, in 10 objects is

$$\frac{11 . 12 . 13 . 14}{1 . 2 . 3 . 4} \infty \frac{24024}{24} \infty 1001, \text{ among only three things is } \frac{5 . 6 . 7}{1 . 2 . 3}$$

$$\infty \frac{210}{6} \infty 35;$$ but if the zero is excluded the number of combinations is 1000 in the first case, 34 in the second.

GALOIS

On Groups and Equations and Abelian Integrals

(Translated from the French by Dr. Louis Weisner, Hunter College of the City of New York.)

Evariste Galois (1811–1832) was born in Paris, was educated at the Lycée Louis-le-Grand and the École Normale, was a rabid republican, was twice imprisoned for his political views, and lost his life in a stupid, boyish duel before he had reached the age of twenty-one. His most important paper, "Mémoire sur les conditions de résolubilité des équations par radicaux" was not published until 1846, when his works appeared in Liouville's *Journal de Mathématiques.*

The night before the duel in which Galois was killed he wrote a letter to his friend Auguste Chevalier in which he set forth briefly his discovery of the connection of the theory of groups with the solution of equations by radicals. In this letter, written apparently under the impression that the result of the duel would be fatal to himself, he asked that it be published in the *Revue encyclopédique,* a wish that was carried out the same year (1832, page 568). His works were republished in 1897 under the auspices of La Société Mathématique de France with an introduction by E. Picard. Further writings of Galois were published by J. Tannery in the *Bulletin des Sciences Mathématiques* (1906–1907) and reprinted the following year in book form. It being impossible to include in a source book of this kind the mémoire above mentioned, the letter to M. Chevalier is here given in translation.

My dear friend,

I have made some new discoveries in analysis.

Some are concerned with the theory of equations; others with integral functions.

In the theory of equations, I have sought to discover the conditions under which equations are solvable by radicals, and this has given me the opportunity to study the theory and to describe all possible transformations on an equation even when it is not solvable by radicals.

It will be possible to make three memoirs of all this.

The first is written, and, despite what Poisson has said of it, I am keeping it, with the corrections I have made.

The second contains some interesting applications of the theory of equations. The following is a summary of the most important of these:

278

1°. From propositions II and III of the first memoir, we perceive a great difference between adjoining to an equation one of the roots of an auxiliary equation and adjoining all of them.

In both cases the group of the equation breaks up by the adjunction in sets such that one passes from one to the other by the same substitution, but the condition that these sets have the same substitutions holds with certainty only in the second case. This is called the *proper decomposition.*[1]

In other words, when a group G contains another H, the group G can be divided into sets, each of which is obtained by multiplying the permutations of H by the same substitution; so that

$$G = H + HS + HS' + \ldots$$

And it can also be divided into sets which contain the same substitutions, so that

$$G = H + TH + T'H + \ldots$$

These two methods of decomposition are usually not identical. When they are identical, the decomposition is *proper.*

It is easy to see that when the group of an equation is not susceptible of any proper decomposition, then, however, the equation be transformed, the groups of the transformed equations will always have the same number of permutations.

On the other hand, when the group of an equation admits a proper decomposition, in which it has been separated into M groups of N permutations, then we can solve the given equation by means of two equations, one having a group of M permutations, the other N.

When therefore we have exhausted in the group of an equation all the possible proper decompositions, we shall arrive at groups which can be transformed, but whose permutations will always be the same in number.

If each of these groups has a prime number of permutations, the equation will be solvable by radicals; otherwise not.

The smallest number of permutations which an indecomposable group can have, when this number is not a prime, is $5 \cdot 4 \cdot 3$.

2°. The simplest decompositions are those which occur in the method of M. Gauss.

As these decompositions are obvious, even in the actual form of the group of the equation, it is useless to spend time on this matter.

---

[1] [A proper decomposition, in modern parlance, is an arrangement of the permutations of a group into cosets with respect to an invariant subgroup.]

What decompositions are practicable in an equation which is not simplified by the method of M. Gauss?

I have called those equations *primitive* which cannot be simplified by M. Gauss's method; not that the equations are really indecomposable, as they can even be solved by radicals.

As a lemma in the theory of primitive equations solvable by radicals, I made in June 1830, in the *Bulletin de Férussac* an analysis of imaginaries in the theory of numbers.

There will be found herewith[1] the proof of the following theorems:

1°. In order that a primitive equation be solvable by radicals its degree must be p^ν, p being a prime.

2°. All the permutations of such an equation have the form

$$x_{k,l,m,\ldots} \mid x_{ak+bl+cm+\cdots+h,\ a'k+b'l+c'm+\cdots+h',\ a''k+\ldots}, \quad k,\ l,\ m,\ldots$$

being ν indices, which, taking p values each, denote all the roots. The indices are taken with respect to a modulus p; that is to say, the root will be the same if we add a multiple of p to one of the indices.

The group which is obtained on applying all the substitutions of this linear form contains in all

$$p^\nu(p^\nu - 1)(p^\nu - p)\ldots(p^\nu - p^{\nu-1})$$

permutations.

It happens that in general the equations to which they belong are not solvable by radicals.

The condition which I have stated in the *Bulletin de Férussac* for the solvability of the equation by radicals is too restricted; there are few exceptions, but they exist.[2]

The last application of the theory of equations is relative to the modular equations of elliptic functions.

We know that the group of the equation which has for its roots the sines of the amplitude[3] of the $p^3 - 1$ divisions of a period is the following:

$$x_{k,l},\ x_{ak+bl,\ ck+dl};$$

---

[1] [Liouville remarks: "Galois speaks of manuscripts, hitherto unpublished, which we shall publish."]

[2] [Galois stated in the *Bulletin des sciences mathématiques de M. Férussac* (1830), p. 271, that the elliptic modular equation of degree $p + 1$ could not be reduced to one of degree p when p exceeds 5; but $p = 7$ and $p = 11$ are exceptions to this statement, as Galois shows in the next page of his letter.]

[3] [Meaning the elliptic sn-function.]

consequently the corresponding modular equation has for its group

$$x_{k}, \ x_{\frac{ak+bl}{ck+dl}},$$

in which $\frac{k}{l}$ may have the $p + 1$ values

$$\infty, 0, 1, 2, \ldots, p - 1.$$

Thus, by agreeing that k may be infinite, we may write simply

$$x_{k}, \ x_{\frac{ak+b}{ck+d}}.$$

By giving to a, b, c, d all the values, we obtain

$$(p + 1)p(p - 1)$$

permutations.

Now this group decomposes *properly* in two sets, whose substitutions are

$$x_{k}, \ x_{\frac{ak+b}{ck+d}},$$

$ad - bc$ being a quadratic residue of p.

The group thus simplified has $\dfrac{(p + 1)p(p - 1)}{2}$ permutations.

But it is easy to see that it is not further properly decomposable, unless $p = 2$ or $p = 3$.

Thus, in whatever manner we transform the equation, its group will always have the same number of substitutions.

But it is interesting to know whether the degree can be lowered.

First, it cannot be made less than p, as an equation of degree less than p cannot have p as a factor of the number of permutations of its group.

Let us see then whether the equation of degree $p + 1$, whose roots are denoted by x_{k} on giving k all its values, including infinity, and has for its group of substitutions

$$x_{k}, \ x_{\frac{ak+b}{ck+d}}$$

$ad - bc$ being a square, can be lowered to degree p.

Now this can happen only if the group decomposes (improperly, of course) in p sets of $\dfrac{(p + 1)(p - 1)}{2}$ permutations each.

Let 0 and ∞ be two conjoint letters of one of these groups. The substitutions which do not change 0 and ∞ are of the form

$$x_{k}, \ x_{m^{2}k}.$$

Therefore if M is the letter conjoint to 1, the letter conjoint to m^2 will be m^2M. When M is a square, we shall have $M^2 = 1$. But this simplification can be effected only for $p = 5$.

For $p = 7$ we find a group of $\dfrac{(p+1)(p-1)}{2}$ permutations, where

$$\infty, 1, 2, 4$$

have respectively the conjoints

$$0, 3, 6, 5.$$

The substitutions of this group are of the form

$$x_k, \ x_{a \cdot \frac{(k-b)}{k-c}},$$

b being the letter conjoint to c, and a a letter which is a residue or a non-residue simultaneously with c.

For $p = 11$, the same substitutions will occur with the same notations,

$$\infty, 1, 3, 4, 5, 9,$$

having respectively for conjoints

$$0, 2, 6, 8, 10, 7.$$

Thus for the cases $p = 5, 7, 11$, the modular equation can be reduced to degree p.

In all rigor, this equation is not possible in the higher cases.

The third memoir concerns integrals.

We know that a sum of terms of the same elliptic function[1] always reduces to a single term, plus algebraic or logarithmic quantities.

There are no other functions having this property.

But absolutely analogous properties are furnished by all integrals of algebraic functions.

We treat at one time every integral whose differential is a function of a variable and of the same irrational function of the variable, whether this irrationality is or is not a radical, or whether it is expressible or not expressible by means of radicals.

We find that the number of distinct periods of the most general integral relative to a given irrationality is always an even number.

If $2n$ is this number, we have the following theorem:

Any sum of terms whatever reduces to n terms plus algebraic and logarithmic quantities.

---

[1] [Galois presumably means a sum of elliptic integrals of the same species.]

The functions of the first species are those for which the algebraic and logarithmic parts are zero.

There are n distinct functions of the first species.

The functions of the second species are those for which the complementary part is purely algebraic.

There are n distinct functions of the second species.[1]

We may suppose that the differentials of the other functions are never infinite except once for $x = a$, and moreover, that their complementary part reduces to a single logarithm, log P, P being an algebraic quantity. Denoting these functions by $\pi(x, a)$, we have the theorem

$$\pi(x, a) - \pi(a, x) = \Sigma\varphi a \cdot \psi x,$$

$\varphi(a)$ and $\psi(x)$ being functions of the first and of the second species.

We infer, calling $\pi(a)$ and ψ the periods of $\pi(x, a)$ and ψx relative to the same variation of x,

$$\pi(a) = \Sigma\psi \times \varphi a.$$

Thus the periods of the functions of the third species are always expressible in terms of the first and second species.

We can also deduce theorems analogous to the theorem of Legendre [2]

$$FE' + EF' - FF' = \frac{\pi}{2}.$$

The reduction of functions of the third species to definite integrals, which is the most beautiful discovery of M. Jacobi, is not practicable, except in the case of elliptic functions.

The multiplication of integral functions by a whole number is always possible, as is the addition, by means of an equation of degree n whose roots are the values to substitute in the integral to obtain the reduced terms.[3]

The equation which gives the division of the periods in p equal parts is of degree $p^{2n} - 1$. Its group contains in all

$$(p^{2n} - 1)(p^{2n} - p)\ldots(p^{2n} - p^{2n-1}) \text{ permutations.}$$

---

[1] [Picard comments: "We thus acquire the conviction that he (Galois) had in his possession the most essential results concerning Abelian integrals which Riemann was to obtain twenty-five years later."]

[2] [According to Tannery, who collated Galois's manuscripts with Liouville's publication of Galois's Works, Galois wrote Legendre's theorem in the form: $E'F'' - E''F' = \frac{\pi}{2}\sqrt{-1}$. Liouville made other alterations of a minor character.]

[3] [Obscure.]

The equation which gives the division of a sum of n terms in p equal parts is of degree p^{2n}.   It is solvable by radicals.

Concerning the Transformation.—First, by reasoning analogous to that which Abel has indicated in his last memoir, we can show that if, in a given relation among integrals, we have the two functions

$$\int \Phi(x, X)dx, \int \Psi(y, Y)dy,$$

the last integral having $2n$ periods, it will be permissible to suppose that y and Y can be expressed by means of a single equation of degree n in terms of x and X.

Then we may suppose that the transformations are constantly made for two integrals only, since one has evidently, in taking any rational function whatever of y and Y,

$$\Sigma \int\!\int f(y, \quad Y)dy = \int F(x, \quad X)dx + \text{an algebraic and logarithmic}$$
quantity.

There are in this equation obvious reductions in the case where the integrals of the two members do not both have the same number of periods.

Thus we have only to compare those integrals both of which have the same number of periods.

We shall prove that the smallest degree of irrationality of two like integrals cannot be greater for one than for the other.

We shall show subsequenty that one may always transform a given integral into another in which one period of the first is divided by the prime number p, and the other $2n - 1$ remain the same.

It will only remain therefore to compare integrals which have the same periods, and such consequently, for which n terms of the one can be expressed without any other equation than a single one of degree n, by means of two of the others, and reciprocally. We know nothing about this.

You know, my dear Auguste, that these subjects are not the only ones I have explored. My reflections, for some time, have been directed principally to the application of the theory of ambiguity to transcendental analysis.[1] It is desired to see *a priori* in a relation among quantities or transcendental functions, what transformations one may make, what quantities one may

---

[1] [Picard comments: "We could almost guess what he means by this, and in this field, which, as he says, is immense, there still to this day remain discoveries to make."]

substitute for the given quantities, without the relation ceasing to be valid. This enables us to recognize at once the impossibility of many expressions which we might seek. But I have no time, and my ideas are not developed in this field, which is immense.

Print this letter in the *Revue encyclopédique*.

I have often in my life ventured to advance propositions of which I was uncertain; but all that I have written here has been in my head nearly a year, and it is too much to my interest not to deceive myself that I have been suspected of announcing theorems of which I had not the complete demonstration.

Ask Jacobi or Gauss publicly to give their opinion, not as to the truth, but as to the importance of the theorems.

Subsequently there will be, I hope, some people who will find it to their profit to decipher all this mess.

Je t'embrasse avec effusion.

E. Galois.

May 29, 1832.

ABEL

On the Continuity of Functions Defined by Power Series

(Translated from the German by Professor Albert A. Bennett, Brown University, Providence, R. I.)

This article constitutes part of the opening portion of a paper originally written in French entitled "Investigation of the series: $1 + \frac{m}{1}x + \frac{m}{1}\frac{(m-1)}{2}x^2 + \frac{m(m-1)(m-2)}{1 \cdot 2 \cdot 3}x^3 + \dots$ and so forth." It first appeared, in a faithful German translation, in the *Journal für die reine und angewandte Mathematik* (Crelle) Berlin 1826, pages 311 to 339, and the extract translated below covers pages 312 to 315). It was reprinted with corrections and notes in Ostwald's *Klassiker der Exacten Wissenschaft*, No. 71, Leipzig, 1895. The article in the original French is in Abel, *Œuvres complètes*, Vol. I, Christiania, 1881, pages 219 to 250.

Niels Henrik Abel (Aug. 5, 1802 to April 6, 1829) was born in Findö, Norway. As a youth, his mathematical achievement was phenomenal. He studied some eighteen months in Germany and France under a grant from the Norwegian government and collaborated in founding Crelle's Journal. He returned to Christiania, 1827, and died suddenly at the age of 26 years. The two volumes of the second edition of his collected works bear testimony to his productivity. The classical terms "Abelian group" and "Abelian function" indicate in widely different fields something of his originality, profundity, and still increasing influence.

The theorem (which is fundamental in analytic function theory) may be stated in modern notation as follows. *If a real power series converges for some positive value of the argument, the domain of uniform convergence extends at least up to and including this point, and the continuity of the sum-function extends at least up to and including this point.* The extension to complex values follows readily by the method used previously by Cauchy (noted below) in the special case of the infinite geometric progression, *Cours d'Analyse*, Paris, 1821, p. 275–278.

This theorem is of special interest, in that it was included in the scope of the investigation by Cauchy, referred to above. Cauchy correctly stated and in substance proved the theorem for the trivial case of the infinite geometric progression. Cauchy proceeded at once to state and claimed to prove a much more general theorem of which this would have been a special case. Cauchy's more general theorem is however false. Abel remarks indeed in this paper in a footnote (p. 316):

In the above-mentioned work of Mr. Cauchy (page 131) one finds the following theorem:

"If the different terms of the series

$$u_0 + u_1 + u_2 + u_3 + \ldots \text{etc.}$$

are functions of one and the same variable x, and indeed continuous functions with respect to this variable in the neighborhood of a particular value for which the series converges, then the sum s of the series is also a continuous function of x in the neighborhood of this particular value."

It appears to me that this theorem suffers exceptions. Thus for example the series

$$\sin \phi - \frac{1}{2} \sin 2\phi + \frac{1}{3} \sin 3\phi - \ldots \text{etc.}$$

is discontinuous for each value $(2m + 1)\pi$ of ϕ, where m is a whole number. It is well-known that there are many series with similar properties.

Abel was the first to note that Cauchy's announced theorem is not in general valid, and to prove the correct theorem for general power series.

This paper appeared at a critical time in the theory of infinite series. (For reference, see *Enc. des Sci. Math.* I, 1, 2. (1907) p. 213 to 221.) Archimedes used the infinite series $1 + \frac{1}{4} + (\frac{1}{4})^2 + \ldots$ Prop. 22, 23, Quadrature of the Parabola, *Works of Archimedes*, T. L. Heath, 1897. p. 249–251. N. Mercator, and Lord Brouncker simultaneously in 1668 introduced the infinite logarithmic series. Sir Isaac Newton (*De analysi per aequationes numero terminorum infinitas*, (1669; London, 1711), used infinite series systematically. Leibniz 1673 remarked upon the divergence of the harmonic series in connection with his harmonic triangle. (J. M. Child, *Early mathematical manuscripts of Leibniz*, Chicago, Open Court, 1920, Page 50.) Both Jacques and Jean Bernoulli considered the same problem in 1689. Even Lagrange (1768) was content to establish the fact that the successive terms of a convergent series approach zero, apparently assuming the converse theorem in such use as he made of series in his *Théorie des fonctions analytiques* (Paris, year V, 1797, p. 50; *Oeuvres*, vol. 9, Paris 1881, p. 85).

The outstanding general discussion of convergence of series prior to this paper was the *Cours d'Analyse de l'École Royale Polytechnique*, (Paris, 1821) of Augustin-Louis Cauchy (Part One is "Analyse Algébrique"). Chap. 6 (pages 123–172) deals with convergence of real series; Chap. 9 (pages 274–328), with convergence of series with complex terms. Gauss (*Commentationes Soc. Gottingen. math.*, 1812, Mém. no. 1; *Werke*, Göttingen, Vol. III, 1876, see pages 139–143) had considered rigorously a particular series (the hypergeometric series) but stated however no general theorem on convergence such as given in the extract here translated. Cauchy's completed theorem on the circle of convergence of a Taylor-series expansion for a holomorphic function was not published until 1832, three years after Abel's death.

Abel's own preface suggests the state of the theory of infinite series at Abel's time. Quoting page 312:

One of the most remarkable series of algebraic analysis is the following:

$$1 + \frac{m}{1}x + \frac{m(m-1)}{1.2}x^2 + \frac{m(m-1)(m-2)}{1.2.3}x^3 + \dots$$
$$+ \frac{m(m-1)\dots[m-(n-1)]}{1.2\dots\dots\dots n}x^n + \dots$$

When m is a positive whole number the sum of the series which is then finite can be expressed, as is known, by $(1+x)^n$. When m is not an integer, the series goes on to infinity, and it will converge or diverge according as the quantities m and x have this or that value. In this case, one writes the same equality

$$(1+x)^m = 1 + \frac{m}{1}x + \frac{m(m-1)}{1.2}x^2 + \dots \text{etc.},$$

but then the equation shows nothing more than that the two expressions $(1+x)^m$, $1 + \frac{m}{1} \cdot x + \frac{m \cdot (m-1)}{1.2} \cdot x^2 + \dots$ have certain common properties, upon which for certain values of m and x, depends the *numerical* equality of the expressions. It is assumed that the numerical equality will always occur whenever the series is convergent, but this has never yet been proved. Not even have all the cases in which the series converges been examined as yet. Even if the existence of the equality mentioned above be *assumed*, it would still remain to find the *value* of $(1+x)^n$, for this expression has in general infinitely many different values, while the series $1 + mx + \dots$ has but a single one only.

The following is the translation of the extract referred to above:

We will first establish some necessary theorems on series. The excellent work of Cauchy, *Cours d'analyse de l'École Polytechnique*, which should be read by every analyst who loves rigor in mathematical investigations, will serve to guide us.

Definition.—An arbitrary series

$$v_0 + v_1 + v_2 + \dots + v_m + \dots$$

will be called *convergent*, if for constantly increasing values of m, the sum, $v_0 + v_1 + \dots + v_m$, approaches arbitrarily near a certain limit. This limit will be called the *sum of the series*. In the contrary case the series will be called *divergent*, and then it has no sum. From this definition it follows that if a series is to be convergent, it is necessary and sufficient that for continually increasing values of m, the sum $v_m + v_{m+1} + \ldots + v_{m+n}$ shall approach arbitrarily close to zero, no matter what be the value of n.

In any convergent series, therefore, the general term v_m approaches arbitrarily close to zero.[1]

THEOREM I.—If a series of positive quantities is denoted by ρ_0, ρ_1, ρ_2, ..., and if for continually increasing values of m, the quotient ρ_{m+1}/ρ_m approaches a limit α, which is greater than 1, then the series

$$\epsilon_0\rho_0 + \epsilon_1\rho_1 + \epsilon_2\rho_2 + \ldots + \epsilon_m\rho_m + \ldots,$$

where ϵ_m is a quantity which for continually increasing values of *m does not approach arbitrarily close to zero*, will necessarily *diverge*.

THEOREM II.—If in a series of positive quantities such as $\rho_0 + \rho_1 + \rho_2 + \ldots + \rho_m + \ldots$, the quotient ρ_{m+1}/ρ_m, for continually increasing values of m, approaches arbitrarily close to a limit[2] which *is smaller than* 1,[3] then the series

$$\epsilon_0\rho_0 + \epsilon_1\rho_1 + \epsilon_2\rho_2 + \ldots + \epsilon_m\rho_m + \ldots,$$

where ϵ_0, ϵ_1, ϵ_2, ..., are quantities which *do not exceed* 1, will necessarily converge.

Indeed by hypothesis, m can always be taken sufficiently large so that one shall have $\rho_{m+1} < \alpha\rho_m$, $\rho_{m+2} < \alpha\rho_{m+1}\ldots$, $\rho_{m+n} < \alpha\rho_{m+n-1}$. Thence it follows that $\rho_{m+k} < \alpha^k\rho_m$, and hence

$$\rho_m + \rho_{m+1} + \ldots + \rho_{m+n} < \rho_m(1 + \alpha + \ldots + \alpha^n) < \frac{\rho_m}{(1-\alpha)},$$

and hence a fortiori

$$\epsilon_m\rho_m + \epsilon_{m+1}\rho_{m+1} + \ldots + \epsilon_{m+n}\rho_{m+n} < \frac{\rho_m}{(1-\alpha)}.$$

Since however $\rho_{m+k} < a^k\rho_m$ and $\alpha < 1$, it is clear that ρ_m and consequently also the sum

$$\epsilon_m\rho_m + \epsilon_{m+1}\rho_{m+1} + \ldots + \epsilon_{m+n}\rho_{m+n}$$

will have zero as limit.[4]

Hence the series given above is convergent.

---

[1] For brevity, in this article, by ω will be meant a quantity which can be smaller than any given quantity no matter how small.

[2] [The text reads "to a limit α which..." This is somewhat inexact in view of the use made of α.]

[3] [And hence smaller than some constant α itself smaller than 1.]

[4] [The context shows that this somewhat ambiguous statement is to be understood in the required sense of $\lim_{m\to\infty}[\lim_{n\to\infty}(\epsilon_m\rho_m + \epsilon_{m+1}\rho_{m+1} + \ldots + \epsilon_{m+n}\rho_{m+n})] = 0.$]

THEOREM III.—If by $t_0, t_1, t_2, \ldots, t_m, \ldots$ is denoted a series of arbitrary quantities, and if the quantity

$$p_m = t_0 + t_1 + t_2 + \ldots + t_m$$

is always less than a definite quantity, δ, then one has

$$r = \epsilon_0 t_0 + \epsilon_1 t_1 + \epsilon_2 t_2 + \ldots + \epsilon_m t_m < \delta \epsilon_0,$$

where $\epsilon_0, \epsilon_1, \epsilon_2, \ldots$ are positive decreasing quantities.

In fact one has

$$t_0 = p_0, \ t_1 = p_1 - p_0, \ t_2 = p_2 - p_1, \ldots$$

hence

$$r = \epsilon_0 p_0 + \epsilon_1(p_1 - p_0) + \epsilon_2(p_2 - p_1) + \ldots + \epsilon_m(p_m - p_{m-1}),$$

or also

$$r = p_0(\epsilon_0 - \epsilon_1) + p_1(\epsilon_1 - \epsilon_2) + \ldots + p_{m-1}(\epsilon_{m-1} - \epsilon_m) + p_m \epsilon_m.$$

Since however $\epsilon_0 - \epsilon_1, \epsilon_1 - \epsilon_2, \ldots$, are positive, the quantity r is obviously smaller than $\delta \cdot \epsilon_0$.

Definition.—A function $f(x)$ is called a continuous function of x between the limits $x = a$, and $x = b$, if for an arbitrary value of x between these limits, the quantity $f(x - \beta)$ approaches arbitrarily close to the limit $f(x)$ for continually decreasing values of β.

THEOREM IV.—If the series

$$f(\alpha) = v_0 + v_1\alpha + v_2\alpha^2 + \ldots + v_m\alpha^m + \ldots$$

converges for a certain value of δ of α, it will also converge for every *smaller* value of α, and in such a way that $f(\alpha - \beta)$, for continually decreasing values of β, approaches arbitrarily close to the limit $f(\alpha)$, it being understood that α is equal to or smaller than δ.

Let

$$v_0 + v_1\alpha + \ldots + v_{m-1}\alpha^{m-1} = \phi(\alpha),$$
$$v_m\alpha^m + v_{m+1}\alpha^{m+1} + \ldots = \psi(\alpha);$$

then one has

$$\psi(\alpha) = \left(\frac{\alpha}{\delta}\right)^m \cdot v_m\delta^m + \left(\frac{\alpha}{\delta}\right)^{m+1} \cdot v_{m+1}\delta^{m+1} + \ldots;$$

hence by means of Theorem III, $\psi(\alpha) < (\alpha/\delta)^m p$, if p denotes the largest of the quantities $v_m\delta^m$, $v_m\delta^m + v_{m+1}\delta^{m+1}$, $v_m\delta^m + v_{m+1}\delta^{m+1} + v_{m+2}\delta^{m+2}, \ldots$ Then for each value of α which is equal to or smaller than δ, one can take m sufficiently large so that one has

$$\psi(\alpha) = \omega.$$

Now $f(\alpha) = \phi(\alpha) + \psi(\alpha)$ holds, and therefore

$$f(\alpha) - f(\alpha - \beta) = \phi(\alpha) - \phi(\alpha - \beta) + \omega.$$

Since further, $\phi(\alpha)$ is an entire function o. α, one can take β so small that

$$\phi(\alpha) - \phi(\alpha - \beta) = \omega$$

holds, and therefore also

$$f(\alpha) - f(\alpha - \beta) = \omega,$$

proving the theorem.

The paper continues, giving an imperfect discussion of power series with variable coefficients, Theorem V, and in Theorem VI disposes of the product of two convergent series: Parts III and IV which form the main substance of the paper deal strictly with the binomial series.

GAUSS

Second Proof of the Fundamental Theorem of Algebra

(Translated from the Latin by Professor C. Raymond Adams, Brown
University, Providence, R. I.)

Carl Friedrich Gauss was born in Braunschweig, Germany, on April 30,
1777. At an early age he displayed marked abilities which brought him to the
notice of the Duke of Braunschweig and secured for him an education. While
a student at Göttingen from 1795 to 1798 he made numerous important discoveries in several fields of mathematics. From 1807 until his death in 1855
he held the post of professor of astronomy at Göttingen, which allowed him to
devote all his time to scientific investigation. He made contributions of
fundamental significance not only in almost every leading field of pure mathematics, but also in astronomy, geodesy, electricity, and magnetism. No other
mathematician of the nineteenth century exerted so profound an influence on
the development of the science as did Gauss.

Gauss gave four proofs[1] of the fundamental theorem of algebra, which
may be stated in the form: *every algebraic equation of degree m has exactly m
roots.*[2] The significance of his first proof in the development of mathematics is

---

[1] The first was discovered in the autumn of 1797 and constituted his Dissertation; it was
published at Helmstädt in 1799 under the title "Demonstratio nova theorematis omnem
functionem algebraicam rationalem integram unius variabilis in factores reales primi vel
secondi gradus resolvi posse;" *Werke*, vol. 3 (1876), pp. 3–30. The second and third proofs,
"Demonstratio nova altera theorematis..." and "Theorematis de resolubilitate...
demonstratio tertia" appeared in 1816 in *Commentationes Societatis regiae scientiarum
Gottingensis recentiores.* vol. 3, (class. math.) pp. 107–134 and pp. 135–142 respectively;
Werke, vol. 3 (1876), pp. 33–56, 59–64. The fourth proof was published in 1850 as
"Beiträge zur Theorie der algebraichen Gleichungen" (erste Abtheilung), *Abhandlungen
der Könilgiden Gesellschaft der Wissenschaften zu Göttingen.* vol. 4, (math Klasse) pp. 3–15;
Werke, vol. 3 (1876), pp. 73–85.

[2] It is not quite certain to whom the credit belongs for first stating this theorem. That an
algebraic equation of the mth degree *may* have *m* roots was recognized by Peter Rothe
(*Arithmetica Philosophica*, Nürnberg, 1608). Albert Girard (*Invention Nouvelle en l'Algebre*,
Amsterdam, 1629) asserted that "every algebraic equation has as many solutions as the
exponent of the highest term indicates;" unfortunately he added the qualification "unless
the equation is incomplete" [i. e., does not contain all powers of x from m down to zero], but
he pointed out that if an equation admits fewer roots than its degree indicates, it is useful
to introduce as many *impossible* [i. e., complex] *solutions* as will make the total number of
roots and impossible solutions equal the degree of the equation. The clear-cut statement
of the theorem used by Gauss seems to be due to Euler in a letter dated December 15, 1742
(*Correspondence Mathématique et Physique*, ed. by Fuss, St. Petersburg, 1845, vol. 1, p. 171).
Before Gauss several attempts to prove the theorem had been made, notably by d'Alembert
(1746), whose proof was so widely accepted that the theorem came to be known, at least in
France, as d'Alembert's theorem; by Euler (1749); by Foncenex (1759); by Lagrange (1772);
and by Laplace (lectures given at the École Polytechnique in 1795 but published in its
Journal only in 1812). The term *fundamental theorem of algebra* appears to have been
introduced by Gauss.

292

made clear by his own words in the introduction to the fourth proof: "[the first proof]...had a double purpose, first to show that all the proofs previously attempted of this most important theorem of the theory of algebraic equations are unsatisfactory and illusory, and secondly to give a newly constructed rigorous proof." In the first three proofs (but not in the fourth) the restriction is made that the coefficients in the equation be real; this, however, is not a serious defect since it is readily shown[1] that the case in which the coefficients are complex can be reduced to that in which they are real. While the first proof is based in part on geometrical considerations, the second is entirely algebraic and has been described[2] as "the most ingenious in conception and the most far-reaching in method" of the four. It is appropriate, therefore, to give here the second proof.

Because of the limitations of space we shall not present the entire paper, but shall pass over the introduction (§1) and give a brief resumé of §§2–6, which contain the proofs of certain theorems, now well known, on the primality of rational integral functions and on symmetric functions. From this point on the translation will be given in full except for one section.

In §2 it is proved that if Y and Y' are any two integral functions[3] of x, a necessary and sufficient condition that they have no common factor other than a constant is that there exist two other integral functions of x, Z and Z', satisfying the identity

$$ZY + Z'Y' \equiv 1.$$

In §3 it is pointed out that if a, b, c,...is any set of m constants and if we define

$$v \equiv (x - a)(x - b)(x - c)\ldots \equiv x^m - \lambda' x^{m-1} + \lambda'' x^{m-2} - \ldots,$$

each λ, or any function of the λ's is a symmetric function of a, b, c,...

§4 is devoted to proving that any integral symmetric function of a, b, c,... is an integral function of the λ's; the uniqueness of this function of the λ's is established in §5.

In §6 the product

$$\pi = (a - b)(a - c)(a - d)\ldots \times (b - a)(b - c)(b - d)\ldots \times$$
$$(c - a)(c - b)(c - d)\ldots \times \ldots$$

is introduced. By §§4, 5 this is a certain integral function of λ', λ'',...; the same function of l', l'',...is denoted by p and is defined as the discriminant[4] of the function

$$y = x^m - l' x^{m-1} + l'' x^{m-2} - \ldots$$

This is regarded as *any* integral function of x of the mth degree with the leading coefficient 1, without regard to the question of factorability, and the l's are to be thought of as variables. On the other hand the function

$$Y = x^m - L' x^{m-1} + L'' x^{m-2} - \ldots$$

is regarded as a *particular*, though arbitrary, function of the same type, with no restrictions on the coefficients, which are to be thought of as arbitrary

---

[1] Cf. Netto, "Rationale Funktionen einer Veränderlichen; ihre Nullstellen," *Encyklopädie der Mathematischen Wissenschaften.*, vol. I, p. 233.

[2] Netto, *Die vier Gauss'schen Beweise...*, Leipzig, 1913, p. 81.

[3] [Throughout the paper Gauss uses the term *integral function* in the sense of *rational integral function*.]

[4] [*Determinant* is the term used by Gauss in common with other writers of the time.]

constants. The value of p for $l' = L'$, $l'' = L''$,...is denoted by P. It is with the factorability of Y that this paper is concerned. On the assumption that Y can be broken up into linear factors,

$$Y = (x - A)(x - B)(x - C)\ldots,$$

the following theorems are proved

I. *If P, the discriminant of Y, is zero, Y and $Y' = \dfrac{dY}{dx}$ have a common factor.*

II. *If P, the discriminant of Y, is not zero, Y and Y' have no common factor.*

7.

It is well to observe, however, that the entire strength of this very simple proof rests on the assumption that the function Y can be reduced to linear factors; but this amounts, at least in the present connection, where we are concerned with the general proof of this reducibility, to no less than assuming what is to be proved. Yet not all of those who have attempted analytic proofs of our principal theorem have been on their guard against this sort of deduction. The source of such an obvious error can be perceived in the very title of their investigations, since all have studied only the *form* of the roots of the equation while the *existence* of the roots, rashly taken for granted, should have been the object of the demonstration. But about this sort of procedure which is entirely at odds with rigor and clarity, enough has already been said in the paper referred to above.[1]   Therefore we will now establish on a more sure foundation the theorems of the preceding section, of which at least a part is essential to our purpose; with the second, and simpler, we begin.

8.

We will denote by ρ the function

$$\frac{\pi(x - b)(x - c)(x - d)\ldots}{(a - b)^2(a - c)^2(a - d)^2\ldots} + \frac{\pi(x - a)(x - c)(x - d)\ldots}{(b - a)^2(b - c)^2(b - d)^2\ldots}$$
$$+ \frac{\pi(x - a)(x - b)(x - d)\ldots}{(c - a)^2(c - b)^2(c - d)^2\ldots} + \ldots,$$

which, since π is divisible by the individual denominators, is an integral function of the unknowns $x, a, b, c,\ldots$ Furthermore we set $dv/dx = v'$, obtaining

$$v' = (x - b)(x - c)(x - d)\ldots + (x - a)(x - c)(x - d)\ldots$$
$$+ (x - a)(x - b)(x - d)\ldots + \ldots$$

---

[1] [Gauss's first proof, to which reference is made in the introduction, §1.]

For $x = a$ we clearly have $\rho . v' = \pi$, from which we conclude that the function $\pi - \rho v'$ is exactly divisible[1] by $x - a$, likewise by $x - b$, $x - c$,...and consequently also by the product v. If then we set

$$\frac{\pi - \rho v'}{v} = \sigma,$$

σ is an integral function of the unknowns x, a, b, c,...and indeed, like ρ, symmetric in the unknowns a, b, c,... Accordingly there can be found two integral functions r and s of the unknowns x, l', l'',...which by the substitutions $l' = \lambda'$, $l'' = \lambda''$,...go over respectively into ρ and σ. If analogously we denote the function

$$mx^{m-1} - (m - 1)l'x^{m-2} + (m - 2)l''x^{m-3} - \ldots,$$

i. e., the derivative dy/dx, by y', so that y' also goes over by those substitutions into v', then clearly by those same substitutions $p - sy - ry'$ goes over into $\pi - \sigma v - \rho v'$, i. e., into zero, and must therefore vanish identically (§5). Hence we have the identity

$$p = sy + ry'$$

If we assume that by the substitutions $l' = L'$, $l'' = L''$,...r and s become respectively R and S, we have also the identity

$$P = SY + RY';$$

and since S and R are integral functions of x, and P is a definite quantity or number, it follows at once that Y and Y' can have no common factor if P is not zero. This is exactly the second theorem of §6.

9.

The proof of the first theorem we will construct by showing that if Y and Y' have no common factor, P can certainly not be zero. To this end we determine by the method of §2 two integral functions of the unknown x, say $f(x)$ and $\varphi(x)$, such that the identity

$$f(x) . Y + \varphi(x) . Y' = 1$$

holds; this we can also write as

$$f(x) . v + \varphi(x) . v' = 1 + f(x) . (v - Y) + \varphi(x) . \frac{d(v - Y)}{dx},$$

or, since we have

$$v' = (x - b)(x - c)(x - d)\ldots$$
$$+ (x - a)\frac{d[(x - b)(x - c)(x - d)\ldots]}{dx},$$

---

[1] [An integral function will be said to be exactly divisible by a second integral function of the same variables if the quotient of the first by the second is a third integral function of these variables.]

in the form

$$\varphi(x) \cdot (x - b)(x - c)(x - d)\ldots$$
$$+ \varphi(x) \cdot (x - a)\frac{d[(x - b)(x - c)(x - d)\ldots]}{dx} + f(x) \cdot (x - a)(x - b)$$
$$(x - c)\ldots = 1 + f(x) \cdot (v - Y) + (x) \cdot \frac{d(v - Y)}{dx}.$$

For brevity we will denote the expression

$$f(x) \cdot (y - Y) + \varphi(x) \cdot \frac{d(y - Y)}{dx},$$

which is an integral function of the unknowns x, l', l'',..., by

$$F(x, l', l'',\ldots);$$

hence we have identically

$$1 + f(x) \cdot (v - Y) + \varphi(x) \cdot \frac{d(v - Y)}{dx} = 1 + F(x, \lambda', \lambda'',\ldots),$$

and therefore the identities

(1) $\varphi(a) \cdot (a - b)(a - c)(a - d)\ldots = 1 + F(a, \lambda', \lambda'',\ldots),$
  $\varphi(b) \cdot (b - a)(b - c)(b - d)\ldots = 1 + F(b, \lambda', \lambda'',\ldots),$
$\ldots\ldots\ldots\ldots$

If then we assume that the product of all the functions

$$1 + F(a, l', l'',\ldots), 1 + F(b, l', l'',\ldots), \ldots,$$

which is an integral function of the unknowns a, b, c,..., l', l'',...
and indeed a symmetric function of a, b, c,..., is denoted by

$$\psi(\lambda', \lambda'',\ldots, l', l'',\ldots),$$

there follows from the multiplication of all the equations (1)
the new identity

(2) $\pi\varphi a \cdot \varphi b \cdot \varphi c\ldots = \psi(\lambda', \lambda'',\ldots, \lambda', \lambda'',\ldots).$

It is furthermore clear that since the product $\varphi a \cdot \varphi b \cdot \varphi c\ldots$
involves the unknowns a, b, c,...symmetrically, an integral
function of the unknowns l', l'',... can be found which by the
substitutions $l' = \lambda'$, $l'' = \lambda''$,... goes over into $\varphi a. \varphi b. \varphi c\ldots$If
t is this function we have identically

(3) $pt = \psi(l', l'',\ldots, l', l'',\ldots),$

for by the substitution $l' = \lambda'$, $l'' = \lambda''$,... this equation becomes
the identity (2).

From the definition of the function F follows immediately the
identity

$$F(x, L', L'',\ldots) = 0.$$

Hence we have successively the following identities.

$$1 + F(a, L', L'', \ldots) = 1, \, 1 + F(b, L', L'', \ldots) = 1, \ldots,$$
$$\psi(\lambda', \lambda'', \ldots, L', L'', \ldots) = 1,$$

and

(4) $$\psi(l', l'', \ldots, L', L'', \ldots) = 1.$$

From equations (3) and (4) jointly, if we set $l' = L'$, $l'' = L''$, ..., follows the relation

(5) $$PT = 1,$$

where T denotes the value of the function t that corresponds to those substitutions. Since this value must be finite, P can certainly not be zero.

10.

From the foregoing it is apparent that every integral function Y of an unknown x whose discriminant is zero can be broken up into factors of which none has a vanishing discriminant. In fact if we find the greatest common divisor of the functions Y and $\dfrac{dY}{dx}$, Y is thereby broken into two factors. If one of these factors[1] again has the discriminant zero, it may in the same way be broken into two factors, and so we shall proceed until Y is finally reduced to factors no one of which has the discriminant zero.

Moreover one easily perceives that of those factors into which Y has been broken, at least one has the property that among the factors of its degree index the factor 2 is present no more frequently than it occurs among the factors of m, the degree index of Y; accordingly if we set $m = k \cdot 2^\mu$, where k is odd, there will be among the factors of Y at least one whose degree is $k' \cdot 2^\nu$, k' being odd and $\nu = \mu$ or $\nu < \mu$. The validity of this assertion follows immediately from the fact that m is the sum of the numbers which indicate the degree of the individual factors of Y.

11.

Before proceeding further we will explain an expression whose introduction is of the greatest use in all investigations of symmetric functions and which will be exceedingly convenient also for our

---

[1] [As a matter of fact only that factor which is the greatest common divisor can have a vanishing discriminant. But the proof of this statement would lead us into various digressions; moreover it is not necessary here, since we should be able to treat the other factor, in case its discriminant should vanish, in the same way and reduce it to factors.]

purposes. We assume that M is a function of some of the unknowns $a, b, c, \ldots$; let μ be the number of those which enter into the expression M, without reference to other unknowns which perhaps are present in M. If these μ unknowns are permuted in all possible ways, not only among themselves but also with the $m - \mu$ remaining unknowns of the set a, b, c,..., there arise from M other expressions similar to M, so that we have in all

$$m(m - 1)(m - 2)\ldots(m - \mu + 1)$$

expressions, including M itself; the set of these we call simply the *set of all M*. From this it is clear what is to be understood by the sum of all M, the product of all $M, \ldots$ Thus, for example, π can be called the product of all $a - b$, v the product of all $x - a$, v' the sum of all $\dfrac{v}{x - a}$, etc.

If perchance M should be a symmetric function of some of the μ unknowns which it contains, the permutations of these among themselves will not alter the function M; hence in the set of all M every term is multiple and in fact is present $1 . 2 \ldots v$ times if v stands for the number of unknowns in which M is symmetric. But if M is symmetric not only in v unknowns but also in v' others, and in v'' still different unknowns, etc., then M is unchanged if any two of the first v unknowns are permuted among themselves, or any two of the following v' among themselves, or any two of the next v'' among themselves, etc., so that identical terms always correspond to

$$1 . 2 \ldots v . 1 . 2 \ldots v' . 1 . 2 \ldots v'' \ldots$$

permutations. If then from these identical terms we retain only one of each, we have in all

$$\frac{m(m - 1)(m - 2)\ldots(m - \mu + 1)}{1 . 2 \ldots v . 1 . 2 \ldots v' . 1 . 2 \ldots v'' \ldots}$$

terms, the set of which we call the *set of all M without repetitions* to distinguish it from the *set of all M with repetitions*. Unless otherwise expressly stated we shall always admit the repetitions.

One further sees easily that the sum of all M, or the product of all M, or more generally any symmetric function whatever of all M is always a symmetric function of the unknowns $a, b, c, \ldots$, whether repetitions are admitted or excluded.

12.

We will now consider the product of all $u - (a + b)x + ab$ without repetitions, where u and x indicate unknowns, and denote the same by ζ. Then ζ will be the product of the following $\frac{1}{2}m(m-1)$ factors:

$$u - (a + b)x + ab, \; u - (a + c)x + ac, \; u - (a + d)x + ad, \ldots ;$$
$$u - (b + c)x + bc, \; u - (b + d)x + bd, \ldots ;$$
$$u - (c + d)x + cd, \ldots ; \ldots$$

Since this function involves the unknowns a, b, c, ... symmetrically, it determines an integral function of the unknowns u, x, l', l'', ..., which shall be denoted by z, with the property that it goes over into ζ if the unknowns l', l'', ... are replaced by λ', λ'', ...   Finally we will denote by Z the function of the unknowns u and x alone to which z reduces if we assign to the unknowns l', l'', ... the particular values L', L'', ...

These three functions ζ, z, and Z can be regarded as integral functions of degree $\frac{1}{2}m(m - 1)$ of the unknown u with undetermined coefficients; these coefficients are

for ζ, functions of the unknowns x, a, b, c, ...
for z, functions of the unknowns x, l', l'', ...
for Z, functions of the single unknown x.

The individual coefficients of z will go over into the coefficients of ζ by the substitutions $l' = \lambda'$, $l'' = \lambda''$, ... and likewise into the coefficients of Z by the substitutions $l' = L'$, $l'' = L''$, ...   The statements made here for the coefficients hold also for the discriminants of the functions ζ, z, and Z. These we will examine more closely for the purpose of obtaining a proof of the

THEOREM.—*Whenever P is not zero, the discriminant of the function Z certainly cannot vanish identically.*

14.[1]

The discriminant of the function ζ is the product of all differences between pairs of quantities $(a + b)x - ab$, the total number of which is

$$\tfrac{1}{2}m(m - 1)[\tfrac{1}{2}m(m - 1) - 1] = \tfrac{1}{4}(m + 1)m(m - 1)(m - 2).$$

This number also expresses the degree in x of the discriminant of the function ζ.   The discriminant of the function z will be of the same degree, while the discriminant of the function Z can be of lower degree if some of the coefficients of the highest power of x

---

[1] [We omit §13 which, containing a proof of the above theorem for the restricted case in which Y is reducible to linear factors, is not essential to the further developments of the paper.]

vanish. Our problem is to prove that in the discriminant of the function Z certainly not *all* the coefficients can be zero.

If we examine more closely the differences whose product is the discriminant of the function ζ, we notice that a part of them (that is, those differences between two quantities $(a + b)x - ab$ which have a common element) provides the *product of all* $(a - b)(x - c)$; from the others (that is, those differences between two quantities $(a + b)x - ab$ which have no common element) arises *the product of all*

$$(a + b - c - d)x - ab + cd$$

without repetitions. The first product contains each factor $a - b$ clearly $m - 2$ times, whereas each factor $x - c$ is contained $(m - 1)(m - 2)$ times; from this it is easily seen that the value of this product is

$$\pi^{m-2}v^{(m-1)(m-2)}.$$

If we denote the second product by ρ, the discriminant of the function ζ becomes equal to

$$\pi^{m-2}v^{(m-1)(m-2)}\rho.$$

If further we indicate by r that function of the unknowns x, l', l'',…which by the substitutions $l' = \lambda'$, $l'' = \lambda''$,…goes over into ρ, and by R that function of x alone into which r goes over by the substitutions $l' = L'$, $l'' = L''$,…, the discriminant of the function z manifestly will be equal to

$$p^{m-2}y^{(m-1)(m-2)}r,$$

while the discriminant of the function Z will be

$$P^{m-2}Y^{(m-1)(m-2)}R.$$

Since by hypothesis P is not zero, it now remains to be shown that R cannot vanish identically.

15.

To this end we introduce another unknown w and will consider the product of all

$$(a + b - c - d)w + (a - c)(a - d)$$

without repetitions; since this involves the $a, b, c,…$ symmetrically, it can be expressed as an integral function of the unknowns w, λ', λ'',… We denote this function by $f(w, \lambda', \lambda'',…)$. The number of the factors $(a + b - c - d)w + (a - c)(a - d)$ will be

$$\tfrac{1}{2}m(m - 1)(m - 2)(m - 3),$$

from which easily follow in succession the equalities

$$f(0, \lambda', \lambda'', \ldots) = \pi^{(m-2)(m-3)}$$
$$f(0, l', l'', \ldots) = p^{(m-2)(m-3)}$$

and

$$f(0, L', L'', \ldots) = P^{(m-2)(m-3)}.$$

The function $f(w, L', L'', \ldots)$ must in general be of degree

$$\tfrac{1}{2}m(m-1)(m-2)(m-3);$$

only in particular cases can it well reduce to lower degree, if perchance some coefficients of the highest power of w vanish; it is however impossible for it to be identically zero, since as the above equation shows, at least the last term of the function does not vanish. We will assume that the highest term of the function $f(w, L', L'', \ldots)$ to have a non-vanishing coefficient is Nw^ν. If we make the substitution $w = x - a$, it is clear that $f(x - a, L', L'', \ldots)$ is an integral function of the unknowns x and a, or what is the same thing, an integral function of x whose coefficients depend upon the unknown a; its highest term is Nx^ν and it therefore has a coefficient that is independent of a and different from zero. In the same way $f(x - b, L', L'', \ldots), f(x - c, L', L'', \ldots),$ $\ldots$ are integral functions of the unknown x which individually have Nx^ν as highest term, while the coefficients of the remaining terms depend upon $a, b, c, \ldots$ Hence the product of the m factors $f(x - a, L', L'', \ldots), f(x - b, L', L'', \ldots), f(x - c, L', L'', \ldots), \ldots$ will be an integral function of x whose highest term is $N^m x^{m\nu}$, whereas the coefficients of the subsequent terms depend upon $a, b, c, \ldots$

We now consider the product of the m factors

$$f(x - a, l', l'', \ldots), f(x - b, l', l'', \ldots), f(x - c, l', l'', \ldots), \ldots,$$

which as a function of the unknowns $x, a, b, c, \ldots, l', l'', \ldots$, symmetric in the $a, b, c, \ldots$, can be expressed in terms of the unknowns $x, \lambda', \lambda'', \ldots, l', l'', \ldots$ and denoted by

$$\varphi(x, \lambda', \lambda'', \ldots, l', l'', \ldots).$$

Thus

$$\varphi(x, \lambda', \lambda'', \ldots, \lambda', \lambda'', \ldots)$$

becomes the product of the factors

$$f(x - a, \lambda', \lambda'', \ldots), f(x - b, \lambda', \lambda'', \ldots), f(x - c, \lambda', \lambda'', \ldots), \ldots$$

and is exactly divisible by ρ, since as is easily seen each factor of ρ is contained in one of these factors. We will therefore set

$$\varphi(x, \lambda', \lambda'', \ldots, \lambda', \lambda'', \ldots) = \rho\psi(x, \lambda', \lambda'', \ldots),$$

ψ indicating an integral function.   From this follows at once the identity

$$\varphi(x, L', L'',\ldots, L', L'',\ldots) = R\psi(x, L', L'',\ldots).$$

We have proved above, however, that the product of the factors $f(x - a, L', L'',\ldots), f(x - b, L', L'',\ldots), f(x - c, L', L'',\ldots),\ldots,$ which is $\varphi(x, \lambda', \lambda'',\ldots, L', L'',\ldots)$ has $N^m x^{m\nu}$ as its highest term; hence the function $\varphi(x, L', L'',\ldots, L', L'',\ldots)$ will have the same highest term and accordingly will not be identically zero.   Therefore R, and likewise the discriminant of the function Z, cannot be identically zero.   Q. E. D.

16.

THEOREM.—*If*[1] $\varphi(u, x)$ *denotes the product of an arbitrary number of factors which are linear in u and x and so of the form*

$$\alpha + \beta u + \gamma x, \alpha' + \beta'u + \gamma'x, \alpha'' + \beta''u + \gamma''x,\ldots,$$

and if w is another unknown, the function[2]

$$\left(u + w\frac{d\varphi(u, x)}{dx}, x - w\frac{d\varphi(u, x)}{du}\right) = \Omega$$

will be exactly divisible by $\varphi(u, x)$.

Proof.—If we set

$$\varphi(u, x) = (\alpha + \beta u + \gamma x)Q = (\alpha' + \beta'u + \gamma'x)Q' = \ldots,$$

then $Q, Q',\ldots$ will be integral functions of the unknowns $u, x, \alpha, \beta, \gamma, \alpha', \beta', \gamma',\ldots$ and we shall have

$$\frac{d\varphi(u, x)}{dx} = \gamma Q + (\alpha + \beta u + \gamma x)\frac{dQ}{dx}$$
$$= \gamma'Q' + (\alpha' + \beta'u + \gamma'x)\frac{dQ'}{dx} = \ldots,$$

$$\frac{d\varphi(u, x)}{du} = \beta Q + (\alpha + \beta u + \gamma x)\frac{dQ}{du}$$
$$= \beta'Q' + (\alpha' + \beta'u + \gamma x)\frac{dQ'}{du} = \ldots$$

[1] [It is hardly necessary to state that the symbols introduced in the preceding section are restricted to that section, in particular that the present meaning of φ and w is not to be confused with the former.]

[2] [The usual notation for total derivative is used here and in §19 to designate partial derivatives.]

If we introduce these values into the factors of the product Ω, that is, into

$$\alpha + \beta u + \gamma x + \beta w \frac{d\varphi(u, x)}{dx} - \gamma w \frac{d\varphi(u, x)}{du},$$

$$\alpha' + \beta' u + \gamma' x + \beta' w \frac{d\varphi(u, x)}{dx} - \gamma' w \frac{d\varphi(u, x)}{du}, \ldots,$$

we obtain the expressions

$$(\alpha + \beta u + \gamma x)\left(1 + \beta w \frac{dQ}{dx} - \gamma w \frac{dQ}{du}\right),$$

$$(\alpha' + \beta' u + \gamma' x)\left(1 + \beta' w \frac{dQ'}{dx} - \gamma' w \frac{dQ'}{du}\right), \ldots,$$

so that Ω becomes the product of $\varphi(u, x)$ and the factors

$$1 + \beta w \frac{dQ}{dx} - \gamma w \frac{dQ}{du}, \; 1 + \beta' w \frac{dQ'}{dx} - \gamma' w \frac{dQ'}{du}, \ldots,$$

i. e., of $\varphi(u, x)$ and an integral function of the unknowns $u, x, w,$ $\alpha, \beta, \gamma, \alpha', \beta', \gamma', \ldots$ Q. E. D.

17.

The theorem of the foregoing paragraph is clearly applicable to the function ς, which from now on we will denote by

$$f(u, x, \lambda', \lambda'', \ldots),$$

so that

$$f\left(u + w \frac{d\varsigma}{dx}, x - w \frac{d\varsigma}{du}, \lambda', \lambda'', \ldots\right)$$

is exactly divisible by ς; the quotient, which is an integral function of the unknowns $u, x, w, a, b, c, \ldots$ and is symmetric in $a, b, c, \ldots$, we will denote by

$$\psi(u, x, w, \lambda', \lambda'', \ldots).$$

From this follow the identities

$$f\left(u + w \frac{dz}{dx}, x - w \frac{dz}{du}, l', l'', \ldots\right) = z\psi(u, x, w, l', l'', \ldots),$$

$$f\left(u + w \frac{dZ}{dx}, x - w \frac{dZ}{du}, L', L'', \ldots\right) = Z\psi(u, x, w, L', L'', \ldots).$$

If then we indicate the function Z simply by $F(u, x)$, i. e., set

$$f(u, x, L', L'', \ldots) = F(u, x),$$

we shall have the identity

$$F\left(u + w \frac{dZ}{dx}, x - w \frac{dZ}{du}\right) = Z\psi(u, x, w, L', L'', \ldots).$$

18.

Assuming that particular values of u and x, say $u = U$ and $x = X$, give

$$\frac{dZ}{dx} = X', \frac{dZ}{du} = U',$$

we have identically

$$F(U + wX', X - wU') = F(U, X)\psi(U, X, w, L', L'', \ldots).$$

Whenever U' does not vanish we can set

$$w = \frac{X - x}{U'}$$

and obtain

$$F\left(U + \frac{XX'}{U'} - \frac{X'x}{U'}, x\right) = F(U, X)\psi\left(U, X, \frac{X - x}{U'}, L', L'', \ldots\right).$$

If we set $u = U + \dfrac{XX'}{U'} - \dfrac{X'x}{U'}$, the function Z therefore becomes

$$F(U, X)\psi\left(U, X, \frac{X - x}{U'}, L', L'', \ldots\right).$$

19.

Since in case P is not zero the discriminant of the function Z is a function of the unknown x that is not identically zero, the number of particular values of x for which this discriminant can vanish is finite; accordingly an infinite number of values of the unknown x can be assigned which give this discriminant a value different from zero. Let X be such a value of x (which moreover we may assume *real*). Then the discriminant of the function $F(u, X)$ will not be zero and it follows by Theorem II, §6 that the functions

$$F(u, X) \text{ and } \frac{dF(u, X)}{du}$$

can have no common divisor. We will further assume that there is a particular value U of u, which may be real or imaginary, *i. e.*, of the form $g + b\sqrt{-1}$, and which makes $F(u, X) = 0$, so that $F(U, X) = 0$. Then $u - U$ will be an undetermined factor of the function $F(u, X)$ and hence the function $\dfrac{dF(u, X)}{du}$ is certainly not divisible by $u - U$. If then we assume that this function $\dfrac{dF(u, X)}{du}$ takes on the value U' for $u = U$, surely U' cannot be zero. Clearly, however, U' is the value of the partial derivative

$\frac{dZ}{du}$ for $u = U$, $x = X$; if then we denote by X' the value of the

partial derivative $\frac{dZ}{dx}$ for the same values of u and x, it is clear from

the proof in the foregoing section that by the substitution

$$u = U + \frac{XX'}{U'} - \frac{X'x}{U'}$$

the function Z vanishes identically and so is exactly divisible by the factor

$$u + \frac{X'}{U'}x - \left(U + \frac{XX'}{U'}\right).$$

If we set $u = x^2$, clearly $F(x^2, x)$ is divisible by

$$x^2 + \frac{X'}{U'}x - \left(U + \frac{XX'}{U'}\right)$$

and thus takes on the value zero if for x we take a root of the equation

$$x^2 + \frac{X'}{U'}x - \left(U + \frac{XX'}{U'}\right) = 0,$$

i. e.,

$$x = \frac{-X' \pm \sqrt{(4UU'U' + 4XX'U' + X'X')}}{2U'}$$

These values are manifestly either real or of the form $g + b\sqrt{-1}$.

Now it can be easily shown that for these same values of x the function Y also must vanish. For it is clear that $f(xx, x, \lambda', \lambda'', \ldots)$ is the product of all $(x - a)(x - b)$ without repetitions and so equals v^{m-1}. From this follow immediately

$$f(xx, x, l', l'', \ldots) = y^{m-1},$$
$$f(xx, x, L', L'', \ldots) = Y^{m-1},$$

or $F(xx, x) = Y^{m-1}$; accordingly a particular value of this function F cannot be zero unless at the same time the value of Y is zero.

20.

By the above investigations the solution of the equation $Y = 0$, that is the determination of a particular value of x which satisfies the equation and is either real or of the form $g + b\sqrt{-1}$, is made to depend upon the solution of the equation $F(u, X) = 0$, provided the discriminant of the function Y is not zero. It may be remarked that if all the coefficients in Y, i. e., the numbers

L', L'', ..., are real, and if as is permissible we take a real value for X, all the coefficients in $F(u, X)$ are also real. The degree of the auxiliary equation $F(u, X) = 0$ is expressed by the number $\frac{1}{2}m(m-1)$; if then m is an even number of the form $2^{\mu}k$, k designating an odd number, the degree of the second equation is expressed by a number of the form $2^{\mu-1}k$.

In case the discriminant of the function Y is zero, it will be possible by §10 to find another function $\mathfrak{Y}$ which is a divisor of Y, whose discriminant is not zero, and whose degree is expressed by a number $2^{\nu}k$ with $\gamma < \mu$. Every solution of the equation $\mathfrak{Y} = 0$ will also satisfy the equation $Y = 0$; the solution of the equation $\mathfrak{Y} = 0$ is again made to depend upon the solution of another equation whose degree is expressed by a number of the form $2^{\nu-1}k$.

From this we conclude that in general the solution of every equation whose degree is expressed by an even number of the form $2^{\mu}k$ can be made to depend upon the solution of another equation whose degree is expressed by a number of the form $2^{\mu'}k$ with $\mu' < \mu$. In case this number also is even,—i. e., if μ' is not zero,—this method can be applied again, and so we proceed until we come to an equation whose degree is expressed by an odd number; the coefficients of this equation are all real if all the coefficients of the original equation are real. It is known, however, that such an equation of odd degree is surely solvable and indeed has a real root. Hence each of the preceding equations is solvable, having either real roots or roots of the form $g + b\sqrt{-1}$.

Thus it has been proved that every function Y of the form $x^m - L'x^{m-1} + L''x^{m-2} - \ldots$, in which L', L'', ... are particular real numbers, has a factor $x - A$ where A is real or of the form $g + b\sqrt{-1}$. In the second case it is easily seen that Y is also zero for $x = g - b\sqrt{-1}$ and therefore divisible by $x - (g - b\sqrt{-1})$ and so by the product $xx - 2gx + gg + bb$. Consequently every function Y certainly has a real factor of the first or second degree. Since the same is true of the quotient [of Y by this factor], it is clear that Y can be reduced to real factors of the first or second degree. To prove this fact was the object of this paper.

III. FIELD OF GEOMETRY

DESARGUES ON PERSPECTIVE TRIANGLES

(Translated from the French by Professor Lao G. Simons, Hunter College, New York City.)

Gérard Desargues was born at Lyons in 1593 and died there in1662. Little is known of his life. For a time, he was an engineer but later he devoted himself to geometry and its applications to art, architecture, and perspective. His abilities were recognized by Descartes, between whom and Desargues there existed an unusual friendship, and by Pascal. Desargues was a geometer in a period when geometry was being discarded for the methods of analytic geometry. The work by which he is known today is *Brouillon projet d'une atteinte aux éuénemens des rencontres d'un cone avec un plan*, Paris, 1639. In this he laid the foundation of Projective Geometry.

A statement made by Chasles[1] may well be used as the reason for including this proposition in any list of contributions to the progress of mathematics. He says in effect that we owe to Desargues a property of triangles which has become fundamental and, in its applications, invaluable in recent geometry, that M. Poncelet made it the basis of his beautiful theory of homologic figures, giving the name *homologiques* to the two triangles in question, *centre d'homologie* to the point of concurrency of the three lines joining the corresponding vertices, and *axe d'homologie* to the line on which the corresponding sides intersect. The theorem is a basic one in the present day theory of projective geometry.

The source of this article is the *Oeuvres de Desargues*, réunies et analysées par M. Poudra, Tome I, Paris, 1864, pp. 413–415. On page 399, Poudra gives the original source of this particular proposition as follows: "Note:—Extrait de la Perspective de Bosse 1648, et faisant suite à la Perspective de Desargues de 1636." And so this important truth is found with two others, one of which is its reciprocal, at the end of a work written by one of the pupils and followers of Desargues.

Geometric Proposition

When the lines *HDa*, *HEb*, *cED*, *lga*, *lfb*, *HlK*, *DgK*, *EfK*,[2] in different planes or the same plane, having any order or direction

---

[1] *Aperçu historique sur l'origine et développement des méthodes en géométrie*, Paris, 1837, 2d ed., Paris, 1875, 3d ed., Paris, 1889. Notes taken from 2d ed., pp. 82–83.

[2] [Desargues omits mention of the line *abc* which is necessary in the determination of *c*. In the figure as given by Poudra, the *a* appears as *o* but for greater clarity the accompanying figure has been lettered to conform with the proof as given by Desargues.]

whatsoever meet in like points[1] the points c, $\int$, g lie in one line $c\int g$.
For whatever form the figure takes, and in all possible cases, the
lines being in different planes, abc, lga, $l\int b$ are in one plane; DEc,
DgK, $K\int E$ in another; and the points c, $\int$, g are in each of the two

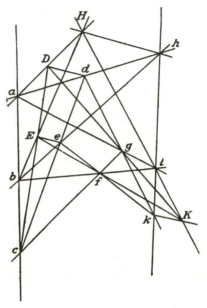

planes; hence they are in one straight line $c\int g$, and the lines in the
same plane are:

$$gD - gK \begin{cases} aD - aH \\ lH - lK \end{cases} \begin{cases} cD - cE \\ bE - bH \end{cases} \quad\Bigg| \quad cD - cE \begin{cases} gD - gK \\ \int K - \int E \end{cases} \quad\Bigg| \quad \begin{array}{l} \text{Hence} \\ c, g, \int \end{array}$$

$$\int K - \int E \begin{cases} lK - lH \\ bH - bE \end{cases} bH - bE \qquad\qquad\qquad\qquad \begin{array}{l} \text{are in} \\ \text{one line} \end{array}$$

And, conversely, if the lines abc, HDa, HEb, DEc, HK, DKg,
$KE\int$[2] in different planes or the same plane, having any direction
and formation whatever intersect in like points, the lines agb[3]
(sic) $b\int l$ tend to meet at the same goal l as HK.[4]   When these lines
are in different planes, $HKgDag$ is one plane, $HK\int Eb\int$ is another

---

[1] [Da, Eb, lK meet at H, etc.]

[2] [Desargues again omits from the given sets of collinear points $ag\int$ and $b\int l$
which are named in his conclusion and $c\int g$ which is implied in the explanation
that follows.]

[3] [This should be $ag\int$ as indicated later in the discussion.]

[4] [The idea of two lines stretching toward a goal is an excellent one.]

and *cbagf* still another, and *HlK*, *bfl*, *agl* are the intersections of these three planes; hence they must meet at the same point *l*. And if the lines are in a single plane, producing the given line *agl* from *a* to meet the line *HK*, and then drawing *lb*, it can be proved that it meets *EK* at a point such as *f* which is in a line with *c* and *g*, that is to say that it goes to *f*, and hence that the two lines *ag*, *bf* intersect at a point *l*, in *HK*. And taking the same lines in different planes once more, if through the points *H*, *D*, *E*, *K* other lines *Hb*, *Dd*, *Ee*, *Kk* are passed tending to an infinitely distant goal, otherwise parallel to one another, and intersecting one of the planes *cbagfl* in the points *b*, *d*, *e*, *k* respectively; then *b*, *l*, *k* are in one straight line; *b*, *d*, *a* in one; *b*, *e*, *b* in one; *k*, *g*, *d* in one; *k*, *f*, *e* in one; and *c*, *e*, *d* in one, because by construction the lines *Hb*, *Kk*, *HlK* are in one plane; *abc*, *bfl*, *klb* in another; and the points *b*, *l*, *k* are in each of the planes; hence they are in one straight line; and so with each of the other sets of three: and all of these lines in the same plane *cbagfl* are divided by means of the parallels through the points *H*, *D*, *E*, *K*, each in the same manner as its corresponding one in the figure of the several planes. Thus the figure which the parallels are used to determine in a single plane *bdabcedgfkl* corresponds line to line, point to point, and proof to proof to that of *abcEHlkgf* in different planes, and the properties of the figures can be reasoned about from either one or the other, and by this means there may be substituted for a figure in relief one in a single plane.

The following note is given by Poudra, the editor of Desargues's works, vol. I, pp. 430–433.

ANALYSIS
OF THE FIRST GEOMETRICAL PROPORTION
OF DESARGUES

Note: The small letters *a*, *b*, *c* of the figure denote points situated in the plane of the paper while the capitals *E*, *D*, *H*, *K* denote points which may be outside the paper.

————

The proposition contains three distinct parts:

1. If two triangles *abl*, *DH*(sic)*K*, in space or in the same plane, are such that the three straight lines *aD*, *bE*, *lK* which connect the corresponding vertices of the two triangles meet at a point *H*, it follows; that the sides of the two triangles meet in three points *c*, *f*, *g* which are in one straight line.

2. If the corresponding sides of two triangles meet in three collinear points c, f, g it follows conversely, not only that the three straight lines aD, bE, lK which connect the corresponding vertices are concurrent at the point H; but also that the three lines ag, bf, HK pass through the point l because c may be regarded as the apex of a pyramid passing through the vertices of the triangles bfE, agD, from which, etc.

Likewise, considering f as the apex of another pyramid passing through the vertices of the two triangles bcE, lgK, it can be demonstrated that the corresponding sides give the collinear points A(sic), D, H.

And again, taking g for the apex, the two triangles acD, lfK have their corresponding sides meeting in collinear points b, E, H.

3. If from the three vertices D, E, K of the triangle DEK and from the vertex H, the vertical lines Dd, Ee, Kk, Hb are drawn, these lines intersect the plane of the paper in the points d, e, k, b which are such that the line bd passes through the point a of the line HD, likewise bk passes through l, de through c, be through b, dk through g. Thus there is determined in the plane of the paper a figure which corresponds point to point, line to line and proof to proof to that in different planes and then the properties of the figures may be reasoned about from either one or the other, and by this means there may be substituted for a figure in relief one in a single plane.

An important remark which reveals the end which Desargues had in view in this proposition.

DESARGUES

On the 4-Rayed Pencil

(Translated from the French by Professor Vera Sanford, Western Reserve University, Cleveland, Ohio.)

The first work to deal with the harmonic properties of points and of pencils of lines was written by Gérard Desargues (1593–1662) with the title *Brouillon proiect d'une atteinte aux éuénemens des rencontres d'un cone auec un plan* (1639), referred to in the preceding article. This title might be translated "Proposed draft of an investigation into the results of the intersection of a cone and a plane."

This work makes use of many terms of fanciful origin, and the definition of involution differs from the ordinary definition of today. Six points on a line are said to be in involution when the product of the segments cut off from a given point by one pair of the points is equal to the corresponding products for each of the other two pairs. The line on which these points are situated is called a tree (*arbre*). The common intersection of several lines is called a knot (*noeud*), each of the intersecting lines is a bough (*rameau*), but in this translation, the terms point and ray will be used since their connotation is clearer. Any of the segments between two of the points that are in involution is a branch (*branche*). Desargues uses many other terms whose meaning is not obvious from the context and which necessitate constant reference to the discussion of these terms which begins his book. His proofs, as will appear in the translation which follows were given without the use of algebraic symbolism and in consequence they seem prolix and involved.

4 Points in Involution.[1]—We may imagine the words four points in involution as expressing two cases of the same sort, since one or the other of these two cases results: the first where four points on a line each at a finite distance yield three consecutive segments of which either end segment is to the middle segment as the sum of the three segments is to the other end segment, and the second in which three points are at finite distances on a line with a fourth point at an infinite distance and in which case, the points likewise yield three segments of which one end segment is to the middle segment as the sum of the three is to the other end segment. This is incomprehensible and seems at first to imply that the three points at a finite distance yield two segments that are equal to

---

[1] [*Œuvres de Desargues*, edited by M. Poudra, Paris, 1864, Vol. I, p. 135–136.]

each other, with the midpoint as both origin[1] and as the endpoint coupled with an infinite distance.

Thus we should take careful note that a line which is bisected and then produced to infinity is one of the cases of the involution of four points.

. . . .[2]

Mutually Corresponding Rays.—In the case of but four points *B, D, G, F* in involution on a line through which there pass four rays *BK, DK, GK, FK* radiating in a pencil[3] from the point *K*, the pairs of rays *DK, FK* or *GK, BK* that pass through the corresponding points *D, F* or *B, G* are here called corresponding rays.

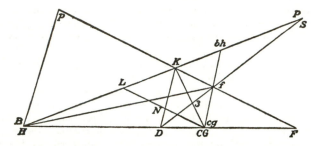

In this case, when the two corresponding rays *BK, GK* are perpendicular to each other, they bisect each of the angles between the other two corresponding rays *DK, FK*.

Since the line *Df* was drawn parallel to any one of the rays *BK* which is perpendicular to its corresponding ray *GK*, the line *Df* is also perpendicular to the ray *GK*.[4]

Furthermore, because of the parallelism of *BK* and *Df*, the ray *GK* bisects *Df* at the point 3.

---

[1] The term *souche*, here translated "origin," is the one used for the point from which the line segments of the points in involution are measured.]

[2] [Here follow many theorems expressed, as has been noted before, in verbose terms. The author follows these theorems with the observation that "For this draft, it is sufficient to note the particular properties with which this case abounds, and if this method of procedure is not satisfactory in geometry, it is easier to suppress it than to develop it clearly and to give it complete form."]

[3] [P. 153, *Ordonnance.*]

[4] [The previous work to which this refers is summarized by Poudra in his commentary as: "Whenever four lines of a pencil are in involution, every parallel to one of the rays of the pencil cuts the three others in three points, and the ray which is conjugate to this parallel bisects the segment formed by the two others. Conversely, if one of the rays bisects a segment taken on a line, this line is parallel to the fourth ray of the involution."]

Thus the two triangles $K3D$, $K3f$ each have a right angle at the point 3, and they have the sides $3K$, $3D$, and $3K$, $3f$ which include the equal angles $K3d$, $K3f$ equal to each other.

Since the two triangles $K3D$, $K3f$ are equal and similar, then the ray GK bisects one of the angles DKF between the corresponding rays DK, FK, and the ray BK clearly bisects the other of the angles made by the same corresponding rays DK, FK.

When anyone of the rays GK bisects one of the angles DKF between the two other mutually corresponding rays DK, FK, this ray GK is perpendicular to its corresponding ray BK which also bisects the other angle included by the corresponding rays DK, FK.

[This is true] since when the line Df was drawn perpendicular to any ray GK, the two triangles $K3D$ and $K3f$ each have a right angle at the point 3 and furthermore each has an equal angle at the point K and also a common side $K3$, consequently they are similar and equal and the ray GK bisects Df at the point 3.

Therefore, the ray BK is parallel to the line Df and it also is perpendicular to the corresponding ray GK.

When in a plane, there is a pencil of four lines BK, DK, GK, FK from the vertex K, and when two of these lines as BK and GK are mutually perpendicular and bisect each of the angles which the two others FK, DK make, it follows that these four lines cut any other line $BDGF$ lying in their plane in four points B, D, G, F which are arranged in involution.

When a line FK in a plane bisects one of the sides Gb of the triangle BGb at f, and when through the point K which is thus determined on one of the other two sides Bb there passes another line KD parallel to the bisected side Gb, the four points B, D, G, F determined by this construction on the third side BG of the triangle are in involution.

When from the angle B which subtends the bisected side Gb there passes another line Bp parallel to the bisected side Gb, the four points F, f, K, p determined on the line FK by the three sides BG, Gb, Bb of the triangle BGb and the line Bp are themselves in involution.

And in the second case, by drawing the line Bf in a way similar to the line GH, the three points G, f, b at a finite distance and the (point at a) infinite distance are in involution (since) to these pass four branches of a pencil whose vertex is at B and which consequently determines on the line FK four points F, f, K, p which are in involution.

Then as the line *FGB* cuts on the line *bf* a segment *Gf* equal to the segment *bf*, the side of a triangle such as *bfK*, this is equivalent to saying that this line *FGH* is double the side *bf* of the triangle *bfK*.

When in a plane a line *FGB* is double one of the sides *bf* of a triangle *bfK*, and when from the point *B* which it determines on either side *bK* of the two other sides of the same triangle, there passes a line *Bp* parallel to the double side *bf*, then this construction yields on the third side *Kf* of the triangle *bfK*, four points *F*, *f*, *K*, *p* which are in involution.

This is evident when the line *BF* is drawn.

When from the angle *K* subtended by the double side *bf* there passes a line *KD* parallel to the double side *bf*, this construction gives on a line the double of *FB*, four points *F*, *B*, *D*, *B* which are in involution as is evident when the line *KG* is drawn.

This theory swarms with similar means of developing theorems when four or even three points are in involution on a line, but these are sufficient to open the mine of that which follows.

PONCELET

On Projective Geometry

(Translated from the French by Professor Vera Sanford, Western Reserve
University, Cleveland, Ohio.)

The contribution of Jean-Victor Poncelet (1788–1867) to the field of geometry was his theory of projections. The idea of projections had been used by
Desargues, Pascal, Newton, and Lambert, but Poncelet's formulation of the
principle of continuity added greatly to the usefulness of the method. This
work was developed under adverse circumstances as Poncelet himself states
in the preface to his *Traité des propriétés projectives des figures:*

This work is the result of researches which I began as early
as the spring of 1813 in the prisons of Russia.[1] Deprived of books
and comforts of all sorts, distressed above all by the misfortunes
of my country and of my own lot, I was not able to bring these
studies to a proper perfection. I discovered the fundamental
theorems, however, that is to say, the principles of the central
projection of figures in general and of conic sections in particular,
the properties of secants and of common tangents to curves, the
properties of polygons that are inscribed or circumscribed in them
etc.[2]

Introduction[3]
· · · · · · · · · ·

Let us consider any figure whatever, in a general position and
undetermined in some way among all those (positions) which it

---

[1] [Poncelet held a commission in the French army and was taken prisoner by
the Russians during Napoleon's retreat from Moscow.]

[2] [*Traité des propriétés projectives des figures*, preface to the first edition (1822)
quoted in the second edition, Paris 1865, Vol. I. The author states that the
first volume of the second edition is identical with the first except for notes
inserted at its end. The second volume is largely made up of material that
had not been printed earlier.]

[3] [The introduction begins with a statement of the work of Monge in geometry, saying,—"The same works of Monge and of his pupils...have proved
that descriptive geometry 'the language of the artist and of the man of genius,'
is sufficient in itself and attains all the height of the concepts of algebraic
analysis." He says, however, that there are still lacunae to be filled in spite of
the many discoveries in regard to the properties of lines and surfaces of the
second order that have been discovered by the principles of rational
geometry. He then passes to a statement of his concept of continuity.]

may assume without violating the laws, conditions, or relations which exist between the divers parts of the system (of figures). Let us suppose that after these are given, we have found one or more relations or properties whether they be *metric* or *descriptive* belonging to the figure which depend on ordinary explicit reasoning, that is to say, on steps which in a particular case we regard as purely rigorous.   Is it not evident that if in keeping the same given properties, we begin to vary the original figure by insensible degrees, or if we subject certain parts of this figure to a continuous motion of any sort, is it not evident that the properties and relations found for the first system remain applicable to successive stages of this system, provided always that we have regard for the particular modifications which may occur as when certain magnitudes vanish or change in direction or sign, etc., modifications which will be easy to recognize *a priori* and by trustworthy rules?

This is at least what one may conclude without the trouble of implicit reasoning, and it is a thing which in our day is generally admitted as a sort of axiom whose truth is manifest, incontestable, and which need not be proved.   Witness the principle of the *correlation of figures* assumed by M. Carnot in his *Geometry of Position* to establish the rule of signs.   Witness also the principle of *functions* used by our greatest geometers to establish the bases of geometry and mechanics.   Witness finally the *infinitesimal calculus*, the *theory of limits*, the *general theory of equations*, and all the writings of our days in which a degree of generality is attached to the concepts.

This principle, regarded as an axiom by the most learned geometers, is the one which we may call the *principle* or the *law of continuity* of the mathematical relations of abstract and figurative size [grandeur abstraite et figurée].

. . . .

CHAPTER I

Preliminary Concepts of Central Projection

1. In the work that follows, we will use the word *projection* in the same sense as the word *perspective*, with almost no exception. Thus projection will be *conical* or *central*.

In this type of projection, the surface on which the given figure is projected may be any surface and the figure itself may be arbitrarily placed in space; but this great generalization is useless for the particular purpose of the researches that follow, and we will assume in general that the given figure and the surface of projection are both planes. Whenever it happens that we are obliged to use the word *projection* in a sense that is more extended, or, on the other hand, when its meaning is still more restricted, we will take pains to state this expressly in advance or else we will use terms that are convenient and precisely defined.

Now let us imagine that from a given point which is taken as a center of projection, there radiates a pencil of straight lines directed toward all of the points of the figure which is traced on a plane. Then if one should cut this pencil of lines [droites projetantes] by a plane located arbitrarily in space, a new figure will appear on this plane which will be the *projection* of the first.

2. It is evident that this projection does not change the relationships [correlation] nor the degree nor order of the lines of the original figure, nor in general, any type of graphical dependence between the parts of the figure that are concerned only with the undetermined direction of lines, their mutual intersection, their contact, etc. It can only change the form or the particular types of lines, and in general all the dependencies that concern absolute and determined quantities such as the opening of angles, constant parameters, etc. Thus, for example, if one line is perpendicular to another in the original figure, we cannot conclude that it will be so in the projection of the figure on a new plane.

3. All these properties of central projection result in a purely geometric manner from its very nature and from concepts that are most commonly admitted, and there is no need to have recourse to algebraic analysis to discover and to prove them. Thus to prove that a line of degree m remains of the same degree in its projection, it is sufficient to notice that the first line cannot be cut in more than m points by a line drawn at will in the plane. This must necessarily be the same in the other plane since the projection of a straight line is always a straight line which must pass through all the points which correspond to those of the original.

4. According to the definition of Apollonius which is generally accepted in geometry, a conic section or simply a conic is a line along which an arbitrary plane meets any cone that has a circular base. A conic thus is nothing else than the projection of a circle, and according to what precedes, it is also a line of second order since the circumference of a circle cannot be cut in more than two points by a straight line arbitrarily drawn in a plane.

5. A figure whose parts are only graphically dependent in the way that has been stated above, that is to say, whose dependence is not changed as a result of projection, will be called a projective figure in the work that follows.

These relations themselves and in general all the relations or properties that subsist at the same time, in the given figure and in its projections are called *projective relations* or *projective properties*.

6. After what we shall say of the projective properties of the position of a figure or its graphic properties, it will always be easy to tell whether the properties are of this sort by their simple statement or from an inspection of the figure. This results at once from their particular nature so that it will suffice to establish and prove these [results] for any projection whatever of the figure to which they belong because they are applicable in general to that figure itself and to all its possible projections.

7. It is certain that nothing can be indicated *a priori* as to the projective properties that concern the relations of size which we call *metric*, and whether they will persist in all the projections of the figure to which they belong. For example, the known relation between the segments of the secants of a circle which concerns only undetermined quantities is not for that reason a projective relation, for we know well that it does not persist for any conic section, the projection of this circle. The reason for this is that this relation depends implicitly on the parameter or radius.

On the other hand, in the case where the given figure includes lines of a particular kind as for example the circumference of a circle, it is not necessary to conclude in consequence that all the relations that belong to it cease to exist as before in the general projections of the figure for if these relations are not dependent on any determined and constant quantity, and if they are all of a type, the contrary will evidently take place.

If then a figure of a particular type enjoys certain metric properties, we cannot state *a priori* and without a preliminary examination either that the properties persist or cease to persist in the diverse projections of the original figure. We constantly realize, however, the importance of recognizing in advance whether such or such relations are or are not projective in nature for it follows that having shown this relation for a particular figure, we can at once extend it to all possible projections of the figure.

8. It does not seem easy to establish a simple rule for all cases. Trigonometric methods and the analysis by coordinates only lead

to results that are forbidding because of their prolix calculations. In view of its importance, however, the question is worthy of attracting the attention of geometers. While waiting for them to solve this in a convenient way for projective relations in general, let us busy ourselves with a particular class of relations less extensive, whose character is remarkable for its simplicity which facilitates the verification and recognition of the relations with which it deals.[1]

. . . .

CHAPTER III

Principles Relating to the Projection of One Plane Figure into Another

99. From the very nature of the properties of a projection as they were defined (5), it follows that when we wish to establish a certain property regarding a given figure, it is sufficient to prove that the property exists in the case of any one of its projections. Among all the possible projections of a figure, some one may exist that may be reduced to the simplest conditions and from which the proof or the investigations which we have proposed may be made with the greatest ease. It may require but a brief glance or at most the knowledge of certain elementary properties of geometry to perceive it or to know it. For example, to take a particular case, suppose the figure contains a conic section, this may be considered the projection of another [figure] in which the conic section is replaced by the circumference of a circle and this single statement is sufficient to shift the most general questions regarding conic sections to others that are purely elementary.

100. From this, we realize the importance of the theory of projections in all research in geometry and [we see] how much the considerations which it presents abridge and facilitate these researches.

With any given figure, all this amounts to as one sees, is to discover that projection which offers conditions that are the most elementary and that are the best suited by their simplicity to

---

[1] [Poncelet here passes to a discussion of the harmonic division of lines, the properties of a harmonic pencil, the similarity of conics, and the projective relations of the areas of plane figures. His second chapter deals with the secants and ideal chords of conic sections, and their properties, including the concept of poles and polars. An ideal chord is a line whose points of intersection with a curve are imaginary.]

disclose the particular relations which we have in mind. The theory of projections has already furnished several methods of attaining this,[1] but it lacks much which it could furnish, and our actual aim is to discover these other methods and to make them known in a purely geometric manner by the aid of ideas established in the foregoing work.

101. Let us first recall the principles that are generally known, the proof of which is the simplest possible.

Any plane figure which comprises a system of lines or of curves which have a common point of intersection may be regarded as the projection of another of the same kind or order in which the point of intersection has passed to infinity and in which the corresponding lines have become parallel.

Clearly it is sufficient in order that this take place that the planes of projection be taken parallel to the line that joins the point of intersection of the first system to the center of projection which is arbitrarily chosen in other respects.

102. Conversely, *a plane figure which comprises a system of straight lines or curves which are parallel or concurrent at infinity, has in general for its projection on any plane a figure of the same order in which the corresponding lines are concurrent at a point at a finite distance, the projection of that of the first system.*

When the plane of projection is parallel to the line that joins the point of intersection to the center of projection, it is evident that the lines of the system are parallel or concurrent at infinity and if we assume, furthermore, that it is parallel to the plane of the first figure, the projection becomes similar to this figure and it is similarly placed.[2]

103. These theorems, giving a geometric interpretation to this concept, generally adopt the idea that *parallel lines meet in a single point at infinity.* We shall see as a consequence that the points of intersection at an infinite distance and at a finite distance are reciprocally interchangeable as a result of projection.

---

[1] [Poncelet is here probably referring to the work of previous writers who used the method of projections to some extent.   Among these are Desargues, Pascal, and Newton.]

[2] [Poncelet here refers to an article of his second chapter in which he shows that when a conic surface is cut by two planes, the resulting sections in general have two points in common thus determining their common secant, and certain metrical relations obtain.   When the planes are parallel, and consequently similar, the common chord becomes an "ideal chord" and the same relations hold true.]

104. If the point of intersection which we are considering were at the same time a point of contact for certain lines of the original figure, it will, according to the nature of central projection, be equally a point of contact of the same lines. Consequently, when this point passes to infinity, the lines in question become tangents at infinity instead of having their corresponding branches merely parallel, a thing which we ordinarily express by saying that they are *asymptotic*.

Furthermore, they may be asymptotes of the first, second... order, if the *osculation* of the original curves was of that order.

Thus asymptotic lines and lines whose course is parallel in certain regions or which have asymptotes that are parallel, enjoy the same properties as lines of the same order which intersect or which touch at a given point so that they differ from these lines only in the fact that their point of intersection or of contact lies at infinity.

105. *Any plane figure whatever, which involves a given straight line, may be considered as the projection of another in which the corresponding line has passed to infinity. Consequently, every system of lines or curves meeting in a point on the first line in the original figure becomes a system of lines that are parallel or concurrent at infinity in the projection derived from it.*

It evidently is sufficient that this may take place that the plane of projection be taken parallel to the plane which encloses the straight line of the original figure and the center of projection which is otherwise arbitrary.

106. Conversely, *any figure which includes an arbitrary number of systems of lines whether straight or curved, that are respectively parallel or asymptotic, that is to say, lines meeting at infinity in each system, has in general for its projection on any plane, another figure in which the points of concurrence at infinity in the first are arranged on one and the same line at a distance which is given and finite.*

107. These last considerations, deduced wholly from the elementary principles of central projection, give an interpretation of the concept of metaphysics that we have already mentioned:

All points situated at infinity on a plane, may be considered ideally as distributed on a single straight line, itself located at infinity on this plane.

We see by what preceeds that all these points are represented in projection by the points of a single straight line situated in general at a given finite distance.

This paradoxical concept thus receives a definite and natural meaning when it is applied to a given figure in a plane and when we suppose that this figure is put in projection on any other plane. The lack of determination involved in the direction of the line at infinity becomes precisely the lack of determination that exists in the case of the plane which projects this line at the moment when it becomes parallel to the plane of the given figure. But we also see that this lack of determination has no place except because we persist in giving mentally a real existence to the line of their common intersection when they [the planes] have become parallel. Besides the lack of determination does not really exist except in the law or in the original construction which gives this line when it is at a finite distance, and not in the same direction, when it ceases to exist in an absolute and geometric manner.

108. In all the preceding theorems, nothing has determined the position of the center of projection in space. This is wholly arbitrary, and for any given point, one can always fulfil one or another of the prescribed conditions. This is not the case in the theorems which follow. They cannot take place except for a series of particular positions of the center of projection and as the proof of this is difficult, and as it is not known to geometers as yet, it is timely that we pause here and devote ourselves to it.

109. *Any plane figure which involves a given line and a conic section may in general be regarded as the projection of another (figure) in which the line has wholly passed to infinity and in which the conic section has become the circumference of a circle.*

To prove this principle in a way which leaves absolutely nothing to be desired from the point of view of geometry, let us suppose that we are required to solve the following question:

110. *Having given a conic section (C) and a straight line MN situated at will in a plane, to find a center and a plane of projection such that the given line MN shall be projected to infinity on this plane and that at the same time, the conic section shall be represented by a circle.*

Let S be the unknown center of projection. According to the conditions of the problem, the plane which passes through this point and through the line *MN* should be parallel to the plane of projection and this last line should cut the conical surface of which (C) is the base and S the vertex in the circumference of a circle. In the first place, it follows from this that the line *MN* should be entirely exterior to the conic section (C), that is to say, [it should be] the *ideal secant* of the curve.

In the second place, if one determines the ideal chord MN which corresponds to this line and to the conic (C) when we join its midpoint O to the center of projection S by the line SO, this line should become equal to half the ideal chord OM and it should make with it an angle MOS which should be a right angle. That is to say:

The auxiliary center of projection should be located on the circumference of a circle described on the midpoint of the ideal chord which corresponds to the given line as center, with a radius equal to half of this chord,[1] and in a plane which is perpendicular to it.

As this is no other condition to be fulfilled, we may conclude that there exists an infinity of centers and planes of projection which satisfy the conditions of the question. But for this [to be true] it is necessary that the line MN shall not meet the curve for otherwise the distance OS or OM will become imaginary.

. . . . [2]

140. It seems actually useless to develop these ideas further, the more so because the course of this work which has as its object the application of the preceeding concepts to the study of the projective properties of conic sections, will serve as a natural elucidation of everything which these concepts may have retained which is obscure or difficult. Besides, one will then see the simplicity with which these concepts lead to properties already known and to an infinity of others which ordinary geometry does not seem to touch easily. And this [is true] without the use of any auxiliary construction and by the use of the simplest theorems, that is those that concern only the direction and the size of the lines of elementary figures, and which require for the most part only a quick glance to make them seen and recognized. Also, I content myself very often by citing the theorems without obliging myself to prove them since they are self-evident or else the simple consequences of theorems already known.

---

[1] [Algebraically, this reduces to saying that the square of the radius is equal to the square of half the chord intercepted in the conic by the corresponding line, but taken with the contrary sign. Consequently, this radius may be either real or imaginary.]

[2] [Here follow many theorems showing how the various types of figures may be regarded as the projections of figures in which the conics are circumferences of circles etc.]

PEAUCELLIER'S CELL

(Translated from the French by Professor Jekuthiel Ginsburg, Yeshiva College, New York City.)

It is possible to draw a circle without having another circle around which to trace it, because a pair of compasses permits of this being done.   Until comparatively recent times, however, it has not been possible to draw a theoretically straight line without having another straight line (ruler, straight-edge) along which to trace it.   The instrument commonly known as the Peaucellier Cell permits of the drawing of such a line by means of a linkage illustrated in the following article.   It was first described in a communication to the Société Philomathique of Paris in 1867.   It was later discovered independently by a Russian mathematician, Lipkin and a description of his instrument is given in the *Fortschritte der Physik* for 1871, p. 40.

Peaucellier's matured description is given in the *Journal de Physique*, vol. II (1873), and is translated below.

It is known that in practical mechanics it is often necessary to transform circular motion into continuous rectilinear motion. Watt realized this with a high degree of perfection.   The transmission invented by him offers many advantages in that it gives very continuous motion without considerable shocks or friction. On account of certain circumstances Watt's invention has serious faults.[1]

· · · · · · · · · ·

The solution that we offer is derived directly from a geometrical principle to which we were led while searching for a solution of the

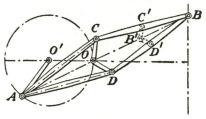

problem which Watt put to himself when he created his parallelogram.   This principle gives a rigorous solution of the problem. It was communicated in the year 1867 to the Philomathic Society of Paris.   The Russian mathematician Lipkin discovered it

---

[1] [A paragraph containing a criticism of Watt's invention is omitted here.]

independently but at a later period.[1]   Here is how the problem is solved.

Conceive of a balancer composed of six movable bars $AC = CB = BD = AD$, $OC = OD$, of which the center of rotation O is fixed. If the extremity A be made to describe a circle passing through C (which can be easily obtained by linking it to the bar OA) of which the radius is OO',—then the opposite extremity B will describe a line perpendicular to the direction OO' and will guide therefore the bar of the piston.

It may be seen among other things that if on the links BC, BD, equal lengths BC', BD' be taken, and if one links the points C', D' with the pieces CB, DB, also linked in B', this point B' will describe a straight line parallel with that described by B, because of the proportion $BC' : B'C' = BC : OC$.

---

[1] [Lippman Lipkin, born at Salaty, Russia, in 1846; died at St. Petersburg (Leningrad), Feb. 21, 1876.]

PASCAL

"Essay pour les Coniques"

(Translated from the French by Dr. Frances Marguerite Clarke, Bryn Mawr College, Bryn Mawr, Penna.)

When Pascal (see page 67) was only sixteen years old, he wrote a brief statement which was doubtless intended by him as the first step in an extended study of conics to be undertaken at some future time. In the following year it was printed in the form of a broadside and bore the simple title, "Essay povr les coniqves. Par B. P." Of this single page only two copies are known, one at Hannover among the papers of Leibniz, and the other in the Bibliothèque nationale at Paris. The illustrations here given appeared at the top of the original broadside. The third lemma involves essentially the "Mystic Hexagram" of Pascal. This translation first appeared in *Isis*, X, 33, with a facsimile of the entire essay, and is reproduced in revised form with the consent of the editors.

Essay on Conics

First Definition

When several straight lines meet at the same point, or are parallel to each other, all these lines are said to be of the same order or of the same *ordonnance*, and the totality of these lines is termed an order of lines, or an *ordonnance* of lines.[1]

Definition II

By the expression "conic section," we mean the circle, ellipse, hyperbola, parabola, and an angle; since a cone cut parallel to its base, or through its vertex, or in the three other directions which produce respectively an ellipse, a hyperbola, and a parabola, produces in the conic surface, either the circumference of a circle, or an angle, or an ellipse, a hyperbola, or a parabola.

---

[1] [This definition is taken almost word for word from Desargues. See the notes to the Brunschvicg and Boutroux edition, t. I., Paris, 1908. This translation is made from the facsimile of the original as given in this edition, and acknowledgment is hereby made of the assistance rendered by these notes in determining the meaning of several passages. It should also be said that the text of this edition is in marked contrast to the imperfect one given in the Paris edition of 1819.]

Definition III

By the word "droite" (straight) used alone, we mean "ligne droite" (straight line).[1]

Lemma I

If in the plane M, S, Q, two straight lines MK, MV, are drawn from point M and two lines SK, SV from point S; and if K be the point of intersection of the lines MK, SK; V, the point of intersection of the lines MV, SV; A, the point of intersection of the lines MA, SA; and μ, the point of intersection of the lines

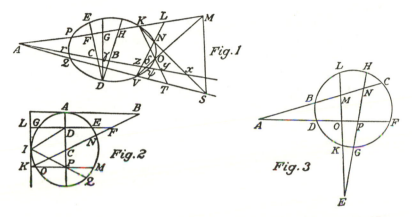

Fig. 1

Fig. 2

Fig. 3

MV, SK; and if through two of the four points A, K, μ, V, which can not lie in the same line with points M, S, and also through points K, V, a circle passes cutting the lines MV, MP, SV, SK at points O, P, Q, N, then I say that the lines MS, NO, PQ, are of the same order.

Lemma II

If through the same line several planes are passed, and are cut by another plane, all lines of intersection of these planes are of the same order as the line through which these planes pass.

On the basis of these two lemmas and several easy deductions from them, we can demonstrate that if the same things are granted as for the first lemma, that is, through points K, V, any conic section whatever passes cutting the lines MK, MV, SK, SV in

---

[1] [In this translation, the word "line," meaning a straight line-segment, will be used for "droite."]

points P, O, N, Q, then the lines MS, NO, PQ will be of the same order. This constitutes a third lemma.[1]

By means of these three lemmas and certain deductions, therefrom, we propose to derive a complete ordered sequence of conics,[2] that is to say, all the properties of diameters and other straight lines,[3] of tangents, &c., the construction of the cone from substantially these data, the description of conic sections by points, etc.

Having done this, we shall state the properties which follow, doing this in a more general manner than usual. Take for example, the following: If in the plane MSQ, in the conic PKV, there are drawn the lines AK, AV, cutting the conic in points P, K, Q, V; and if from two of these four points, which do not lie in the same line with point A,—say the points K, V, and through two points N, O, taken on the conic, there are produced four lines KN, KO, VN, VO, cutting the lines AV, AP at points L, M, T, S,—then I maintain that the proportion composed of the ratios of the line PM to the line MA, and of the line AS to the line SQ, is the same as the proportion composed of the ratio of the line PL to the line LA, and of the line AT to the line TQ.

We can also demonstrate that if there are three lines DE, DG, DH that are cut by the lines AP, AR at points F, G, H, C, γ, B, and if the point E be fixed in the line DC, the proportion composed of the ratios of the rectangle $EF.FG$ to the rectangle $EC.C\gamma$, and of the line $A\gamma$ to the line AG, is the same as the ratio of the rectangle $EF.FH$ to the rectangle $EC.CB$, and of the line AB to the line AH. The same is also true with respect to the ratio of the rectangle $FE.FD$ to the rectangle $CE.CD$. Consequently, if a conic section passes through the points E, D, cutting the lines AH, AB in points P, K, R, ψ, the proportion composed of the ratio of the rectangle of these lines EF, FC, to the rectangle of the lines EC, $C\gamma$, and of the line γA to the line AG, will be the same as the ratio of the rectangle of the lines FK, FP, to the rectangle of the lines CR, $C\psi$, and of the rectangle of the lines AR, $A\psi$, to the rectangle of the lines AK, AP.

We can also show that if four lines AC, AF, EH, EL intersect in points N, P, M, O, and if a conic section cuts these lines in

---

[1] [This involves the so-called "Mystic Hexagram," the dual of Brianchon's Theorem given on page 331. Pascal did not state the hexagram theorem in the form commonly seen in textbooks.]

[2] ...des éléments coniques complets.

[3] ...et côtés droits.

points *C, B, F, D, H, G, L, K,* the proportion consisting of the ratios of rectangle *MC.MB* to rectangle *PF.PD,* and of rectangle *AD.AF* to rectangle *AB.AC,* is the same as the proportion composed of the ratios of rectangle *ML.MK* to the rectangle *PH.PG,* and of rectangle *EH.EG* to rectangle *EK.EL.*

We can also demonstrate a property stated below, due to *M.* DESARGUES of Lyons, one of the great geniuses of this time and well versed in mathematics, particularly in conics, whose writings on this subject although few in number give abundant proof of his knowledge to those who seek for information. I should like to say that I owe the little that I have found on this subject to his writings, and that I have tried to imitate his method, as far as possible, in which he has treated the subject without making use of the triangle through the axis.

Giving a general treatment of conic sections, the following is the remarkable property under discussion: If in the plane *MSQ* there is a conic section *PQN,* on which are taken four points *K, N, O, V* from which are drawn the lines *KN, KO, VN, VO,* in such a way that through the same four points only two lines may pass, and if another line cuts the conic at points *R, ψ,* and the lines *KN, KO, VN, VO,* in points *X, Y, Z, δ,* then as the rectangle *ZR.Zψ* is to the rectangle *γR.γψ,* so the rectangle *δR.δψ* is to the rectangle *XR.Xψ.*

We can also prove that, if in the plane of the hyperbola, the ellipse, or the circle *AGE* of which the center is *C,* the line *AB* is drawn touching the section at *A,* and if having drawn the diameter we take line *AB* such that its square shall be equal to the square of the figure,[1] and if *CB* is drawn, then any line such as *DE,* parallel to line *AB* and cutting the section in *E,* and the lines *AC, CB* in points *D, F,* then if the section *AGE* is an ellipse or a circle, the sum of the squares of the lines *DE, DF* will be equal to the square of the line *AB*; and in the hyperbola, the difference between the same squares of the lines *DE, DF* will be equal to the square of the line *AB.*

---

[1] [In order that the square of segment *AB,* which is equal to *DE + DF,* shall be equal to one fourth of the circumscribed rectangle, the conic must be a circle. If the conic is an ellipse, *AB* will be taken equal to the axis which is perpendicular to *CA.*

DESARGUES treated analogous questions in his *Brouillon Projet* (Œuvres de DESARGUES, I, p. 202 et p. 284).]

Attention should be called to the fact that the statements of both DESARGUES and PASCAL immediately lead up to the equation of conics.

We can also deduce [from this] several problems; for example:

From a given point to draw a tangent to a given conic section.

To find two diameters that meet in a given angle.

To find two diameters cutting at a given angle and having a given ratio.

There are many other problems and theorems, and many deductions which can be made from what has been stated above, but the distrust which I have, due to my little experience and capacity, does not allow me to go further into the subject until it has passed the examination of able men who may be willing to take this trouble. After that if someone thinks the subject worth continuing, I shall endeavor to extend it as far as God gives me the strength.

At Paris, M.DC.XL.

BRIANCHON'S THEOREM

(Translated from the French by Professor Nathan Altshiller-Court, University of Oklahoma, Norman, Okla.)

Charles Julien Brianchon was born in Sèvres (France) in 1785. In 1804 he entered the École Polytechnique where he studied under Gaspard Monge. In 1808 he was appointed lieutenant of artillery and in this capacity took part in the campaigns in Spain and Portugal. Later he became professor of applied sciences in the artillery school of the Royal Guard. He died at Versailles in 1864.

His paper on curved surfaces of the second degree was published in the *Journal de l'École Polytechnique*, cahier 13, 1806. Only the first part, pp. 297–302, of it is given here in translation.

This paper contains the famous theorem, known under the author's name, which, together with Pascal's theorem (see page 326), is at the very foundation of the projective theory of conic sections. This theorem is one of the earliest and most remarkable examples of the principle of duality and it played an important role in the establishing of this far-reaching principle. The paper is also one of the earliest to make use of the theory of poles and polars to obtain new geometrical results. It is interesting to note that this article, which made the author's name familiar to every student of geometry, was written by him at the age of 21, while he was still in school.

Lemma

Given a line AA' (Fig. A.) of known length, if on this line or on the line produced a point O is taken arbitrarily, dividing the line into two segments OA, OA', it is always possible to determine on this line or on the line produced a point P which forms two new segments PA, PA' proportional to the first two.

It is evident that of the two points O and P, one will be situated on the line AA' itself, and the other on this line produced.

I

(Fig. 1). Consider in space three arbitrary lines AA', BB', CC', which, produced if necessary, meet in the same point P.[1]

---

[1] As an example take the three edges AA', BB', CC', of a truncated triangular pyramid.

Mémoire sur les surfaces du 2.ᵉ Degré.

Journal de l'École Polytechnique, N.º 15. An 1806.   Planche 2.ᵉ

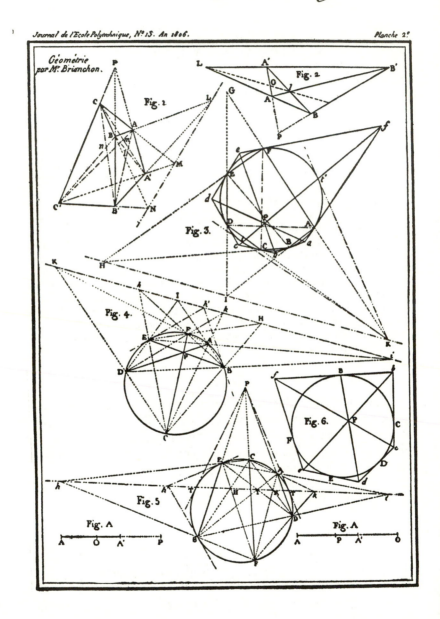

Name	the point of intersection of		produced if need be.
L	AB and $A'B'$		
l	AB'	$A'B$	
M	AC	$A'C'$	
m	AC'	$A'C$	
N	BC	$B'C'$	
n	BC'	$B'C$	

It is evident that the three points		are situated on the line of intersection of the plane which passes through the three points		with the plane which passes through the three points	
L, M, N		A, B, C		A', B', C'	
L, m, n		A, B, C'		A', B', C	
l, M, n		A, B', C		A', B, C'	
l, m, N		A', B, C		A, B', C'	

Now it is seen by taking any two of these four lines of intersection that they have a point in common; hence each one of them will meet the remaining three, and consequently they are all situated in the same plane, which I shall designate by XY.

II

This plane XY in which the six points L, M, N, l, m, n, are situated has the property that it divides each of the three lines AA', BB', CC', produced if necessary, into two segments proportional to those which the point P forms on these lines.

This property is based on the following proposition, taken from the *Géométrie de position*, p. 282.[1]

"In any complete quadrilateral[2] having its three diagonals, each of these diagonals is cut by the other two in proportional segments."

(Fig. 2). Consider, for example, what happens in the plane of the two lines AA', BB'; this plane is met by the plane XY along the line Ll, and B, B' are the points of concurrence of the opposite sides of the quadrilateral $ALA'l$; so that the three diagonals of this complete quadrilateral are BB', Ll, AA'.

---

[1] [The reference is to the book of L. M. N. Carnot, Paris, 1803. It is noteworthy that, writing three years after the publication of this work, Brianchon considers the book so well known that in quoting it he does not find it necessary to give the author's name.]

[2] A complete quadrilateral is the aggregate of four lines produced until they meet; and the line joining the point of intersection of two of these lines to the point of intersection of two others is called a diagonal.

Thus, according to the preceding theorem, any one of these three diagonals, say AA', is cut by the other two, Ll and BB', into proportional segments OA, OA', PA, PA'; that is to say, these four segments satisfy the relation

$$OA : OA' = PA : PA'.$$

Let us imagine that we project,[1] upon any plane whatever, the system of three lines AA', BB', CC', and all the construction lines, and let us denote the projection of a point by the inverted letter denoting the point itself. According to this convention, $\rotatebox{180}{L}$ represents the projection of the point L, and the same for the others.

This granted, those of the six points L, M, N, l, m, n, which lie on the same line, will have their projections also in a straight line, so that the six points $\rotatebox{180}{L}$, $\rotatebox{180}{M}$, $\rotatebox{180}{N}$, $\rotatebox{180}{l}$, $\rotatebox{180}{m}$, $\rotatebox{180}{n}$, will be arranged on four straight lines, in the same way in which are arranged, in space, the points of which they are the projections.

III

It follows from the above that when the three lines AA', BB', CC' are drawn in the same plane, the six points L, M, N, l, m, n are still arranged on this plane in such a way that when taken by threes in the indicated order (1), each of these groups of three points belongs to the same line.

IV

Should it happen that three of the six points $\rotatebox{180}{L}$, $\rotatebox{180}{M}$, $\rotatebox{180}{N}$, $\rotatebox{180}{l}$, $\rotatebox{180}{m}$, $\rotatebox{180}{n}$, (for example $\rotatebox{180}{l}$, $\rotatebox{180}{m}$, $\rotatebox{180}{n}$), which, in general, do not lie on a straight line, were to do so, as this would indicate that the plane of projection is perpendicular to the plane XY, one would conclude that all the six points lie on the same straight line, and then (II) this latter line would cut each of the three lines AA', BB', CC', produced if need be, into two segments proportional to those formed by the point P on the same lines.

V

One can, with the aid of the preceding considerations, demonstrate several remarkable properties belonging to the curves of the second degree. In order to succeed, let us recall the following proposition:

---

[1] [It is understood "orthogonally."]

(Figs. 3.) "In any hexagon (*ABCDEF*) inscribed in a conic section, the three points of intersection (*H*, *I*, *K*) of the opposite sides always lie on a straight line."

Or more generally:

"If on the perimeter of any conic section six points *A*, *B*, *C*, *D*, *E*, *F* are taken at random, and if the lines *AB*, *AF* are produced, if necessary, until they meet the lines *DE*, *DC*, in *I*, *K* respectively, the three lines *IK*, *BC*, *FE* cross each other in the same point." (*Géométrie de position*, p. 452.)

VI

(Figs. 4 and 5).   Again having three lines *AD*, *BE*, *CF*, inscribed in a curve of the second degree in such a way that they concur or cut each other in the same point *P*, if we carry out the constructions indicated in the figure, we see, according to the last theorem, that the points *H*, *I*, *K* are situated in a straight line; now this does not take place when the three lines *AD*, *BE*, *CF*, being subject to crossing each other in the same point *P*, have no other relation to each other: hence (IV) the six points *H*, *I*, *K*, *b*, *i*, *k* are all situated on the same line which divides each of the three chords *AD*, *BE*, *CF*, produced if need be, into two segments proportional to those which the point *P* forms on the same chords.

VII

Suppose now that one of the three chords, for example *CF*, changes in length, but in such a way as to retain the point *P* on its direction; the two points *I*, *i* will remain fixed, and the four remaining points *H*, *b*, *K*, *k* will still be situated on the indefinite line *Ii*. Hence when this variable chord *CF* coincides with one, say *BE*, of those which remain fixed, the lines *BF*, *CE* will become tangent to the curve and will have their point of intersection *b'* situated on *Ii*.

VIII

When the point *P* is outside the area of the conic section, there is an instant when the two extremities of the moving chord unite in a single point *T*, situated on the perimeter of the curve and on the line *Ii*.

IX

(Fig. 3). Let *abcdef* be an arbitrary hexagon circumscribed about a conic section, and *B*, *C*, *D*, *E*, *F*, *A*, the respective points of contact of the sides *ab*, *bc*, *cd*, *de*, *ef*, *fa*:

1°. The points of intersection H, I, K of the opposite sides of the inscribed hexagon $ABCDEF$, are three points situated on the same straight line (V);

2°. If we draw the diagonal which meets the curve in the two points t, t', the lines KT, Kt', will be tangent at t, t' respectively (VIII) and the same holds for the other diagonals.

3°. If from any point K of the line HIK the two tangents Kt, Kt' are drawn to the conic section, the chord tt', which joins the two points of contact, passes constantly through the same point P. (VIII).

Hence the three diagonals fc, be, ad cut each other in the same point P, that is to say:

"In any hexagon circumscribed about a conic section, the three diagonals cross each other in the same point."

This last theorem is pregnant with curious consequences; here is an example.

X

(Fig. 6.) Suppose that two of the six points of contact, say A and B, become united in a single point B, the vertex a will also coincide with B, and the figure will be reduced to a circumscribed pentagon $bcdef$; then applying the preceding theorem to this special case, we see that the three lines fc, be, db must cut each other in the same point P, that is to say,

"If in an arbitrary pentagon ($bcdef$), circumscribed about a curve of the second degree, the diagonals (be, cf) are drawn, which do not issue from the same angle, they meet in a point (P) situated on the line (dB) which joins the fifth angle (d) to the point of contact (B) of the opposite side."

This proposition gives at once the solution of the following problem...Determine the points where five known lines are touched by a curve of the second degree...These points once found, we can obtain all the other points of the curve by a very simple construction, which requires, just as the first, no instrument other than the ruler (VI).

The conic section having been constructed, we may propose to ourselves to draw to it a tangent through a point taken outside of or on the perimeter of the curve. The construction is carried out, the same as the two preceding ones, without the intervention of the compasses and, moreover, without its being necessary to know anything but the trace of the curve (VII), (VIII).

BRIANCHON AND PONCELET

On the Nine-point Circle Theorem

(Translated from the French by Dr. Morris Miller Slotnick, Harvard University, Cambridge, Mass.)

The nine-point circle was discovered by Brianchon and Poncelet, who published the proof of the fact that the circle passes through the nine points in a joint paper: "Recherches sur la détermination d'une hyperbole équilatère, au moyen de quatre conditions données," appearing in Georgonne's *Annales de Mathématiques*, vol. 11 (1820–1821), pp. 205–220. The theorem, of which the translation appears below, occurs as *Théorème IX*, (p. 215) in a sequence of theorems, and is there used merely as a lemma.

The theorem is known as Feuerbach's theorem, although the latter published his *Eigenschaften einiger merkwürdigen Punkte des geradlinigen Dreiecks*, in which the theorem occurs, in Nürnberg in 1822. However, Feuerbach proved the remarkable fact that the nine-point circle is tangent to the inscribed and the three escribed circles of the triangle. This booklet has been reprinted in 1908 (Berlin, Mayer und Müller), and there the theorems in question are found on pages 38–46. The proofs of Feuerbach are all of a quantitative nature, based as they are on the numerical relations existing between the radii of the various circles associated with a triangle and the distances between the centers of pairs of these circles.

The circle which passes through the feet of the perpendiculars dropped from the vertices of any triangle on the sides opposite them, passes also through the midpoints of these sides as well as through the midpoints of the segments which join the vertices to the point of intersection of the perpendiculars.

Proof.—Let P, Q, R be the feet of the perpendiculars dropped from the vertices of the triangle ABC on the opposite sides; and let K, I, L be the midpoints of these sides.[1]
The right triangles CBQ and ABR being similar,

$$BC : BQ = AB : BR;$$

from which, since K and L are the midpoints of BC and AB,

$$BK \cdot BR = BL \cdot BQ;$$

that is to say, the four points K, R, L, Q lie on one circle.[2]

---

[1] [The figure which appears here is similar to that which is published with the manuscript.]

[2] [The text here reads: "appartiennent à une même circonférence."]

337

It may similarly be shown that the four points K, R, I, P lie on a circle, as well as the four points P, I, Q, L.

This done, if it were possible that the three circles in question not be one and the same circle, it would follow that the common chords of the circles taken two at a time would pass through a point; now, these chords are precisely the sides of the triangle ABC, whoch do not pass through a common point; it is equally impossible to suppose that the three circles are different from one another; thus they must coincide in one and the same circle.

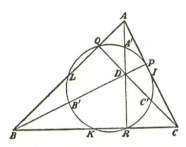

Now, let C', A', B' be the midpoints of the segments DC, DA, DB which join the point of intersection D of the three altitudes of the triangle ABC with each of the respective vertices. The right triangles CDR and CQB being similar, we have

$$CD : CR = CB : CQ;$$

from which, since C' and K are the midpoints of the segments CD and CB,

$$CC' \cdot CQ = CR \cdot CK;$$

that is to say, the circle which passes through K, R, Q passes also through C'.

It may be shown in the same manner that the circle passes through the other two points A', B'; thus it passes through the nine points $P, Q, R, I, K, L, A', B', C'$; *which was to be proven.*

FEUERBACH

On the Theorem Which Bears His Name

(Translated from the German by Professor Roger A. Johnson, Hunter College
New York City.)

Karl Wilhelm Feuerbach (1800–1834) was a professor of mathematics in the Gymnasium at Erlangen, Germany. He is known chiefly for the theorem which bears his name and which is reproduced in this article. The preceding article calls attention to the theorem and to the earlier work upon the subject by Brianchon and Poncelet. The present translation is made from the *Eigenschaften einiger merkwürdigen Punkte des geradlingen Dreiecks, und mehrerer durch Sie bestimmten Linien und Figuren,* published by Feuerbach in 1822 (2d ed., 1908). It includes certain passages which lead up to the theorem, but it omits such parts of the work as are not used in the proof. As is the case with the preceding article, this furnishes a source of some of the interesting work on the modern geometry of the triangle.

CHAPTER I

On the Centers of the Circles, Which Touch the Three Sides of a Triangle

§1. [It is stated to be known, that there are four circles which are tangent to the sides of a triangle ABC, one internal to the triangle and the others external. Their centers S, S', S'', S''', are the points of intersection of the bisectors of the angles of the triangle.]

§2. Let the radii of the circles about the centers S, S', S'', S''' be denoted by r, r', r'', r''' respectively; then we know that their values are

$$r = \frac{2\Delta}{a+b+c}, r' = \frac{2\Delta}{-a+b+c}, r'' = \frac{2\Delta}{a-b+c}, r''' = \frac{2\Delta}{a+b-c}$$

where, as usual, a, b, c denote the sides BC, CA, AB, and Δ the area of the triangle ABC.

. . . .

CHAPTER II

ON THE POINT OF INTERSECTION OF THE PERPENDICULARS
DROPPED FROM THE VERTICES OF A TRIANGLE ON THE
OPPOSITE SIDES

§19. If from the vertices of the angles of a triangle ABC, perpendiculars AM, BN, CP are dropped on the opposite sides, intersecting, as is well known, in a point O, the feet of these perpendiculars determine a triangle MNP, which is notable as having the least perimeter of all triangles inscribed in triangle ABC...

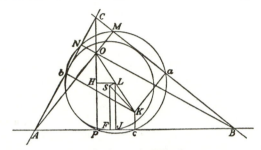

[Denoting the angles of the triangle by α, β, γ respectively], we have $AP = b \cos \alpha$, $AN = c \cos \alpha$, hence

$$NP^2 = (b^2 + c^2 - 2bc \cos \alpha)\cos^2 \alpha;$$

but $b^2 + c^2 - 2bc \cos \alpha = a^2$, whence $\overline{NP^2} = a^2 \cos^2 \alpha$ and $NP = a \cos \alpha$; similarly $MP = b \cos \beta$, and $MN = c \cos \gamma$; or, if the cosines are expressed in terms of the sides,

$$MN = \frac{c(a^2 + b^2 - c^2)}{2ab}$$

[and similarly for MP and NP]. If we add these three values, then since

$$a^2(-a^2 + b^2 + c^2) + b^2(a^2 - b^2 + c^2) + c^2(a^2 + b^2 - c^2) = 16\Delta^2 \,[1]$$

$MN + MP + NP = \dfrac{8\Delta^2}{abc}$, and since $\dfrac{abc}{4\Delta} = R$,[2] where R, as

previously, represents the radius of the circle circumscribed about triangle ABC, the desired result is obtained:

$$MN + NP + MP = \frac{2\Delta}{R}.$$

. . . .

---

[1] [This is the familiar formula $\Delta^2 = (a + b + c)(-a + b + c)(a - b + c)(a + b - c)$ in a modified form.]

[2] [Easily established, since $ab \sin \gamma = 2\Delta$, and $2R \sin \gamma = c$.]

If the triangle has an obtuse angle, say at A, then the corresponding term NP has the negative sign in this relation:

$$MN + MP - NP = 2\Delta/R.$$

The theorems which follow are subject to similar modification.

. . . .

§23. Since $AN = c \cos \alpha$, $AP = b \cos \alpha$, and $\tfrac{1}{2}bc \sin \alpha = \Delta$, it follows that $\triangle ANP = \Delta.\cos^2 \alpha$; likewise $\triangle BMP = \Delta.\cos^2 \beta$ and $\triangle CMN = \Delta.\cos^2 \lambda$; hence

$$\triangle MNP = \Delta(1 - \cos^2 \alpha - \cos^2 \beta - \cos^2 \gamma);$$

but since $\cos^2 \gamma = (\cos \alpha \cos \beta - \sin \alpha \sin \beta)^2$ and $\sin^2 \alpha \sin^2 \beta = 1 - \cos^2 \alpha - \cos^2 \beta + \cos^2 \alpha \cos^2 \beta$, then

$$\cos^2 \alpha + \cos^2 \beta + \cos^2 \gamma = 1 - 2 \cos \alpha \cos \beta \cos \gamma,$$

and accordingly

$$\triangle MNP = 2\Delta \cos \alpha \cos \beta \cos \gamma =$$
$$\frac{(-a^2 + b^2 + c^2)(a^2 - b^2 + c^2)(a^2 + b^2 - c^2)}{4a^2b^2c^2}\Delta.$$

§24. If ρ designates the radius of the inscribed circle of triangle MNP and $\rho^{(1)}$, $\rho^{(2)}$, $\rho^{(3)}$ those of the escribed circles, then, by virtue of §2 and §23, we have for the acute triangle ABC:

$$\rho = \frac{4\Delta \cos \alpha \cos \beta \cos \gamma}{a \cos \alpha + b \cos \beta + c \cos \gamma}$$

$$\rho^{(1)} = \frac{4\Delta \cos \alpha \cos \beta \cos \gamma}{-a \cos \alpha + b \cos \beta + c \cos \gamma}$$

[and if angle α is obtuse, these equations are modified by changing the sign of each term containing $\cos \alpha$.]

Now since in general, by §19, $a \cos \alpha + b \cos \beta + c \cos \gamma = 2\Delta/R$, we have for the acute triangle ABC:

$$\rho = 2R \cos \alpha \cos \beta \cos \gamma$$

and on the other hand, if A is obtuse,

$$\rho^{(1)} = -2R \cos \alpha \cos \beta \cos \gamma.$$

. . . .

26. The radius of the circle circumscribed about the triangle MNP is equal to

$$\frac{MN.MP.NP}{4 \triangle MNP} = \frac{abc \cos \alpha \cos \beta \cos \gamma}{8\Delta \cos \alpha \cos \beta \cos \gamma} = \tfrac{1}{2}R$$

that is, to half the radius of the circle circumscribed about the triangle.

. . . .

§32. Since angle AOP equals ABC, then $AO = \dfrac{AP}{\sin \beta}$, and

since $AP = b \cos \alpha$, and $\sin \beta = \dfrac{b}{2R}$, therefore:

$AO = 2R \cos \alpha$; similarly $BO = 2R \cos \beta$ and $CO = 2R \cos \gamma$;
hence
$$AO + BO + CO = 2R(\cos \alpha + \cos \beta + \cos \gamma)$$
[and by substitution of the formulas for the cosines and algebraic reduction, we find for any acute triangle]
$$\cos \alpha + \cos \beta + \cos \gamma = \frac{r + R}{R}, \; AO + BO + CO = 2(r + R)$$
[If the triangle has an obtuse angle, e. g., at C, then]
$$AO + BO - CO = 2(r + R).$$
. . . .

§35. We have $OM = BO \cos \gamma$, and since by §32 $BO = 2R \cos \beta$, therefore
$$OM = 2R \cos \beta \cos \gamma;$$
similarly $ON = 2R \cos \alpha \cos \gamma$, and $OP = 2R \cos \alpha \cos \beta$. If one multiplies these expressions respectively by those of AO, BO, CO

(§32), then, since $\cos \alpha \cos \beta \cos \gamma = \dfrac{\rho}{2R}$ (§24)
$$AO.OM = BO.ON = CO.OP = 2\rho R$$
That is, *the point of intersection of the three perpendiculars of triangle ABC divides each into two parts, whose rectangle equals double the rectangle of the radius of the circle inscribed in triangle MNP and that of the circle circumscribed about triangle ABC.*
. . . .

CHAPTER III

On the Center of the Circle, Which Is Circumscribed about a Triangle

§45. If K is the center of the circle circumscribed about a triangle ABC, and if perpendiculars Ka, Kb, Kc are dropped from this point on the sides, BC, CA, AB; then if we draw AK, $Kc = AK \cos AKc$; and because $AK = R$ and angle AKc equals ACB, therefore
$$Kc = R \cos \gamma;$$
and similarly $Kb = R \cos \beta$ and $Ka = R \cos \alpha$. If we compare these expressions with those found in §32 for AO, BO, CO,... we have at once
$$AO = 2Ka, \; BO = 2Kb, \; CO = 2Kc,$$

In any triangle the distance from the center of the circumscribed circle to any side is half the distance from the common point of the altitudes to the opposite vertex.

. . . .

CHAPTER IV

DETERMINATION OF THE RELATIVE POSITIONS OF THE POINTS PREVIOUSLY DISCUSSED. [Bestimmung der Gegenseitigen Lage der Vornehmsten bisher betrachteten Punkte.]

§49. If K and S are the centers of the circles circumscribed and inscribed to the triangle ABC, and perpendiculars Kc and SF are dropped from these to the side AB, then

$$\overline{KS}^2 = (Ac - AF)^2 + (SF - Kc)^2$$

Now we have $Ac = \tfrac{1}{2}c$ and $AF = \tfrac{1}{2}(-a + b + c)$, whence:

$$Ac - AF = \tfrac{1}{2}(a - b);$$

further, because (§2)

$$SF = \frac{2\Delta}{a + b + c},$$

and (§45)

$$Kc = \frac{c(a^2 + b^2 - c^2)}{8\Delta},$$

therefore
$SF - Kc =$

$$\frac{(-a + b + c)(a - b + c)(a + b - c) - c(a^2 + b^2 - c^2)}{8\Delta}$$

If now we substitute in the above expression for $\overline{KS}^2$, after the necessary reductions we have

$$\overline{KS}^2 = \frac{a^2b^2c^2 - abc(-a + b + c)(a - b + c)(a + b - c)}{16\Delta^2},$$

whence by means of the known values of the radii r and R we arrive at the result:

$$\overline{KS}^2 = R^2 - 2rR$$

In any triangle the square of the distance between the centers of the in- and circumscribed circles equals the square of the radius of the circumscribed circle, diminished by twice the rectangle of this radius and the radius of the inscribed circle.

[By similar methods we find that if S' is the center and r' the radius of an escribed circle,

$$\overline{KS'^2} = R^2 + 2r'R^1$$

. . . .

§51, 53. [By exactly the same methods, we derive the relations]

$$\overline{OS^2} = 2r^2 - 2\rho R$$
$$\overline{KO^2} = R^2 - 4\rho R$$

§54. If L is the center of the circle circumscribed about the circle MNP, whose radius (§26) has been found to be $\frac{1}{2}R$, and if OL is drawn, then since O is also the center of the inscribed circle of triangle MNP, then (§49) $\overline{OL^2} = \frac{1}{4}R^2 - \rho R$, and since we have just seen that $\overline{KO^2} = R^2 - 4\rho R$, it follows that $\overline{KO^2} = 4\overline{OL^2}$, or

$$KO = 2OL.$$

[If the triangle is obtuse, a slight modification of the proof leads to the same result, viz:]

In any triangle the common point of the altitudes is twice as far from the center of the circumscribed circle as from the center of the circle through the feet of the altitudes.

§55. If perpendiculars LJ, LH are dropped from the center L on the lines AB, CP, $LJ = PH$, and since H is a right angle, it is known that in triangle OPL, $PH = \dfrac{-\overline{OL}^2 + \overline{OP}^2 + \overline{LP}^2}{2OP}$. But $LP = \frac{1}{2}R$, and (§54) $\overline{OL}^2 = \frac{1}{4}R^2 - \rho R$; further (§35, 45) $OP.Kc = \rho R$, whence $\overline{LP}^2 - \overline{OL}^2 = OP.Kc$. If this expression is substituted in $PH = LJ$, the result is

$$LJ = \frac{1}{2}(OP + Kc).$$

From this property it follows at once that the points O, L, K are on one and the same line, and the theorem shines out (erhellet):

In any triangle the center of the circumscribed circle, the common point of the altitudes, and the center of the circle through feet of the altitudes lie on one and the same straight line, whose mid-point is the last named point.

§56. Because the point L, then, lies at the center of the line KO, therefore the point J is also the center of the line Pc, whence

---

[1] In a historical note, the theorem is attributed to Euler. The history of this theorem has been investigated in detail by Mackay, *Proceedings of the Edinburgh Math. Society*, V. 1886-7, p. 62.

it follows that $Lc = LP = \frac{1}{2}R$; and similarly on each of the other sides AC, BC. Thus we have the theorem:

In any triangle the circle which passes through the feet of the altitudes also cuts the sides at their mid-points.

§57. If the line LS is drawn, we know that in triangle KOS, since L is the mid-point of KO, $2\overline{SL}^2 + 2\overline{OL}^2 = \overline{KS}^2 + \overline{OS}^2$. If we substitute in this equation the values of the squares of OL KS, OS, as found in 54, 49, 51, thus there comes

$$\overline{LS}^2 = \frac{1}{4}R^2 - rR + r^2 = (\frac{1}{2}R - r)^2,$$

or:

$$LS = \frac{1}{2}R - r.$$

Similarly, setting a, b, c in turn negative,

$$LS' = \frac{1}{2}R + r',\ \ LS'' = \frac{1}{2}R + r'',\ \ LS''' = \frac{1}{2}R + r'''.$$

Since now (§26)$\frac{1}{2}R$ is the radius of the circle circumscribed about triangle MNP, we deduce from a known property of circles which are tangent, the following theorem:

The circle which passes through the feet of the altitudes of a triangle touches all four of the circles which are tangent to the three sides of the triangle and specifically, it touches the inscribed circle internally and the escribed circles externally.

WILLIAM JONES

The First Use of π for the Circle Ratio

(Selections Made by David Eugene Smith from the Original Work.)

William Jones (1675–1749) was largely a self-made mathematician. He had considerable genius and wrote on navigation and general mathematics. He edited some of Newton's tracts. The two passages given below are taken from the *Synopsis Palmariorum Matheseos: or, a New Introduction to the Mathematics*, London, 1706. The work was intended "for the Vse of some Friends, who had neither Leisure, Conveniency, nor, perhaps, Patience, to search into so many different Authors, and turn over so many tedious volumes, as is unavoidably required to make but a tolerable Progress in the Mathematics." It was a very ingenious compendium of mathematics as then known. The symbol π first appears on page 243, and again on p. 263. The transcendence of π was proved by Lindemann in 1882. For the transcendence of e, which proved earlier (1873), see page 99.

Taking a as an arc of 30°, aud t as a tangent in a figure given, he states (p. 243):

$$6a, \text{ or } 6 \times t - \frac{1}{3}t^2 + \frac{1}{5}t^5, \text{ \&c.} = \frac{1}{2} \text{ Periphery } (\pi) \dots$$

Let

$$\alpha = 2\sqrt{3}, \ \beta = \frac{1}{3}\alpha, \ \gamma = \frac{1}{3}\beta, \ \delta = \frac{1}{3}\gamma, \text{ \&c.}$$

Then

$$\alpha - \frac{1}{3}\beta + \frac{1}{5}\gamma - \frac{1}{7}\delta + \frac{1}{9}\epsilon, \text{ \&c.} = \frac{1}{2}\pi,$$

or

$$\alpha - \frac{1}{3}\frac{3\alpha}{9} + \frac{1}{5}\frac{\alpha}{9} - \frac{1}{7}\frac{3\alpha}{9^2} + \frac{1}{9}\frac{\alpha}{9^2} - \frac{1}{11}\frac{3\alpha}{9^3} + \frac{1}{13}\frac{\alpha}{9^3}, \text{ \&c.}$$

Theref. the (Radius is to ½ Periphery, or) Diameter is to the Periphery, as 1,000, &c to 3.141592653 . 5897932384 . 6264338327 . 9502884197 . 1693993751 . 0582097494 . 4592307816 . 4062862089 . 9862803482 . 5342117067. 9+ True to above a 100 Places; as Computed by the accurate and Ready Pen of the Truly Ingenious Mr. *John Machin*.

On p. 263 he states:

There are various other ways of finding the *Lengths*, or *Areas* of particular *Curve Lines*, or *Planes*, which may very much facilitate the Practice; as for Instance, in the *Circle*, the *Diameter* is to *Circumference* as 1 to

$$\frac{16}{3} - \frac{4}{239} - \frac{1}{3}\frac{16}{5^3} - \frac{4}{239^3} + \frac{1}{5}\frac{16}{5^5} - \frac{4}{239^5} -, \&c. =$$

$$3.14159, \&c. = \pi \ldots$$

Whence in the *Circle*, any one of these three, a, c, d, being given, the other two are found, as, $d = c \div \pi = \overline{a \div \frac{1}{4}\pi}\Big|^{\frac{1}{2}}$, $c = d \times \pi$

$= \overline{a \times 4\pi}\Big|^{\frac{1}{2}}$, $a = \frac{1}{4}\pi d^2 = c^2 \div 4\pi$.

GAUSS

On the Division of a Circle into n Equal Parts

(Translated from the Latin by Professor J. S. Turner, University of Iowa, Ames, Iowa.)

Carl Friedrich Gauss (1777–1855) was a student at Göttingen from 1795 to 1798, and during this period he conceived the idea of least squares, began his great work on the theory of numbers (*Disqvisitiones arithmeticae*, Leipzig, 1801), and embodied in the latter his celebrated proposition that a circle can be divided into n equal parts for various values of n not theretofore known. This proposition is considered in the *Disqvisitiones*, pages 662–665. From this edition the following translation has been made, the portion selected appearing in sections 365 and 366.

For further notes upon Gauss and his works see pages 107 and 292.

(365.) We have therefore, by the preceding investigations, reduced the division of the circle into n parts, if n is a prime number, to the solution of as many equations as there are factors into which $n - 1$ can be resolved, the degrees of these equations being determined by the magnitude of the factors. As often therefore as $n - 1$ is a power of the number 2, which happens for these values of n: 3, 5, 17, 257, 65537 *etc.*, the division of the circle is reduced to quadratic equations alone, and the trigonometric functions of the angles $\dfrac{P}{n}$, $\dfrac{2P}{n}$ *etc.* can be expressed by square roots more or less complicated (according to the magnitude of n); hence in these cases the division of the circle into n parts, or the description of a regular polygon of n sides, can evidently be effected by geometrical constructions. Thus for example for $n = 17$, by arts. 354, 361, the following[1] expression is easily derived for the cosine of the angle $\frac{1}{17}P$:

$$-\frac{1}{16} + \frac{1}{16}\sqrt{17} + \frac{1}{16}\sqrt{(34 - 2\sqrt{17})} - \frac{1}{8}\sqrt{(17 + 3\sqrt{17}}$$
$$- \sqrt{(34 - 2\sqrt{17})} - 2\sqrt{(34 + 2\sqrt{17})});$$

[1] [An elegant presentation of Gauss's method will be found on p. 220 of Casey's *Plane Trigonometry* (Dublin, 1888), where, however, the last terms of equations (550), (551), (552) should be c_1, c_2, b_2 respectively.]

the cosines of the multiples of this angle have a similar form, but the sines have one more radical sign. Truly it is greatly to be wondered at, that, although the geometric divisibility of the circle into three and five parts was already known in the times of Euclid, nothing has been added to these discoveries in the interval of 2000 years, and all geometers have pronounced it as certain, that beyond the divisions referred to and those which readily follow, namely divisions into 15, 3.2^μ, 5.2^μ, 15.2^μ and also into 2^μ parts, no others can be effected by geometrical constructions. Moreover it is easily proved, if the prime number n is equal to $2^m + 1$, that the exponent m cannot involve other prime factors than 2, and so must be either equal to 1 or 2 or a higher power of 2; for if m were divisible by any odd number ζ (greater than unity), and $m = \zeta\eta$, $2^m + 1$ would be divisible by $2^\eta + 1$, and therefore necessarily composite. Consequently all values of n by which we are led to none but quadratic equations are contained in the form $2^{2^\nu} + 1$; thus the 5 numbers 3, 5, 17, 257, 65,537 result by setting $\nu = 0$, 1, 2, 3, 4 or $m = 1, 2, 4, 8, 16$. By no means for all numbers contained in that form, however, can the division of the circle be performed geometrically, but only for those which are prime numbers. Fermat indeed, misled by induction, had affirmed that all numbers contained in that form are necessarily primes; but the celebrated Euler first remarked that this rule is erroneous even for $\nu = 5$, or $m = 32$, the number $2^{32} + 1 = 4294967297$ involving the factor 641.

But as often as $n - 1$ involves other prime factors than 2, we are led to higher equations; namely to one or more cubics when 3 is found once or more frequently among the factors of $n - 1$; to equations of the 5th degree when $n - 1$ is divisible by 5 etc., AND WE CAN DEMONSTRATE WITH ALL RIGOR THAT THESE HIGHER EQUATIONS CAN IN NO WAY BE EITHER AVOIDED OR REDUCED TO LOWER, although the limits of this work do not permit this demonstration to be given, which nevertheless we effected since a warning must be given lest anyone may still hope to reduce to geometrical constructions other divisions beyond those which our theory furnishes, for example divisions into 7, 11, 13, 19 etc. parts, and waste his time uselessly.

(366.) If the circle is to be divided into a^α parts, where a denotes a prime number, this can clearly be effected geometrically when $a = 2$, but for no other value of a, provided $a > 1$; for then besides those equations which are required for the division into a

parts it is also necessary to solve $\alpha - 1$ others of the ath degree; moreover these can in no way be either avoided or depressed. Consequently the degrees of the necessary equations can be as certained generally (evidently also for the case where $\alpha = 1$) from the prime factors of the number $(a - 1)a^{\alpha-1}$.

Finally, if the circle is to be divided into $N = a^\alpha b^\beta c^\lambda \ldots$ parts, *a, b, c etc.* denoting unequal prime numbers, it suffices to effect the divisions into a^α, b^β, c^γ *etc.* parts (art. 336); and therefore, to ascertain the degrees of the equations required for this purpose, it is necessary to examine the prime factors of the numbers $(a - 1)a^{\alpha-1}$, $(b - 1)b^{\gamma-1}$, $(c - 1)c^{\gamma-1}$ *etc.*, or which amounts to the same thing, of the product of these numbers. It may be observed that this product expresses the number of numbers prime to N and less than N (art. 38). Therefore the division is effected geometrically only when this number is a power of two; indeed when it involves prime factors other than 2, for instance p, p' *etc.*, equations of degree p, p' *etc.* can in no way be avoided. Hence it is deduced generally that, in order that a circle may be geometrically divisible into N parts, N must be *either* 2 or a higher power of 2, *or* a prime number of the form $2^m + 1$, *or* the product of several such prime numbers, *or* the product of one or more such prime numbers into 2 or a higher power of 2; or briefly, it is necessary that N should involve neither any odd prime factor which is not of the form $2^m + 1$, nor even any prime factor of the form $2^m + 1$ more than once. The following 38 such values of N are found below 300: 2, 3, 4, 5, 6, 8, 10, 12, 15, 16, 17, 20, 24, 30, 32, 34, 40, 48, 51, 60, 64, 68, 80, 85, 96, 102, 120, 128, 136, 160, 170, 192, 204, 240, 255, 256, 257, 272.

SACCHERI

On Non-Euclidean Geometry[1]

(Translated from the Latin by Professor Henry P. Manning, Brown University, Providence, R. I.)

Geronimo Saccheri was born in 1667 and died in 1733. He was a Jesuit and taught in two or three of the Jesuit colleges in Italy. His chief work, published about the time of his death, is an attempt to prove Euclid's parallel postulate as a theorem by showing that the supposition that it is not true leads to a contradiction. The path to his "contradiction" consists of a series of propositions which actually constitute the main part of the elementary non-Euclidean geometry, published in this way about a hundred years before it was published as such.

The final discovery of the non-Euclidean geometry was not based on the work of Saccheri. Neither Lobachevsky nor Bolyai seems to have ever heard of him. But Saccheri is the most important figure in the preparation for this discovery in the period that precedes it, and after his relation to it was pointed out in 1889 his work took its place as standing at the head of the literature of the subject.

Euclid Freed from Every Flaw[2]

Book I

in which is demonstrated: Any two straight lines lying in the same plane, on which a straight line makes the two interior angles on the same side less than two right angles, will at length meet each other on the same side if they are produced to infinity.

---

[1] On the general subject of non-Euclidean geometry, including biographical notes, see Engel and Stäckel, *Die Theorie der Parallellinien* (Leipzig, 1895); *Urkunden zur Geschichte der Nichteuklidischen Geometrie* (Leipzig, 1898, 1913).

[2 This book was written in Latin and the Latin text has been published along with the translation by Professor Halsted (Chicago, 1920). We have checked this with the German translation given in *Die Theorie der Parallellinien* by Engel and Stäckel, Leipzig, 1895, pages 41–135. The text itself of which we are translating a part is preceded by a "Preface to the Reader" and a summary of the contents "added in place of an index."]

Part I

Proposition I.—*If two equal straight lines, AC and BD, make equal angles on the same side with a line AB, I say that the angles with the joining line CD will also be equal.*

Proof.[1]—Let *A* and *D* be joined, and *C* and *B*. Then let the triangles *CAB* and *DBA* be considered. It follows (I, 4)[2] that the bases *CB* and *AD* will be equal. Then consider the triangles *ACD* and *BDC*. It follows (I, 8) that the angles *ACD* and *BDC* will be equal. Q. E. D.[3]

Proposition II.—*In the same quadrilateral ABCD[4] let the sides AB and CD be bisected at the points M and H. I say that the angles with the joining line MH will then be right angles.*

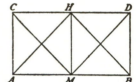

Proof.—Let the joining lines *AH* and *BH* be drawn, also *CM* and *DM*. Since in this quadrilateral the angles *A* and *B* are given equal, and also (from the preceding) *C* and *D*, it follows from I, 4 (since also the equality of the sides is known) that in the triangles *CAM* and *DBM* the bases *CM* and *DM* will be equal; also in the triangles *ACH* and *BDH* the bases *AH* and *BH*. Therefore, from a comparison of the triangles *CHM* and *DHM*, and again of the triangles *AMH* and *BMH*, it will follow (I, 8) that the angles in these at the points *M* and *H* will be equal to each other, and so right angles. Q. E. D.

Proposition III.—*If two equal straight lines AC and BD stand perpendicularly to a straight line AB, I say that the joining line CD will be equal to, or less than, or greater than AB, according as the angles with the same CD are right or obtuse or acute.*

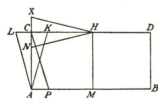

Proof of the first part. Each angle *C* and *D* being a right angle, if possible let one of those, say *CD*, be greater than the other, *AB*. On *DC*

---

[1] ["Demonstratur," *It is proved.*]

[2] [..."ex quarta primi." This is a reference to Euclid. For such references we shall give only the numbers of the book and proposition.]

[3] ["Quod erat demonstrandum," written out in full in the original as published with Halsted's translation.]

[4] [Literally, "The uniform quadrilateral remaining."]

let a portion *DK* be taken equal to *BA*, and let *A* and *K* be joined. Since therefore on *BD* stand two equal perpendicular lines *BA* and *DK* the angles *BAK* and *DKA* will be equal (1).[1] But this is absurd, since the angle *BAK* is by construction less than the assumed right angle *BAC*, and the angle *DKA* is an exterior angle by construction, and therefore (I, 16) greater than the interior opposite angle *DCA*, which is supposed to be a right angle. Therefore neither of the given lines *DC* and *BA* is greater than the other if the angles with the joining line *CD* are right angles, and therefore they are equal to each other. Q. E. D. for the first part.

Proof of the second part. But if the angles with the joining line *CD* are obtuse, let *AB* and *CD* be bisected at the points *M* and *H*, and let *M* and *H* be joined. Since therefore on the straight line *MH* stand two perpendicular lines *AM* and *CH* (from what precedes) and we have with the joining line *AC* the right angle at *A*, *CH* will not be equal to *AM* (1) since the angle at *C* is not a right angle. But neither will it be greater: otherwise, taking on *HC* a portion *KH* equal to *AM*, we shall have equal angles with the joining line *AK* (1). But this is absurd as above. For the angle *MAK* is less than a right angle and the angle *HKA* is greater than the obtuse angle *HCA* which is interior and opposite (I, 16). It results therefore that *CH*, while the angles with the joining line *CD* are obtuse, is less than *AM*, and therefore the double of the former, *CD*, is less than the double of the latter, *AB*. Q. E. D. for the second part.

Proof of the third part. But, finally, if the angles with the joining line *CD* were acute, the perpendicular *MH* being drawn as before, we proceed thus: Since on the line *MH* stand perpendicularly two straight lines *AM* and *CH*, and with the joining line *AC* there is a right angle at *A*, the line *CH* will not be equal to *AM*, since there is lacking a right angle at *C*. But neither will it be less: otherwise, if on *HC* produced we take *HL = AM*, the angles formed with the joining line *AL* will be equal (as above). But this is absurd. For the angle *MAL* is by construction greater than the angle *MAC* supposed a right angle, and the angle *HLA* is by construction interior and opposite, and so less than the exterior angle *HCA* (I, 16), which is supposed acute. It remains, therefore, that *CH*, while the angles with the joining line *CD* are acute, is greater

---

[1] [... "ex prima hujus." These are references to previous theorems, which we will indicate simply by putting the number in the parentheses.]

than *AM*, and so *CD*, the double of the former, is greater than *AB*, the double of the latter. Q. E. D. for the third part.

Thus it follows that the joining line *CD* will be equal to, or less than, or greater than *AB*, according as the angles with the same *CD* are right or obtuse or acute. Q. E. D.

Corollary 1.—Hence in every quadrilateral containing three right angles and one obtuse or acute, the sides adjacent to the angle which is not a right angle are less, respectively, than the opposite sides if the angle is obtuse, but greater if it is acute. For it has already been demonstrated of the side *CH* with respect to the opposite side *AM*, and in a similar way it is shown of the side *AC* with respect to the opposite side *MH*. For as the lines *AC* and *MH* are perpendicular to *AM*, they cannot be equal to each other (1), because of the unequal angles with the joining line *CH*. But neither (in the hypothesis of the obtuse angle at *C*) can a certain portion *AN* of *AC* be equal to *MH*, than which certainly *AC* is greater; otherwise (1) the angles with the joining line *HN* would be equal which is absurd as above. But again (in the hypothesis of the acute angle at *C*) if we wish that a certain *AX*, taken on *AC* produced, shall be equal to *MH*, than which certainly *AC* is smaller, the angles with the joining line *HX* will be equal for the same reason, which is absurd in the same way as above. It remains therefore that in the hypothesis indeed of the obtuse angle at the point *C* the side *AC* will be less than the opposite side *MH*, but in the hypothesis of the acute angle it will be greater. Q. E. I.[1]

Corollary 2.—By much more will CH be greater than any portion of AM, as say *PM*, since the joining line *CP* makes a more acute angle with *CH* on the side towards the point *H*, and an obtuse angle (I, 16) with *PM* towards the point *M*.

Corollary 3.—Again it follows that all these statements are true if the perpendiculars AC and BD are of a certain finite length fixed by us, or are supposed to be infinitely small. This indeed ought to be noted in the rest of the propositions that follow.

Proposition IV.—But conversely (in the figure of the preceding proposition), the angles with the joining line CD will be right or obtuse or acute according as CD is equal to, or less than, or greater than, the opposite AB.

Proof.—For if the line *CD* is equal to the opposite *AB*, and nevertheless the angles with the same are obtuse or acute, already such angles will prove it (from the preceding) not equal to, but less

---

[1] ["Quod erat intentum," *What was asserted.*]

than, or greater than the opposite *AB*, which is absurd, contrary to hypothesis. The same holds in the other cases. It stands therefore that the angles with the joining line *CD* are right or obtuse or acute according as the line *CD* is equal to, or less than, or greater than, the opposite *AB*. Q. E. D.

Definitions.—Since (from 1) a straight line joining the extremities of equal lines standing perpendicularly to the same line (which we shall call base) makes equal angles with them, therefore there are three hypotheses to be distinguished in regard to the nature of these angles. And the first indeed I will call *the hypothesis of the right angle*, but the second and third I will call *the hypothesis of the obtuse angle* and *the hypothesis of the acute angle*.

Proposition V.—*The hypothesis of the right angle, if true in a single case, is always in every case the only true hypothesis.*

Proof.—Let the joining line *CD* make right angles with any two equal lines, *AC* and *BD*, standing perpendicularly to any *AB*. *CD* will be equal to *AB* (3). Take on *AC* and *BD* produced the two *CR* and *DX*, equal to *AC* and *BD*, and join *R* and *X*. We shall easily show that the joining line *RX* will be equal to *AB* and the angles with it right angles; and first indeed by superposition of the quadrilateral *ABDC* upon the quadrilateral *CDXR*, with the common base *CD*. But then we can proceed more elegantly thus: join *A* and *D* and *R* and *D*. It follows (I, 4) that the triangles *ACD* and *RCD* will be equal, the bases *AD* and *RD*, and also the angles *CDA* and *CDR* and, because they are equal remainders to a right angle, *ADB* and *RDX*. Wherefore again (from the same I, 4) will be equal in the triangles *ADB* and *RDX* the base *AB* to the base *RX*. Therefore (from the preceding) the angles with the joining line *RX* will be right angles, and therefore we shall persist in the same hypothesis of the right angle.

And since the length of the perpendiculars can be increased to infinity on the same base, with the hypothesis of the right angle always holding, it must be proved that the same hypothesis will remain in the case of any diminution of the same perpendiculars, which is proved as follows.

Take in *AR* and *BX* any two equal perpendiculars *AL* and *BK*, and join *L* and *K*. Even if the angles with the joining line are not right angles, yet they will be equal to each other (1). They will therefore be obtuse on one side, say towards AB, and towards RX acute, as the angles at each of these points are equal to two right angles (I, 13). But it follows also that LR and KX are perpendiculars equal to each other standing on RX. Therefore (3) LK will be greater than the opposite RX and less than the opposite AB.

But this is absurd, since AB and RX have been shown equal. Therefore the hypothesis of the right angle will not be changed under any diminution of the perpendiculars while the given AB remains the base.

But neither will the hypothesis of the right angle be changed under any diminution or greater amplitude of the base, since it is evident that any perpendicular BK or BX can be considered as base, and so in turn AB and the equal opposite line KL or XR can be considered as the perpendicular.

It follows therefore that the hypothesis of the right angle, if true in any case, is always in every case the only true hypothesis. Q. E. D.

Proposition VI.—The hypothesis of the obtuse angle, if it is true in one case, is always in every case the only true hypothesis.

Proof.—Let the joining line CD make obtuse angles with any two equal perpendiculars AC and BD standing on any straight line AB. CD will be less than AB (3). Take on AC and BD produced any two portions CR and DX, equal to each other, and join R and X. Now I seek in regard to the angles with the joining line RX, which will be equal to each other (1). If they are obtuse we have the theorem as asserted. But they are not right, because we should then have a case of the hypothesis of the right angle, which (from the preceding) would leave no place for the hypothesis of the obtuse angle. But neither are they acute. For in that case RX would be greater than AB (3), and therefore still greater than CD. But that this cannot be true is shown as follows. If the quadrilateral $CDXR$ is known to be filled with straight lines cutting off portions from CR and DX equal to each other, this implies a passing from the straight line CD which is less than AB, to RX greater than the same, and so a passing through a certain ST equal to AB. But that this is absurd in our present hypothesis follows from it, because then there would be one case of the hypothesis of the right angle (4), which would leave no place for the hypothesis of the obtuse angle (from the preceding). Therefore the angles with the joining line *RX* ought to be obtuse.

Then taking on *AC* and *BD* equal portions *AL* and *BK*, we shall show in a similar manner that the angles with the joining line *LK* cannot be acute towards *AB*, because then it would be greater than *AB*, and therefore still greater than *CD*. But from this there ought to be found as above a certain intermediate line between *CD* which is smaller, and *LK* which is larger than *AB*, intermediate I say, and equal to *AB*, which certainly, from what has just been noted, would take away all place for the hypothesis of the obtuse angle. Finally, for this same reason, the angles with the joining line *LK* cannot be right angles. Therefore they will be obtuse. Therefore on the same base *AB*, the perpendiculars being increased or diminished at will, there will always remain the hypothesis of the obtuse angle.

But the same ought to be demonstrated on the assumption of any base. Let there be chosen for base one of the above mentioned perpendiculars, say *BX*. Bisect *AB* and *RX* at the points *M* and *H* and join *M* and *H*. *MH* will be perpendicular to *AB* and *RX* (2). But the angle at *B* is a right angle by hypothesis, and the angle at *X* is obtuse as just proved. Therefore make the right angle *BXP* on the side of *MH*, *XP* will cut *MH* at a certain point *P* situated between the points *M* and *H*, since, on the one hand, the angle *BXH* is obtuse, and, on the other hand, if we join *X* and

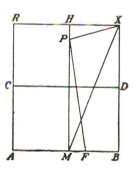

M, the angle *BXM* is acute (I, 17). Then indeed since the quadrilateral *XBMP* contains three right angles from what is already known and one obtuse at the point *P*, because it is exterior with respect to the interior opposite right angle at the point *H* of the triangle *PHX* (I, 16), the side *XP* will be less than the opposite side *BM* (3, Cor. 1). Therefore, taking in *BM* the portion *BF* equal to *XP*, the angles with the joining line *PF* will be equal to each other, and even obtuse, since the angle *BFP* is obtuse on account of the interior opposite angle *FMP* (I, 16). Therefore under any base *BX* the hypothesis of the obtuse angle holds true.

Moreover there stands as above the same hypothesis on the same base *BX,* however the equal perpendiculars are increased at will or diminished.   And so it follows that the hypothesis of the obtuse angle, if it is true in one case, is always in every case the only true hypothesis.   Q. E. D.

Proposition VII.—The hypothesis of the acute angle, if it is true in one case, is always in every case the only true hypothesis.

It is proved most easily.   For if the hypothesis of the acute angle should permit any case of either the hypothesis of the right angle or of the obtuse angle, already (from the two preceding propositions) no place will be left for the hypothesis of the acute angle, which is absurd.   Therefore the hypothesis of the acute angle, if it is true in one case, is always in every case the only true hypothesis.

The following propositions, given here without the proofs, show in part the course pursued by Saccheri.

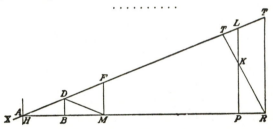

Proposition XI.—Let the straight line AP (as long as you please) cut two straight lines PL and AD, the first in a right angle at P, but the second at A in an angle that is acute towards the side of PL.   I say that the straight lines AD and PL (in the hypothesis of the right angle) will at length meet in some point, and indeed at a finite or terminated distance, if produced towards those parts on which they make with the base line AP the two angles less than two right angles.

Proposition XII.—Again I say that the straight line AD will meet the line PL somewhere towards these parts (and indeed at a finite or terminated distance) even in the hypothesis of the obtuse angle.

Proposition XIII.—If the straight line XA (of any designated length) intersecting two straight lines AD and XL, makes with them on the same side interior angles XAD and AXL less than two right angles, I say that those two (even if neither of those angles is a right angle) will meet each other at length in some point on the side of those angles, and indeed at a finite or terminated distance, if holds true the hypothesis of the right angle or of the obtuse angle.

· · · · · · · · · ·

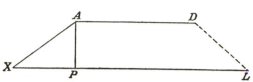

Proposition XIV.—The hypothesis of the obtuse angle is absolutely false, because it destroys itself.[1]

· · · · · · · · · ·

Proposition XXIII.—If two lines AX and BX lie in the same plane, either they have one common perpendicular (even in the hypothesis of the acute angle), or prolonged, both towards one side or towards the other, unless somewhere one meets the other at a finite distance, they will always more and more nearly approach each other.

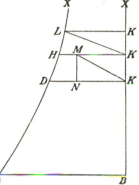

· · · · · · · · · ·

Proposition XXXIII.—The hypothesis of the acute angle is absolutely false, because repugnant to the nature of a straight line.[2]

---

[1] [Propositions XII and XIII lead to Euclid's postulate, and so to the hypothesis of the right angle, "even in the hypothesis of the obtuse angle."]

[2] [Up to this point his proofs are clear and logical. But he fails to find his contradiction and falls back on vague illogical reasoning. In the proof of Proposition XXXIII he says, "For then we have two lines which produced must run together into the same line and have at one and the same infinitely distant point a common perpendicular." Then he says he will go into first principles most carefully in order not to omit any objection. Finally in Part II he comes to

Proposition XXXVIII. *The hypothesis of the acute angle is absolutely false, because it destroys itself.*

In the summary at the beginning he says that after the falsity of the hypothesis of the obtuse angle is shown "begins a long battle against the hypothesis of the acute angle," which alone denies the truth of that axiom.]

LOBACHEVSKY

On Non-Euclidean Geometry

(Translated from the French by Professor Henry P. Manning, Brown
University, Providence, R. I.)

Nicholas Ivanovich Lobachevsky was born in 1793 and died in 1856. For
almost his entire life he was connected with the University of Kasan where he
was professor of mathematics and finally rector. He wrote several memoirs
and books on the theory of parallels, of which three may be mentioned as the
most important: (1) *New Foundations of Geometry*, published first in Russian
in 1835–1838. A German translation is given by Engel and Stäckel *Urkunden*,
vol. I, pages 67–236. There was an English translation made by Halsted in
1897, and a French translation made in 1901. (2) *Geometrical Investigations
on the Theory of Parallels*, written in German and published as a book in Berlin
in 1840. This was translated into French by Höuel in 1866, and into English
in 1891 by Halsted (Chicago, 1914). (3) *Pangeometry*, published simultaneously in Russian and French in 1855, translated into German in 1858 and
again in 1902, and into Italian in 1867. This is more condensed, many proofs
being omitted with references to *Geometrical Investigations*, and as it was
written near the end of Lobachevsky's life it may be regarded as representing
the final development of his ideas. When this was written he had become
blind and had to dictate whatever he wrote to his pupils.

Pangeometry

or a Summary of Geometry Founded upon a General and Rigorous
Theory of Parallels.[1]

The notions upon which the elementary geometry is founded
are not sufficient for a deduction from them of a demonstration
of the theorem that the sum of the three angles of a rectilinear
triangle is always equal to two right angles, a theorem the truth

---

[1] [*Collection complète des œuvres géométriques de N. I. Lobatcheffsky*, volume
II, Kasan, 1886, pages 617–680. This translation has been compared with
the Russian edition by Mrs. D. H. Lehmer of Brown University. There are
a few differences. Some superfluous words are omitted in the French and
obscure passages in the Russian are explained more fully and so made clearer.
Apparently the Russian edition was printed first and the French shows some
slight revision. Some of these differences will be pointed out below.]

of which no one has doubted to the present time because we meet
no contradiction in the consequences which we have deduced from
it, and because direct measures of angles of rectilinear triangles
agree with this theorem within the limits of error of the most
perfect measures.

The insufficiency of the fundamental notions for the demonstration of this theorem has forced geometers to adopt explicitly or
implicitly auxiliary suppositions, which, however simple they
appear, are no less arbitrary and therefore inadmissible. Thus,
for example, one assumes that a circle of infinite radius becomes
a straight line, and a sphere of infinite radius a plane, that the
angles of a rectilinear triangle always depend only on the ratios
of the sides and not on the sides themselves, or, finally, as it is
ordinarily done in the elements of geometry, that through a given
point of a plane we can draw only a single line parallel to another
given line in the plane, while all other lines drawn through the
same point and in the same plane ought necessarily to cut the
given line if sufficiently prolonged. We understand by the term
"line parallel to a given line" a line which, however far it is prolonged in both directions, never cuts the one to which it is parallel.
This definition is of itself insufficient, because it does not sufficiently characterize a single straight line. We may say the same
thing of most of the definitions given ordinarily in the elements of
geometry, for these definitions not only do not indicate the generation of the magnitudes which they define, but they do not even
show that these magnitudes can exist. Thus we define the straight
line and the plane by one of their properties. We say that
straight lines are those which always coincide when they have
two points in common, and that a plane is a surface in which a
line lies entirely when the line has two points in common with it.

Instead of commencing geometry with the plane and the
straight line as we do ordinarily, I have preferred to commence it
with the sphere and the circle, whose definitions are not subject
to the reproach of being incomplete, since they contain the generation of the magnitudes which they define.

Then I define the plane as the geometrical locus of the intersections of equal spheres described around two fixed points as centers.
Finally I define the straight line as the geometrical locus of the
intersections of equal circles, all situated in a single plane and
described around two fixed points of this plane as centers. If
these definitions of the plane and straight line are accepted all the

theory of perpendicular planes and lines can be explained and demonstrated with much simplicity and brevity.[1]

Being given a straight line and a point in a plane, I define as parallel through the given point to the given line, the limiting line between those drawn in the same plane through the same point and prolonged on one side of the perpendicular from the point to the given line, which cut it, and those which do not cut it.[2]

I have published a complete theory of parallels under the title *Geometrical Investigations on the Theory of Parallels*, Berlin, 1840, in the Finck publishing house. In this work I have stated first all the theorems which can be demonstrated without the aid of the theory of parallels. Among these theorems the theorem which gives the ratio of the area of a spherical triangle to the entire area of the sphere upon which it is traced, is particularly remarkable (*Geometrical Investigations*, §27.)[3] If A, B, and C are the angles of a spherical triangle and π represents 2 right angles, the ratio of the area of the triangle to the area of the sphere to which it belongs will be equal to the ratio of

$$\tfrac{1}{2}(A + B + C - \pi)$$

to four right angles.

Then I demonstrate that the sum of the three angles of a rectilinear triangle can never surpass two right angles (§19), and that, if the sum is equal to two right angles in any triangle, it will be so in all (§20). Thus there are only two suppositions possible: Either the sum of the three angles of a rectilinear triangle is always equal to two right angles, the supposition which gives the known geometry, or in every rectilinear triangle this sum is less than two right angles, and this supposition serves as the basis of another geometry, to which I had given the name of *imaginary geometry*, but which it is perhaps more fitting to call *pangeometry* because this name indicates a general geometrical theory which includes the ordinary geometry as a particular case. It follows from the principles adopted in the pangeometry that a perpendicular p let fall from a

---

[1] [He seems to refer to work elsewhere, or perhaps to his teaching. These ideas are not developed further in this book.]

[2] [The Russian adds: That side on which the intersection occurs I call *the side of parallelism*.]

[3] [The *Geometrical Investigations* is the work translated by Halsted. See page 360. In further references to this work only the section number will be given.]

point of a straight line upon one of the parallels makes with the first line two angles of which one is acute. I call this angle the *angle of parallelism* and the side of the first line where it is found,[1] side which is the same for all the points of this line, the *side of parallelism*. I denote this angle by $\Pi(p)$, since it depends upon the length of the perpendicular. In the ordinary geometry we have always $\Pi(p) = $ a right angle for every length of p. In the pangeometry the angle $\Pi(p)$ passes through all values from zero, which corresponds to $p = \infty$, to $\Pi(p) = $ a right angle for $p = 0$ (§23). In order to give the function $\Pi(p)$ a more general analytical value I assume that the value of this function for p negative, case which the original definition does not cover, is fixed by the equation

$$\Pi(p) + \Pi(-p) = \pi.$$

Thus for every angle $A > 0$ and $< \pi$ we can find a line p such that $\Pi(p) = A$, where the line p will be positive if $A < \pi/2$. Reciprocally there exists for every line p an angle A such that $A = \Pi(p)$.

I call *limit circle*[2] the circle whose radius is infinite. It can be traced approximately by constructing in the following manner as many points as we wish. Take a point on an indefinite straight line, call this point *vertex* and the line *axis* of the limit circle, and construct an angle $A > 0$ and $< \pi/2$ with vertex at the vertex of the limit circle and the axis of the limit circle as one of its sides. Then let a be the line which gives $\Pi(a) = A$ and lay off on the second side of the angle from the vertex a length equal to $2a$. The extremity of this length will be found on the limit circle. To continue the tracing of the limit circle on the other side of the axis it will be necessary to repeat the construction on that side. It follows that all the lines parallel to the axis of the limit circle can be taken as axes.

The rotation of the limit circle around one of its axes produces a surface which I call *limit sphere*,[3] surface which is, therefore, the limit which the sphere approaches if the radius increases to infinity. We shall call the axis of rotation, and therefore all the lines parallel to the axis of rotation, *axes of the limit sphere*, and we shall call *diametral plane* every plane which contains one or several axes of the limit sphere. The intersections of the limit sphere by its

---

[1] [Russian: Where the acute angle is found.]

[2] [This is the *oricycle* or *boundry-curve*.]

[3] [*Orispbere* or *boundary-surface*.]

diametral planes are limit circles. A part of the surface of the limit sphere bounded by three limit circle arcs will be called a *limit sphere triangle*. The limit circle arcs will be called the *sides*, and the dihedral angles between the planes of these arcs the *angles* of the limit sphere triangle.

Two lines parallel to a third are parallel to each other (§25). It follows that all the axes of a limit circle and of a limit sphere are parallel to one another. If three planes two by two intersect in three parallel lines and if we limit each plane to the part which is between these parallels, the sum of the three dihedral angles which these planes form will be equal to two right angles (§28). It follows from this theorem that the sum of the angles of a limit sphere triangle is always equal to two right angles, and everything that is demonstrated in the ordinary geometry of the proportionality of the sides of rectilinear triangles can therefore be demonstrated in the same manner in the pangeometry of the limit sphere triangles if only we will replace the lines parallel to the sides of the rectilinear triangle by limit circle arcs drawn through the points of one of the sides of the limit sphere triangle and all making the same angle with this side.[1] Thus, for example, if p, q, and r are the sides of a limit sphere right triangle and P, Q, and $\pi/2$ the angles opposite these sides, it is necessary to assume, as for right angled rectilinear right triangles of the ordinary geometry, the equations

$$p = r \sin P = r \cos Q,$$
$$q = r \cos P = r \sin Q,$$
$$P + Q = \frac{\pi}{2}.$$

In the ordinary geometry we demonstrate that the distance between two parallel lines is constant. In pangeometry, on the contrary, the distance p from a point of a line to the parallel line diminishes on the side of parallelism, that is to say, on the side towards which is turned the angle of parallelism $\Pi(p)$.

Now let s, s', s'',... be a series of limit circle arcs lying between two parallel lines which serve as axes to all these limit circles, and suppose that the parts of these parallel lines between

---

[1] [Apparently he would say that two limit circle arcs cutting a third so that corresponding angles are equal would be like the parallels of ordinary geometry. Thus a limit circle arc cutting one side of a limit sphere triangle may be "parallel" to one of the other sides.]

two consecutive arcs are all equal to one another and equal to x. Denote by E the ratio of s to s',[1]

$$\frac{s}{s'} = E,$$

where E is a number greater than unity.[2]

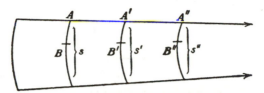

Suppose[3] first that $E = n/m$, m and n being two integer numbers, and divide the arc s into m equal parts. Through the points of division draw lines parallel to the axes of the limit circles. These parallels will divide each of the arcs s', s'', etc., into m parts equal to one another. Let[4] AB be the first part of s, $A'B'$ the first part of s', $A''B''$ the first part of s'' etc., A, A', A'',...the points situated upon one of the given parallels, and put $A'B'$ upon AB so that A and A' will coincide and $A'B'$ fall along AB. Repeat this superposition n times. Since by supposition $s/s' = n/m$, it will be necessary that $nA'B' = mAB$, and therefore that the second extremity of $A'B'$ will coincide after the nth superposition with the second extremity of s, which will be divided into n equal parts. s', s'',... will also be divided into m equal parts each by the lines parallel to the two given parallels. But if we consider that in making the superposition indicated above, $A'B'$ carries the part of the plane limited by this arc and the two parallels drawn through its extremi-

---

[1] [The Russian adds: when x is equal to 1.]

[2] [Russian: positive and greater than unity.]

[3] [There are no figures for the pangeometry in the Œuvres from which this translation is made. The figures that we are using are taken from the German translation made by Heinrich Liebmann, Leipzig, 1902.]

[4] [Instead of the rest of this paragraph the Russian says: We superimpose the area between the arcs s' and s'' over the area between s and s', and the arc s' on the arc s, and hence s'' on s'. We repeat the arc s'/m. It has to go n times in the arc s. Parallelism of lines makes the arc s''/m go n times in s'. Hence

$$s/s' = s'/s''.$$

And a few lines below: This implies that for every line x, $s' = sE^{-x}$, etc.]

ties, it is clear that at the same time while n times $A'B'$ covers all of the arc s, $nA''B''$ will cover all of the arc s', and so on, because in this case the parallels ought to coincide in all their extent, so that we have

$$nA''B'' = mA'B',$$

or, what is the same thing,

$$\frac{s'}{s''} = \frac{n}{m} = E, \frac{s'}{s''} = E, \text{ etc.,}$$

which is what we had to demonstrate.

To demonstrate the same thing in the case where E is an incommensurable number we can employ one of the methods used for similar cases in ordinary geometry. For the sake of brevity I will omit these details. Thus

$$\frac{s}{s'} = \frac{s'}{s''} = \frac{s''}{s'''} = \ldots = E.$$

After this it is not difficult to conclude that

$$s' = sE^{-x},$$

where E is the value of s/s' for x, the distance between the arcs s and s', equal to unity.

It is necessary to remark that this ratio E does not depend on the length of the arc s, and remains the same if the two given parallel lines are moved away from each other or approach each other. The number E, which is necesarily greater than unity, depends only on the unit of length, which is the distance between two successive arcs, and which is entirely arbitrary. The property which we have just demonstrated with respect to the arcs s, s', s'' ... subsists for the areas P, P', P'', ..., limited by two successive arcs and the two parallels. We have then

$$P' = PE^{-x}.$$

If we unite n such areas P, P', P'', ... $P^{(n-1)}$, the sum will be

$$P\frac{1 - E^{-nx}}{1 - E^{-x}}.$$

For $n = \infty$ this expression gives the area of the part of the plane between two parallel lines, limited on one side by the arc s, and unlimited on the side of the parallelism, and the value of this will be

$$\frac{P}{1 - E^{-x}}.$$

If we choose for unit of area the area P which corresponds to an arc s also a unit, and to $x = 1$, we shall have in general for any arc s

$$\frac{Es}{E-1}.$$

In the ordinary geometry the ratio designated by E is constant and equal to unity. It follows that in the ordinary geometry two parallel lines are everywhere equidistant and that the area of the part of the plane situated between two parallel lines and limited on one side only by a perpendicular common to them is infinite.

Consider for the present a right angled rectilinear triangle in which a, b, and c are the sides, and A, B, and $\pi/2$ the angles opposite these sides. For the angles A and B can be taken the angles of parallelism $\Pi(\alpha)$ and $\Pi(\beta)$ corresponding to lines of positive length α and β. Let us agree also to denote hereafter by a letter with an accent a line whose length corresponds to an angle of parallelism which is the complement to a right angle of the angle of parallelism corresponding to the line whose length is denoted by the same letter without accent, so that we have always

$$\Pi(\alpha) + \Pi(\alpha') = \frac{\pi}{2},$$

$$\Pi(b) + \Pi(b') = \frac{\pi}{2}.$$

Denote[1] by $f(a)$ the part of a parallel to an axis of a limit circle intercepted between the perpendicular to the axis through the vertex of the limit circle and the limit circle itself, if this parallel passes through a point of the perpendicular whose distance from the vertex is a, and let $L(a)$ be the length of the arc from the vertex to this parallel.

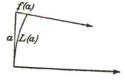

In the ordinary geometry we have

$$f(a) = 0, \; L(a) = a,$$

for every length a.

---

[1] [The Russian says: Erect a perpendicular a to an axis of a limit circle at the vertex. Through the apex of the perpendicular draw a line parallel to the axis on the side of parallelism. Designate by $f(a)$ the part of the parallel between the perpendicular and the limit circle itself and by $L(a)$ the length of the arc from the vertex to this parallel.]

Draw a perpendicular AA' to the plane of the right angled triangle whose sides have been denoted by a, b, and c, perpendicular 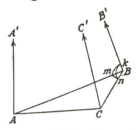 through the vertex A of the angle $\Pi(\alpha)$. Pass through this perpendicular two planes, of which one, which we will call the first plane, passes also through the side b, and the other, the second plane, through the side c. Construct in the second plane the line BB' parallel to AA' which passes through the vertex B of the angle $\Pi(\beta)$, and pass a third plane through BB' and the side a of the triangle. This third plane will cut the first in a line CC' parallel to AA'. Conceive now a sphere described from the point B as center with a radius arbitrary, but smaller than a, a sphere which will therefore cut the two sides a and c of the triangle and the line BB' in three points which we will call, the first n, the second m, and the third k. The arcs of great circles, intersections of this sphere by the three planes passing through B, which unite two by two the points n, m, and k, will form a spherical triangle right angled at m, whose sides will be $mn = \Pi(\beta)$, $km = \Pi(c)$, and $kn = \Pi(a)$. The spherical angle knm will be equal to $\Pi(b)$ and the angle kmn will be a right angle. The three lines being parallel to one another, the sum of the three dihedral angles which the parts of the planes $AA'BB'$, $AA'CC'$, and $BB'CC'$ situated between the lines AA', BB', and CC' form with one another will be equal to two right angles.[1]   It follows that the third angle of the spherical triangle will be $mkn = \Pi(\alpha')$. We see then that to every right angled rectilinear triangle whose sides are a, b, and c, and the opposite angles $\Pi(\alpha)$, $\Pi(\beta)$, and $\pi/2$ corresponds a right angled spherical triangle whose sides are $\pi(\beta)$, $\Pi(c)$, and $\Pi(a)$, and the opposite angles $\Pi(\alpha')$, $\Pi(b)$, and $\pi/2$. Construct another right angled rectilinear triangle whose sides perpendicular to each other are[2] α' and a, whose hypotenuse is g, and in which $\Pi(\lambda)$ is the angle opposite the side a, and $\Pi(\mu)$ the angle opposite the side α'. Pass from this triangle to the spherical triangle which corresponds in the same manner as the spherical triangle

---

[1] [The Russian says: The three lines AA', BB', and CC', being parallel to one another, give a sum of dihedral angles equal to π.]

[2] [It would be better to put a before α'.]

kmn corresponds to the triangle ABC.   The sides of this spherical triangle will then be

$$\Pi(\mu), \Pi(g), \Pi(a),$$

and the opposite angles

$$\Pi(\lambda'), \Pi(\alpha'), \frac{\pi}{2},$$

and it will have its parts equal to the corresponding parts of the spherical triangle kmn, for the sides of the latter were

$$\Pi(c), \Pi(\beta), \Pi(a),$$

and the opposite angles

$$\Pi(b), \Pi(\alpha'), \frac{\pi}{2},$$

which shows that these spherical triangles have the hypotenuse and an adjacent angle the same.

It follows that

$$\mu = c, \; g = \beta, \; b = \lambda',$$

and thus the existence of the right angled rectilinear triangle with the sides

$$a, \; b, \; c,$$

and the opposite angles

$$\Pi(\alpha), \Pi(\beta), \frac{\pi}{2},$$

supposes the existence of a right angled rectilinear triangle with the sides

$$a, \; \alpha', \; \beta,$$

and the opposite angles

$$\Pi(b'), \Pi(c), \frac{\pi}{2}.$$

We can express the same thing by saying that if

$$a, \; b, \; c, \; \alpha, \; \beta$$

are the parts of a right angled rectilinear triangle,

$$a, \; \alpha', \; \beta, \; b', \; c$$

will be the parts of another right angled rectilinear triangle.[1]

---

[1] [There seems to be no simple geometrical relation between the two rectilinear triangles nor anything more than an empirical law for deriving the parts of the spherical triangle from those of the rectilinear triangle.   There are two ways in which the parts of two right triangles may correspond and there is

If we construct the limit sphere of which the perpendicular AA' to the plane of the given right angled rectilinear triangle is an axis and

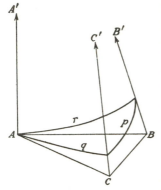

the point A the vertex, we shall have a triangle situated upon the limit sphere and produced by its intersections with the planes drawn through the three sides of the given triangle. Denote the three sides of this limit sphere triangle by p, q, and r, p the intersection of the limit sphere by the plane which passes through a, q the intersection by the plane which passes through b, and r the intersection by the plane which passes through c. The angles opposite these sides will be $\Pi(\alpha)$ opposite p, $\Pi(\alpha')$ opposite q, and a right angle opposite r. From the conventions adopted above $q = L(b)$ and $r = L(c)$. The limit sphere will cut the line CC' at a point whose distance from C will be, from the same conventions, $f(b)$. In the same manner we shall have $f(c)$ for the distance from the point of intersection of the limit sphere with the line BB' to the point B.

It is easy to see that we shall have

$$f(b) + f(a) = f(c).$$

some confusion here because the correspondence of the two spherical triangles as derived from corresponding rectilinear triangles is not the correspondence in which the parts of one are equal to the corresponding parts of the other.

If we indicate the parts of the first rectilinear triangle by writing

$$a, b, c, \alpha, \beta,$$

and the corresponding parts of the spherical triangle by writing

$$\beta, c, a, \alpha', b,$$

we can write for the second rectilinear triangle

$$a, \alpha', g, \lambda, \mu,$$

or, substituting for g, λ, and μ their values,

$$a, \alpha,' \beta, b,' c,$$

and then for the second spherical triangle

$$c, \beta, a, b, \alpha',$$

and we find that the parts of the two spherical triangles are not arranged according to the way in which they are equal.

When he writes the parts of the first spherical triangle for the purpose of comparing the two he changes the order so that the parts of the two are arranged in this way.]

In the triangle whose sides are the limit circle arcs p, q, and r we shall have

$$p = r \sin \Pi(\alpha), \quad q = r \cos \Pi(\alpha).$$

Multiplying the first of these two equations by $E^{f(b)}$ we have

$$pE^{f(b)} = r \sin \Pi(\alpha)E^{f(b)}.$$

But

$$pE^{f(b)} = L(a),$$

and therefore

$$L(a) = r \sin \Pi(\alpha)E^{f(b)}.$$

In the same way

$$L(b) = r \sin \Pi(\beta)E^{f(a)}.$$

At the same time $q = r \cos \Pi(\alpha)$, or, what is the same thing, $L(b) = r \cos \Pi(\alpha)$. A comparison of the two values of $L(b)$ gives the equation

$$\cos \Pi(\alpha) = \sin \Pi(\beta)E^{f(a)}. \tag{1}$$

Substituting b' for α and c for β without changing a, which is permitted from what we have demonstrated above, we shall have

$$\cos \Pi(b') = \sin \Pi(c)E^{f(a)},$$

or, since

$$\Pi(b) + \Pi(b') = \frac{\pi}{2},$$

$$\sin \Pi(b) = \sin \Pi(c)E^{f(a)}.$$

In the same way we ought to have

$$\sin \Pi(a) = \sin \Pi(c)E^{f(b)}.$$

Multiply the last equation by $E^{f(a)}$ and substitute $f(c)$ for $f(a) + f(b)$. This will give

$$\sin \Pi(a)E^{f(a)} = \sin \Pi(c)E^{f(c)}.$$

But as in a right angled rectilinear triangle the perpendicular sides can vary while the hypotenuse remains constant, we can put in this equation $a = 0$ without changing c. This will give, since $f(0) = 0$ and $\Pi(0) = \pi/2$,

$$1 = \sin \Pi(c)E^{f(c)},$$

or

$$E^{f(c)} = \frac{1}{\sin \Pi(c)}$$

for every line c.

Now take equation (1)

$$\cos \Pi(\alpha) = \sin \Pi(\beta) E^{f(a)}$$

and substitute $1/\sin \Pi(a)$ for $E^{f(a)}$. It will take the following form

$$\cos \Pi(\alpha) \sin \Pi(a) = \sin \Pi(\beta). \qquad (2)$$

Changing α and β to b' and c without changing a we find

$$\sin \Pi(b) \sin \Pi(a) = \sin \Pi(c).$$

Equation (2) with a change of letters gives

$$\cos \Pi(\beta) \sin \Pi(b) = \sin \Pi(\alpha).$$

If in this equation we change β, b, and α to c, α', and b' we shall get

$$\cos \Pi(c) \cos \Pi(\alpha) = \cos \Pi(b). \qquad (3)$$

In the same way we shall have

$$\cos \Pi(c) \cos \Pi(\beta) = \cos \Pi(a) \qquad (4)$$

. . . .

It follows[1] from what precedes that spherical trigonometry remains the same, whether we adopt the supposition that the sum of the three angles of a rectilinear triangle is always equal to two right angles, or adopt the supposition that this sum is always less than two right angles, which is very remarkable and does not hold for rectilinear trigonometry.

Before demonstrating the equations which express in pangeometry the relations between the sides and angles of any rectilinear triangle we shall seek for every line x the form of the function which we have denoted hitherto by $\Pi(x)$.

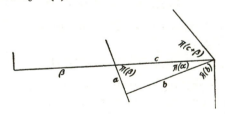

Consider[2] for this purpose a right angled rectilinear triangle whose sides are a, b, c, and the opposite angles $\Pi(\alpha)$, $\Pi(\beta)$, $\pi/2$. Prolong c beyond the vertex of the angle $\Pi(\beta)$ and make the

---

[1] [In the part omitted the ordinary equations of spherical trigonometry are derived from the preceding equations.]

[2] [We have combined and changed a little the figures given here by Liebmann.]

prolongation equal to β. The perpendicular to β erected at the extremity of this line and on the side opposite to that of the angle $\Pi(\beta)$ will be parallel to a and its prolongation beyond the vertex of $\Pi(\beta)$. Draw also through the vertex of $\Pi(\alpha)$ a line parallel

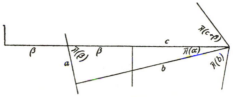

to this same prolongation of a. The angle which this line will make with c will be $\Pi(c + \beta)$ and the angle which it will make with b will be $\Pi(b)$, and we shall have the equation

$$\Pi(b) = \Pi(c + \beta) + \Pi(\alpha). \tag{II}$$

If we take the length β from the vertex of the angle $\Pi(\beta)$ on the side c itself and erect at its extremity a perpendicular to β on the side of the angle $\Pi(\beta)$, this line will be parallel to the prolongation of a beyond the vertex of the right angle. Draw through the vertex of the angle $\Pi(\alpha)$ a line parallel to this last perpendicular, which will therefore also be parallel to the second prolongation of a. The angle of this parallel with c will be in all cases $\Pi(c - \beta)$ and the angle which it makes with b will be $\Pi(b)$. Therefore

$$\Pi(b) = \Pi(c - \beta) - \Pi(\alpha). \tag{II'}$$

It is easy to convince ourselves that this equation is true not only if $c > \beta$, but also if $c = \beta$ and if $c < \beta$. In fact, if $c = \beta$ we have, on the one hand, $\Pi(c - \beta) = \Pi(0) = \pi/2$, and, on the other hand, the perpendicular to c drawn through the vertex of the angle $\Pi(\alpha)$ becomes parallel to a, whence it follows that $\Pi(b) = \frac{\pi}{2} - \Pi(\alpha)$, which agrees with our equation.

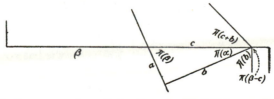

If $c < \beta$ the extremity of the line β will fall beyond the vertex of the angle $\Pi(\alpha)$ at a distance equal to $\beta - c$. The perpendicular to β at this extremity of β will be parallel to a and to the line through the vertex of the angle $\Pi(\alpha)$ parallel to a, whence it follows that the two adjacent angles which this parallel makes with c will be, the acute equal to $\Pi(\beta - c)$, the obtuse equal to $\Pi(\alpha) + \Pi(b)$. But the sum of two adjacent angles is always equal to two right angles. Thus

$$\Pi(\beta - c) + \Pi(\alpha) + \Pi(b) = \pi,$$

or

$$\Pi(b) = \pi - \Pi(\beta - c) - \Pi(\alpha).$$

But from the definition of the function $\Pi(x)$

$$\pi - \Pi(\beta - c) = \Pi(c - \beta),$$

which gives

$$\Pi(b) = \Pi(c - \beta) - \Pi(\alpha),$$

that is to say, the equation found above, which is thus demonstrated for all cases.

The two equations (Π) and (Π') can be replaced by the following two

$$\Pi(b) = \tfrac{1}{2}\Pi(c + \beta) + \tfrac{1}{2}\Pi(c - \beta)$$
$$\Pi(\alpha) = \tfrac{1}{2}\Pi(c - \beta) - \tfrac{1}{2}\Pi(c + \beta).$$

But equation (3) gives us

$$\cos \Pi(c) = \cos \Pi(b)/\cos \Pi(\alpha),$$

and in substituting in this equation in place of $\Pi(b)$ and $\Pi(\alpha)$ their values we get

$$\cos \Pi(c) = \frac{\cos \left[\tfrac{1}{2}\Pi(c + \beta) + \tfrac{1}{2}\Pi(c - \beta)\right]}{\cos \left[\tfrac{1}{2}\Pi(c - \beta) - \tfrac{1}{2}\Pi(c + \beta)\right]}.$$

From this equation we deduce the following

$$\tan^2 \tfrac{1}{2}\Pi(c) = \tan \tfrac{1}{2}\Pi(c - \beta) \tan \tfrac{1}{2}\Pi(c + \beta).$$

As the lines c and β can vary independently of each other in a right angled rectilinear triangle, we can put successively in the last equation $c = \beta$, $c = 2\beta, \ldots c = n\beta$, and we conclude from the equations thus deduced that in general for every line c and for every positive integer n

$$\tan^n \tfrac{1}{2}\Pi(c) = \tan \tfrac{1}{2}\Pi(nc).$$

It is easy to demonstrate the truth of this equation for n negative or fractional, whence it follows that in choosing the unit of length so that we have $\tan \tfrac{1}{2}\Pi(1) = e^{-1}$, where e is the base of Naperian logarithms, we shall have for every line x

$$\tan \tfrac{1}{2}\Pi(x) = e^{-x}.$$

This expression gives $\Pi(x) = \pi/2$ for $x = 0$, $\Pi(x) = 0$ for $x = \infty$, and $\Pi(x) = \pi$ for $x = -\infty$, agreeing with what we have adopted and demonstrated above.

BOLYAI

On Non-Euclidean Geometry

(Translated from the Latin by Professor Henry P. Manning, Brown University, Providence, R. I.)

János Bolyai (1802–1860) was the son of Farkas Bolyai, a fellow student of Gauss's at Göttingen. Farkas wrote to Gauss in 1816 that his son, then a boy of fourteen, had already a good knowledge of the calculus and its applications to mechanics. János went to the engineering school at Vienna at the age of sixteen and entered the army at the age of twenty-one. About 1825 or 1826 he worked out his theory of parallels and published it in 1832 as an appendix to the first part of a work by his father, the book having the imprimatur of 1829. It was in Latin, but was later translated into French (1867), Italian (1868), German (1872), and English (by Halsted, 1891). The title of the father's work is *Tentamen Juventutem Studiosam in Elementa Matheseos Purae...introducendi.* It appeared in two parts at Maros-Vasarhely, in 1832, 1833. It is the appendix to the first part that is here translated.

APPENDIX[1]

exhibiting the absolutely true science of space,[2] independent of Axiom XI[3] of Euclid (not to be decided a priori), with the geometrical quadrature of a circle in the case of its falsity.

Explanation of signs[4]

$\overline{AB}$[5] denotes the complex of all the points on a line with the points A and B.

---

[1] [Ioannis Bolyai de Bolya, *Appendix*, editio nova, published by the Hungarian Academy of Science, Budapest, 1902, in honor of the centennial aniversary of the author's birth. This was published, as originally, along with the *Tentamen* of his father, and also separately.]

[2] [We may remark that "absolutely true science of space" is a different thing from "absolute science of space," or "absolute geometry," which are the terms often used and seem to have been used sometimes by Bolyai himself.]

[3] [Euclid's axiom of parallels which the best authorities now call "Postulate V."]

[4] [In the edition of the Hungarian Academy points are denoted by small letters in German type. We shall use capital italic letters as is customary in modern textbooks in geometry. Also in that edition parentheses are used much more frequently than with us, often a clause that is an essential part of a sentence is inclosed in parentheses. We shall omit most of these parentheses.]

[5] [Two or more letters without any mark over them denote a limited figure, while a figure is unlimited in a part denoted by a letter with a mark over it.]

$\overline{AB}$ denotes that half of the line AB cut at A which contains the
point B.

$\overline{ABC}$ denotes the complex of all the points which are in the same
plane with the points A, B, and C (these not lying in the same
straight line).

$AB\overline{C}$ denotes the half of the plane ABC cut apart by $\overline{AB}$ that
contains the point C.

ABC denotes the smaller of the portions into which $\overline{ABC}$ is
divided by the complex of $B\overline{A}$ and $B\overline{C}$, or the angle whose
sides[1] are $B\overline{A}$ and $B\overline{C}$.

$ABCD$[2] denotes (if D is in ABC and $B\overline{A}$ and $C\overline{D}$ do not cut each
other) the portion of ABC enclosed by $B\overline{A}$, BC, and $C\overline{D}$.
But $BACD$ is the portion of the plane $\overline{ABC}$ between $\overline{AB}$ and
$\overline{CD}$.

R denotes right angle.

$AB \simeq CD$[3] denotes $CAB = ACD$.

AB (not given in the list) denotes the segment from A to B of the line $\overline{AB}$; and
so, in the definition of $ABCD$ below, BC denotes a segment. But we may
note that ABC denotes an angle and not a triangle. When the author wishes
to name a triangle he says "triangle ABC" or inserts the sign "$\triangle$" (see, for
example, §13). We may note also that an angle with him is a portion of a
plane, and that two angles having a side in common but lying in different
planes will form a dihedral angle. See, for example, §7.]

[1] [He calls them *legs*.]

[2] [If two lines in a plane are cut by a third, we can say that the half-lines on
one side of this third line lie in one direction along the
two given lines, and the half-lines on the other side lie
in the other direction. Now in the first of the two definitions given here we read the two pairs of points taken
on the two lines in opposite directions, and in the second
definition we read them in the same direction. We have
an illustration of the first definition in $MACN$ at the
beginning of §2, and an illustration of the second in $BNCP$
at the beginning of §7.]

[3] In the relation represented by this sign AB and CD
are two lines in a plane cut by a third at A and C, with
the points A and B taken on one line and the points C
and D on the other in the same direction. This sign
is used when $A\overline{B}$ and $C\overline{D}$ intersect as well as when they
do not intersect. See, for example, §5, where $EC \simeq$
BC.

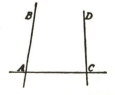

$\equiv$ denotes congruent.[1]

$x \rightarrow a^2$ denotes x tends to a as limit.

$\bigcirc r$ denotes circumference of circle of radius r.

$\odot r$ denotes area of circle of radius r.

§1

Given $A\overline{M}$, if $B\overline{N}$, lying in the same plane with it, does not cut it, but every half-line $B\overline{P}$ in ABN[3] does cut it, let this be denoted by

$$BN\|AM.[4]$$

It is evident that there is *given* such a $B\overline{N}$ and indeed from any point B outside of $\overline{AM}$ *only one*, and that

$$BAM + ABN \text{ is not } > 2R;$$

for when BC[5] is moved around B until

$$BAM + ABC = 2R,$$

at some point $B\overline{C}$ *first* does not cut $A\overline{M}$, and then $BC\|AM$. And it is evident that $BN\|EM$, wherever E may be on $\overline{AM}$ (supposing in all these cases that $AM > AE$).[6]

And if, with the point C on $A\overline{M}$ going off to infinity, we always have $CD = CB$, always we shall have

$$CDB = CBD < NBC.$$

But $NBC \rightarrow 0$. Therefore $ADB \rightarrow 0$.

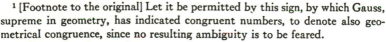

---

[1] [Footnote to the original] Let it be permitted by this sign, by which Gauss, supreme in geometry, has indicated congruent numbers, to denote also geometrical congruence, since no resulting ambiguity is to be feared.

[2] [Bolyai uses the sign "⌒."]

[3] [In the angle ABN, the words "in ABN" are in parentheses in the original. See page 375, footnote 4.]

[4] [This is a relation of half-lines, but in writing it the author always leaves out the mark over the second letter.]

[5] [Here he speaks of the segment BC because he thinks of the point C moving along $A\overline{M}$, but for the limiting position he writes $B\overline{C}$.]

[6] [This seems to mean that E is not to lie beyond M on $A\overline{M}$, or that M has been taken far enough out on this half-line to be beyond where we wish to take E.]

§2

If BN‖AM, we shall have also CN‖AM.[1]

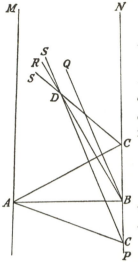

For let *D* be somewhere in *MACN* If *C* lies on $B\overline{N}$, $B\overline{D}$ will cut $\overline{AM}$ because *BN‖AM*, and so also $C\overline{D}$ will cut $\overline{AM}$. But if *C* is in $B\overline{P}$, let $B\overline{Q}‖CD$. $B\overline{Q}$ falls in *ABN* (§1)[2] and cuts $\overline{AM}$. and so $C\overline{D}$ cuts $\overline{AM}$. Therefore $C\overline{D}$ cuts $\overline{AM}$ in both cases. But $C\overline{N}$ does not cut $\overline{AM}$. Therefore always *CN‖AM*.

§3

If BR and CS are both ‖AM and C is not in $\overline{BR}$, then $B\overline{R}$ and $C\overline{S}$ do not intersect each other.

For if $B\overline{R}$ and $C\overline{S}$ had a point *D* in common, *DR* and *DS* would at the same time ‖AM (§2), and $D\overline{S}$ would fall on $D\overline{R}$ (§1) and *C* on $\overline{BR}$, contrary to hypothesis.

§4

If MAN > MAB, for every point B in $A\overline{B}$ is given a point C in $A\overline{M}$ such that BCM = NAM.

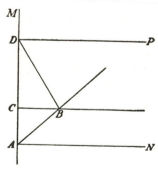

---

[1] [It is left to the reader to see from the figure that *C* is a point of *BN*. The author often omits details in this way when they are shown in the figure.]

[2] [BN does not cut $C\overline{D}$, nor will it do so, even if it rotate towards $B\overline{A}$, as long as the two lines intersect below *B*.]

For there is given a $BDM > NAM$ (§1)[1], and also an $MDP = MAN$, and B falls in $NADP$. If therefore we move NAM along AM until $A\overline{N}$ comes to $D\overline{P}$, sometime $A\overline{N}$ will have passed through B and there will be a $BCM = NAM$.

<h2 style="text-align:center">§5</h2>

If $BN|||AM$ (p. 377) there is a point F in $A\overline{M}$ such that $FM \backsimeq BN$.
For there is a $BCM > CBN$ (§1), and if $CE = CB$, and so $EC \backsimeq BC$, it is evident that $BEM < EBN$. Let P traverse EC, the angle BPM always called u and the angle PBN always called[2] v. It is evident that u is at first less than the corresponding value of v, but afterwards greater. But u increases from BEM to BCM *continuously*, since there is *no* angle $> BEM$ and $< BCM$ to which u is not at some time equal (§4). Likewise v decreases from EBN to CBN continuously. And so there is given on EC a point F such that $BFM = FBN$.

<h2 style="text-align:center">§6</h2>

If $BN|||AM$ and E is anywhere in $\overline{AM}$ and G in $\overline{BN}$, then $GN|||$
EM and $EM|||GN$.
For $BN|||EM$ (§1), and hence $GN|||EM$(§2). If then $FM \backsimeq BN$ (§5), then $MFBN \equiv NBFM$, and so, since $BN|||FM$, also $FM|||BN$, and by what precedes $EM|||GN$.

<h2 style="text-align:center">§7</h2>

If BN and CP are both $|||AM$, and C is not in $\overline{BN}$, then $BN|||CP$.

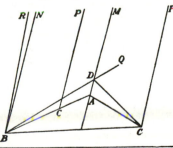

---

[1] [If we let D move off on $A\overline{M}$, the angle ADB will approach zero, and therefore at some time become less than the supplement of NAM. Then if M is taken beyond this position of D, we shall have BDM the supplement of ADB greater than NAM.]

[2] [To show this in the figure on p. 377, P ought to be put where A is.]

For $B\overline{N}$ and $C\overline{P}$ do not cut each other (§3). Moreover, AM, BN, and CP are in a plane or not. And in the first case AM lies in $BNCP$[1] or not.

If AM, BN, and CP are in a plane and AM falls in $BNCP$, then any $B\overline{Q}$ in NBC cuts $\overline{AM}$ in a point D because $BN|||AM$; and then since $DM|||CP$ (§6) it is evident that $D\overline{Q}$ cuts $C\overline{P}$, and so $BN|||CP$.

But if BN and CP lie on the same side of AM, then one of them, for example CP, will fall *between* the other two, $\overline{BN}$ and $\overline{AM}$, and any $B\overline{Q}$ in NBA will cut $A\overline{M}$, and so also $C\overline{P}$. Therefore $BN|||CP$.

If MAB and MAC form an *angle*,[2] then CBN has in common with ABN only $B\overline{N}$, but $A\overline{M}$ in ABN with $B\overline{N}$, and so NBC with $A\overline{M}$, have in common nothing. But a $BC\overline{D}$ drawn through any $B\overline{D}$ in NBA will cut $A\overline{M}$ because $B\overline{D}$ cuts $A\overline{M}$, BN being$|||AM$. Therefore if $BC\overline{D}$ is moved about BC[3] until *first* it leaves $A\overline{M}$, at last $BC\overline{D}$ will fall in $BC\overline{N}$. For the same reason it will fall in $BC\overline{P}$. Therefore BN falls in $BC\overline{P}$. Then if $BR|||CP$, because also $AM|||CP$, BR will fall in BAM for the same reason, and in BCP because $BR|||CP$. And so $B\overline{R}$ is common to MAB and PCB, and is therefore $B\overline{N}$ itself.[4] Therefore $BN|||CP$.

If therefore $CP|||AM$ and B is outside of $\overline{CAM}$, then the intersection of BAM and BCP, that is, $B\overline{N}$, is $|||$ both to AM and to CP.[5]

---

[1] [Notice that the points B and N are taken on one line and the points C and P on the other in the same direction. Thus "in $BNCP$" means in the entire strip between $\overline{BN}$ and $\overline{CP}$, and not simply in that portion of this strip which is above BC. In this paragraph we should take the CP that is to the right in Figure 58 and regard the entire figure as lying in one plane. In the second paragraph we take the CP at the left, while in the third paragraph we take the CP again at the right, but this time the three lines not in one plane.]

[2] [A dihedral angle. See page 375, end of footnote 5.]

[3] [The point D moving off indefinitely on $A\overline{M}$. This brings $B\overline{D}$ into coincidence with $B\overline{N}$ and $C\overline{D}$ with $C\overline{P}$.]

[4] [The argument is this: BN, which is $|||$ AM, lies in BCP as well as in BAM, and is therefore their intersection. But in the same way we can say that BR, which is $|||$ CP lies in BAM as well as in BCP and is also their intersection.]

[5] [Footnote to the original] If the third case had been taken first, the other two could have been solved as in §10 more briefly and more elegantly (from edition I, volume I, Errata of the Appendix).

§8

If BN||| and ≕ CP (or more briefly BN||| ≕ CP), and if AM in NBCP bisects BC at right angles, then BN|||AM.

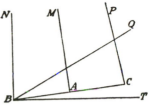

For if $B\overline{N}$ should cut $A\overline{M}$, also $C\overline{P}$ would cut $A\overline{M}$ at the same point, since $MABN \equiv MACP$, which would be common to $B\overline{N}$ and $C\overline{P}$, although $BN|||CP$. But if $B\overline{Q}$ in CBN cuts $C\overline{P}$ then also $B\overline{Q}$ cuts $A\overline{M}$. Therefore $BN|||AM$.

§9[1]

If BN |||AM, if MAP ⊥ MAB, and if the angle which NBD makes with NBA on the side of MABN where MAP is, is < R, then MAP and NBD cut each other.

For let $BAM = R$, let $AC \perp BN$ (whether B falls at C or not), and let $CE \perp BN$ in NBD. ACE will be $< R$ by hypothesis and $AF \perp CE$ will fall in ACE. Let $A\overline{P}$ be the intersection of $AB\overline{F}$ and $AM\overline{P}$ (these having the point A in common). Then $BAP = BAM = R$ (since $BAM \perp MAP$). If, finally, $AB\overline{F}$ be put upon $AB\overline{M}$, A and B remaining fixed,[2] $A\overline{P}$ will fall on $A\overline{M}$, and since $AC \perp BN$ and $AF < AC$, it is evident that AF will end on this side of $B\overline{N}$ and so BF will fall in ABN. But $B\overline{F}$ will cut $A\overline{P}$ in this position because $BN|||AM$, and so also in their *first* positions $A\overline{P}$ and $B\overline{F}$ will cut each other. The point of intersection is a point common to $MA\overline{P}$ and $NB\overline{D}$, and so $MA\overline{P}$ and $NB\overline{D}$ cut each other.

---

[1] [It will be noticed particularly in this section that he sometimes uses a letter in naming a line or plane and later defines the letter more specifically. Thus at the beginning $\overline{BN}$ and $\overline{AM}$ are arbitrarily given, $BN ||| AM$, but A and B are not both arbitrarily given on these lines, for later they are taken so that $AB \perp AM$. Then he speaks of the half-plane $MA\overline{P}$ although later he takes $A\overline{P}$ as the intersection of this half-plane and another, $AB\overline{F}$, and $NB\overline{D}$ is mentioned twice before he draws $AF \perp CE$, thus determining $B\overline{D}$ apparently as drawn through F.]

[2] [We should say, If $AB\overline{F}$ is revolved on AB so as to fall upon $AB\overline{M}$.]

Then it follows easily that $MA\overline{P}$ and $NB\overline{D}$ mutually intersect, if the sum of the interior angles which they make with $MABN$ is $< 2R.$[1]

§10

If BN and CP are both $|||| \simeq AM$, then also $BN|||| \simeq CP.$[2]

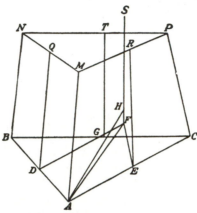

For MAB and MAC either make an *angle* or are in a plane.

If the former let $\overline{QDF}$ bisect at right angles the line AB. DQ will be $\perp AB$, and so $DQ||||AM$ (§8). Likewise, if $\overline{ERS}$ bisect AC at right angles, then $ER||||AM$, and therefore $DQ||||ER$ (§7). Easily (through §9) it follows that $\overline{QDF}$ and $\overline{ERS}$ mutually intersect,[3] and the intersection $\overline{FS}$ is $||||DQ$ (§7), and since $BN||||DQ$

---

[1] [If he means when neither plane is $\perp MABN$, then at least we can say that he does not prove it. See footnote 3 below.]

[2] [We should remember that this theorem has already been proved so far as the first sign, $||||$, is concerned (§7). It is only the equality of angles represented by the sign $\simeq$ that has to be proved here.]

[3] [The theorem of §9 as proved does not seem to apply here directly, for neither of these two planes is perpendicular to the plane of $D\overline{Q}$ and $E\overline{R}$, but a proof can easily be given analogous to the proof of §9. $\overline{QDF}$ is perpendicular to ABC, and if $\overline{DH}$ is their intersection (not drawn in the figure) the perpendicular from E to $\overline{DH}$, which will fall upon $\overline{DH}$ since it cannot intersect the perpendicular $D\overline{A}$ (Euclid I, 28), will be shorter than any other line from E to $\overline{QDF}$, and so shorter than any line from E to $\overline{DQ}$. If then we revolve HDE around DE until it falls into the plane of $QDER$, $D\overline{H}$ will fall within the angle QDE and will intersect $E\overline{R}$. In the same way the intersection of $RE\overline{S}$ with ABC, if carried in this revolution into the plane of $QDER$, will fall in the angle

is also $FS\|\|BN$. Then for every point of $\overline{FS}$ is $FB = FA = FC$,[1]
and FS falls in the plane TGF bisecting the line BC at right angles.
But (by §7), since $FS\|\|BN$, also $GT\|\|BN$. In the same way we
prove $GT\|\|CP$. Moreover GT bisects the line BC at right angles,
and so $TGBN \equiv TGCP$ (§1)[2] and $BN\|\|\| \simeq CP$.

If BN, AM, and CP are in a plane let FS, falling *outside* of this
plane, be $\|\|\| \simeq AM$. Then (by the preceding) $FS\|\|\| \simeq$ both BN
and CP, and so $BN\|\|\| \simeq CP$.

§11

Let the complex of the point A and of *all* the points of which
any one B is such that f $BN\|\|\|AM$, then $BN \simeq AM$, be called F,
and let the section of F by any plane containing the line AM be
called L.

In any line which is $\|\|\| AM$ F has one point and only one, and
it is evident that L is divided by AM into two congruent parts.
Let $\overline{AM}$ be called the *axis* of L. It is evident also that in any
plane containing AM there will be one L *with axis* $A\overline{M}$. Any
such L will be called the *L of the axis* $A\overline{M}$, in the plane considered.
It is evident that if L is revolved about AM the F will be described
of which $A\overline{M}$ is called the *axis*, and conversely the F may be
attributed to the axis $A\overline{M}$.

§12

*If B is anywhere in the L of $A\overline{M}$ and $BN\|\|\| \simeq AM$ (§11), then
the L of $A\overline{M}$ and the L of $B\overline{N}$ coincide.*

For let the L of $B\overline{N}$ be called for distinction l, and let C be any-
where in l, and $CP\|\|\| \simeq BN$ (§11). Then, since also $BN\|\|\| \simeq AM$
will CP be $\|\|\| \simeq AM$ (§10), and so C will fall also in L. And if C
is anywhere in L and $CP\|\|\| \simeq AM$, then $CP\|\|\| \simeq BN$ (§10), and C
will fall also in l (§11). Therefore L and l are the same, and any
$B\overline{N}$ is also axis of L and is $\simeq$ among all the axes of L.

The same in the same manner is evident of F.

RED and will intersect $D\overline{Q}$. In the plane of $QDER$ those two half-lines must
intersect each other, and so before the revolution they must have intersected
each other, and their intersection was a point common to $QD\overline{F}$ and $RE\overline{S}$.]

[1] [Every point of FS is equidistant from A, B, and C.]

[2] [If $TGCP$ is placed upon $TGBN$ (revolved about TG) so that GC falls on GB,
CP and BN drawn from the same point $\|\|\| GT$ must coincide by §1.]

§13

If BN|||AM and CP|||DQ, and BAM + ABN = 2R, then also DCP + CDQ = 2R.

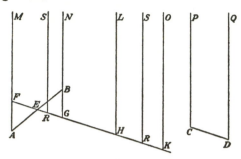

Let $EA = EB$ and $EFM = DCP$ (§4), then, since

$$BAM + ABN = 2R = ABN + ABG,$$

will

$$EBG = EAF,$$

and if also $BG = AF$,

$$\Delta EBG = \Delta EAF,$$

and

$$BEG = AEF$$

and G falls in $\overline{FE}$.   Then is $GFM + FGN = 2R$ (because $EGB = EFA$).   Also $GN|||FM$ (§6), and so if $MFRS \equiv PCDQ$, then $RS |||GN$ (§7) and R falls in or outside of FG (if CD is not $= FG$, where the thing is evident).

I. In the first case FRS is not $> 2R - RFM = FGN$ because $RS|||FM$.   But since $RS|||GN$ also FRS is not $< FGN$, and so $FRS = FGN$, and

$$RFM + FRS = GFM + FGN = 2R.$$

Therefore $DCP + CDQ = 2R$.

II. If R falls outside of FG, then $NGR = MFR$, and we can let $MFGN = NGHL = LHKO$, and so on until FK first becomes $=$ or $> FR$.   This $KO|||HL|||FM$ (§7).   If K falls on R then KO falls on RS (§1), and so

$$RFM + FRS = KFM + FKO = KFM + FGN = 2R;$$

but if R falls in HK, then (from I)

$$RHL + HRS = 2R = RFM + FRS = DCP + CDQ.$$

§14

If $BN|||AM$ and $CP|||DQ$ and $BAM + ABN < 2R$, then also $DCP + CDQ < 2R$.

For if $DCP + CDQ$ is not $<$, and so (§1) is $= 2R$, then (by §13) also $BAM + ABN = 2R$, contrary to hypothesis.

§15

Weighing carefully §§13 and 14, *let the system of geometry resting on the hypothesis of the truth of Euclid's Axiom XI be called* Σ, *and let that one built on the contrary hypothesis be S. All things which are not expressly declared be in* Σ *or S are to be understood to be announced absolutely, that is, to be true whether* Σ *or S is true.*

§16

If AM is the axis of any L, then L in Σ is a straight line $\perp AM$.

For at any point B of L let the axis be BN. In Σ

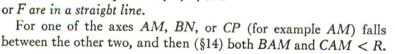

$$BAM + ABN = 2BAM = 2R,$$

and so $BAM = R$. And if C is any point in $\overline{AB}$ and $CP|||AM$, then (by §13) $CP \eqcirc AM$ and C is in L (§11).

But in S no three points A, B, C, of L or F are in a straight line.

For one of the axes AM, BN, or CP (for example AM) falls between the other two, and then (§14) both BAM and $CAM < R$.

§17

L is also in S a line and F a surface.

For (from §11) any plane perpendicular to the axis $\overline{AM}$ through any point of F will cut F in the circumference of a circle whose plane is not perpendicular to any other axis $\overline{BN}$ (§14). Let F revolve about BN. Every point of F will remain in F (§12) and the section of F by a plane not perpendicular to $\overline{BN}$ will describe a surface. And F (by §12), whatever are the points A and B in it, can be made congruent to *itself* in such a way that A will fall at B. Therefore F is a *uniform surface.*

Hence it is evident (§11 and §12) that L is a *uniform line.*[1]

---

[1] [Footnote to the original] It is not necessary to restrict the demonstration to S, since the statement may easily be made so as to hold absolutely (for S and Σ). (Edition I, volume I, Errata of the Appendix.)

§18

The section of any plane through a point A of F, oblique to the axis AM, with F in S is the circumference of a circle.

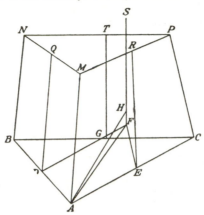

For let A, B, and C be three points of this section, and BN and CP axes. $AMBN$ and $AMCP$ will make an angle. For otherwise the plane determined by A, B, and C (§16) will contain AM, contrary to hypothesis. Therefore the planes bisecting at right angles AB and AC will intersect each other (§10) in an axis FS of F, and $FB = FA = FC$. Let AH be $\perp FS$, and revolve FAH about FS. A will describe a circumference of radius HA going through B and C, lying *at the same time* in F and in $\overline{ABC}$, nor will F and $\overline{ABC}$ have anything in common except $\circ HA$.

It is evident also that $\circ HA$ will be described by the extremity of the portion FA of the line L (like a radius) rotating in F about F.[1]

§19

The perpendicular BT to the axis BN of L, falling in the plane of L, is in S tangent to L. [See the figure on p. 385.]

For L has no point in $B\overline{T}$ except B (§14), but if BQ falls in TBN, then the center of the plane section through BQ perpendicular to TBN with the F of $B\overline{N}$ is manifestly located in $B\overline{Q}$,[2] and if

---

[1] [In the surface F about the point F. In the original there is not this confusion because points are denoted by small German letters. The point F is here taken on FS so that $AM \simeq FS$.]

[2] [Apparently because the entire figure is symmetrical with respect to the plane TBN, and therefore the section is symmetrical with respect to the line BQ.]

BQ is the diameter it is evident that $\overline{BQ}$ cuts the L of $B\overline{N}$ in Q.

§20

*Through any two points of F a line L is determined (§11 and §18),
and, since from §16 and §19 L is perpendicular to all its axes,
any L-angle in F is equal to the angle of the planes through its sides
perpendicular to F.*

§21

*Two L-lines $A\overline{P}$ and $B\overline{D}$ in the same F making with a third L-line
AB a sum of interior-angles $< 2R$ intersect each other.* [By $\overline{AP}$ in F
is meant the L drawn through A and P, and by $A\overline{P}$ that half of it
beginning at A in which P falls. See the second figure, p. 381.]

For if AM and BN are axes of F, then $AM\overline{P}$ and $BN\overline{D}$ cut
each other (§9),[1] and F cuts their intersection (§7 and §11), and
so $A\overline{P}$ and $B\overline{D}$ mutually intersect.

It is evident from this that Axiom XI and all the things which
are asserted in plane geometry and trigonometry follow *absolutely*
on F, L-lines taking the place of straight lines. Therefore the
trigonometrical functions are to be ac-
cepted in the same sense as in Σ, and the
circumference of the circle in F whose
radius is the L-line $= r$, is $= 2\pi r$, and
likewise $\odot r$ in F is $= \pi r^2$ (π being $\frac{1}{2}\odot 1$ in
F, or $3.1415926\ldots$).

§22

*If $\overline{AB}$ is the L of $A\overline{M}$ and C is in $A\overline{M}$,
and the angle CAB formed from the straight
line $A\overline{M}$ and the L-line $A\overline{B}$ is moved first
along $A\overline{B}$ and then along $B\overline{A}$ to infinity, the
path $\overline{CD}$ of C will be the L of $C\overline{M}$.*

For (calling the latter l), let D be any point in $\overline{CD}$, $DN\|CM$,
and B the point of L falling in $\overline{DN}$. $BN \approx AM$, and $AC = BD$,
and so also $DN \approx CM$, and therefore D will be in l. But if D
is in[2] l and $DN\|CM$, and B is the point of L common to it and

---

[1] [Proved in §9 only when one of the angles is a right angle.]

[2] [This is a new D. First he takes any point of the path of C and proves
that it is on l, and then he takes any point on l and proves that it is a point of
the path of C. Here $\overline{CD}$ is not necessarily straight.]

$\overline{DN}$, then $AM \backsimeq BN$ and $CM \backsimeq DN$, whence it is clear that $BD = AC$, and D falls in the path of the point C and l and $\overline{CD}$ are the same.   We designate such an l by $l\|L$.

§23

If the L-line $CDF\|ABE$ (§22), and $AB = BE$, and AM, $B\overline{N}$, and $E\overline{P}$ are axes, plainly $CD = DF$; and if any three points A, B, and E belong to AB and $AB = n . CD$, then will $AE = n . CF$, and therefore (plainly also for incommensurables AB, AE, and CD),

$$AB : CD = AE : CF,$$

and $AB : CD$ is *independent* of AB and *directly determined* by AC. Let this quantity be denoted by the capital letter (as X) of the same name as the small letter (as x) by which AC is denoted.

§24

Whatever be x and y, $Y = X^{\frac{y}{x}}$ (§23).

For one of the two letters x and y will be a multiple of the other (for example, y of x) or not.

If $y = nx$, let $x = AC = CG = GH$ etc., until is made $AH = y$. Then let $CD\|GK\|HL$.   Then (§23)

$$X = AB : CD = CD : GK = GK : HL,$$

and so

$$\frac{AB}{HL} = \left(\frac{AB}{CD}\right)^{n},$$

or

$$Y = X^n = X^{\frac{y}{x}}.$$

If x and y are multiples of i, say $x = mi$ and $y = ni$, then by the preceding $X = I^m$, $Y = I^n$, and therefore $Y = X^{\frac{n}{m}} = X^{\frac{y}{x}}$.

The same is easily extended to the case of incommensurability of x and y.   But if $q = y - x$, clearly $Q = Y : X$.

Now it is manifest that in Σ for any x is $X = 1$; but in S, $X > 1$, and for *any* AB and ABE there is a $CDF\|ABE$ such that $CDF = AB$, whence $AMBN \equiv AMEP$[1] although the latter is a multiple of the former, which is indeed singular, but evidently does not prove the absurdity of S.

---

[1] [$AMBN$ means that portion of the plane that lies between the complete lines AM and BN, and $AMEP$ means that portion that lies between the complete lines AM and EP.   See page 375 footnote 5.]

FERMAT

On Analytic Geometry

(Translated from the French by Professor Joseph Seidlin, Alfred College, Alfred, N. Y.)

The following extract is from Fermat's *Introduction aux Lieux Plans et Solides*. It appears in the the *Varia Opera Mathematica* of Fermat in 1679, and in the *Œuvres de Fermat*, ed. Tannery and Henry, Paris, 1896. It shows how clearly Fermat understood the connection between algebra and geometry. It will be observed that Fermat uses the terms "plane and solid loci" in an older sense, somewhat different from the one now recognized.
The French text will be found in the *Œuvres*, vol. III, pp. 85–96.

Introduction to Plane and Solid Loci

None can doubt that the ancients wrote on loci. We know this from Pappus, who, at the beginning of Book VII, affirms that Apollonius had written on plane loci and Aristæus on solid loci. But, if we do not deceive ourselves, the treatment of loci was not an easy matter for them. We can conclude this from the fact that, despite the great number of loci, they hardly formulated a single generalization, as will be seen later on. We therefore submit this theory to an apt and particular analysis which opens the general field for the study of loci.

Whenever two unknown magnitudes appear in a final equation, we have a locus, the extremity of one of the unknown magnitudes describing a straight line or a curve. The straight line is simple and unique; the classes of curves are indefinitely many,—circle, parabola, hyperbola, ellipse, etc.

When the extremity of the unknown magnitude which traces the locus, follows a straight line or a circle, the locus is said to be plane; when the extremity describes a parabola, a hyperbola, or an ellipse, the locus is said to be solid....

It is desirable, in order to aid the concept of equation, to let the two unknown magnitudes form an angle, which usually we would suppose to be a right angle, with the position and the extreme point of one of the unknown magnitudes established. If neither of the two unknowns is greater than a quadratic, the locus will be plane or solid, as can be clearly seen from the following:

Let NZM be a straight line of given position with point N

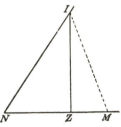

fixed. Let NZ be the unknown quantity a and ZI (the line drawn to form the angle NZI) the other unknown quantity e.

If $da = be$, the point I will describe a line of fixed position. Indeed, we would have $\frac{b}{d} = \frac{a}{e}$. Consequently the ratio $a:e$ is given, as is also the angle at Z. Therefore both the triangle NIZ and the angle INZ are determined. But the point N and the position of the line NZ are given, and so the position of NI is determined. The synthesis is easy.

To this equation we can reduce all those whose terms are either known or combined with the unknowns a and e, which may enter simply or may be multiplied by given magnitudes.

$$z^{\mathrm{II}} - da = be.$$

Suppose that $z^{\mathrm{II}} = dr$. We then have

$$\frac{b}{d} = \frac{r - a}{e}.$$

If we let $MN = r$, point M will be fixed and we shall have $MZ = r - a$.

The ratio $\frac{MZ}{ZI}$ therefore becomes fixed. With the angle at Z given, the triangle IZM will be determined, and in drawing MI it follows that this line is fixed. Thus point I will be on a line of determined position. A ike conclusion can be reached without difficulty for any equation containing the terms a or e.

Here is the first and simplest equation of a locus, from which all the loci of a straight line may be found; for example, the proposition 7 of Book I of Apollonius "On Plane Loci," which has since, however, found a more general expression and mode of construction. This equation yields the following interesting proposition: "Assume any number of lines of given position. From a given point draw lines forming given angles. If the sum of the products of the lines thus drawn by the given lines equals a given area, then the given point will trace a line of determined position."

We omit a great number of other propositions, which could be considered as corollaries to those of Apollonius.

The second species of equations of this kind are of the form $ae = z^{ii}$, in which case point I traces a hyperbola. Draw NR parallel to ZI; through any point, such as M, on the line NZ, draw MO parallel to ZI. Construct the rectangle NMO equal in area to z^{ii}. Through the point O, between the asymptotes NR, NM, describe a hyperbola; its position is determined and it will pass through point I, having assumed, as it were, ae,—that is to say the rectangle NZI,—equivalent to the rectangle NMO. To this equation we may reduce all those whose terms are in part constant, or in part contain a or e or ae.

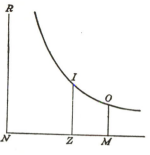

If we let

$$d^{ii} + ae = ra + se$$

we obtain by fundamental principles $ra + se - ae = d^{ii}$. Construct a rectangle of such dimensions as shall contain the terms $ra + se - ae$. The two sides will be $a - s$ and $r - e$, and their rectangle, $ra + se - ae - rs$,

If from d^{ii} we subtract rs, the rectangle

$$(a - s)(r - e) = d^{ii} - rs.$$

Take NO equal to s, and ND, parallel to ZI, equal to r.

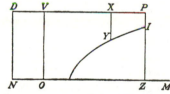

Through point D, draw DP parallel to NM; through point O, OV parallel to ND; prolong ZI to P.

Since $NO = s$ and $NZ = a$, we have $a - s = OZ = VP$. Similarly, since $ND = ZP = r$ and $ZI = e$, we have $r - e = PI$. The rectangle $PV \times PI$ is therefore equal to the given area $d^{ii} - rs$; the point I is therefore on a hyperbola having PV, VO as asymptotes.

If we take any point X, the parallel XY, and construct the rectangle $VXY = d^{ii} - rs$, and through point Y we describe a hyperbola between the asymptotes PV, VO, it will pass through point I. The analysis and construction are easy in every case.

The following species of loci equations arises if we have $a^2 = e^2$ or if a^2 is in a given relation to e^2, or, again, if $a^2 + ae$ is in a given relation to e^2. Finally this type includes all the equations whose terms are of the second degree containing a^2, e^2, or ae. In all these cases point I traces a straight line, which is easily demonstrated.

If the ratio $\dfrac{NZ^2 + NZ.ZI}{ZI^2}$ is given, and any parallel OR is

drawn, then it is easy to show that $\dfrac{NO^2 + NO.OR}{OR^2}$ has the value

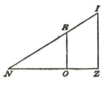

of the given ratio. The point I will therefore be on a line of determined position. The same will be true of all equations whose terms are either the squares of the unknowns or their product. It is needless to enumerate additional specific instances.

If to the squares of the unknowns, with or without their product, are added absolute terms or terms which are the products of one of the unknowns by a given magnitude, the construction is more difficult. We shall indicate the construction and give the proof for several cases.

If $a^2 = de$, point I is on a parabola.

Let NP be parallel to ZI; with NP as diameter, construct the parabola whose parameter is the given line d and whose ordinates are parallel to NZ. The point I will be on the parabola whose position is defined. In fact, it follows from the construction that the rectangle $d \times NP = PI^2$, that is, $d \times IZ = NZ^2$ and, consequent y, $de = a^2$.

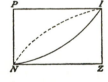

To this equation we can easily reduce all those in which, with a^2, appear the products of the given magnitudes and e, or with e^2 appear the products of the given magnitudes with a. The same would hold true were the equation to contain absolute terms.

If, however, $e^2 = da$, then, in the preceding figure, with N as vertex and with NZ as diameter, construct the parabola whose parameter is d and whose ordinates are parallel to the line NP. It is plain that the imposed condition is satisfied.

If we let $b^2 - a^2 = de$, we have $b^2 - de = a^2$. Divide b^2 by d; let $b^2 = dr$, and we have $dr - de = a^2$ or $d(r - e) = a^2$.

We shall have reduced this equation to the former [—that is, $a^2 = de$,—] by replacing $r - e$ by e.

Let us assume MN (p. 393) parallel to ZI and equal to r; through the point M draw MO parallel to NZ. Point M and the position of the line MO are now given. It follows from the construction that $OI = r - e$. Therefore $d \times OI = NZ^2 = MO^2$.

The parabola drawn with M as vertex, diameter MN, d as parameter, and the ordinates parallel to NZ, satis-
fies the condition as is clearly shown by the
construction.

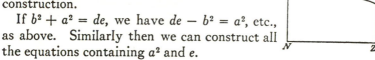

If $b^2 + a^2 = de$, we have $de - b^2 = a^2$, etc.,
as above. Similarly then we can construct all
the equations containing a^2 and e.

But a^2 is often found with e^2 and with absolute terms. Let
$b^2 - a^2 = e^2$.

The point I will be on a circle of determined position if the angle
NZI is a right angle.

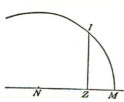

Assume MN equal to b. The circle described with N as
center and with NM as radius will satisfy
the condition. That is to say, that no matter which point I is taken, anywhere on the
circumference, it is clear that ZI^2 (or e^2) will
equal NM^2 (or b^2) $- NZ^2$ (or a^2).

To this equation may be reduced all those
containing terms in a^2, e^2, and in a or e multiplied by given magnitudes, provided angle NZI be a right angle,
and, moreover, that the coefficient of a^2 be equal to that of e^2.

Let
$$b^2 - 2da - a^2 = e^2 + 2re.$$

Adding r^2 to both sides and, thus replacing e by $e + r$, we have
$$r^2 + b^2 - 2da - a^2 = e^2 + r^2 + 2re.$$

Adding d^2 to $r^2 + b^2$, thus replacing a by $d + a$, and denoting
the sum of the squares $r^2 + b^2 + d^2$ by p^2, we get
$$p^2 - d^2 - 2da - a^2 = r^2 + b^2 - 2da - a^2,$$
which leads to
$$p^2 - d^2 = r^2 + b^2.$$

If now we replace $a + d$ by a and $e + r$ by e, we shall have
$$p^2 - a^2 = e^2,$$
which equation is reduced to the preceding.

By like reasoning we are able to reduce all similar equations.
Based on this method we have built up all of the propositions of
the Second Book of Apollonius "On Plane Loci" and we have
proved that the six first cases have loci for any points whatever,
which is quite remarkable and which was probably unknown to
Apollonius.

When $\dfrac{b^2 - a^2}{e^2}$ is a given ratio, the point I will be on an ellipse.

Let MN equal b.   With M as vertex, NM as diameter, and N as center describe an ellipse whose ordinates are parallel to ZI, so that the squares of the ordinates shall be in a given ratio to the product of the segment of the diameter.   The point I will be on that ellipse. That is, $NM^2 - NZ^2$ is equal to the product of the segments of the diameter.

To this equation can be reduced all those in which a^2 is on one side of the equation and e^2 with an opposite sign and a different coefficient on the other side.   If the coefficients are the same and the angle a right angle, the locus will be a circle, as we have said. If the coefficients are the same but the angle is not a right angle, the locus will be an ellipse.

Moreover, though the equations include terms which are products of a or e by given magnitudes, the reduction may nevertheless be made by the method which we have already employed.

If $(a^2 + b^2):e^2$ is a given ratio, the point I will be on a hyperbola. Draw NO parallel to ZI; let the given ratio be equal

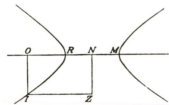

to $b^2:NR^2$.   Point R will then be fixed.   With R as vertex, RO as diameter, and N as center, construct an hyperbola whose ordinates are parallel to NZ, such that the product of the whole diameter (MR) by RO together with RO^2 shall be to OI^2 as $NR^2:b^2$.   It follows, letting $MN = NR$, that $(MO \times OR + NR^2):(OI^2 + b^2)$ is equal to $NR^2:b^2$, the given ratio.

But

$$MO \times OR + NR^2 = NO^2 = ZI^2 = e^2$$

and

$$OI^2 + b^2 = NZ^2 \text{ (or } a^2\text{)} + b^2.$$

Therefore $e^2:(b^2 + a^2) = NR^2:b^2$ and, inverting, $(b^2 + a^2):e^2$ is the given ratio.   Therefore point I is on an hyperbola of determined position.

By the scheme we have already employed we may reduce to this equation all those in which a^2 and e^2 are contained with given terms (separately) or with expressions involving the products of a or e by the given terms, and in which a^2 and e^2 have the same sign and appear on the opposite sides of the equation.   If the signs were different the locus would be a circle or an ellipse.

The most difficult type of equation is that containing, along with a^2 and e^2, terms involving ae, other given magnitudes, etc.
Let

$$b^2 - 2a^2 = 2ae + e^2.$$

Add a^2 to both sides so as to have $a + e$ as a factor of one of the members. Then

$$b^2 - a^2 = a^2 + 2ae + e^2.$$

Replace $a + e$ by, say, e; then, according to the preceding development, the circle MI will satisfy the equation; that is to say, $MN^2 \ (= b^2) - NZ^2$ $(= a^2) = ZI^2 (= [a + e]^2)$. Letting $VI = NZ$ $= a$, we have $ZV = e$.

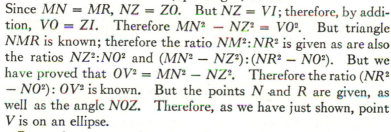

In this problem, however, we are looking for the point V or the extremity of the line e. It is therefore necessary to find, and to indicate, the line upon which the point V is located.

Let MR be parallel to ZI and equal to MN. Draw NR which meets IZ, prolonged, at O. Since $MN = MR$, $NZ = ZO$. But $NZ = VI$; therefore, by addition, $VO = ZI$. Therefore $MN^2 - NZ^2 = VO^2$. But triangle NMR is known; therefore the ratio $NM^2 : NR^2$ is given as are also the ratios $NZ^2 : NO^2$ and $(MN^2 - NZ^2) : (NR^2 - NO^2)$. But we have proved that $OV^2 = MN^2 - NZ^2$. Therefore the ratio $(NR^2 - NO^2) : OV^2$ is known. But the points N and R are given, as well as the angle NOZ. Therefore, as we have just shown, point V is on an ellipse.

By analogous procedure we reduce to the preceding cases all the others in which along with the terms containing ae and a^2 or e^2 are also terms consisting of products of a and e by given magnitudes. The discussion of these different cases is very easy. The problem may always be solved by means of a triangle of known configuration.

We have therefore included in a brief and clear exposition all that the ancients have left unexplained concerning plane and solid loci. Consequently one can recognize at once which loci apply to all cases of the final proposition of Book I of Apollonius "On Plane Loci," and one can generally discover without great difficulty all which pertains to that matter.

As a culminating point to this treatise, we shall add a very interesting proposition of almost obvious simplicity:

"Given the position of any number of lines; if from some definite point lines be drawn forming given angles with the given lines, and the sum of the squares of all the segments is equal to a given area, the point will describe a solid locus of determined position."

A single example will suffice to indicate the general method

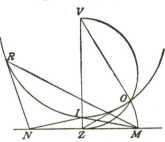

of construction. Given two points N and M, required the locus of the points such that the sum of the squares of IN, IM, shall be in a given ratio to the triangle INM.

Let $NM = b$. Let e be the line ZI drawn at right angles to NM, and let a be the distance NZ. In accordance with fundamental principles, $(2a^2 + b^2 - 2ba + 2e^2)$: be is a given ratio. Following in treatment the procedures previously explained we have the suggested construction.

Bisect NM at Z; erect at Z the perpendicular ZV; make the ratio $4ZV:NM$ equal to the given ratio. On VZ draw the semicircle VOZ, inscribe $ZO = ZM$, and draw VO. With V as center and VO as radius draw the circle OIR. If from any point R on this circle, we draw RN, RM, I say that $RN^2 + RM^2$ is in the given ratio to the triangle RNM.

The constructions of the theorems on loci could have been much more elegantly presented if this discovery had preceded our already old revision of the two books on plane loci. Yet, we do not regret this work, however precocious or insufficiently ripe it may be. In fact, there is for science a certain fascination in not exposing to posterity works which are as yet spiritually incomplete; the labor of the work at first simple and clumsy gains strength as well as stature through new inventions. It is quite important that the student should be able to discern clearly the progress which appears veiled as well as the spontaneous development of the science.

DESCARTES

On Analytic Geometry

(Translated from the French by Professor David Eugene Smith, Teachers College, Columbia University, New York City, and the late Marcia L. Latham, Hunter College, New York City.)[1]

René Descartes (1596–1650), philosopher, mathematician, physicist, soldier, and littérateur, published the first book that may properly be called a treatise on analytic geometry. This appeared as the third appendix to his *Discours de la méthode pour bien conduire sa raison et chercher la vérité dans les sciences,* which was published at Leyden in 1637. Pierre de Fermat (c. 1608–1665) had already conceived the idea as early as 1629, as is shown by a letter written by him to Roberval on Sept. 22, 1636, but he published nothing upon the subject. For the posthumous publication see p. 389.

The following extract constitutes the first eight pages of the first edition (pages 297–304, inclusive) of the *Discours.*

La Geometrie

Book I

Problems the Construction of Which Requires Only Straight Lines and Circles

Any problem in geometry can easily be reduced to such terms that a knowledge of the lengths of certain straight lines is sufficient for its construction.[2] Just as arithmetic consists of only four or five operations, namely, addition, subtraction, multiplication, division, and the extraction of roots, which may be considered a kind of division, so in geometry, to find required lines it is merely necessary to add or subtract other lines; or else, taking one line which I shall call unity in order to relate it as closely as possible

---

[1] From the edition of *La Géométrie* published in facsimile and translation by The Open Court Publishing Co., Chicago, 1925, and here reprinted with the permission of the publisher.

[2] [Large collections of problems of this nature are contained in the following works: Vincenzo Riccati and Girolamo Saladino, *Institutiones Analyticae,* Bologna, 1765; Maria Gaetana Agnesi, *Istituzioni Analitiche,* Milan, 1748; Claude Rabuel, *Commentaires sur la Géométrie de M. Descartes,* Lyons, 1730; and other books of the same period or earlier.]

to numbers,[1] and which can in general be chosen arbitrarily, and having given two other lines, to find a fourth line which shall be to one of the given lines as the other is to unity (which is the same as multiplication); or, again, to find a fourth line which is to one of the given lines as unity is to the other (which is equivalent to division); or, finally, to find one, two, or several mean proportionals between unity and some other line (which is the same as extracting the square root, cube root, etc., of the given line). And I shall not hesitate to introduce these arithmetical terms into geometry, for the sake of greater clearness.

For example, let AB be taken as unity, and let it be required to multiply BD by BC. I have

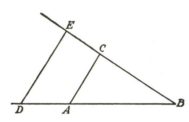

only to join the points A and C, and draw DE parallel to CA; then BE is the product of BD and BC.

If it be required to divide BE by BD, I join E and D, and draw AC parallel to DE; then BC is the result of the division.

If the square root of GH is desired, I add, along the same straight line, FG equal to unity; then, bisect-

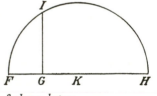

ing FH at K, I describe the circle FIH about K as a center, and draw from G a perpendicular and extend it to I, and GL is the required root. I do not speak here of cube roots, or other roots, since I shall speak more conveniently of them later.

Often it is not necessary thus to draw the lines on paper, but it is sufficient to designate each by a single letter. Thus, to add the lines BD and GH, I call one a and the other b, and write $a + b$. Then $a - b$ will indicate that b is subtracted from a; ab that a is multiplied by b; $\frac{a}{b}$ that a is divided by b; aa or a^2 that a is multiplied by itself; a^3 that this result is multiplied by a, and so on, indefinitely.[2] Again, if I wish to extract the square root of $a^2 + b^2$,

---

[1] [Van Schooten, in his Latin edition of 1683, has this note: "*Per unitatem intellige lineam quandam determinatam qua ad quamvis reliquarum linearum talem relationem babeat, qualem unitas ad certum aliquem numerum.*"]

[2] [Descartes uses a^3, a^4, a^5, a^6, and so on, to represent the respective powers of a, but he uses both aa and a^2 without distinction. For example, he often has $aabb$, but he also uses $3a^2/4b^2$.]

I write $\sqrt{a^2 + b^2}$; if I with to extract the cube root of $a^3 - b^3$ $+ ab^2$, I write $\sqrt[3]{a^3 - b^3 + ab^2}$, and similarly for other roots.[1] Here it must be observed that by a^2, b^3, and similar expressions, I ordinarily mean only simple lines, which, however, I name squares, cubes, etc., so that I may make use of the terms employed in algebra.

It should also be noted that all parts of a single line should always be expressed by the same number of dimensions, provided unity is not determined by the conditions of the problem. Thus, a^3 contains as many dimensions as ab^2 or b^3, these being the component parts of the line which I have called $\sqrt[3]{a^3 - b^3 + ab^2}$. It is not, however, the same thing when unity is determined, because unity can always be understood, even when there are too many or too few dimensions; thus, if it be required to extract the cube root of $a^2b^2 - b$, we must consider the quantity a^2b^2 divided once by unity, and the quantity b multiplied twice by unity.[2]

Finally, so that we may be sure to remember the names of these lines, a separate list should always be made as often as names are assigned or changed. For example, we may write, $AB = 1$, that is AB equal to 1;[3] $GH = a$, $BD = b$, and so on.

If, then, we wish to solve any problem, we first suppose the solution already effected, and give names to all the lines that seem needful for its construction,—to those that are unknown as well as to those that are known. Then, making no distinction between known and unknown lines, we must unravel the difficulty in any way that shows most naturally the relations between these lines, until we find it possible to express a single quantity in two ways.[4] This will constitute an equation, since the terms of one of these two expressions are together equal to the terms of the other.

---

[1] [Descartes writes: $\sqrt{C.a^3 - b^3 + abb}$.]

[2] [Descartes seems to say that each term must be of the third degree, and that therefore we must conceive of both a^2b^2 and b as reduced to the proper dimension.]

[3] [Van Schooten adds "seu unitati," p. 3. Descartes writes, $AB \infty 1$. He seems to have been the first to use this symbol. Among the few writers who followed him, was Hudde (1633–1704). It is very commonly supposed that ∞ is a ligature representing the first two letters (or diphthong) of "aequale." See, for example, M. Aubry's note in W. R. Ball's *Recréations Mathématiques et Problèmes des Temps Anciens et Modernes*, French edition, Paris, 1909, Part III, p. 164. See also F. Cajori, *Hist. of Math. Notations*, vol. I, p. 301.]

[4] [That is, we must solve the remaining simultaneous equations.]

We must find as many such equations as there are supposed to
be unknown lines;[1] but if, after considering everything involved,
so many cannot be found, it is evident that the question is not
entirely determined.   In such a case we may choose arbitrarily
lines of known length for each unbroken line to which there
corresponds no equation.

If there are several equations, we must use each in order, either
considering it alone or comparing it with the others, so as to obtain
a value for each of the unknown lines; and so we must combine
them until there remains a single unknown line which is equal to
some known line, or whose square, cube, fourth power, fifth power,
sixth power, etc., is equal to the sum or difference of two or more
quantities, one of which is known, while the others consist of
mean proportionals between unity and this square, or cube, or
fourth power, etc., multiplied by other known lines.   I may
express this as follows:

$$z = b,$$

or

$$z^2 = -az + b^2,$$

or

$$z^3 = az^2 + b^2z - c^3,$$

or

$$z^4 = az^3 - c^3z + d^4, \text{ etc.}$$

That is, z, which I take for the unknown quantity, is equal to b;
or, the square of z is equal to the square of b diminished by a
multiplied by z; or, the cube of z is equal to a multiplied by the
square of z, plus the square of b multiplied by z, diminished by the
cube of c; and similarly for the others.

Thus, all the unknown quantities can be expressed in terms of a
single quantity, whenever the problem can be constructed by
means of circles and straight lines, or by conic sections, or even by
some other curve of degree not greater than the third or fourth.

But I shall not stop to explain this in more detail, because I
should deprive you of the pleasure of mastering it yourself, as well

---

[1] [Van Schooten (p. 149) gives two problems to illustrate this statement.
Of these, the first is as follows: Given a line segment AB containing any point
C, required to produce AB to D so that the rectangle $AD.DB$ shall be equal to
the square on CD. He lets $AC = a$, $CB = b$, and $BD = x$.   Then $AD =
a + b + x$, and $CD = b + x$, whence $ax + bx + x^2 = b^2 + 2bx + x^2$ and
$x = \dfrac{b^2}{a - b}$.]

as of the advantage of training your mind by working over it, which is in my opinion the principal benefit to be derived from this science. Because, I find nothing here so difficult that it cannot be worked out by any one at all familiar with ordinary geometry and with algebra, who will consider carefully all that is set forth in this treatise.

I shall therefore content myself with the statement that if the student, in solving these equations, does not fail to make use of division wherever possible, he will surely reach the simplest terms to which the problem can be reduced.

And if it can be solved by ordinary geometry, that is, by the use of straight lines and circles traced on a plane surface, when the last equation shall have been entirely solved there will remain at most only the square of an unknown quantity, equal to the product of its root by some known quantity, increased or diminished by some other quantity also known.[1] Then this root or unknown line can easily be found. For example, if I have $z^2 = az + b^2$,[2] I construct a right triangle NLM with one side LM, equal to b,

the square root of the known quantity b^2, and the other side, LN, equal to $\frac{1}{2}a$, that is to half the other known quantity which was multiplied by z, which I suppose to be the un- known line. Then prolonging

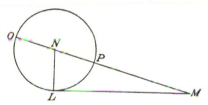

MN, the hypotenuse[3] of this triangle, to O, so that NO is equal to NL, the whole line OM is the required line z. This is expressed in the following way:[4]

$$z = \tfrac{1}{2}a + \sqrt{\tfrac{1}{4}a^2 + b^2}.$$

But if I have $y^2 = -ay + b^2$, where y is the quantity whose value is desired, I construct the same right triangle NLM, and

---

[1] [That is, an expression of the form $z^2 = az \pm b$. "Esgal a ce qui se produit de l'Addition, ou soustraction de sa racine multiplée par quelque quantité connue, & de quelque autre quantité aussy connue."]

[2] [Descar es proposes to show how a quadratic may be solved geomet- rically.]

[3] [Descartes says "prolongeant MN la baze de ce triangle," because the hypotenuse was commonly taken as the base in earlier times.]

[4] [From the figure $OM.PM = LM^2$. If $OM = z$, $PM = z - a$, and since $LM = b$, we have $z(z - a) = b^2$ or $z^2 = az + b^2$. Again, MN $= \sqrt{\tfrac{1}{4}a^2 + b^2}$, whence $OM = z = ON + MN = \tfrac{1}{2}a + \sqrt{\tfrac{1}{4}a^2 + b^2}$. Descartes ignores the second root, which is negative.]

on the hypotenuse MN lay off NP equal to NL, and the remainder PM is y, the desired root. Thus I have

$$y = -\tfrac{1}{2}a + \sqrt{\tfrac{1}{4}a^2 + b^2}$$

In the same way if I had

$$x^4 = -ax^2 + b^2,$$

PM would be x^2 and I should have

$$x = \sqrt{-\tfrac{1}{2}a + \sqrt{\tfrac{1}{4}a^2 + b^2}},$$

and so for other cases.

Finally, if I have $z^2 = az - b^2$, I make NL equal to $\tfrac{1}{2}a$ and LM equal to b as before; then, instead of joining the points M and N, I draw MQR parallel to LN, and with N as a center describe a circle through L cutting MQR in the points Q and R; then z, the line sought, is either MQ or MR, for in this case it can be expressed in two ways, namely,

$$z = \tfrac{1}{2}a + \sqrt{\tfrac{1}{4}a^2 - b^2},$$

and

$$z = \tfrac{1}{2}a - \sqrt{\tfrac{1}{4}a^2 - b^2}.$$

And if the circle described about N and passing through L neither cuts nor touches the line MQR, the equation has no root, so that we may say that the construction of the problem is impossible.

These same roots can be found by many other methods. I have given these very simple ones to show that it is possible to construct all the problems of ordinary geometry by doing no more than the little covered in the four figures that I have explained.[1] This is one thing which I believe the ancient mathematicians did not observe, for otherwise they would not have put so much labor into writing so many books in which the very sequence of the propositions shows that they did not have a sure method of finding all,[2] but rather gathered those propositions on which they had happened by accident.

---

[1] [It will be seen that Descartes considers only three types of the quadratic equation in z, namely $z^2 + az - b^2 = 0$, $z^2 - az - b^2 = 0$, and $z^2 - az + b^2 = 0$. It thus appears that he has not been able to free himself from the old traditions to the extent of generalizing the meaning of the coefficients,— as negative and fractional as well as positive. He does not consider the type $z^2 + az + b^2 = 0$, because it has no positive roots.]

[2] ["Qu'ils n'ont point eu la vraye methode pour les trouuer toutes."]

POHLKE'S THEOREM

(Translated from the German by Professor Arnold Emch, University of Illinois, Urbana, Ill.)

Karl Pohlke was born in Berlin on January 28, 1810, and died there November 27, 1876. He taught in various engineering schools, closing his work in the Technische Hochschule in Charlottenburg. In his *Darstellende Geometrie* (Berlin, 1859; 4th ed., 1876, p. 109) is found "the principal theorem of axonometry," now generally known as Pohlke's Theorem. It is here translated as an important piece of source material, but without proof, the demonstration being available in an article by Arnold Emch, *American Journal of Mathematics*, vol. 40 (1918). On the general development of orthogonal axonometry see F. J. Obenrauch, *Geschichte der darstellenden und projectiven Geometrie*, Brünn,1897, pp. 385 seq.

Three segments of arbitrary length a_1x_1, a_1y_1, a_1z_1, which are drawn in a plane from a point a_1 under arbitrary angles, form a parallel projection of three equal segments ax, ay, az from the origin on three perpendicular coordinate axes; however, only one of the segments $a_1x_1, \ldots$, or one of the angles may vanish.

RIEMANN

On Riemann's Surfaces and Analysis Situs

(Translated from the German by Dr. James Singer, Princeton University, Princeton, N. J.)

Georg Friedrich Bernhard Riemann (1826–1866) was born at Breselenz in Hannover and died at Selasca on his third trip to Italy. He studied theology at Göttingen and also attended some mathematical lectures there. He soon gave up theology for mathematics and studied under Gauss and Stern. In 1847 he went to Berlin, drawn by the fame of Dirichlet, Jacobi, Steiner, and Eisenstein. He returned to Göttingen in 1850 to study physics under Weber and there he received his doctorate the following year. He became a Privat-dozent at Göttingen in 1854, a professor in 1857, and in 1859 succeeded Dirichlet as ordinary professor. His work on the differential equations of physics, a series of lectures edited by Hattendorf and later by H. Weber, is still a standard textbook. In prime numbers also he opened a new field. The first section of this article is a translation of part of Riemann's paper entitled "Allgemeine Voraussetzungen und Hülfsmittel für die Untersuchung von Functionen unbeschränkt veränderlicher Grössen," which appeared in Crelle's *Journal für reine und angewandte Mathematik*, Bd. 54, 1857, pp. 103–104. The second division is a translation of part of his paper, "Lehrsätze aus der Analysis Situs für die Theorie der Integrale von zweigliedrigen vollständigen Differentialien," which appeared in the same issue, pp. 105–110. The papers can also be found in his *Mathematische Werke* collected by H. Weber, first edition, pp. 83–89; second edition, pp. 90–96.

The importance of these two contributions of Riemann can scarcely be exaggerated. Thanks to the Riemann surface the theory of single-valued analytic functions of one variable can largely be extended to multiple-valued functions. Riemann introduced his surface for the purpose of studying algebraic functions, in which field it plays a fundamental part. In Riemann's work we find also the real beginning of modern Analysis Situs, which in a larger sense was however created by Poincaré (1895).

1. For many investigations, especially in the investigation of algebraic and Abelian functions, it is advantageous to represent geometrically the modes of branching of a multivalued function in the following way: We imagine another surface spread over the (x, y)-plane and coincident with it (or an infinitely thin body spread over the plane) which extends as far and only as far as the function is defined. By a continuation of this function this surface will

404

be likewise further extended. In a portion of the plane in which there exist two or more continuations of the function, the surface will be double or multifold; it will consist there of two or more sheets, each one of which represents one branch of the function. Around a branch point of the function a sheet of the surface will be continued into another, so that in the neighborhood of such a point the surface can be considered as a helicoid with its axis perpendicular to the (x, y)-plane at this point, and with infinitely small pitch. If the function takes on again its original value after several revolutions of z around a branch point (for example, as $(z - a)^{\frac{m}{n}}$, where m and n are relatively prime numbers, after n revolutions of z around a), we must then of course assume that the topmost sheet of the surface is continued into the lowermost by means of the remaining ones. The multiple-valued function has only one definite value for each point of the surface representing its modes of branching and therefore can be regarded as a fully determined function of the position in this surface.

· · · · · · · · · ·

2. In the investigation of functions which arise from the integration of total differentials several theorems belonging to Analysis Situs are almost indispensable. This name, used by Leibniz, although perhaps not entirely with the same significance, may well designate a part of the theory of continuous entities which treats them not as existing independently of their positions and measurable by one another but, on the contrary, entirely disregarding the metrical relations, investigates their local and regional properties. While I propose to present a treatment entirely free from metric considerations, I will here present in a geometric form only the theorems necessary for the integration of two-termed total differentials.

· · · · · · · · · ·

If upon a surface F two systems of curves, a and b, together completely bound a part of this surface, then every other system of curves which together with a, completely bounds a part of F also constitutes with b the complete boundary of a part of the surface; which part is composed of both of the first partitions of the surface joined along a (by addition or subtraction, according as they lie upon opposite or upon the same side of a). Both systems of curves serve equally well for the complete boundary of a part of F and can be interchanged as far as the satisfying of this requirement.[1]

If upon the surface F there can be drawn n closed curves $a_1, a_2, \ldots, a_n$ which neither by themselves nor with one another completely bound a part of this surface F, but with whose aid every other closed curve does form the complete boundary of a part of F, the surface is said to be $(n + 1)$-fold connected.

This character of the surface is independent of the choice of the system of curves $a_1, a_2, \ldots, a_n$ since any other n closed curves $b_1, b_2, \ldots, b_n$ which are not sufficient to bound completely a part of this surface, do likewise completely bound a part of F when taken together with any other closed curve.

Indeed, since b_1 completely bounds a part of F when taken together with curves a, one of these curves a can be replaced by b_2 and the remaining curves a. Therefore, any other curve, and consequently also b_2, together with b_1 and these $n - 1$ curves a is sufficient for the complete boundary of a part of F, and hence one of these $n - 1$ curves a can be replaced by b_1, b_2 and the remaining $n - 2$ curves a. If, as assumed, the curves b are not sufficient for the complete boundary of a part of F, this process can clearly be continued until all the a's have been replaced by the b's.

---

[1] [Note by H. Weber.] The theorem stated here needs to be somewhat restricted and made more precise, as was pointed out by Tonelli (*Atti della R. accademia dei Lincei*, Ser. II, vol. 2, 1875. In an extract from the *Nachrichten der Gesselschaft der Wissenschaften zu Göttingen*, 1875.)

If the system of curves a completely bounds a part of the surface F when taken together with a system of curves b as well as with a second system of curves c, it is generally necessary, in order that the systems of curves b and c taken together likewise bound a part of the surface, that no subset of the curves a together with b or with c already bounds a part of the surface. The part of the surface bounded by the systems of curves b, c which, even when the parts of the surface a, b, and a, c are simple, can consist of several separate pieces, are described by Tonelli in the following fashion: It consists of the totality of the parts of the surface a, b, and a, c when those parts which are bounded by the curves a are taken away from the parts common to both of these surface partitions.

The example given by Tonelli of a closed five-fold connected double anchor ring bounded by a point illustrates this relation and makes it intuitive.

These remarks have no influence on the use which Riemann makes of this theorem for the definition of the $(n + 1)$-fold connectivity, since the system here denoted by a always consists of only one curve, namely the curve a which is replaced by b.

By means of a crosscut,—i. e., a line lying in the interior of the surface and going from a boundary point to a boundary point,—an $(n + 1)$-fold connected surface F can be changed into an n-fold connected one, F'. The parts of the boundary arising from the cutting play the role of boundary even during the further cutting so that a crosscut can pass through no point more than once but can end in one of its earlier points.

Since the lines $a_1, a_2, \ldots, a_n$ are not sufficient for the complete boundary of a part of F, if one imagines F cut up by these lines, then the piece of the surface lying on the right of a_n as well as that lying on the left must contain boundary elements other than the lines a and which belong therefore to the boundary of F. One can therefore draw a line in the one as well as the other of these pieces of surface not cutting the curves a from a point of a_n to the boundary of F. Both of these two lines q' and q'' taken together then constitute a crosscut q of the surface F which satisfies the requirement.

Indeed, on the surface F' arising from this crosscut of F the curves $a_1, a_2, \ldots, a_{n-1}$ are closed curves lying in the interior of F' which are not sufficient to bound a part of F and hence also not a part of F'. However, every other closed curve l lying in the interior of F' constitutes with them the complete boundary of a part of F'. For the line l forms with a complex of the lines $a_1, a_2, \ldots, a_n$ the complete boundary of a part f of F. However, it can be shown that a_n cannot occur in the boundary of f; because then, according as f lies on the left or right side of a_n, q' or q'' would go from the interior of f to a boundary point of F, hence to a point lying outside of f, and therefore would cut the boundary of f contrary to the hypothesis that l as well as the lines a, excepting for the point of intersection of a_n and q, always lie in the interior of F'.

The surface F' into which F is decomposed by the crosscut q is therefore n-fold connected, as required.

It shall now be shown that the surface F is changed into an n-fold connected one F' by any crosscut p which does not decompose it into separate pieces. If the pieces of surface adjacent to the two sides of the crosscut p are connected, a line b can be drawn from one side of p through the interior of F' back to the starting point on the other side. This line b forms a line in the interior of F leading back into itself; and since the crosscut issuing from it on both sides goes to a boundary point, b cannot constitute the complete boundary of either of the two pieces of surface into which it separates F.   We can therfore replace one of the curves a by the curve b and each of the remaining $n - 1$ curves a by a curve in the interior of F' and the curve b, if necessary; wherefrom we can deduce by the same means as above the proof that F' is n-fold connected.

An $(n + 1)$-fold connected surface will therefore be changed into an n-fold connected one by means of any crosscut which does not separate it into pieces.

The surface arising from a crosscut can be divided again by a new crosscut, and after n repetitions of this operation an $(n + 1)$-fold connected surface will be changed into a simply connected one by means of n successive non-interesting crosscuts.   To apply these considerations to a surface without boundary, a closed surface, we must change it into a bounded one by the specialization of an arbitrary point; so that the first division is made by means of this point and a crosscut beginning and ending in it, hence by a closed curve.   For example, the surface of an anchor ring, which is 3-fold connected, will be changed into a simply connected surface by means of a closed curve and a crosscut.

Simply-connected Surface

It will be decomposed into parts by any crosscut, and any closed curve in it constitutes the complete boundary of a part of the surface.

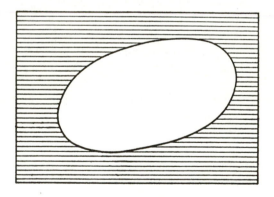

Doubly-connected Surface

It will be reduced to a simply-connected one by any crosscut q that does not disconnect it. Any closed curve in it can, with the aid of a, constitute the complete boundary of a part of the surface.

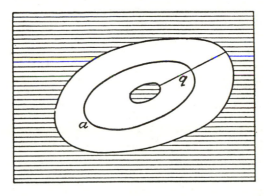

Triply-connected Surface

In this surface any closed curve can constitute the complete boundary of a part of the surface with the aid of the curves a_1 and a_2. It is decomposed into a doubly connected surface by any crosscut that does not disconnect it and into a simply connected one by two such crosscuts, q_1 and q_2.

This surface is double in the regions α, β, γ, δ of the plane.   The arm of the surface containing a_1 is imagined as lying under the other and is therefore represented by dotted lines.

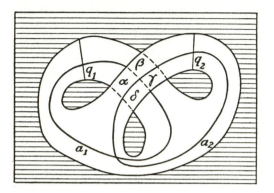

RIEMANN

On the Hypotheses which Lie at the Foundations of Geometry

(Translated from the German by Professor Henry S. White, Vassar College, Poughkeepsie, N. Y.)

For a biographical sketch of Riemann see page 404.
The paper here translated is Riemann's *Probe-Vorlesung,* or formal initial lecture on becoming Privat-Docent. It is extraordinary in scope and originality and it paved the way for the now current theories of hyperspace and relativity. It was read on the 10th of June, 1854, for the purpose of Riemann's "Habilitation" with the philosophical faculty of Göttingen. This explains the form of presentation, in which analytic investigations could be only indicated; some elaborations of them are to be found in the *"Commentatio mathematica, qua respondere tentatur quaestioni ab Illma Academia Parisiensi propositae"* etc., and in the appendix to that paper. It appears in vol. XIII of the *Abbandlungen* of the Royal Society of Sciences of Göttingen.

Plan of the Investigation

It is well known that geometry presupposes not only the concept of space but also the first fundamental notions for constructions in space as given in advance. It gives only nominal definitions for them, while the essential means of determining them appear in the form of axioms. The relation of these presuppositions is left in the dark; one sees neither whether and in how far their connection is necessary, nor a priori whether it is possible.

From Euclid to Legendre, to name the most renowned of modern writers on geometry, this darkness has been lifted neither by the mathematicians nor by the philosophers who have labored upon it. The reason of this lay perhaps in the fact that the general concept of multiply extended magnitudes, in which spatial magnitudes are comprehended, has not been elaborated at all. Accordingly I have proposed to myself at first the problem of constructing the concept of a multiply extended magnitude out of general notions of quantity. From this it will result that a multiply extended magnitude is susceptible of various metric relations and that space accordingly constitutes only a particular case of a

411

triply extended magnitude. A necessary sequel of this is that the propositions of geometry are not derivable from general concepts of quantity, but that those properties by which space is distinguished from other conceivable triply extended magnitudes can be gathered only from experience. There arises from this the problem of searching out the simplest facts by which the metric relations of space can be determined, a problem which in nature of things is not quite definite; for several systems of simple facts can be stated which would suffice for determining the metric relations of space; the most important for present purposes is that laid down for foundations by Euclid. These facts are, like all facts, not necessary but of a merely empirical certainty; they are hypotheses; one may therefore inquire into their probability, which is truly very great within the bounds of observation, and thereafter decide concerning the admissibility of protracting them outside the limits of observation, not only toward the immeasurably large, but also toward the immeasurably small.

I. The Concept of n-fold Extended Manifold

While I now attempt in the first place to solve the first of these problems, the development of the concept of manifolds multiply extended, I think myself the more entitled to ask considerate judgment inasmuch as I have had little practise in such matters of a philosophical nature, where the difficulty lies more in the concepts than in the construction, and because I have not been able to make use of any preliminary studies whatever aside from some very brief hints which Privy Councillor Gauss has given on the subject in his second essay on biquadratic residues and in his Jubilee booklet, and some philosophical investigations of Herbart.

1

Notions of quantity are possible only where there exists already a general concept which allows various modes of determination. According as there is or is not found among these modes of determination a continuous transition from one to another, they form a continuous or a discrete manifold; the individual modes are called in the first case points, in the latter case elements of the manifold. Concepts whose modes of determination form a discrete manifold are so numerous, that for things arbitrarily given there can always be found a concept, at least in the more highly developed languages, under which they are comprehended (and mathematicians have been able therefore in the doctrine of discrete quantities to set out without scruple from the postulate that given things are to be considered as all of one kind); on the other hand there are in common life only such infrequent occasions to form concepts whose modes of determination form a continuous manifold, that the positions of objects of sense, and the colors, are probably the only simple notions whose modes of determination form a multiply extended manifold. More frequent occasion for the birth and development of these notions is first found in higher mathematics.

Determinate parts of a manifold, distinguished by a mark or by a boundary, are called quanta. Their comparison as to quantity comes in discrete magnitudes by counting, in continuous magnitude by measurement. Measuring consists in superposition of the magnitudes to be compared; for measurement there is requisite some means of carrying forward one magnitude as a measure for the other. In default of this, one can compare two magnitudes only when the one is a part of the other, and even then one can only decide upon the question of more and less, not upon the question of how many. The investigations which can be set on foot about them in this case form a general part of the doctrine of quantity independent of metric determinations, where magnitudes are thought of not as existing independent of position and not as expressible by a unit, but only as regions in a manifold. Such inquiries have become a necessity for several parts of mathematics, namely for the treatment of many-valued analytic functions, and the lack of them is likely a principal reason why the celebrated theorem of Abel and the contributions of Langrange, Pfaff, and Jacobi to the theory of differential equations have remained so long unfruitful. For the present purpose it will be sufficient to bring forward conspicuously two points out of this general part of the doctrine of extended magnitudes, wherein nothing further is assumed than what was already contained in the concept of it. The first of these will make plain how the notion of a multiply extended manifold came to exist; the second, the reference of the determination of place in a given manifold to determinations of quantity and the essential mark of an n-fold extension.

2

In a concept whose various modes of determination form a continuous manifold, if one passes in a definite way from one mode of determination to another, the modes of determination which are traversed constitute a simply extended manifold and its essential mark is this, that in it a continuous progress is possible from any point only in two directions, forward or backward.   If now one forms the thought of this manifold again passing over into another entirely different, here again in a definite way, that is, in such a way that every point goes over into a definite point of the other, then will all the modes of determination thus obtained form a doubly extended manifold.   In similar procedure one obtains a triply extended manifold when one represents to oneself that a double extension passes over in a definite way into one entirely different, and it is easy to see how one can prolong this construction indefinitely.   If one considers his object of thought as variable instead of regarding the concept as determinable, then this construction can be characterized as a synthesis of a variability of $n + 1$ dimensions out of a variability of n dimensions and a variability of one dimension.

3

I shall now show how one can conversely split up a variability, whose domain is given, into a variability of one dimension and a variability of fewer dimensions.   To this end let one think of a variable portion of a manifold of one dimension,—reckoning from a fixed starting-point or origin, so that its values are comparable one with another—which has for every point of the given manifold a definite value changing continuously with that point; or in other words, let one assume within the given manifold a continuous function of place, and indeed a function such that it is not constant along any portion of this manifold.   Every system of points in which the function has a constant value constitutes now a continuous manifold of fewer dimensions than that which was given.   By change in the value of the function these manifolds pass over, one into another, continuously; hence one may assume that from one of them all the rest emanate, and this will come about, speaking generally, in such a way that every point of one passes over into a definite point of the other.   Exceptional cases, and it is important to investigate them,—can be left out of consideration here.   By this means the fixing of position in the given manifold is referred to the determination of one quantity and the fixing of position in a manifold of fewer dimensions.   It is easy now to show that this latter has $n - 1$ dimensions if the given manifold was n-fold extended.   Hence by repetition of this procedure, to n times, the fixing of position in an n-dimensional manifold is reduced to n determinations of quantities, and therefore the fixing of position in a given manifold is reduced, whenever this is possible, to the determination of a finite number of quantities.   There are however manifolds in which the fixing of position requires not a finite number but either an infinite series or a continuous manifold of determinations of quantity.   Such manifolds are constituted for example by the possible determinations of a function for a given domain, the possible shapes of a figure in space, et cetera.

II. Relations of Measure, of Which an n-dimensional Manifold is Susceptible, on the Assumption that Lines Possess a Length Independent of Their Position; that is, that Every Line Can Be Measured by Every Other

Now that the concept of an n-fold extended manifold has been constructed and its essential mark has been found to be this, that the determination of position therein can be referred to n determinations of magnitude, there follows as second of the problems proposed above, an investigation into the relations of measure that such a manifold is susceptible of, also into the conditions which suffice for determining these metric relations.   These relations of measure can be investigated only in abstract notions of magnitude and can be exhibited connectedly only in formulae; upon certain assumptions, however, one is able to resolve them into relations which are separately capable of being represented geometrically, and by this means it becomes possible to express geometrically the results of the calculation.   Therefore if one is to reach solid ground, an abstract investigation in formulae is indeed unavoidable, but its results will allow an exhibition in the clothing of geometry.   For both parts the foundations are contained in the celebrated treatise of Privy Councillor Gauss upon curved surfaces.

1

Determinations of measure require magnitude to be independent of location, a state of things which can occur in more than one way. The assumption that first offers itself, which I intend here to follow out, is perhaps this, that the length of lines be independent of their situation, that therefore every line be measurable by every other.   If the fixing of the location is referred to determinations of magnitudes, that is, if the location of a point in the n-dimensional manifold be expressed by n variable quantities x_1, x_2, x_3, and so on to x_n, then the determination of a line will reduce to this, that the quantities x be given as functions of a single variable.   The problem is then, to set up a mathematical expression for the length of lines, and for this purpose the quantities x must be thought of as expressible in units.   This problem I shall treat only under certain restrictions, and limit myself first to such lines as have the ratios of the quantities dx—the corresponding changes in the quantities x—changing continuously; one can in that case think of the lines as laid off into elements within which the ratios of the quantities dx may be regarded as constant, and the problem reduces then to this: to set up for every point a general expression for a line-element which begins there, an expression which will therefore contain the quantities x and the quantities dx.   In the second place I now assume that the length of the line-element, neglecting quantities of the second order, remains unchanged when all its points undergo infinitely small changes of position; in this it is implied that if all the quantities dx increase in the same ratio, the line-element likewise changes in this ratio.   Upon these assumptions it will be possible for the line-element to be an arbitrary homogeneous function of the first degree in the quantities dx which remains unchanged when all the dx change sign, and in which the arbitrary constants are continuous functions of the quantities x.   To find the simplest cases, I look first for an expression for the $(n-1)$-fold extended manifolds which are everywhere equally distant from the initial point of the line-element, that is, I look for a continuous function of place, which renders them distinct from one another.   This will have to diminish or increase from the initial point out in all directions; I shall assume that it increases in all directions and therefore has a minimum in that point.   If then its first and second differential quotients are finite, the differential of the first order must vanish and that of the second order must never be negative; I assume that it is always positive.   This differential expression of the second order accordingly remains constant if ds remains constant, and increases in squared ratio when the quantities dx and hence also ds all change in the same ratio.   That expression is therefore $=$ const. ds^2, and consequently $ds =$ the square root of an everywhere positive entire homogeneous function of the second degree in quantities dx having as coefficients continuous functions of the quantities x. For space this is, when one expresses the position of a point by rectangular coordinates, $ds = \sqrt{\Sigma(dx)^2}$; space is therefore comprised under this simplest case. The next case in order of simplicity would probably contain the manifolds in which the line-element can be expressed by the fourth root of a differential expression of the fourth degree. Investigation of this more general class indeed would require no essentially different principles, but would consume considerable time and throw relatively little new light upon the theory of space, particularly since the results cannot be expressed geometrically. I limit myself therefore to those manifolds in whlch the line-element is expressed by the square root of a differential expression of the second degree. Such an expression one can transform into another similar one by substituting for the n independent variables functions of n new independent variables. By this means however one cannot transform every expression into every other; for the expression contains $n \cdot \dfrac{n+1}{2}$ coefficients which are arbitrary functions of the independent variables; but by introducing new variables one can satisfy only n relations (conditions), and so can make only n of the coefficients equal to given quantities. There remain then $n \cdot \dfrac{n-1}{2}$ others completely determined by the nature of the manifold that is to be represented, and therefore for determining its metric relations $n \cdot \dfrac{n-1}{2}$ functions of position are requisite. The manifolds in which, as in the plane and in space, the line-element can be reduced to the form $\sqrt{\Sigma(dx)^2}$ constitute therefore only a particular case of the manifolds under consideration here. They deserve a particular name, and I will therefore term *flat* these manifolds in which the square of the line-element can be reduced to the sum of squares of total differentials. Now in order to obtain a conspectus of the essential differences of the manifolds representable in this prescribed form it is necessary to remove those that spring from the mode of representation, and this is accomplished by choosing the variable quantities according to a definite principle.

2

For this purpose suppose the system of shortest lines emanating from an arbitrary point to have been constructed. The position of an undetermined point will then be determinable by specifying the direction of that shortest line in which it lies and its distance, in that line, from the starting-point; and it can therefore be expressed by the ratios of the quantities dx^0, that is the limiting ratios of the dx at the starting point of this shortest line and by the length s of this line. Introduce now instead of the dx^0 such linear expressions $d\alpha$ formed from them, that the initial value of the square of the line-element equals the sum of the squares of these expressions, so that the independent variables are: the quantity s and the ratios of quantities $d\alpha$. Finally, set in place of the $d\alpha$ such quantities proportional to them, $x_1, x_2, \ldots, x_n$, that the sum of their squares $= s^2$. After introducing these quantities, the square of the line-element for indefinitely small values of x becomes $= \Sigma(dx)^2$, and the term of next order in that $(ds)^2$ will be equal to a homogeneous expression of the second degree in the $n\dfrac{n-1}{2}$ quantities $(x_1 dx_2 - x_2 dx_1)$, $(x_1 dx_3 - x_3 dx_1), \ldots$, that is, an indefinitely small quantity of dimension four; so that one obtains a finite magnitude when one divides it by the square of the indefinitely small triangle-area in whose vertices the values of the variables are $(0, 0, 0, \ldots)$, $(x_1, x_2, x_3, \ldots)$, $(dx_1, dx_2, dx_3, \ldots)$. This quantity retains the same value, so long as the quantities x and dx are contained in the same binary linear forms, or so long as the two shortest lines from the values 0 to the values x and from the values 0 to the values dx stay in the same element of surface, and it depends therefore only upon the place and the direction of that element. Plainly it is $= 0$ if the manifold represented is flat, that is if the square of the line-element is reducible to $\Sigma(dx)^2$, and it can accordingly be regarded as the measure of the divergence of the manifold from flatness in this point and in this direction of surface. Multiplied by $-\tfrac{3}{4}$ it becomes equal to the quantity which Privy Councillor Gauss has named the measure of curvature of a surface.

For determining the metric relations of an n-fold extended manifold representable in the prescribed form, in the foregoing discussion $n \cdot \dfrac{n-1}{2}$ functions of position were found needful; hence when the measure of curvature in every point in $n \cdot \dfrac{n-1}{2}$ surface-directions is given, from them can be determined the metric relations of the manifold, provided no identical relations exist among these values, and indeed in general this does not occur. The metric relations of these manifolds that have the line-element represented by the square root of a differential expression of the second degree can thus be expressed in a manner entirely independent of the choice of the variable quantities. A quite similar path to this goal can be laid out also in case of the manifolds in which the line-element is given in a less simple expression; *e. g.*, as the fourth root of a differential expression of the fourth degree. In that case the line-element, speaking generally, would no longer be reducible to the form of a square root of a sum of squares of differential expressions; and therefore in the expression for the square of the line-element the divergence from flatness would be an indefinitely small quantity of the dimension two, while in the former manifolds it was indefinitely small of the dimension four. This peculiarity of the latter manifolds may therefore well be called flatness in smallest parts. The most important peculiarity of these manifolds, for present purposes, on whose account solely they have been investigated here, is however this, that the relations of those doubly extended can be represented geometrically by surfaces, and those of more dimensions can be referred to those of the surfaces contained in them; and this requires still a brief elucidation.

3

In the conception of surfaces, along with the interior metric relations, in which only the length of the paths lying in them comes into consideration, there is always mixed also their situation with respect to points lying outside them. One can abstract however from external relations by carrying out such changes in the surfaces as leave unchanged the length of lines in them; *i. e.*, by thinking of them as bent in any arbitrary fashion,—without stretching— and by regarding all surfaces arising in this way one out of another as equivalent. For example, arbitrary cylindrical or conical surfaces are counted as equivalent to a plane, because they can be formed out of it by mere bending, while interior metric relations remain unchanged; and all theorems regarding them—the whole of planimetry—retain their validity; on the other hand they count as essentially distinct from the sphere, which cannot be converted into a plane without stretching. According to the above investigation in every point the interior metric relations of a doubly extended manifold are characterized by the measure of curvature if the line-element can be expressed by the square root of a differential expression of the second degree, as is the case with surfaces. An intuitional significance can be given to this quantity in the case of surfaces, namely that it is the product of the two curvatures of the surface in this point; or also, that its product into an indefinitely small triangle-area formed of shortest lines is equal to half the excess of its angle-sum above two right angles, when measured in radians. The first definition would presuppose the theorem that the product of the two radii of curvature is not changed by merely bending a surface; the second, the theorem that at one and the same point the excess of the angle-sum of an indefinitely small triangle above two right angles is proportional to its area. To give a tangible meaning to measure of curvature of an n-dimensional manifold at a given point and in a surface direction passing through that point, it is necessary to start out from the principle that a shortest line, originating in a point, is fully determined when its initial direction is given. According to this, a determinate surface is obtained when one prolongs into shortest lines all the initial directions going out from a point and lying in the given surface element; and this surface has in the given point a determinate measure of curvature, which is also the measure of curvature of the n-dimensional manifold in the given point and the given direction of surface.

4

Now before applications to space some considerations are needful regarding flat manifolds in general, *i. e.*, regarding those in which the square of the line-element is representable by a sum of squares of total differentials.

In a flat n-dimensional manifold the measure of curvature at every point is in every direction zero; but by the preceding investigation it suffices for determining the metric relations to know that at every point, in $n \cdot \dfrac{n-1}{2}$ surface directions whose measures of curvature are independent of one another, that measure is zero. Manifolds whose measure of curvature is everywhere zero may be regarded as a particular case of those manifolds whose curvature is everywhere constant. The common character of those manifolds of constant curvature can also be expressed thus: that the figures lying in them can be moved without stretching. For it is evident that the figures in them could not be pushed along and

rotated at pleasure unless in every point the measure of curvature
were the same in all directions. Upon the other hand, the metric
relations of the manifold are completely determined by the
measure of curvature. About any point, therefore, the metric
relations in all directions are exactly the same as about any other
point, and so the same constructions can be carried out from it,
and consequently in manifolds with constant curvature every
arbitrary position can be given to the figures. The metric relations
of these manifolds depend only upon the value of the measure of
curvature, and it may be mentioned, with reference to analytical
presentation, that if one denotes this value by α, the expression
for the line element can be given the form

$$\frac{1}{1 + \frac{\alpha}{4}\sqrt{\Sigma dx^2}}$$

5

Consideration of surfaces with constant measure of curvature
can help toward a geometric exposition. It is easy to see that
those surfaces whose curvature is positive will always permit themselves to be fitted upon a sphere whose radius is unity divided by
the square root of the measure of curvature; but to visualize the
complete manifold of these surfaces one should give to one of them
the form of a sphere and to the rest the form of surfaces of rotation
which touch it along the equator. Such surfaces as have greater
curvature than this sphere will then touch the sphere from the
inner side and take on a form like that exterior part of the surface
of a ring which is turned away from the axis (remote from the
axis); they could be shaped upon zones of spheres having a smaller
radius, but would reach more than once around. Surfaces with
lesser positive measure of curvature will be obtained by cutting
out of spherical surfaces of greater radius a portion bounded by
two halves of great circles, and making its edges adhere together.
The surface with zero curvature will be simply a cylindrical surface
standing upon the equator; the surfaces with negative curvature
will be tangent to this cylinder externally and will be formed like
the inner part of the surface of a ring, the part turned toward the
axis.

If one thinks of these surfaces as loci for fragments of surface
movable in them, as space is for bodies, then the fragments are
movable in all these surfaces without stretching. Surfaces with

positive curvature can always be formed in such wise that those fragments can be moved about without even bending, namely as spherical surfaces, not so however those with negative curvature. Beside this independence of position shown by fragments of surface, it is found in the surface with zero curvature that direction is independent of position, as is not true in the rest of the surfaces.

III. Application to Space

1

Following these investigations concerning the mode of fixing metric relations in an n-fold extended magnitude, the conditions can now be stated which are sufficient and necessary for determining metric relations in space, when it is assumed in advance that lines are independent of position and that the linear element is representable by the square root of a differential expression of the second degree; that is if flatness in smallest parts is assumed.

These conditions in the first place can be expressed thus: that the measure of the curvature in every point is equal to zero in three directions of surface; and therefore the metric relations of the space are determined when the sum of the angles in a triangle is everywhere equal to two right angles.

In the second place if one assumes at the start, like Euclid, an existence independent of situation not only for lines but also for bodies, then it follows that the measure of curvature is everywhere constant; and then the sum of the angles in all triangles is determined as soon as it is fixed for one triangle.

In the third place, finally, instead of assuming the length of lines to be independent of place and direction, one might even assume their length and direction to be independent of place. Upon this understanding the changes in place or differences in position are complex quantities expressible in three independent units.

2

In the course of preceding discussions, in the first place relations of extension (or of domain) were distinguished from those of measurement, and it was found that different relations of measure were conceivable along with identical relations of extension. Then were sought systems of simple determinations of measure by means of which the metric relations of space are completely deter-

mined and of which all theorems about such relations are a necessary consequence. It remains now to examine the question how,
in what degree and to what extent these assumptions are guaranteed by experience. In this connection there subsists an essential
difference between mere relations of extension and those of
measurement: in the former, where the possible cases form a
discrete manifold the declarations of experience are indeed never
quite sure, but they are not lacking in exactness; while in the latter,
where possible cases form a continuum, every determination
based on experience remains always inexact, be the probability
that it is nearly correct ever so great. This antithesis becomes
important when these empirical determinations are extended
beyond the limits of observation into the immeasurably great and
the immeasurably small; for the second kind of relations obviously
might become ever more inexact, beyond the bounds of observation, but not so the first kind.

When constructions in space are extended into the immeasurably
great, unlimitedness must be distinguished from infiniteness; the
one belongs to relations of extension, the other to those of measure.
That space is an unlimited, triply extended manifold is an assumption applied in every conception of the external world; by it at
every moment the domain of real perceptions is supplemented and
the possible locations of an object that is sought for are constructed,
and in these applications the assumption is continually being
verified. The unlimitedness of space has therefore a greater
certainty, empirically, than any experience of the external. From
this, however, follows in no wise its infiniteness, but on the contrary
space would necessarily be finite, if one assumes that bodies are
independent of situation and so ascribes to space a constant
measure of curvature, provided this measure of curvature had any
positive value however small. If one were to prolong the elements
of direction, that lie in any element of surface, into shortest lines
(geodetics), one would obtain an unlimited surface with constant
positive measure of curvature, consequently a surface which would
take on, in a triply extended manifold, the form of a spherical
surface, and would therefore be finite.

3

Questions concerning the immeasurably large area, for the
explanation of Nature, useless questions. Quite otherwise is it
however with questions concerning the immeasurably small.

Knowledge of the causal connection of phenomena is based essentially upon the precision with which we follow them down into the infinitely small. The progress of recent centuries in knowledge of the mechanism of Nature has come about almost solely by the exactness of the syntheses rendered possible by the invention of Analysis of the infinite and by the simple fundamental concepts devised by Archimedes, Galileo, and Newton, and effectively employed by modern Physics. In the natural sciences however, where simple fundamental concepts are still lacking for such syntheses, one pursues phenomen into the spatially small, in order to perceive causal connections, just as far as the microscope permits. Questions concerning spatial relations of measure in the indefinitely small are therefore not useless.

If one premise that bodies exist independently of position, then the measure of curvature is everywhere constant; then from astronomical measurments it follows that it cannot differ from zero; at any rate its reciprocal value would have to be a surface in comparison with which the region accessible to our telescopes would vanish. If however bodies have no such non-dependence upon position, then one cannot conclude to relations of measure in the indefinitely small from those in the large. In that case the curvature can have at every point arbitrary values in three directions, provided only the total curvature of every metric portion of space be not appreciably different from zero. Even greater complications may arise in case the line element is not representable, as has been premised, by the square root of a differential expression of the second degree. Now however the empirical notions on which spatial measurements are based appear to lose their validity when applied to the indefinitely small, namely the concept of a fixed body and that of a light-ray; accordingly it is entirely conceivable that in the indefinitely small the spatial relations of size are not in accord with the postulates of geometry, and one would indeed be forced to this assumption as soon as it would permit a simpler explanation of the phenomena.

The question of the validity of the postulates of geometry in the indefinitely small is involved in the question concerning the ultimate basis of relations of size in space. In connection with this question, which may well be assigned to the philosophy of space, the above remark is applicable, namely that while in a discrete manifold the principle of metric relations is implicit in the notion of this manifold, it must come from somewhere else in the case of a continuous manifold. Either then the actual things forming the groundwork of a space must constitute a discrete manifold, or else the basis of metric relations must be sought for outside that actuality, in colligating forces that operate upon it.

A decision upon these questions can be found only by starting from the structure of phenomena that has been approved in experience hitherto, for which Newton laid the foundation, and by modifying this structure gradually under the compulsion of facts which it cannot explain. Such investigations as start out, like this present one, from general notions, can promote only the purpose that this task shall not be hindered by too restricted conceptions, and that progress in perceiving the connection of things shall not be obstructed by the prejudices of tradition.

This path leads out into the domain of another science, into the realm of physics, into which the nature of this present occasion forbids us to penetrate.

"Riemann, who was logically the immediate predecessor of Einstein, brought in a new idea of which the importance was not perceived for half a century. He considered that geometry ought to start from the infinitesimal, and depend upon integration for statements about finite lengths, areas, or volumes. This requires, inter alia, the replacement of the straight line by the geodesic: the latter has a definition depending upon infinitesimal distances, while the former has not. The traditional view was that, while the length of a curve could, in general, only be defined by integration, the length of the straight line between two points could be defined as a whole, not as the limit or a sum of little bits. Riemann's view was that a straight line does not differ from a curve in this respect. Moreover, measurement, being performed by means of bodies, is a physical operation, and its results depend for their interpretation upon the laws of physics. This point of view has turned out to be of very great importance. Its scope has been extended by the theory of relativity, but in essence it is to be found in Riemann's dissertation." (Bertrand Russell, *The Analysis of Matter*, p. 21, New York, 1927, Harcourt, Brace and Company, Quoted by permission of the publishers.)

MONGE

On the Purpose of Descriptive Geometry

(Translated from the French by Professor Arnold Emch, University of Illinois, Urbana, Ill.)

Gaspard Monge (1746–1818) was the son of an itinerant tradesman. At twenty-two he was professor of mathematics in the military school at Mézières and finally held a similar position in the École Polytechnique in Paris. He is known chiefly for his elaboration of descriptive geometry, a theory which had been suggested by Frézier in 1738. He lectured upon the subject at Paris in "l'an 3 de la République" (1794–1795) and his *Géométrie Descriptive* was published in "l'an 7" (1798–1799). He had already laid the foundations for the theory when teaching at Mézières, and on January 11, 1775 he had presented a memoir before the Académie des Sciences in which he made use of two planes of projection. He was one of the leaders in the foundation and organization of the École Normale and the École Polytechnique. The following brief quotation from his treatise (5th ed., Paris, 1927, pp. 1–2) will serve to set forth the purpose which he had in view and which the government guarded as a secret for some years because of its value in the construction of fortifications:

Descriptive geometry has two objects: the first is to establish methods to represent on drawing paper which has only two dimensions,—namely, length and width,—all solids of nature which have three dimensions,—length, width, and depth,—provided, however, that these solids are capable of rigorous definition.

The second object is to furnish means to recognize accordingly an exact description of the forms of solids and to derive thereby all truths which result from their forms and their respective positions.

REGIOMONTANUS

ON THE LAW OF SINES FOR SPHERICAL TRIANGLES

(Translated from the Latin by Professor Eva M. Sanford, College for Women, Western Reserve University, Cleveland, Ohio.)

Johann Müller (1436–1476), known as Regiomontanus, was the first to write a treatise devoted wholly to trigonometry. This appeared in manuscript about 1464, and had the title *De triangulis omnimodis*. The completeness of this work may be judged from the author's treatment of the Law of Sines for Spherical Triangles, a theorem which was probably of his own invention.

In every right-angled triangle, the ratio of the sine of each side to the sine of the angle which it subtends is the same.[1]

Given the triangle *abg* having the angle *b* a right angle. I say that the ratio of the sine of the side *ab* to the sine of the angle *agb* is the same as the ratio of the sine of the side *bg* to the sine of the angle *bag*, and also as the ratio of the sine of the side *ag* to the sine of the angle *abg*, which we shall prove as follows.

It is inevitable that each of the angles *a* and *g* is a right angle, or that one or other of them is a right angle, or that neither is a right angle. If each of them is a right angle, then, by hypothesis, the point *a* is the pole of the circle *bg*, and moreover the point *b* is the pole of the circle *ag* and *g* is the pole of the circle *ab*. Thus, by definition, each of the three arcs will measure[2] its respective angle. Therefore, the sine of any one of the three sides will be the same as that of the angle opposite, and accordingly the sine of each side has the same ratio to the sine of its respective angle, this ratio being that of equality.

If, however, but one of the angles *a* and *g* is a right angle, let this one be the angle *g*. Since the hypothesis made *b* a right angle also, then, on this supposition, *a* is the pole of the circle *bg*, and each of the arcs *ba* and *ag* is a fourth of a great circle. Thus by definition, each of the arcs *ab*, *bg*, and *ga* will determine the size

---

[1] [*De triangulis omnimodis*, Lib. IIII, XVI, pp. 103–105, Nürnberg, 1533. The proof as given in this work has been divided into paragraphs for greater clarity in the translation.]

[2] [Literally "will determine the size of."]

of its respective angle, and the sine of any side will be the same as that of the corresponding angle by applying the definition of the sine of the angle. It will then be evident that the sine of each side has the same ratio, namely that of equality, to the sine of its corresponding angle.

But if neither of the angles *a* and *g* be a right angle, no one of the three sides will be a quadrant of a great circle, but they will be found in three-fold variety.[1] If each of the angles *a* and *g* should be acute, each of the arcs *ab* and *bg* will be less than a quadrant, and accordingly the arc *ag* will be less than a fourth of a great circle. Then let the arc *ga* be produced[2] toward *a* until it becomes the quadrant *gd*, and taking the chord which is the side of a great square[3] as a radius and the point *g* as a center, describe a great circle cutting the arc *gb* produced in the point *e*.

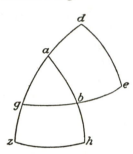

Finally, let the arc *ag* be extended to the point *z* thus obtaining the quadrant *az* whose chord, swung about the pole *a* generates a circle which meets the arc *ab* extended in the point *b*. We have drawn a diagram illustrating these conditions.

But if each of the angles be obtuse, each of the arcs *ab* and *gb* will exceed a quadrant, and we know that the arc *ag* is less than a quadrant. Therefore, prolonging the arc *ag* on both sides as before until the fourth arc *gd* is formed and the arc *az* also, let two great circles be described with the centers at *g* and *a*. The circumference of the one described with *g* as a center will necessarily cut the arc *gb*, which is greater than a quadrant. Let this happen at the point *e*. The other circle described with *a* as a center will cut the arc *ab* at the point *b*. Thus another figure will be produced.

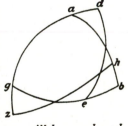

[1] [That is, all three sides will be less than a quadrant, or *ab* and *bg* will each be greater and *ag* less, or *bg* and *ag* will each be greater than a quadrant and *ab* will be less. It should be noted that Regiomontanus uses the letters in his diagrams in the order in which they appear in the Greek alphabet which is a natural outcome of his familiarity with mathematical classics in Greek.]

[2] [Literally "increased."]

[3] [Literally, "costa quadrati magni." This is evidently the chord of the quadrant of a great circle, the pole being used as a center in describing the circle on the sphere.]

But if one of the angles *a* and *g* is obtuse and the other acute,
let *a* be obtuse and let the other be acute. Then, according to
the cases cited, each of the arcs *bg* and *ga* is greater than a quadrant,
but the arc *ab* is less than a quadrant. Therefore let two quad-
rants *gd* and *az*, which share the arc *dz*, be cut off from the arc
ag. Then the circumference of a circle described as before with
g as a pole will cut the arc *bg* which is greater than a quadrant.
Let *e* be this point of intersection. Moreover the circumference
of the circle described about *a* will not cut the arc *ab*, since this
arc is less than a quadrant, but it will meet it if it is prolonged
sufficiently, as at *b*. Therefore when
neither of the angles *a* and *g* is a right
angle, although we use a triple diagram,
yet a single syllogism will result.

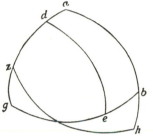

Since the two circles *gd* and *ge* meet
obliquely,[1] and since two points are
marked on the circumference of the
circle *gd* with the perpendiculars *ab* and
de drawn at these points, then accord-
ing to the preceding demonstration the ratio of the sine of the arc
ga to the sine of the arc *ab* will be as the sine of the arc *gd* to the
sine of the arc *de*, and, by interchanging these terms,[2] the ratio
of the sine *ga* to the sine *gd* will be that of the sine *ab* to the sine
de. In like fashion, the two circles *az* and *ab* meet obliquely, and
two points *g* and *z* are marked on the circumference of the circle
az from which are drawn two perpendicular arcs *gb* and *zb*. Therefore, according to the foregoing proofs, the ratio of the sine *ag*
to the sine *gb* is as that of the sine *az* to the sine *zb*; and by alternation, the sine *ag* is to the sine *az* as the sine *gb* is to the sine *zb*.
Moreover, the sine *ag* is to the sine *az* as the sine *ga* is to the sine
gd. Each of the arcs *az* and *ga* is a quadrant. Therefore the
sine of the side *ab* has the same ratio to the sine *de* as the sine of the
side *gb* has to the sine *zb*, which is that of the sine of the side *ag*
to the sine of the quadrant. Moreover, the sine *de* is the sine of
the angle *agb*, for the arc *de* measures the angle *agb* with the point
g acting as the pole of the circle *de*. In like manner, the sine *zb*
is the sine of the angle *bag*. Furthermore, the sine of the quadrant
is the sine of a right angle, therefore the ratio of the sine of the
side *ab* to the sine of the angle *agb*, and that of the sine of the side

---

[1] [Literally, "are inclined toward each other."]
[2] [Literally, "by permuting the terms."]

bg to the sine of the angle *bag*, and also the ratio of the sine of the side *ag* to the sine of the right angle *abg* are one and the same, which was to be shown.

In every triangle, not right-angled, the sines of the sides have the same ratio as the sines of the angles opposite.[1]

The statement which the preceding proposition demonstrated for right-angled triangles may be proved for triangles that are not right-angled. Suppose that the triangle *abg* has no right angle. I say that the ratio of the sine of the side *ab* to the sine of the angle *g*, and that of the sine of the side *bg* to the sine of the angle *a*, and of the sine of the side *ga* to the sine of the angle *b* are one and the same.

I draw a perpendicular *ad* from *a* cutting the arc *bg* if it remains inside the triangle, or meeting the arc *bg* opportunely prolonged if it falls outside the triangle but being coterminus with neither *ab* nor *ag*; for in such a case, one of the angles *b* and *g* would be considered to be a right angle which our hypothesis has stated is not a right angle. Therefore; let it fall first within the triangle,

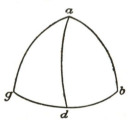

marking out two triangles *abd* and *agd*. According to the preceding proof, but alternating the terms, the ratio of the sine *ab* to the sine *ad* is the same as that of the sine of the angle *adb*, a right angle, to that of the angle *abd*. But by the same previous proof, the ratio of the sine *ad* to the sine *ag* is the same as that of the sine of the angle *agd* to the sine of a right angle *adg*, since the sine of the angle *adg* is the same as that of the angle *adb* and since each of them is a right angle. Then[2] the sine of *ab* will be to the sine of *ag* as the sine of the angle *agb* is to the sine of the angle *abg*; and by alternation, the sine of the side *ab* will be to the sine of the angle *agb* as the sine of the side *ag* is to the sine of the angle *abg*.

Finally, you will conclude that the ratio of the sine of the side *bg* to the sine of the angle *bag* is the same, if from one of the vertices *b* or *g* you draw an arc perpendicular to the side opposite it.

But if the perpendicular *ad* falls outside the triangle, thus changing the figure a little, let us seek the original syllogism; for reasoning by alternation from the preceding proof, the sine *ab* will be to the sine *ad* as the sine of the right angle *adb* s to the sine

---

[1] [Lib. IIII, XVII.]

[2] ["By reason of the equal indirect proportion."]

of the angle *abd*. Likewise, the sine of *ad* will be to the sine *ag* as the sine of the angle *agb* is to the sine of the right angle *adg*. Therefore, the sine of the side *ab* will be to the sine of the side *ag* as the sine of the angle *agb* is to the sine of the angle *abd*. Moreover, the sine of the angle *abd* is also the sine of the angle *abg* by common knowledge.[1] Therefore the sine *ab* is to the sine *ag* as the sine of the angle *agb* is to the sine of the angle *abd*, and thus,

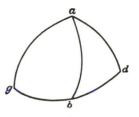

changing the terms, the sine of the side *ab* is to the sine of the angle *agb* as the sine of the side *ag* is to the sine of the angle *abg*. Finally, we shall prove that this is the ratio of the sine of the side *bg* to the sine of the angle *bag*, by the method which we have used above. Therefore the statement which was demonstrated in these theorems in regard to right-angled and non-right-angled triangles, respectively, we are at last free to state in general in regard to all triangles of whatever sort they may be, and we shall now consider step by step the great and jocund fruits which this study is to yield.

---

[1] ["Per communem scientiam" a direct translation from the Greek name for axiom.]

REGIOMONTANUS

On the Relations of the Parts of a Triangle

(Translated from the Latin by Professor Vera Sanford, Western Reserve University, Cleveland, Ohio.)

Regiomontanus is the Latin name assumed by Johann Müller (1436–1476), being derived from his birthplace, Königsberg, in Lower Franconia. In a block-book almanac prepared by him his name appears as Magister Johann van Kunsperck. He was known in Italy, where he spent some years, as Joannes de Monteregio. He wrote *De triangulis omnimodis c.* 1464, but it was not printed until 1533. It was the first work that may be said to have been devoted solely to trigonometry. The following extract is from this work, lib. II, p. 58. In it Regiomontanus shows the relations of the parts of a triangle, and from it is easily derived the formula which, in our present symbols, would appears as $\Delta = \frac{1}{2}bc \sin A$.

XXVI

Given the area of a triangle and the rectangle[1] of the two sides, then the angle opposite the base will either be known or with the known angle will equal two right angles.[2]

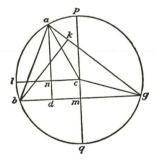

Using again the diagrams of the preceding proposition, if the perpendicular *bk* meeting the line *ag* falls outside the triangle, then by the first case, the ratio of *bk* to *ba* will be known, and so

---

[1] [*I.e.*, the product.]

[2] [This enables one to find sin A, having given the area of the triangle and the product *bc*. Regiomontanus, however, does not seem to have changed this into the form: Given A, b, and c to find the area. The theorem determines the acute angle at the vertex, whether this be interior or exterior to the triangle.]

by the angle of this first figure, we shall assume *bak* as known, accordingly the angle *bag* with the known angle *bak* will equal two right angles. But if the perpendicular *bk* falls inside the triangle as is seen in the third diagram [the one here shown] of the preceeding proposition, then as before *ab* will have a known ratio to *bk*, and therefore the angle *bak* or *bag* will be known. But if the perpendicular *bk* coincides with the side *ab*, the angle *bag* must have been a right angle and therefore must be known, which indeed happens when the area of the proposed triangle equals that of the rectangle which is inclosed by the two sides.

PITISCUS

On the Laws of Sines and Cosines

(Translated from the Latin by Professor Jekuthiel Ginsburg, Yeshiva College, New York City.)

Bartholemäus Pitiscus (1561–1613), a German clergyman, wrote the first satisfactory textbook on trigonometry, and the first book to bear this title, —the *Trigonometriae sive de dimensione triangulorum libri quinque* (Frankfort, 1595, with later editions in 1599, 1600, 1608, and 1612, and an English edition in 1630). The selections here translated are from the 1612 edition, pages 95 and 105, and set forth the laws of sines and cosines. The translation makes use of modern symbols.

Fragment I

The ratio of the sides of a triangle to each other is the same as the ratio of the sines of the opposite angles.

The sines are halves of the [corresponding] chords. The sides of a triangle have the same ratio as the chords of the opposite angles, hence the ratio of the sides will be equal to the ratio of the sines, because the ratio of the whole quantity to another whole quantity is the same as the ratio of a half to a half, according to proposition 19 of Book 2, and it lies in the nature of the thing itself.[1]

The sides of the plane triangle will be the chords of the opposite angles or of the arcs by half of which the angles are measured.

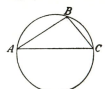

Thus: If the circle *ABC* be circumscribed around the triangle *ABC*, the side *AB* will be the chord of angle *ACB*; that is, of the arc *AB* which measures the angle *ACB*. The side *BC* will be the chord of the angle *BAC*; that is, the chord of the arc *BC* which measures the angle *BAC*. Similarly the side *AC* will be the chord of the angle *ABC*; that is, of the arc *AC* which determines the angle *ABC*.

---

[1] [To prove this Pitiscus uses the circumscribed circle in the following way.]

Hence the side AB has the same ratio to the side BC as the chord of the angle ACB to the chord of the angle BAC which was to be proved.[1]

. . . .

Fragment II

When the three sides of an oblique triangle are given, the segments made by the altitude drawn from the vertex of the greatest angle are given.[2] . . . .

Subtract the square of one of the lateral sides of a triangle from the sum of the squares of the other two. Divide the remainder by twice the base and you will get the segment between the altitude and the other lateral side.[3]

---

[1] [According to Tropfke in his *Geschichte der Elementar-Mathematik* (V, p. 74) there were two methods of proving the Law of Sines: one used by Vieta (1540–1603) and traced back to Levi Ben Gerson (1288–1344), who was the first to formulate it in the West; the other to Nasîr ed-din al-Tûsî (1201–1274) and used by Regiomontanus, Pitiscus, and others. This is the method here given. It is equivalent to the modern method of expressing the sides a, b, c, as $2r \sin A$, $2r \sin B$, $2r \sin C$ respectively.]

[2] [In the $\triangle ABC$, AG is $\perp$ to BC.

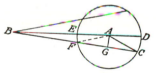

To find CG, Pitiscus describes a circle with a radius AC and uses the known proportion

$$BC:BD = BE:BF,$$

in which $BC = a$, $AC = b$, $BA = c$. Then $BD = BA + AD = BA + AC = c + b$. Also, $BE = BA - AE = c - AC = c - b$, and $BF = BC - CF = a - 2x$.

Hence

$$a:c + b = c - b:a - 2x,$$

or

$$c^2 - b^2 = a^2 - 2ax;$$

therefore

$$2ax = a^2 + b^2 - c^2$$
$$x = \frac{a^2 + b^2 - c^2}{2a}$$

From this he derives the scholium which follows.]

[3] [From this there is only one step to the general form of the Law of Cosines. Pitiscus did not make that step, perhaps because he considered it self-evident; but he used the theorem above given in exactly the same way as we now use the Law of Cosines; that is, he used it in finding the values of the angles from the given sides.]

PITISCUS

On Bürgi's Method of Trisecting an Arc

(Translated from the Latin by Professor Jekuthiel Ginsburg, Yeshiva College, New York City.)

Jobst Bürgi's (1552–1632) solution of the equation used in the trisection of an arc was given by Bartholomeus Pitiscus (1561–1613) in his *Trigonometria*, (1595; 1612 edition, pages 50–54). Whether Bürgi obtained it from Arabic sources or discovered it independently is an interesting question that has not as yet been answered satisfactorily.

The material in the translated "fragments" is interesting on account of the bearing it has on questions of both algebra and trigonometry. The explanation consists of two fragments, one of which is introductory to the other.

Fragment 1 [p. 38], Problem 3. Given the chord (*subtensa*) of an arc less than half the circumference, and the chord of double the given arc, required to find the chord of the triple arc.[1]

Solution ("rule").—Subtract the square of the chord of the given arc from the square of the chord of double the arc. The remainder divide by the chord of the given arc. The quotient will be the chord of the triple arc.[2]...

Fragment 2 [p. 50], Problem 6. Given the chord of an arc, find the chord of a third of the same arc.

Solution.—Take a third of the given chord; add something to it, and assuming the result to be the required chord compute the

---

[1] In modern notation: given $2 \sin a$ and $2 \sin 2a$, to find $2 \sin 3a$.

[2] To prove this Pitiscus makes use of the fact that the chords of the three arcs form the sides and diagonals of an inscribed quadrilateral. If arc AB = arc BC = arc CD, $AB = BC = CB$ = chord of given arc. $AC = BD$ = chord of double the arc, and AD = chord of triple arc. According to a well-known theorem, $AC.BD = AB.CD + AD.BC$ or

$$\overline{AC}^2 = \overline{AB}^2 + AD.AB$$

Hence AD, or the chord of the triple arc, equals $\dfrac{\overline{AC}^2 - \overline{AB}^2}{AB}.$

Hence the proposition is proved.

given chord using the method of problem 3.   Note the difference
by plus or minus and, repeating the same operation on another
assumed value of the required chord, mark the new difference by
plus or minus.   Having done this you will find the truth infallibly
by the Rule of False.

Example.—Let the given arc *AD* or 30° be taken as 5176381.
Required to find the chord of a third of the arc, namely, of the
arc of 10°.

The given chord	= 5176381
One third of it	= 1725460
Increased value of the third	= 1730000
or	= 1740000
or	= 1750000

The first assumption is	= 1730000
The chord of the triple arc (30⁰) computed from this according to the method of problem 3	= 5138223
But it should be	= 5176381
Hence the difference is minus	= 38158

The

The second assumption is	= 1740000
The value of the chord of the triple arc computed by the method of problem 3	= 5167320
But it should be	= 5176381

Hence the difference is minus 9161

Now according to the Rule of False multiply across: that is,
the first difference by the second assumption and the second
difference by the first assumption.   And since they are both
negative subtract the products and you will have the number
to be divided.

The first product	= 66394920000
The second product	= 15675530000

The number to be divided is 50719390000

Also from one of the minus numbers subtract the other and you will get the
divisor.

One of them is	38158
The other is	9061

The divisor is 29097

The performed division will give for the chord AB the number 1743114. On this number perform an operation similar to that performed on each of the two assumed values, and again there will be a difference,—but very small, namely 3. Taking a number slightly greater than 1743114, namely the number 1743115, and repeating on it the above operation you will find that the chord AD will be almost equal to the given value 5176381 but in the end it will be a little greater. Therefore, the chord 1743115 will also be a little greater but nearer the truth than 1743114, as will appear from the computation; hence there will not be an appreciable difference between the given value of AD and the computed one.

Another Method by Algebra. Solution. Divide the given chord by $3x - x^3$.[1] The quotient will be the chord of a third of the given arc.

Proof of the Rule. The chord of any arc is equal to three roots less one cube, the root being equal to the chord of a third of this arc.[2]

This is demonstrated as follows: Let AD be the given chord of the arc $ABCD$. It is required to find the chord of AB, BC, or

CD, [each of which is] a third of the arc. Let x be the chord of the third of the arc. Hence each of the chords AC and BD of the double arcs will be $l4q - 1bq$,[3] as has been demonstrated in the solution of the preceding problem. Since $ABCD$ is an inscribed quadrilateral, the product of the diagonals AC and BD is equal to the sum of the products of the opposite sides, by proposition 54 of the first book [of the *Trigonometria*]. Multiply the diagonals and the square is $4x^2 - x^4$.[4]

Then multiply the side AB by the side CD,[5] that is x by x; x^2 is obtained. This, being subtracted from the square [made by]

---

[1] [Pitiscus uses l instead of x, and c for x^3. The chord of the lesser arc is equal to the root of the equation $3x - x^3 = AD$. In the translation we have used modern symbols.]

[2] [In modern notation, $2 \sin 3A = 3(2 \sin A) - (2 \sin A)^3$, which reduces to $\sin 3A = 3 \sin A - 4 \sin^3 A$.]

[3] The Pitiscus notation for $\sqrt{4x^2 - x^4}$.

[4] [Pitiscus here adds in parentheses the following characteristic remark: "Because to multiply a surd number by itself is nothing else than removing the sign l" i. e. the radical sign.]

[5] [Pitiscus retains the coefficient 1, writing $1l$ for $1x$, $1q$ for q or x^2, etc. In the translation this coefficient is omitted.]

the diagonals, that is, $4x^2 - x^4$ leaves $3x^2 - x^4$ for the rectangle (or the product) made by BC and AD. This rectangle $3x^2 - x^4$, being divided by BC, that is by x, will give as a result $AD = 3x - x^3$.[1]

Hence $3x - x^3$, where x is a chord of a third of the arc is equal to the chord of the given arc.

In consequence of the above, if the chord $[k]$ of the given arc is equal to $3x - x^3$, the root [of the equation $3x - x^3 = k$] will be the chord of the third part.[2]

---

[1] [In modern notation:
$$AC.BD = AB.CD + AD.BC.$$
But
$$AC = BD = \sqrt{4x^2 - x^4},$$
by a previous demonstration
$$(x = AB = BC = CD.)$$
Hence
$$AC.BD = 4x^2 - x^4 = x^2 + x.AD.$$
$$\therefore AD = \frac{4x^2 - x^4 - x^2}{x} = 3x - x^3.]$$

[2] [This is equivalent, in modern symbols, to saying that
$$\sin 3A = 3 \sin A - 4 \sin^3 A.]$$

DE MOIVRE

On His Formula

(Translated from the Latin and from the French by
Professor Raymond Clare Archibald, Brown University.)

De Moivre's Formula is usually stated in the form

$$(\cos x + i \sin x)^n = \cos nx + i \sin nx,$$

where n is any real number. The equivalent of this form was given by
Euler (in 1748, Extract **E** below) and proved true for all real values of n
(in 1749, Extract **F** below). The result is not explicitly stated in any of
De Moivre's writings. But it will be observed that in more than one of them
(1707–38, Extracts **A–D** below) the formula and its application were thoroughly familiar to him; and that in passages where it is suggested (1722, 1730)
that certain eliminations shall be performed, on carrying these out we are led
to exactly the formula associated with his name. This was made clear in
Braunmühl's historical sketch in *Bibliotheca Mathematica*, series 3, vol. 2, p.
97–102, and in his *Vorlesungen über Geschichte der Trigonometrie*, part 2, 1903,
p. 75–78. While Hutton's translations (*Philosophical Transactions*, abridged,
vols. 5, 6, 8) have been the basis of the translations in Extracts **A**, **B**, and **D**
they have not been slavishly followed as they were carefully compared with
the originals. The original display of formulae and symbolism has been preserved except that in such an expression as y.y, y^2 has been substituted.

Abraham De Moivre was born in Champagne, France, in 1667, studied
mathematics under Ozanam in Paris, and repaired in 1688 to London where he
spent the remaining 66 years of his life. He was an intimate friend of Newton,
and his notable mathematical publications led to his election not only as a
member of the Royal Society, and of the Berlin Academy of Sciences, but also
as a foreign associate of the Paris Academy of Sciences. Of his *Annuity Upon
Lives* there were seven editions, five in English (1725, 1743, 1750, 1752, 1756),
one in Italian (1776) much enlarged with notes of Gaeta and Fontana and the
basis of lectures at the University of Pavia, and one in German (1906) by
Czuber. Among his other publications, which display great analytic power,
skill and invention, were *The Doctrine of Chances* (three editions, 1718. 1738,
1856), *Miscellanea Analytica* (1730) which brought about his election to the
Berlin Academy, a number of papers in the *Philosophical Transactions*, and an
important 8-page pamphlet of 1733 (*Approximatio ad summan terminorum
binomii* $(a + b)^n$ *in seriem expansi*, English editions in the last two editions
of *Doctrine of Chances*) presenting the first treatment of the probability integral
and essentially of the normal curve. For a facsimile of the original edition of
this pamphlet and for references to other discoveries of De Moivre, see *Isis*, vol.
8, 1926, p. 671–683. He was one of the commissioners appointed by the Royal
Society in 1712 to arbitrate on the claims of Newton and Leibniz to the
invention of the infinitesimal calculus.

Miss Clerke has recorded of De Moivre (*D.N.B.*) that he once said that he would rather have been Molière than Newton; and he knew his works and those of Rabelais almost by heart. In Pope's *Essay on Man* one finds (iii, l.103–104):

> Who made the spider parallels design,
> Sure as Demoivre, without rule or line?

A

"*Æquationum quarundam Potestatis tertiæ, quintæ, septimæ, novæ, & superiorum, ad infinitum usque pergendo, in terminis finitis, ad imstar Regularum pro Cubicis quae vocantur,* Cardani, *Resolutio Analytica,*" *Philosophical Transactions,* 1707, no. 309, vol. 25, p. 2368–2371.

"*The analytic solution of certain equations of the third, fifth, seventh, ninth and other higher uneven powers, by rules similar to those called Cardan's.*"

"The analytic solution of certain equations of the third, fifth, seventh, ninth and other higher uneven powers, by rules similar to those called Cardan's."

Let n denote any number whatever, y an unknown quantity or root of this equation, and a the absolute known quantity, or what is called the homogeneum comparationis[1]; let also the relation between these be expressed by the equation

$$ny + \frac{n^2-1}{2\times3}ny^3 + \frac{n^2-1}{2\times3}\times\frac{n^2-9}{4\times5}ny^5$$
$$+ \frac{(n^2-1)}{2\times3}\times\frac{(n^2-9)}{4\times5}\times\frac{(n^2-25)}{6\times7}ny^7, \text{etc.} = a.$$

From the nature of this series it is manifest, that if n be taken as any odd integer, either positive or negative, then the series will term nate and the equation become one of those above mentioned, the root of which is

$$(1) \qquad y = \tfrac{1}{2}\sqrt[n]{\sqrt{1+a^2}+a} - \frac{\tfrac{1}{2}}{\sqrt[n]{\sqrt{1+a^2}+a}}$$

$$\text{or } (2) \qquad y = \tfrac{1}{2}\sqrt[n]{\sqrt{1+a^2}+a} - \tfrac{1}{2}\sqrt[n]{\sqrt{1+a^2}-a}$$

$$\text{or } (3) \qquad y = \frac{\tfrac{1}{2}}{\sqrt[n]{\sqrt{1+a^2}-a}} - \tfrac{1}{2}\sqrt[n]{\sqrt{1+a^2}-a}$$

$$\text{or } (4) \qquad y = \frac{\tfrac{1}{2}}{\sqrt[n]{\sqrt{1+a^2}-a}} - \frac{\tfrac{1}{2}}{\sqrt[n]{\sqrt{1+a^2}+a}}[2]$$

---

[1] ["Homogeneum comparationis" in algebra was a name given by Vieta (1540–1603) to an equation's constant term, which he placed on the right-hand side of the equation and all the other terms on the left.]

[2] [In the denominator of the second term of the original, the second sign was − instead of +.]

For example, let it be required to find the root of the following equation of the fifth degree, $5y + 20y^3 + 16y^5 = 4$; in this case $n = 5$ and $a = 4$. According to (1) the root will then be

$$y = \tfrac{1}{2}\sqrt[5]{\sqrt{17} + 4} - \frac{\tfrac{1}{2}}{\sqrt[5]{\sqrt{17} + 4}}$$

whose numerical value is readily found. First $\sqrt{17} + 4 = 8.1231$, whose logarithm is 0.9097164, of which the fifth part is 0.1819433, the number corresponding to which is $1.5203 = \sqrt[5]{\sqrt{17} + 4}$. Also the arithmetic complement of 0.1819433 is 9.8180567, to which the number $0.6577 = \dfrac{1}{\sqrt[5]{\sqrt{17} + 4}}$. Therefore the half difference of these numbers is $0.4313 = y$.

It may be here observed that, instead of the general root, it may be sufficient to take $y = \tfrac{1}{2}\sqrt[n]{2a} - \dfrac{\tfrac{1}{2}}{\sqrt[n]{2a}}$, whenever the number n is very large in comparison with unity. For example, if the equation were $5y + 20y^3 + 16y^5 = 682$; then $\log.2a = 3.1348143$, of which the fifth part, 0.6269628 corresponds to the number 4.236. Also its arithmetic complement is 9.3730372 which corresponds to the number 0.236. The half difference of these two numbers is $2 = y$.

Again, if the terms of the preceding equation be alternately positive and negative, or which is the same thing, if the series be as follows:

$$ny + \frac{1-n^2}{2\times3}ny^3 + \frac{1-n^2}{2\times3}\times\frac{9-n^2}{4\times5}ny^5 +$$

$$\frac{1-n^2}{2\times3}\times\frac{9-n^2}{4\times5}\times\frac{25-n^2}{6\times7}ny^7,\text{ etc.} = a,\text{[1]}$$

---

[1] [If $y = \sin\phi$, $a = \sin n\phi$ we have $\sin n\phi$ expressed in terms of $\sin\phi$, a result which Newton had already given in a letter of 13 June 1676 (*Commercium Epistolicum J. Collins et Aliorum* ed. Biot and Lefort, p. 106), and which De Moivre derived in an article, "A method of extracting the root of an infinite equation" in *Philosophical Transactions* for 1698, no. 240, vol. 20, 1699, p.190. This relation was derived as a special case of the following result with which the article opens: "If $az + bz^2 + cz^3 + dz^4 + ez^5$ etc. $= gy + by^2 + iy^3 + ky^4 + ly^5$ etc. then will

$$zbe = \frac{g}{a}y + \frac{b - bA^2}{a}y^2 + \frac{i - 2bAB - cA^3}{a}y^3 +$$

$$\frac{k - bB^2 - 2bAC - 3cA^2B - dA^4}{a}y^4 +$$

$$\frac{l - 2bBC - 2bAD - 3cAB^2 - 3cA^2C - 4dA^3B - eA^5}{a}y^5\text{ etc.}"$$

where A, B, C, D, etc.

its root will be

$$(1) \quad y = \tfrac{1}{2}\sqrt[n]{a + \sqrt{a^2-1}} + \frac{\tfrac{1}{2}}{\sqrt[n]{a + \sqrt{a^2-1}}},$$

$$\text{or } (2) \quad y = \tfrac{1}{2}\sqrt[n]{a + \sqrt{a^2-1}} + \tfrac{1}{2}\sqrt[n]{a - \sqrt{a^2-1}},$$

$$\text{or } (3) \quad y = \frac{\tfrac{1}{2}}{\sqrt[n]{a - \sqrt{a^2-1}}} + \tfrac{1}{2}\sqrt[n]{a - \sqrt{a^2-1}},$$

$$\text{or } (4) \quad y = \frac{\tfrac{1}{2}}{\sqrt[n]{a - \sqrt{a^2-1}}} + \frac{\tfrac{1}{2}}{\sqrt[n]{a + \sqrt{a^2-1}}}.$$

It should be here noted that if $\dfrac{n-1}{2}$ is an odd number, the sign of the root when found must be changed to the contrary sign.

If the equation $5y - 20y^3 + 16y^5 = 6$ be proposed, then $n=5$ and $a=6$. The root is equal to

$$\tfrac{1}{2}\sqrt[5]{6 + \sqrt{35}} + \frac{\tfrac{1}{2}}{\sqrt[5]{6 + \sqrt{35}}}.$$

Or, since $6 + \sqrt{35} = 11.916$ of which the logarithm is 1.0761304 and of which the fifth part is 0.2152561, the arithmetic complement being 9.7847439. Hence the numbers corresponding to these logarithms are 1.6415 and 0.6091 respectively whose half sum $1.1253 = y$.

But if it happen that a is less than unity then the second form of the root is rather to be preferred as more convenient for the purpose. Thus if the equation were

$$5y - 20y^3 + 16y^5 = \frac{61}{64}$$

$$y = \frac{1}{2}\sqrt[5]{\frac{61}{64} + \sqrt{\frac{-375}{4096}}} + \frac{1}{2}\sqrt[5]{\frac{61}{64} - \sqrt{\frac{-375}{4096}}}. \quad \text{And if by any}$$

are respectively equal to the coefficients of y, y^2, y^3, y^4, etc., a result to which W. Jones refers (in his *Synopsis Palmariorum Matheseos*, London, 1706, p. 188) as a "*Theorem*" of "that Ingenious Mathematician *Mr. De Moivre*." In a review of this book of Jones in *Acta Eruditorum*, 1707, p. 176; this result is called "Theorema Moivræanum," a term assigned to a theorem which does not necessarily have any connection with trigonometric functions. The terms De Moivre's Formula, De Moivre's Theorem, applied to the formula we are considering, do not seem to have come into general use till the early part of the nineteenth century. Tropfke cites A. L. Crelle, *Lehrbuch der Elemente der Geometrie und der ebenen und sphärischen Trigonometrie*, Berlin, vol. 1, 1826, §335 for the use of the former term.]

means the fifth root of the binomial can be extracted the root will come out true and possible, although the expression seems to include an impossibility. Now the fifth root of the binomial $\frac{61}{64} + \sqrt{\frac{-375}{4096}}$ is $\frac{1}{4} + \frac{1}{4}\sqrt{-15}$, and of the binomial $\frac{61}{64} - \sqrt{\frac{-375}{4096}}$ is $\frac{1}{4} - \frac{1}{4}\sqrt{-15}$ whose semi-sum $\frac{1}{4} = y$.   But if the extraction can not be performed, or should seem too difficult, the thing may always be effected by a table of natural sines in the following manner.

To the radius 1 let $a = \frac{61}{64} = 0.95112$ be the sine of a certain arc which therefore will be 72° 23′, the fifth part of which (because n = 5) is 14°28′; the sine of this is 0.24981, nearly $= \frac{1}{4}$.   So also for equations of higher degree.[1]

B

"De Sectione Anguli," *Philosophical Transactions*, 1722, no. 374, vol. 32. p. 228–230.

"Concerning the section of an angle"

In the beginning of the year 1707, I fell upon a method by which a given equation of these forms

$$ny + \frac{n^2-1}{2\times 3}Ay^3 + \frac{n^2-9}{4\times 5}By^5 + \frac{n^2-25}{6\times 7}Cy^7, \text{ etc. } = a,$$

or

$$ny + \frac{1-n^2}{2\times 3}Ay^3 + \frac{9-n^2}{4\times 5}By^5 + \frac{25-n^2}{6\times 7}Cy^7, \text{ etc. } = a,$$

(where $A, B, C, \ldots$ represent the coefficients of the preceding terms) may have its roots determined in the following manner.

Set $a + \sqrt{a^2+1} = v$ in the first case and $a + \sqrt{a^2-1} = v$ in the second.   Then will in the first case

$$y = \tfrac{1}{2}\sqrt[v]{v} - \frac{\frac{1}{2}}{\sqrt[v]{v}}; \text{ and in the second } y = \tfrac{1}{2}\sqrt[v]{v} + \frac{\frac{1}{2}}{\sqrt[v]{v}}.$$

---

[1] [From this example it is clear that in 1707 De Moivre was in possession of the formula

$$\tfrac{1}{2}\sqrt[v]{\sin n\phi + \sqrt{-1}.\cos n\phi} + \tfrac{1}{2}\sqrt[v]{\sin n\phi - \sqrt{-1}.\cos n\phi} = \sin \phi,$$

where n is an odd integer.   In 1730, as we shall presently see, De Moivre formulated a relation equivalent to the following:

$$\tfrac{1}{2}\sqrt[v]{\cos n\phi + \sqrt{-1}.\sin n\phi} + \tfrac{1}{2}\sqrt[v]{\cos n\phi - \sqrt{-1}.\sin n\phi} = \cos \phi,$$

where *n* is any positive integer.]

These solutions were inserted in the Philosophical Transactions, No. 309, for the months Jan. Feb. March of that year.

Now by what artifices these formulae were discovered will clearly appear from the demonstration of the following theorem.

In a unit circle let x denote the versed sine of any arc, and t that of another; and let the former arc be to the latter as 1 to n. Then, assuming two equations which may be regarded as known[1], namely

$$1 - 2z^n + z^{2n} = -2z^n t \text{ and } 1 - 2z + z^2 = -2zx,$$

on eliminating z, there will arise an equation by which the relation between x and t will be determined.

Corollary I.—If the latter arc be a semicircle, the equations will be

$$1 + z^n = 0, 1 - 2z + z^2 = -2zx,$$

from which if z be eliminated there will arise an equation by which will be determined the versed sines of the arcs which are to the semicircle taken once, or thrice, or five times, etc. as 1 to n.

Corollary II.—If the latter arc is a circumference, the equations are

$$1 - z^n = 0, 1 - 2z + z^2 = -2zx,$$

from which after z is eliminated will arise an equation by which are determined the versed sines of the arcs, which are to the circumference, taken once, twice, thrice, four times, etc., as 1 to n.

Corollary III.—If the latter arc is 60 degrees, the equations are

$$1 - z^n + z^{2n} = 0 \text{ and } 1 - 2z + z^2 = -2zx$$

from which on eliminating z, will arise an equation which determines the versed sines of the arcs which are to the arc of 60°, multiplied by 1,7,13,19,25, etc. or by 5,11,17,23,29,etc., as 1 to n.

If the latter arc be 120° the equations will be

$$1 + z^n + z^{2n} = 0 \text{ and } 1 - 2z + z^2 = -2zx,$$

---

[1] [Let x = versed sin $\phi = 1 - \cos \phi$, t = versed sin $n\phi = 1 - \cos n\phi$ then these equations are $1 - 2\cos n\phi z^n + z^{2n} = 0$ and $1 - 2 \cos \phi z + z^2 = 0$. Compare quotation C, corollary I. The elimination of z gives

$$\sqrt[n]{\cos n\phi \pm \sqrt{\cos^2 n\phi - 1}} = \sqrt[n]{\cos n\phi \pm \sqrt{-1}.\sin n\phi}$$
$$= \cos \phi + \sqrt{-1}. \sin n\phi,$$

or $(\cos\phi + \sqrt{-1}.\sin \phi)^n = \cos n\phi + \sqrt{-1}.\sin n\phi$, De Moivre's formula, for n an odd integer.]

from which if z be eliminated there will arise an equation, by which are determined the versed sines of the arcs, which are to the arcs of 120°, multiplied by 1,4,7,10,13, etc., or by 2,5,8,11,14, etc. as 1 to n.

C

[A. De Moivre], *Miscellanea Analytica*, London, 1730, pp. 1–2.

Lemma 1.—*If l and x are the cosines of two arcs A and B of a circle of radius unity, and if the first arc is to the second as the number* n *is to unity then*

$$x = \tfrac{1}{2}\sqrt[n]{l + \sqrt{l^2-1}} + \sqrt[n]{\frac{\tfrac{1}{2}}{l + \sqrt{l^2-1}}}$$

Corollary I.—Set $\sqrt[n]{l + \sqrt{l^2-1}} = z$; then will $z^n = l + \sqrt{l^2-1}$ or $z^n - l = \sqrt{l^2-1}$, or, squaring both sides $z^{2n} - 2lz^n + l^2 = l^2 - 1$. Cancelling equal terms on each side and having made the proper transposition $z^{2n} - 2lz^n + 1 = 0$. From what was assumed $\sqrt[n]{l + \sqrt{l^2 - 1}} = z$, it follows from the above lemma that

$$x = \tfrac{1}{2}z + \frac{\tfrac{1}{2}}{z}, \text{ or } z^2 - 2xz + 1 = 0.$$

Corollary II.—If between the two equations $1 - 2lz^n + z^{2n} = 0$, $1 - 2xz + z^2 = 0$, the quantity z be eliminated there will arise a new equation defining a relation between the cosines l and x, providing the arc A is less than a quadrant.

Corollary III.—But if the arc A is greater than a quadrant then its cosine will be $-l$, from which it results that the equations will turn out to be $1 + 2lz^n + z^{2n} = 0$, $1 - 2xz + z^2 = 0$; and if z be eliminated between these there will arise a new equation expressing the relation between the cosines l and x.

Corollary IV.—And in particular if z be eliminated between the equations $1 \mp 2lz^n + z^{2n} = 0$, $1 - 2xz + z^2 = 0$ there will arise a new equation expressing a relation between the cosine of the arc A (less or greater than a quadrant according as l has the negative or positive sign) and all the cosines of the arcs $\dfrac{A}{n}$,

$\dfrac{C-A}{n}, \dfrac{C+A}{n}, \dfrac{2C-A}{n}, \dfrac{2C+A}{n}, \dfrac{3C-A}{n}, \dfrac{3C+A}{n}$, etc., in which series of arcs C denotes the entire circumference.[1]

---

[1] [That is $\sqrt[n]{\cos A \pm i \sin A} = \cos \dfrac{2k\pi \pm A}{n} + i \sin \dfrac{2k\pi \pm A}{n}$, $k = 0,1,2,3, \dots$.]

D

"*De Reductione* Radicalium *ad simpliciores terminos, seu de* extrabenda *radice quacunque data ex* Binomio $a + \sqrt{+b}$, *vel* $a + \sqrt{-b}$." Epistola, *Philosophical Transactions*, 1739, no.451, vol.40, p. 463–478.

"*On the reduction* of radicals to simpler terms, or the extraction of roots *of any* binomial $a + \sqrt{+b}$ or $a + \sqrt{-b}$. Letter."

[The paper consists almost wholly in the discussion of four problems. Our quotation is of problems 2–3, p.472–74.]

Problem II.—*To extract the cube root of the impossible binomial* $a + \sqrt{-b}$.

Suppose that root to be $x + \sqrt{-y}$, the cube of which is $x^3 + 3x^2\sqrt{-y} - 3xy - y\sqrt{-y}$. Now put $x^3 - 3xy = a$, and $3x^2\sqrt{-y} - y\sqrt{-y} = \sqrt{-b}$. Then the squares of these will give two new equations, namely

$$x^6 - 6x^4y + 9x^2y^2 = a^2$$
$$-9x^4y + 6x^2y^2 - y^3 = -b.$$

Then the difference of these squares is

$$x^6 + 3x^4y + 3x^2y^2 + y^3 = a^2 + b;$$

the cube root of which is $x^2 + y = \sqrt[3]{a^2 + b} = m$, say. Hence $x^2 + y = m$, or $y = m - x^2$ which value of y substituted in the equation $x^3 - 3xy = a$ gives $x^3 - 3mx + 3x^3 = a$, or $4x^3 - 3mx = a$; which is the very same equation as has been before deduced from the equation $2x = \sqrt[3]{a + \sqrt{-b}} + \sqrt[3]{a - \sqrt{-b}}$. Nevertheless it does not follow that in the equation $4x^3 - 3mx = a$, the value of x can be found by the former equation since it consists of two parts each including the imaginary quantity $\sqrt{-b}$; but this will best be done by means of a table of sines.

Therefore let the cube root be extracted of the binomial $81 + \sqrt{-2700}$. Put $a = 81$, $b = 2700$; then $a^2 + b = 6561 + 2700 = 9261$, the cube root of which is 21, which set equal to m makes $3mx = 63x$. Hence the equation to be solved will be $4x^3 - 63x = 81$, which being compared with the equation for the cosines, namely $4x^3 - 3r^2x = r^2c$[1] gives $r^2 = 21$, hence $r = \sqrt{21}$ and therefore $c = \dfrac{a}{r^2} = \dfrac{81}{21} = \dfrac{27}{7}$.

---

[1] [If this equation be put in the form $4\left(\dfrac{x}{r}\right)^3 - 3\left(\dfrac{x}{r}\right) = \dfrac{c}{r}$ it may be regarded as equivalent to the trigonometric formula $4\cos^3\dfrac{A}{3} - 3\cos\dfrac{A}{3} = \cos A$, if

To find the circular arc corresponding to the radius $\sqrt{21}$ and $c = \frac{27}{7}$, put the whole circumference equal to C, and take the arcs $\frac{A}{3}$, $\frac{C-A}{3}$, $\frac{C+A}{3}$, which will easily be known by a trigonometrical calculation, especially by using logarithms; then the cosines of the arcs to the radius $\sqrt{21}$ will be the three roots of the quantity x; since $y = m - x^2$, there will therefore be as many values of y, and thence a triple value of the cube root of the binomial $81 + \sqrt{-2700}$, which must now be accommodated to numbers.[1]

Make then $\sqrt{21}$ to $\frac{27}{7}$ as the tabular radius is to the cosine of an arc A, which will be nearly 32°42′. Hence the arc C−A will be 327°18′, and C+A 392°42′, of which the third parts will be 10°54′, and 109°6′, and 130°54′. But now as the first of these is less than a quadrant, its cosine, that is, the sine of 79°6′ ought to be considered as positive; and both the other two being greater than a quadrant, their cosines, that is the sines of the arcs 19°6′ and 40°54′, must be considered as negative. Now by trigonometrical calcu-

---

$\frac{c}{r} = \cos A$ and $\frac{x}{r} = \cos \frac{A}{3}$. De Moivre identifies the problem of finding the cube root with that of trisecting an angle.]

1 [In general terms the argument is as follows:

$c = r \cos A$, $x = r \cos \frac{A}{3}$ and by comparing the two cubic equations $m = r^2$, $a = r^2 c$. Therefore $r = \sqrt{m}$, $c = \frac{a}{m} = \sqrt{m} \cos A$. Then $x = \sqrt{m} \cos \frac{A}{3}$,

$$x = \sqrt{m} \cos \frac{C-A}{3},$$

$x = \sqrt{m} \cos \frac{C+A}{3}$ are the three roots of the cubic equation. But since $a = r^2 c = \sqrt{m^3} \cos A$, and $b = m^3 - a^2 = m^3(1 - \cos^2 A) = m^3 \sin^2 A$, $x = \sqrt{m} \cos \frac{A}{3}$, $\dots$, $y = m - x^2 = m\left(1 - \cos^2\frac{A}{3}\right) = m \sin\frac{A}{3} \dots$ Hence on substituting in the equation $\sqrt{a + i \sqrt{b}} = x + i \sqrt{y}$ we get

$$[\sqrt{m^3}(\cos A + i \sin A)]^{\frac{1}{3}} = \sqrt{m}\left\{\cos\frac{A}{3} + i \sin\frac{A}{3}\right\}.$$

or

$$\sqrt{m}\left(\cos\frac{C-A}{3} + i \sin\frac{C-A}{3}\right),$$

or

$$\sqrt{m}\left(\cos\frac{C+A}{3} + i \sin \frac{C+A}{3}\right).]$$

lation it appears, that these sines, to radius $\sqrt{21}$ will be 4.04999 and -1.4999, and -3.0000, or $\frac{9}{2}$, and $-\frac{3}{2}$ and -3. Hence there will be as many values of the quantity y, namely all those represented by $m - x^2$, namely $21 - \frac{81}{4}$, and $21 - \frac{9}{4}$, and $21 - 9$, that is $\frac{3}{4}$, $\frac{75}{4}$, 12 and the square roots of which are $\frac{1}{2}\sqrt{3}, \frac{5}{2}\sqrt{3}, 2\sqrt{3}$. Therefore the values of $\sqrt{-y}$ will be $\frac{1}{2}\sqrt{-3}, \frac{5}{2}\sqrt{-3}, 2\sqrt{-3}$. Hence the values of $\sqrt[3]{81 + \sqrt{-2700}}$ are $\frac{9}{2} + \frac{1}{2}\sqrt{-3}, -\frac{3}{2} + \frac{5}{2}\sqrt{-3}$, and $-3 + \frac{1}{2}\sqrt{-3}$. And by proceeding in the same manner, there will be found the three values of $\sqrt[3]{81 - \sqrt{-2700}}$, which are $\frac{9}{2} - \frac{1}{2}\sqrt{-3}, \frac{3}{2} - \frac{5}{2}\sqrt{3}$, and $-3 - \frac{1}{2}\sqrt{-3}$.

There have been several authors, and among them the eminent Wallis, who have thought that those cubic equations which are referred to the circle, may be solved by the extraction of the cube root of an imaginary quantity, and of $81 + \sqrt{-2700}$, without regard to the table of sines, but that is a mere fiction and a begging of the question. For on attempting it, the result always recurs back again to the same question as that first proposed. And the thing cannot be done directly, without the help of the table of sines, especially when the roots are irrational, as has been observed by many others.

Problem III.—*To extract the nth root of the impossible binomial* $a + \sqrt{-b}$.

Let that root be $x + \sqrt{-y}$; then making $\sqrt[n]{a^2 + b} = m$, and $\frac{n - 1^{[1]}}{n} = p$, n any integer, describe, or conceive to be described, a circle, the radius of which is $\sqrt{m}$, in which take an arc A the cosine of which is $\frac{a}{m^p}$, and let C be the whole circumference. To the same radius take the cosines of the arcs $\frac{A}{n}, \frac{C-A}{n}, \frac{C+A}{n}$, $\frac{2C-A}{n}, \frac{2C+A}{n}, \frac{3C-A}{n}, \frac{3C+A}{n}$, etc., till the number of them be

---

[1] [This should be $\frac{n-1}{2} = p$.]

equal to n. Then all these cosines will be so many values of x; and the quantity y will always be $m - x^2$.[1]

E

Euler, *Introductio in Analysin Infinitorum*, Lausanne, 1748, vol.1, Chapter 8, "De quantitatibus transcendentibus ex circulo ortis" ["On transcendental quantities derived from the circle,"] p.97–98, §§132–133; Reprinted in *Leonhardi Euleri Opera Omnia*, Leipzig, series 1, vol.8, 1922, p.140–141.

132. Since $(\sin.z)^2 + (\cos.z)^2 = 1$, on decomposing into factors we get $(\cos.z + \sqrt{-1}.\sin.z)(\cos.z - \sqrt{-1}\ \sin.z) = 1$. These factors, although imaginary are of great use in the combination and multiplication of arcs. For example, let us seek the product of these factors

$$(\cos.z + \sqrt{-1}.\sin.z)(\cos.y + \sqrt{-1}.\sin.y),$$

we find

$$\cos.y\ \cos.z - \sin.y\ \sin.z + \sqrt{-1}.(\cos.y\ \sin.z + \sin.y\ \cos.z);$$

but since

$$\cos.y\ \cos.z - \sin.y\ \sin.z = \cos.(y+z)$$

and

$$\cos.y\ \sin.z + \sin.y\ \cos.z = \sin.(y+z)$$

we obtain the product

$$(\cos.y + \sqrt{-1}.\sin.y)(\cos.z - \sqrt{-1}.\sin.z) = \cos.(y + z) \\ +\sqrt{-1}.\sin.(y+z).$$

Similarly

$$(\cos.y - \sqrt{-1}.\sin.y)(\cos.z - \sqrt{-1}.\sin.z) = \cos.(y + z) \\ -\sqrt{-1}.\sin.(y+z).$$

In the same way

$$(\cos.x \pm \sqrt{-1}.\sin.x)(\cos.y \pm \sqrt{-1}.\sin.y)(\cos.z + \sqrt{-1}.\sin.z) \\ = \cos.(x + y + z) \pm \sqrt{-1}.\sin.(x + y + z).$$

---

1 [That is,—

$$(\sqrt{m^n}(\cos A + i\ \sin A))^{\frac{1}{n}} = \sqrt{m}\left(\cos\frac{2k\pi \pm A}{n} + i\sin\frac{2k\pi \pm A}{n}\right)$$

$$k = 0,1,2,\ldots\frac{n-1}{2} \text{ if } n \text{ is odd; or } k = 1,2,\ldots\frac{n}{2} \text{ if } n \text{ is even,}$$

De Moivre's theorem for any unit fraction.

Practically all of extract **D** is given by De Moivre in a communication dated April 29, 1740, in N. Saunderson, *The Elements of Algebra*, Cambridge, vol. 2, 1740, p.744–748.]

133. Hence it follows that

$$(\cos.z \pm \sqrt{-1}.\sin.z)^2 = \cos.2z \pm \sqrt{-1}.\sin.2z,$$

and

$$(\cos.z \pm \sqrt{-1}.\sin.z)^3 = \cos.3z \pm \sqrt{-1}.\sin.3z;$$

and in general

$$(\cos.z \pm \sqrt{-1}.\sin.z)^n = \cos.nz \pm \sqrt{-1}.\sin.nz.$$

From these, by virtue of the double signs, we deduce

$$\cos.nz = \frac{(\cos.z + \sqrt{-1}.\sin.z)^n + (\cos.z - \sqrt{-1}.\sin.z)^n}{2}$$

and

$$\sin.nz = \frac{(\cos.z + \sqrt{-1}.\sin.z)^n - (\cos.z - \sqrt{-1}.\sin.z)^n}{2}$$

Developing these binomials into series we get

$$\cos.nz = (\cos.z)^n - \frac{n(n-1)}{1.2}(\cos.z)^{n-2}(\sin.z)^2$$

$$+ \frac{n(n-1)(n-2)(n-3)}{1.2.3.4}(\cos.z)^{n-4}(\sin.z)^4$$

$$- \frac{n(n-1)(n-2)(n-3)(n-4)(n-5)}{1.2.3.4.5.6}(\cos.z^{n-6})(\sin.z)^6 + \text{etc.}$$

and

$$\sin.nz = \frac{n}{1}(\cos.z)^{n-1}\sin.z - \frac{n(n-1)(n-2)}{1.2.3}(\cos.z^{n-3})(\sin.z)^3$$

$$+ \frac{n(n-1)(n-2)(n-3)(n-4)}{1.2.3.4.5}(\cos.z)^{n-5}(\sin.z)^5$$

$$- \text{etc.}[1]$$

---

[1] [A little later in the chapter Euler gives the formula;

$$\cos.v = \frac{e^{+v\sqrt{-1}} + e^{-v\sqrt{-1}}}{2}, \quad \sin.v = \frac{e^{+v\sqrt{-1}} - e^{-v\sqrt{-1}}}{2\sqrt{-1}};$$

and

$$e^{+v\sqrt{-1}} = \cos.v + \sqrt{-1}.\sin.v,$$
$$e^{-v\sqrt{-1}} = \cos.v - \sqrt{-1}.\sin.v.$$

Roger Cotes gave much earlier the equivalent of the formula

$$\log(\cos x + i \sin x) = i x$$

(*Philosophical Transactions*, 1714, vol. 29, 1717, p.32) under the form: Si quadrantis circuli quilibet arcus, radio CE descriptus, sinum habeat CX, sinumque complementi ad quadrantem XE: sumendo radium CE, pro Modulo, arcus erit rationis inter EX + XC$\sqrt{-1}$ & CE mensura ducta in $\sqrt{-1}$.]

F

Euler, "Recherches sur les racines imaginaires des equations", *Histoire de l'Academie Royale des Sciences et Belles Lettres*, Berlin, vol.5 (1749), 1751, p.222–288. The following extract covers §79–85, p.265–268.

§79. Problem I.—*An imaginary quantity being raised to a power which is any real number, determine the form of the imaginary which results.*

Solution. Suppose $a + b\sqrt{-1}$ is the imaginary quantity, and m the real exponent; it is required to determine M and N, such that

$$(a + b\sqrt{-1})^m = M + N\sqrt{-1}.$$

Set $\sqrt{(a^2 + b^2)} = c$; c will always be a real positive quantity since we do not here regard the ambiguity of the sign $\sqrt{\ }$. Further, let us seek the angle ϕ such that its sine is equal to $\dfrac{b}{c}$ and cosine $\dfrac{a}{c}$, here having regard to the nature of the quantities a and b if they are positive or negative. It is certain that one can always find this angle ϕ whatever the quantities a and b are, provided that they are real, as we suppose. Now having found this angle ϕ which will be always real, one will at the same time find other angles whose sine is $\dfrac{b}{c}$ and cosine $\dfrac{a}{c}$ are the same; namely, on setting π for the angle of 180°, all the angles ϕ, $2\pi + \phi$, $4\pi + \phi$, $6\pi + \phi$, $8\pi + \phi$, etc. to which one may add $-2\pi + \phi$, $-4\pi + \phi$, $-6\pi + \phi$, $-8\pi + \phi$, etc. That being said $a + b\sqrt{-1} = c(\cos\phi + \sqrt{-1}.\sin\phi)$ and raising to the proposed power

$$(a + b\sqrt{-1})^m = c^m(\cos\phi + \sqrt{-1}.\sin\phi)^m$$

where c^m will always have a real positive value. In consequence of the demonstration that

$$(\cos\phi + \sqrt{-1}.\sin\phi)^m = \cos m\phi + \sqrt{-1}.\sin m\phi,$$

where it is to be remarked that since m is a real quantity, the angle $m\phi$ will be also real and hence also its sine and cosine, we will have

$$(a + b\sqrt{-1})^m = c^m(\cos m\phi + \sqrt{-1}.\sin m\phi).$$

Or, the power $(a + b\sqrt{-1})^m$ is contained in the form

$$M + N\sqrt{-1},$$

on setting $M = c^m \cos m\phi$ and $N = c^m \sin m\phi$ where $c = \sqrt{(a^2 + b^2)}$, and $\cos\phi = \dfrac{a}{c}$ and $\sin\phi = \dfrac{b}{c}$. C.QF.T.[1]

§80. *Corollary I.*—In the same way that $(\cos\phi + \sqrt{-1}.\sin\phi)^m$ $= \cos m\phi + \sqrt{-1}.\sin m\phi$, is also $(\cos\phi - \sqrt{-1}.\sin\phi)^m = \cos m\phi$ $- \sqrt{-1}.\sin m\phi$; and hence

$$(a - b\sqrt{-1})^m = c^m (\cos m\phi - \sqrt{-1}.\sin m\phi),$$

where ϕ is the same angle as in the preceding case.

§81. *Corollary II.*—If the exponent m is negative, since $\sin -m\phi = -\sin m\phi$ and $\cos -m\phi$ - $\cos m\phi$, then

$$(\cos\phi \pm \sqrt{-1}. \sin\phi)^{-m} = \cos m\phi \mp \sqrt{-1}.\sin m\phi$$

and

$$(a \pm b\sqrt{-1})^{-m} = c^{-m}(\cos m\phi \mp \sqrt{-1}.\sin m\phi).$$

§82. *Corollary III.*—If m is an integer positive or negative the formula $(a + b\sqrt{-1})^m$ has only a single value; for whatever is substituted for ϕ of all the angles $\pm 2\pi + \phi$, $\pm 4\pi + \phi$, $\pm 6\pi + \phi$, etc., one always finds the same values for $\sin m\phi$ and $\cos m\phi$.

§83. *Corollary IV.*—But if the exponent m is a rational number $\dfrac{\mu}{\nu}$, the expression $(a + b\sqrt{-1})^{\frac{\mu}{\nu}}$ will have as many different values as there are units in ν. For, on substituting for ϕ the angle one will obtain as many different values for $\sin m\phi$ and $\cos m\phi$ as the number ν contains of units.

§84. *Corollary V.*—Whence it is clear that if m is an irrational number, or incommensurable to unity, the expression $(a + b\sqrt{-1})^m$ will have an infinite number of different values, since all the angles ϕ, $\pm 2\pi + \phi$, $\pm 4\pi + \phi$, $\pm 6\pi + \phi$, etc., will furnish different values for $\sin m\phi$ and $\cos m\phi$.

§85. *Scholium I.*—The foundation of the solution of this problem is that $(\cos\phi + \sqrt{-1}.\sin\phi)^m = \cos m\phi + \sqrt{-1} \sin m\phi$, whose truth is proved by known theorems regarding the multiplication of angles. For having two angles ϕ and θ,

$$(\cos\phi + \sqrt{-1}.\sin\phi)(\cos\theta + \sqrt{-1}.\sin\theta) = \cos (\phi + \theta)$$
$$+ \sqrt{-1}. \sin(\phi + \theta),$$

which is clear by actual multiplication which gives $\cos\phi \cos\theta - \sin\phi \sin\theta + (\cos\phi.\sin\theta + \sin\phi.\cos\theta)\sqrt{-1}$. But $\cos\phi.\cos\theta - \sin\phi.\sin\theta = \cos (\phi + \theta)$, and $\cos\phi \sin\theta + \sin\phi \cos\theta = \sin (\phi + \theta).$

---

[1] [Ce qu'il fallait trouver, which was to be found.]

Hence one may readily deduce the consequence that

$$(\cos\phi + \sqrt{-1}.\sin\phi)^m = \cos m\phi + \sqrt{-1}.\sin m\phi,$$

when the exponent m is an integer. All doubt that the same formula is also true, when m is any number, is removed by differentiation, after having taken logarithms. For, taking logarithms, there results

$$ml(\cos\phi + \sqrt{-1}.\sin\phi) = l(\cos m\phi + \sqrt{-1} \sin m\phi).$$

Treating the angle ϕ as a variable quantity we will have

$$\frac{-m\ d\phi\ \sin\phi + m\ d\phi\sqrt{-1}.\cos\phi}{\cos\phi + \sqrt{-1}.\sin\phi} =$$
$$\frac{-m\ d\phi\ \sin m\phi + m\ d\phi\sqrt{-1}.\cos m\phi}{\cos m\phi + \sqrt{-1}.\sin m\phi}.$$

On multiplying the numerators by $-\sqrt{-1}$, one obtains

$$\frac{m\ d\phi(\cos\phi + \sqrt{-1}.\sin\phi)}{\cos\phi + \sqrt{-1}.\sin\phi} = \frac{m\ d\phi(\cos m\phi + \sqrt{-1}.\sin m\phi)}{\cos m\phi + \sqrt{-1}.\sin m\phi}$$
$$= m\ d\phi,$$

which is an identical equation.[1]

---

1 [This proves that either $(\cos\phi + \sqrt{-1}.\sin\phi)^m$ and $\cos m\phi + \sqrt{-1}.\sin m\phi$ are equal for all values of ϕ or differ by a constant. For $\phi = 0$ they are equal. Hence the proof is completed.]

CLAVIUS AND PITISCUS

On Prosthaphaeresis

(Translated from the Latin by Professor Jekuthiel Ginsburg, Yeshiva College, New York City.)

In the years immediately preceding the discovery of logarithms, mathematicians made use of a method called *prosthaphaeresis* to replace the operations of multiplication and division by addition and subtraction. The method was based on the equivalent of the formula

$$\cos (A - B) - \cos (A + B) = 2 \sin A \sin B$$

Nicolaus Raymarus Ursus Dithmarsus used it in the solution of spherical triangles where it is necessary to find the fourth proportional to the sinus totus (radius), sin A, and sin B. Christopher Clavius (1537–1612) extended the method to the cases of secants and tangents; in fact he showed how to find the product of any two numbers by this method, thus in a way anticipating the theory of logarithms.

In the first fragment translated below, Clavius shows how the product of two sines may be found by the method of prostaphaeresis. This seems to be the original essence of the discovery as presented by Raymarus Ursus. It is applicable only when each of the two factors is less that the sinus totus and may therefore be considered as the sine of some arc.

In the second fragment Clavius shows how to proceed in the case when one of the numbers to be multiplied is greater than the sinus totus.

Both fragments are from his *Astrolabium* (Rome, 1593), Book I, lemma 53.

Fragment I

Let for example it be required to find the declination[1] of 17° 45′π.

Since it is true that the sinus totus is to the sine of the maximum declination[2] of the point as the sine of the distance of the given

---

[1] [The declination is the distance from a point on the celestial sphere to the equator. (The distance of course is to be measured on the arc of the meridian.)

The point 17°45′π is a point in the third sign of the zodiac. Since every sign of the zodiac is 30° and since the first sign begins at the intersection of the ecliptic with the equator, the distance of the point from that intersection equals 30° + 30° + 17°45′ = 77°45′.]

[2] [The angle formed by the equator and ecliptic.]

ecliptical point from the nearest equinoctial point to the sine of
the declination of the same point.[1]

Hence, by prosthaphaeresis,

Max. declination 23°30′	Complement of greater 12°15′
Distance from	
equinox 77°45′	Smaller Arc 23°30′

Sum of complement and smaller arc 35°45′.   The sine is 3842497
Diff. of complement and smaller arc 11°15′.   The sine is 1950903

The sum of the sines 7793400
Half of the sum   3896700   declination 22°56′[2]

Fragment II

When ratio of the sinus totus to a number less than itself is
equal to the ratio of a number greater than the sinus totus to the
required number[3] proceed as follows: The third number, which
is greater than the sinus totus, should be divided by the sinus
totus.   The quotient will be the number obtained when seven

---

[1] [Let AB be the arc of the equator, AC an arc of the ecliptic, A the nearest
point of the equinox,—that is, the intersection of the equator with the ecliptic,
—C a point in the third sign of the ecliptic whose distance
AC from the equinox is equal to 2 signs plus 17°45′, or
77°45′, CB the distance of the point C from the equator
measured on the arc of the meridian.   We see that CB
is perpendicular to AB and triangle ABC is a right spherical triangle.   We have then

$$\sin B : \sin A = \sin AC : \sin CB$$

where

$$\sin B = \sin 90° = 1 \text{ (sinus totus)}, \quad A = 23°30′$$

(this being the known angle of intersection of the ecliptic and equator).
Sin $AC = \sin 77° 45′$, and CB is the arc to be computed; hence in modern nota-

$$1 : \sin 23°30′ = \sin 77°45′ : \sin CB,$$

tion which is adopted for the purposes of prosthaphaeresis.]

[2] [Explanation.   According to a theorem proved by Clavius on p. 179,

$$1 : \sin A = \sin B : \tfrac{1}{2} [\sin \overline{(90° - A + B)} - \sin \overline{(90° - A - B)}]$$

where

$A = 77°45′$, $90° - A = 12°15′$, $B = 23°30′$; $\overline{90° - A + B} = = 23°30′ +$
12°15′ = 35°45′, $\overline{90° - A} - B = 23°30′ - 12°15′ = = 11°15′$.
Sin $\overline{(90° - A + B)}$ and sin $\overline{(90° - A - B)}$ are to be added or subtracted
according as $90° - A$ is greater or less than B.   Hence the required product
of sin 23° 30′ by sin 77°45′ = $\tfrac{1}{2}$ sin 35°45′ − $\tfrac{1}{2}$ sin 11°15′.]

[3] [*I.e.* 10,000,000: $A = B:x$ where $A < 10^7$ and $B > 10^7$.   The difficulty
here is due to the fact that B cannot be represented as a sine.]

figures are cut off on the right (ad dexterum). These seven figures
form the remainder. Then the proportion "the sinus totus is to
the smaller number as the residue is to the required number" will
be adapted to the use of prosthaphaeresis if the arcs of the smaller
number and the residue considered as sines are taken from the
table. To the fourth proportional thus found should be added
the product of the smaller number by the quotient of the above
division.[1]

PITISCUS ON PROSTHAPHAERESIS[2]

Problem I.—Given a proportion in which three terms are known.
To solve the proportion in which the first term is the radius,
while the second and third terms are
sines, avoiding multiplication and
division.[3]

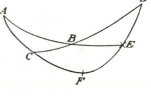

Find the sum of the complements of
the arcs corresponding to these and you
will have a right spherical triangle
which fits in the fourth case of spher-
ical triangles which is solvable by prosthaphaeresis alone.

For example, given the proportion:

"radius *AE* is to sin *EF* as sin *AB* is to sin *BC*."

---

[1] [As an illustration, Clavius considers the product of 3,912,247 by 11,917,537
or the proportion 10^7: 391,247 = 11,917,535: x, in which the third term,
11,917,535, is greater than 10^7. Dividing it by 10^7 the quotient is 1 and
the remainder 1,917,535. Clavius then considers the auxiliary proportion
10^7: 391,247 = 1,917,535:x. From the tables he finds that 391,247 = sin 23°2′,
1,917,535 = sin 11°3′. Hence the proportion becomes

$$10^7: \sin 23°2′ = \sin 11°3′:x.$$

He proceeds then as above

10^7: sin 23°2′ = sin 11°3′: ½[sin (90° − 23°2′ + 11°3′) − sin (90° − 23°2′ −
11° 3′)]

which is equivalent to

10^7: sin 23°2′ = sin 11°3′: ½[sin 78°1′ − sin 55°55′],

which gives

$$x = 749,923.$$

To this result should be added the product of the second term 391,247 by the
quotient 1.]

. . . .

[2] *Trigonometria*, 1612 ed., p. 149.

[3] [*I.e.*, solve 1: sin A = sin B:x, for x, without using either multiplication or
division.]

For the given arcs *EF*, *AB* in the second and third place find the complements *ED*, *BE* and you will have a triangle *BED* which is right-angled at *E* in which the required quantity *BC* is the complement of the side *DB*. This you will find by the fourth axiom of the spherical triangles without multiplication or division.

Let the side *AB* = 42°, and

$$EF = 48°25',$$

Then

$$BE = 48°,$$
$$DE = 41°35',$$

from which it will follow

$$DE = 41°35' = 41°35'$$
$$BE = 48°0' \text{ compl. } 42°0'$$

89°35'	83°35'	sin	9937354
0°25'		sin	72721
			10010075

[Half of this is]   5005037

which is the sine of the required arc *BC*, or 36°2'.

CLAVIUS

On Prosthaphaeresis as Applied to Trigonometry

(Translated from the Latin by Professor Jekuthiel Ginsburg, Yeshiva College, New York City.)

The following fragment is from Christopher Clavius, *Astrolabium*, (Rome, 1593), pp. 179–180. It contains the proof of the formula

$$\cos (A - B) - \cos (A + B) = 2 \sin A, \sin B.$$

Clavius considers three cases:

1) $A + B = 90°$,
2) $A + B < 90°$,
3) $A + B > 90°$.

In the first case the formula becomes

$$\sin 2A = 2 \sin A \cos A.$$

Clavius credits Nicolaus Raymarus Ursus Dithmarsus with the discovery of the theorem, but according to A. Braunmühl (*Vorlesungen über Geschichte der Trigonometrie* p. 173) the latter proved only two of the cases, namely the second and the third. In any case Clavius was the first to publish the theorem and the proof in the complete form.

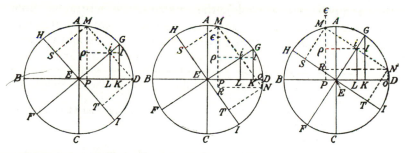

Clavius, as may be seen from his opening note, used the theorem as an introduction to the method of "prosthaphaeresis" which, in the years immediately preceding Napier's discovery of logarithms, was used by mathematicians like Raymarus Ursus, Bürgi, Clavius, and others to replace multiplication and division by the operation of addition and subtraction. The bearing of the subject upon the theory, if not the invention, of logarithms is apparent. The translation is as follows:

Lemma LIII.—Three or four years ago Nicolaus Raymarus Ursus Dithmarsus published a leaflet in which he proposed, among other things, an ingenious device by means of which he solved many spherical triangles by prosthaphaeresis[1] only. But since it is usable only when the sines are assumed in a proportion and when the sinus totus takes the first place, we will attempt here to make the doctrine more general, so that it will hold not only for sines, and [not only] when the sinus totus is in the first place, but also for tangents, secants, versed sines, and other numbers, no matter whether the sinus totus appears at the beginning or in the middle, and even when it does not appear at all. These things are entirely new and full of pleasure and satisfaction [iucunditatis ac voluptatis plena].

THEOREM.—The sinus totus is to the sine of any arc as the sine of another arbitrary arc [sinus alterius cujuspiam arcus] is to a quantity composed of these two arcs in a way required for the purpose of prosthaphaeresis. The smaller is to be added to the complement of the greater and the sine of the sum is to be taken.[2] Then

1. If the minor arc is equal to the complement of the greater (that is, when the two arcs are together equal to a quadrant), half of the computed sine will be the required fourth term of the proportion.[3]

2. If, however, the smaller arc is less than the complement of the greater (which happens when the sum of the two arcs is less than a quadrant of a circle), the smaller arc is subtracted from the complement of the greater so that we now have the difference between the same arcs that have been added before, and the sine of this difference[4] is subtracted from the sine of the arc formed

---

[1] [On page 178 Clavius defines prosthaphaeresis as a method in which only addition and subtraction are used.]

[2] [If the arcs are A and B, then the operation will be equivalent to taking sin $(90° - B + A)$.]

[3] [That is, in modern notation

$$1: \sin A = \sin B: \tfrac{1}{2} \sin (90° - B + A),$$

or

$$1: \sin [(90° - B) = \sin B: \tfrac{1}{2} \sin (90° - B + 90° - B),$$

which reduces to

$$1: \cos B = \sin B: \tfrac{1}{2} \sin 2B,$$

or to

$$\sin 2B = 2 \sin B \cos B.]$$

[4] [*I.e.*, sin $(90° - A - B)$.]

before. Half of the remainder will be the fourth proportional required.[1]

3. If, however, the smaller arc is greater than the complement of the greater (which takes place when the sum of the arcs is greater than the quadrant of a circle), the complement of the greater is subtracted from the lesser arc, so that we again have the difference between the arcs that have been previously added; the sine of this difference is to be added to the sine of the arc previously formed. Half of this sum will be the required fourth proportional.[2]

This is the rule of the above-mentioned author, which will be proven in the following way:

In the first figure we see that EG is the sinus totus. Further, EG is to GK (sine of arc GD), as Ei (sine of arc ID, or HM) to a problematical sine iL. And since the minor arc GD is equal to [itself], DG which is the complement of the greater arc ID (or if GD is the greater and ID the smaller, $ID = DI$, the complement of the greater arc GD), the fourth proportional required will be PQ, which is equal to half the sine MP of the arc MD, composed of the smaller arc DG and of GM, the complement of the greater arc HM.

In the second and third figures we also have the sinus totus EG is to GK (sine of arc GD), as Ei (the sine of the arc IN, or HM) is to the required sine iL. And because in the second figure the smaller arc GD is less than GN, the complement of the greater arc IN (or if by chance GD will be greater and IN smaller, the lesser IN will be smaller than the complement ID of the greater arc GD), the required sine PQ [which is the fourth proportional to I, sin A, sin B] will be obtained in the following way: the sine RP of the difference DN (that is, its equal ME) is to be subtracted from MP, which is the sine of the arc MD composed of the minor arc DG and the complement GM of the greater arc HM. The line

---

[1] [The proportion will then have the form

$$1 : \sin A = \sin B : \tfrac{1}{2} \, [\sin \overline{(90° - A + B)} - \sin \overline{(90° - A - B)}]$$

which is equivalent to

$$1 : \sin A = \sin B : [\tfrac{1}{2} \cos (A - B) - \cos (A + B)].$$

[2] [In modern notation. If $B > 90° - A$ (which means that $A + B > 90°$) we change the way of subtraction, taking $B - (90° - A)$ instead of $(90° - A) - B$.]

PQ will then be half of the remainder PE, just as QR will be half of the total MR.[1]

If, perchance, GD is the greater arc and IN the smaller, MP will nevertheless be the sine of the arc MB, composed of the smaller arc MH and the complement HB of the greater arc GD.

In the third figure, since the smaller arc IN is greater than ID, the complement of the greater arc GD (or, if perchance GD will be the smaller arc and IN the greater, the smaller GD will exceed the complement GN of the greater arc IN), the required fourth proportional will be obtained by adding the sine RP of the difference ND, that is, adding ME = RP to MP, the sine of the arc MB, which consists of the smaller arc (HM) and HB, the complement of the greater. The line PQ, which is half of the total line EP (since QR = ½MR), will be equal to the required line iL.

If the arc GD is, perchance, the smaller, and IN the greater, MP will nevertheless be the sine of arc MD, composed of the lesser arc GD, and of GM, the complement of the greater arc HM.

---

[1] [In modern notation:

$MP = \sin MD = \sin (DG + GM) = \sin (DG + 90° - HM) = \cos (DG - HM)$. Further, triangles $QR = \frac{1}{2}MR$, since in the similar triangles MIQ and MRN we have $MI = \frac{1}{2}MN$.]

GAUSS

On Conformal Representation

(Translated from the German by Dr. Herbert P. Evans, University of
Wisconsin, Madison, Wis.)

Carl Friedrich Gauss was born at Braunschweig April 23, 1777, and died at
Göttingen February 23, 1855. From 1795 until 1798 he was a student at
Göttingen and during the ten years immediately following this period many
of his great fundamental discoveries in pure mathematics and astronomy were
made. As a recognition of his work in astronomy he was made director of
the Göttingen observatory in 1807, and this position he held until his death.
Almost every field of pure and applied mathematics has been enriched by the
genius of Gauss, and his researches were so far reaching and fundamental that
he is considered as the greatest of German mathematicians. The present
memoir was inspired by a prize problem of the Royal Society of Sciences in
Copenhagen and is entitled: "General Solution of the Problem to so Represent
the Parts of One Given Surface upon another Given Surface that the Representation shall be Similar, in its Smallest Parts, to the Surface Represented."
This memoir was written in 1822 and won the prize offered by the Society.
It is found in volume 4, pages 192 to 216, of Gauss's collected works, published
in 1873.

A transformation whereby one surface is represented upon another with
preservation of angles is called today a *conformal*[1] transformation. The earliest
conformal transformation dates back to the Greeks, who were familiar with
the stereographic projection of a sphere upon a plane. Lagrange[2] considered
the conformal representation of surfaces of revolution upon a plane, but it
remained for Gauss, in the memoir herein translated, to solve the general
problem of the conformal representation of one surface upon another. The
memoir may be considered as the basis for the theory of conformal representation and is fundamental to the more modern theory of analytic functions of a
complex variable.

Conformal Representation

General Solution of the Problem to so represent the Parts of
one Given Surface upon another Given Surface that the representation shall be Similar, in its smallest Parts, to the Surface
represented.[3]

---

[1] The term comformal was introduced by Gauss subsquently to the present memoir.
[2] *Collected works* (V. 4, pp. 635–692).
[3] [Gauss, *Werke*, Band 4, p. 193.]

§1. The nature of a curved surface is specified by an equation between the coordinates x, y, z associated with every point on the surface. As a consequence of this equation each of these three variables can be considered as a function of the other two. It is more general to introduce two new variables t, u and to represent each of the variables x, y, z as a function of t and u. By this means definite values of t and u are, at least in general, associated with a definite point of the surface, and conversely.

§2. Let the variables X, Y, Z, T, U have the same significance in reference to a second curved surface as x, y, z, t, u have in reference to the first.

§3. To represent a first surface upon a second is to lay down a law, by which to every point of the first there corresponds a definite point of the second. This will be accomplished by equating T and U to definite functions of the two variables t and u. Insofar as the representation is to satisfy certain conditions these functions can no longer be supposed arbitrary. As thereby X, Y, Z also become functions of t and u, in addition to the conditions which are prescribed by the nature of the two surfaces, these functions must satisfy the conditions which are to be fulfilled in the representation.

§4. The problem of the Royal Society prescribes that the representation should be similar in its smallest parts to the surface represented. First this requirement must be formulated analytically. By differentiation of the functions which express x, y, z, X, Y, Z in terms of t and u there result the following equations:

$$dx = adt + a'du,$$
$$dy = bdt + b'du,$$
$$dz = cdt + c'du,$$
$$dX = Adt + A'du,$$
$$dY = Bdt + B'du,$$
$$dZ = Cdt + C'du.$$

The prescribed condition requires, first, that the lengths of all indefinitely short lines extending from a point in the second surface and contained therein shall be proportional to the lengths of the corresponding lines in the first surface, and secondly, that every angle made between these intersecting lines in the first surface shall be equal to the angle between the corresponding lines in the second surface. **A linear** element on the first surface may be written

$$\sqrt{(a^2+b^2+c^2)dt^2+2(aa'+bb'+cc')dt.du+(a'^2+b'^2+c'^2)du^2},$$

and the corresponding linear element of the second surface is

$$\sqrt{(A^2+B^2+C^2)dt^2+2(AA'+BB'+CC')dt.du+(A'^2+B'^2+C'^2)du^2}.$$

In order that these two lengths shall be in a given ratio independently of dt and du it is obvious that the three quantities

$$a^2+b^2+c^2,\ aa'+bb'+cc',\ a'^2+b'^2+c'^2$$

must be respectively proportional to the three quantities

$$A^2+B^2+C^2,\ AA'+BB'+CC',\ A'^2+B'^2+C'^2.$$

If the endpoints of a second element on the first surface correspond to the values

$$t,\ u \text{ and } t+\delta t,\ u+\delta u,$$

then the cosine of the angle which this element makes with the first element is

$$\frac{(adt+a'du)(a\delta t+a'\delta u)+(bdt+b'du)(b\delta t+b'\delta u)+(cdt+c'du)(c\delta t+c'\delta u)}{\sqrt{\{(adt+a'du)^2+(bdt+b'du)^2+(cdt+c'du)^2\}\{(a\delta t+a'\delta u)^2+(b\delta t+b'\delta u)^2+(c\delta t+c'\delta u)^2\}}}$$

The cosine of the angle between the corresponding elements on the second surface is given by a similar expression, which is obtained if only a, b, c, a', b', c' are replaced by A, B, C, A', B', C'. Obviously the two expressions become equal if the above mentioned proportionality exists, and the second condition is therefore included in the first, which also is clear in itself by a little reflection.

The analytical expression of our problem is, accordingly, that the equations

$$\frac{A^2+B^2+C^2}{a^2+b^2+c^2}=\frac{AA'+BB'+CC'}{aa'+bb'+cc'}=\frac{A'^2+B'^2+C'^2}{a'^2+b'^2+c'^2}$$

shall hold. This ratio will be a finite function of t and u which we will designate by m^2. Then m is the ratio by which linear dimensions on the first surface are increased or diminished in representing them on the second surface (accordingly as m is greater or less than unity). In general this ratio will be different at different points; in the special case for which m is a constant, corresponding finite parts will also be similar, and if moreover $m=1$ there will be complete equality and the one surface is developable upon the other.

§5. If for the sake of brevity we will put

$$(a^2 + b^2 + c^2)dt^2 + 2(aa' + bb' + cc')dt.du$$
$$+ (a'^2 + b'^2 + c'^2)du^2 = \omega,$$

it is to be noted that the differential equation $\omega = 0$ will allow two integrations. For since the trinomial ω may be broken into two factors linear in dt and du, either one factor or the other must vanish, resulting in two different integrations. One integral will correspond to the equation

$$0 = (a^2 + b^2 + c^2)dt + \{(a'^2 + b'^2 + c'^2$$
$$+ i\sqrt{(a^2 + b^2 + c^2)(a'^2 + b'^2 + c'^2) - (aa' + bb' + c')^2}\}du$$

(where i for brevity is written in place of $\sqrt{-1}$, since it is readily seen that the irrational part of the expression must be imaginary); the other integral will correspond to a quite similar equation, obtained by exchanging $-i$ with i. Consequently if the integral of the first equation is

$$p + iq = \text{Const.},$$

where p and q signify real functions of t and u, the other integral will be

$$p - iq = \text{Const.}$$

Consequently, by the nature of the case,

$$(dp + idq)(dp - idq)$$

or

$$dp^2 + dq^2$$

must be a factor of ω, or

$$\omega = n(dp^2 + dq^2),$$

where n will be a finite function of t and u.

We will now designate by Ω the trinomial into which

$$dX^2 + dY^2 + dZ^2$$

changes when for dX, dY, dZ are substituted their values in terms of T, U, dT, dU, and suppose that as in the foregoing the two integrals of the equation $\Omega = 0$ are

$$P + iQ = \text{Const.},$$
$$P - iQ = \text{Const.},$$

and

$$\Omega = N(dP^2 + dQ^2),$$

where P, Q, N are real functions of T and U. These integrations (aside from the general difficulties of integration) can obviously be carried out previous to the solution of our main problem.

If now such functions of t, u be substituted for T, U that the condition of our main problem is satisfied, then Ω may be replaced by $m^2\omega$ and we have

$$\frac{(dP + idQ)(dP - idQ)}{(dp + idq)(dp - idq)} = \frac{m^2n}{N}.$$

It is easily seen however, that the numerator in the left hand side of this equation can be divisible by the denominator only if either

$$dP + idQ \text{ is divisible by } dp + idq$$

and

$$dP - idQ \text{ is divisible by } dp - idq$$

or

$$dP + idQ \text{ is divisible by } dp - idq$$

and

$$dP - idQ \text{ is divisible by } dp + idq.$$

In the first case therefore, $dP + idQ$ will vanish if $dp + idq = 0$, or $P + iQ$ will be constant if $p + iq$ is constant, i. e., $P + iQ$ will be a function only of $p + iq$; and likewise $P - iQ$ will be a function only of $p - iq$. In the other case $P + iQ$ will be a function of $p - iq$ only and $P - iQ$ a function of $p + iq$. It is easily understood that these conclusions also hold conversely; namely that, if $P + iQ$, $P - iQ$ are assumed to be functions of $p + iq$, $p - iq$ (either respectively or reversely), the finite divisibility of Ω by ω follows and accordingly the required proportionality exists.

Moreover, it is easily seen that if, for example, we put

$$P + iQ = f(p + iq),$$

and

$$P - iQ = f'(p - iq),$$

the nature of the function f' is dependent upon that of f. That is, if the constant quantities which the latter may perhaps involve are all real, then f' must be identical with f, in order for real values of P, Q to correspond to real values of p, q. On the contrary supposition the function f' may be obtained from f by merely substituting $-i$ for i therein. Accordingly we have

$$P = \tfrac{1}{2}f(p + iq) + \tfrac{1}{2}f'(p - iq),$$
$$iQ = \tfrac{1}{2}f(p + iq) - \tfrac{1}{2}f'(p - iq),$$

or, what is the same thing, when the function f is assumed quite arbitrary (constant imaginary elements included at pleasure), P is placed equal to the real part and iQ (in the case of the second solution $-iQ$) to the imaginary part of $f(p + iq)$, then by eliminating T and U they will be expressed as functions of t and u. Thus the given problem is completely and generally solved.

§6. If $p' + iq'$ represents an arbitrary function of $p + iq$ (where p', q' are real functions of p, q), it is easily seen that also

$$p' + iq' = \text{Const. and } p' - iq' = \text{Const.}$$

represent integrals of the differential equation $\omega = 0$; in fact, these equations are quite the equivalents of

$$p + iq = \text{Const. and } p - iq = \text{Const.}$$

respectively. Similarly the integrals

$$P' + iQ' = \text{Const. and } P' - iQ' = \text{Const.}$$

of the differential equation $\Omega = 0$ will be the equivalents of

$$P + iQ = \text{Const. and } P - iQ = \text{Const.}$$

respectively, if $P' + iQ'$ represents an arbitrary function of $P + iQ$ (where P', Q' are real functions of P and Q). From this it is clear that in the general solution of our problem, which has been given in the foregoing section, $p'q'$ can take the place of p, q and P', Q' the place of P, Q, respectively. Although the solution gains no greater generality by this substitution, yet occasionally in the application, one form can be more useful than the other.

§7. If the functions which arise from the differentiation of the arbitrary functions f, f' are designated by φ and φ' respectively, so that

$$df(v) = \varphi(v)dv, \qquad df'(v) = \varphi'(v)dv,$$

then as a result of our general solution it follows that

$$\frac{dP + idQ}{dp + idq} = \varphi(p + iq), \qquad \frac{dP - idQ}{dp - idq} = \varphi'(p - iq).$$

Therefore

$$\frac{m^2 n}{N} = \varphi(p + iq) \cdot \varphi'(p - iq).$$

The ratio of magnification is consequently defined by the formula

$$m = \sqrt{\frac{dp^2 + dq^2}{\omega} \cdot \frac{\Omega}{dP^2 + dQ^2} \cdot \varphi(p + iq) \cdot \varphi'(p - iq)}.$$

§8. We will now illustrate our general solution by means of several examples,[1] whereby the kind of application, as well as the

---

[1] [Only the first example is reproduced here, sections 9–13 of the original memoir being omitted.]

nature of several details still to come in for consideration, will best be brought to light.

First consider two plane surfaces, in which case we may write

$$x = t, \quad y = u, \quad z = 0,$$
$$X = T, \quad Y = U, \quad Z = 0.$$

The differential equation

$$\omega = dt^2 + du^2 = 0$$

gives here the two integrals

$$t + iu = \text{Const.}, \qquad t - iu = \text{Const.},$$

and likewise the two integrals of the equation

$$\Omega = dT^2 + dQ^2 = 0$$

are as follows:

$$T + iU = \text{Const.}, \qquad T - iU = \text{Const.}$$

The two general solutions of the problem are accordingly

I. $T + iU = f(t + iu), \qquad T - iU = f'(t - iu).$
II. $T + iU = f(t - iu), \qquad T - iU = f'(t + iu).$

These results may also be expressed thus: If the characteristic f designates an arbitrary function, the real part of $f(x + iy)$ is to be taken for X, and the imaginary part, with omission of the factor i, for either Y or $-Y$.

If the notation φ, φ' are used in the sense of §7, and if we put

$$\varphi(x + iy) = \xi + i\eta, \qquad \varphi'(x - iy) = \xi - i\eta,$$

where obviously ξ and η are to be real functions of x and y, then in the case of the first solution we have

$$dX + idY = (\xi + i\eta)(dx + idy),$$
$$dX - idY = (\xi - i\eta)(dx - idy),$$

and consequently

$$dX = \xi dx - \eta dy,$$
$$dY = \eta dx + \xi dy.$$

Now take

$$\xi = \sigma \cos \gamma, \qquad \eta = \sigma \sin \gamma,$$
$$dx = ds \cdot \cos g, \qquad dy = ds \cdot \sin g,$$
$$dX = dS \cdot \cos G, \qquad dY = dS \cdot \sin G.$$

thereby defining ds as a linear element in the first plane making an angle g with the x-axis and dS as the corresponding linear element in the second plane making an angle G with the X-axis. From these equations there results

$$dS \cdot \cos G = \sigma ds \cos (g + \gamma),$$
$$dS \cdot \sin G = \sigma ds \sin (g + \gamma),$$

and if σ is regarded as positive (which is permissible) it follows that

$$dS = \sigma ds, \; G = g + \gamma$$

It is thus seen (in agreement with §7) that σ represents the ratio of magnification of the element ds in the representation dS, and as requisite, is independent of g; also, since γ is independent[1] of g, it follows that all linear elements extending from a point in the first plane are represented by elements in the second plane, which meet at the same angles, *in the same sense*, as do the corresponding elements in the first plane.

If f is taken to be a linear function, so that $f(v) = A + Bv$, where the constant coefficients are of the form

$$A = a + bi, \; B = c + ei$$

then

$$\varphi(v) = B = c + ei$$

and consequently[2]

$$\sigma = \sqrt{c^2 + e^2}, \; \gamma = \text{arc tan } \frac{e}{c}.$$

The ratio of magnification is therefore the same at all points and the representation is completely similar to the surface represented.[3] For every other function f (as one can easily prove) the ratio of magnification will not be constant, and the similarity will therefore occur only in the smallest parts.

If points in the second plane are prescribed which, in the representation, are to correspond with a certain number of given points in the first plane, then by the common method of interpolation we can easily find the simplest algebraic function f for which this condition is satisfied. Namely, if we designate the values of

---

[1] [That γ is independent of g follows from the fact that σ and ξ, η are independent of g.]

[2] [This follows from the fact that, in this case, $\xi = c$, $\eta = e$, and by definition $\sigma = \sqrt{\xi^2 + \eta^2}$.]

[3] [The similarity is called complete if finite parts of the two surfaces are similar.]

$x + iy$ for the given points by a, b, c, etc., and the corresponding values of $X + iY$ by A, B, C etc., then we have to make

$$f(v) = \frac{(v - b)(v - c)\ldots}{(a - b)(a - c)\ldots} \cdot A + \frac{(v - a)(v - c)\ldots}{(b - a)(b - c)\ldots} \cdot B$$
$$+ \frac{(v - a)(v - b) + \ldots}{(c - a)(c - b) + \ldots} \cdot C + \text{etc.},$$

which is an algebraic function of v whose order is a unit less than the number of given points.[1] In the case of only two points, the function becomes linear and consequently there is complete similarity.

If the second solution is carried through in the same way we find that the similarity is reversed, as all elements in the representation make the same angles with one another as do the corresponding elements in the original surface but in the reverse sense, and so that lies to the right which before lay to the left. This difference is not essential however, and disappears if we take for the under side in one plane the side before regarded as the upper side.

. . . .

§14. It remains to consider more fully one feature occurring in the general solution. We have shown in §5, that there are always just two solutions, since either $P + iQ$ must be a function of $p + iq$, and $P - iQ$ a function of $p - iq$; or $P + iQ$ must be a function of $p - iq$, and $P - iQ$ a function of $p + iq$. We shall now show that always in the case of one solution the parts in the representation are situated similarly as on the surface represented; in the other solution, on the contrary, they lie in the reverse sense; at the same time we shall specify the criterion by means of which this can be settled a priori.

First of all we observe, that of perfect or reversed similarity there can be discussion only insofar as on each of the two surfaces two sides are distinguished, one of which is considered as the upper and the other as the under. Since this in itself is somewhat arbitrary, the two solutions do not differ at all essentially, and a reversed similarity becomes perfect as soon as the side on one surface, considered as the upper side, is taken as the under side. In our solution this distinction cannot present itself, since the surfaces were defined only by the coordinates of their points. If

---

[1] [In the original memoir an application of this process to geodesy is also mentioned.]

one is concerned with this distinction, the nature of the surfaces must first be specified in another manner which includes this in itself. For this purpose we shall assume that the nature of the first surface is defined by the equation $\psi = 0$, where ψ is a given one-valued function of x, y, z. At all points of the surface the value of ψ will thus be zero, and at all points not on the surface it will have a value different than zero. By a passage through the surface, generally speaking, ψ will change from positive to negative or by opposite motion from negative to positive, *i. e.*, on one side the value of ψ will be positive and on the other negative: the first side will be considered as the upper and the other as the under. Likewise the nature of the second surface is to be similarly specified by the equation $\Psi = 0$, where Ψ is a given one-valued function of the coordinates X, Y, Z. Differentiation gives

$$d\psi = edx + gdy + hdz,$$
$$d\Psi = Edx + GdY + HdZ,$$

where e, g, h will be functions of x, y, z and E, G, H functions of X, Y, Z.

Since the considerations through which we attain our aim, although in themselves not difficult, are yet of a somewhat unusual kind, we shall take pains to give them the greatest clarity. We shall assume six intermediate representations in the plane to be inserted between the two corresponding representations on the surfaces whose equations are $\psi = 0$ and $\Psi = 0$, so that eight different representations come in for consideration, namely the surfaces:

The corresponding
points of which have
as coordinates:

1. The original in the surface $\psi = 0$................ x, y, z.
2. Representation in the plane..................... x, y, O.
3. Representation in the plane..................... t, u, O.
4. Representation in the plane..................... p, q, O.
5. Representation in the plane..................... P, Q, O.
6. Representation in the plane..................... T, U, O.
7. Representation in the plane..................... X, Y, O.
8. Representation in the surface $\Psi = 0$............ X, Y, Z.

We shall now compare these different representations solely in relation to the relative positions of their infinitesimal linear elements, disregarding entirely the ratio of their lengths; two representations will be considered as similar, if two linear elements extending from a point are such that the one which lies to the right in the one surface corresponds to the one which lies to the right in the other; in the contrary case the linear elements will be said to be reversely situated. In case of the planes 2–7 the side on which the third coordinate has a positive value will always be considered as the upper side; in case of the first and last surfaces, on the other hand, the distinction between the upper and under sides depends merely upon the positive or negative values of ψ and Ψ, as has already been agreed upon.

First of all, it is clear that at every point of the first surface where one arrives at the upper side by giving z a positive increment, for x and y unchanged, the representation in 2 will be similar to that in 1; this will obviously be the case whenever b is positive; and the contrary will occur when b is negative, in which case the representation 2 will be reversely situated with respect to 1.

In the same way the representations 7 and 8 will be situated similarly or reversely, accordingly as H is positive or negative.

In order to compare the representations in 2 and 3 let ds be the length of an infinitesimal line in the former surface, extending from the point with coordinates x, y to another with coordinates $x + dx$, $y + dy$, and let l denote the angle between this element and the positive x-axis, the angle increasing in the same sense as we pass from the x-axis to the y-axis; thus:

$$dx = ds \cdot \cos l, \qquad dy = ds \cdot \sin l.$$

In the representation in 3 let $d\sigma$ be the length of the line which corresponds to ds and let λ, in the above sense, be the angle it makes with the t-axis, so that

$$dt = d\sigma \cdot \cos \lambda, \qquad du = d\sigma \cdot \sin \lambda.$$

We have therefore, in the notation of §4,

$$ds \cdot \cos l = d\sigma \cdot (a \cos \lambda + a' \sin \lambda),$$
$$ds \cdot \sin l = d\sigma \cdot (b \cos \lambda + b' \sin \lambda),$$

and consequently

$$\tan l = \frac{b \cos \lambda + b' \sin \lambda}{a \cos \lambda + a' \sin \lambda}.$$

If x and y are now considered as fixed and l, λ as variable, it follows by differentiation that

$$\frac{dl}{d\lambda} = \frac{ab' - ba'}{(a \cos \lambda + a' \sin \lambda)^2 + (b \cos \lambda + b' \sin \lambda)^2} =$$
$$(ab' - ba')\left(\frac{d\sigma}{ds}\right)^2.$$

It is thus seen that, accordingly as $ab' - ba'$ is positive or negative, l and λ will increase simultaneously or change in the opposite sense, and therefore in the first case the representations 2 and 3 are similarly situated, while in the second case they are reversely situated.

From the combination of these results with the foregoing it follows that the representations 1 and 3 are similarly or reversely situated, accordingly as $(ab' - ba')/b$ is positive or negative.

Since the equation

$$edx + gdy + hdz = 0$$

as also

$$(ea + gb + hc)dt + (ea' + gb' + hc')du = 0,$$

must hold on the surface $\psi = 0$, irrespective of how the ratio of dt and du is chosen, we have identically

$$ea + gb + hc = 0 \text{ and } ea' + gb' + hc' = 0.$$

Wherefrom it follows that e, g, h must be respectively proportional to the quantities $bc' - cb'$, $ca' - ac'$, $ab' - ba'$, thus

$$\frac{bc' - cb'}{e} = \frac{ca' - ac'}{g} = \frac{ab' - ba'}{h}.$$

We can apply any one of these three expressions, or, on multiplication by the positive quantity $e^2 + g^2 + h^2$, the resulting symmetrical expression,

$$ebc' + gca' + hab' - ecb' - gac' - hba',$$

as a criterion for the similarity or reversal of position of the parts in the representations 1 and 3.

Likewise the similarity or reversal of parts in the representations 6 and 8 depends upon the positive or negative value of the quantity

$$\frac{BC' - CB'}{E} = \frac{CA' - AC'}{G} = \frac{AB' - BA'}{H},$$

or, if we prefer, upon the sign of the symmetrical quantity

$$EBC' + GCA' + HAB' - ECB' - GAC' - HBA'.$$

The comparison of the representations 3 and 4 is based on quite similar grounds as that of 2 and 3, and the similar or reverse situation of the parts depends upon the positive or negative sign of the quantity

$$\frac{\partial p}{\partial t} \cdot \frac{\partial q}{\partial u} - \frac{\partial p}{\partial u} \cdot \frac{\partial q}{\partial t}.$$

Likewise the positive or negative sign of

$$\frac{\partial P}{\partial T} \cdot \frac{\partial Q}{\partial u} - \frac{\partial P}{\partial U} \cdot \frac{\partial Q}{\partial T}$$

determines the similar or reverse situation of the parts in the representations 5 and 6.

Finally, to compare the representations 4 and 5 the analysis of §8 may be employed, from which it is clear that these are similar or reversed in the situation of their smallest parts, accordingly as the first or second solution is chosen, that is, whether

$$P + iQ = f(p + iq) \text{ and } P - iQ = f'(p - iq),$$

or

$$P + iQ = f(p - iq) \text{ and } P - iQ = f'(p + iq).$$

From all this we now conclude that if the representation in the surface $\Psi = 0$ is not only to be similar in its smallest parts to its image on the surface $\psi = 0$, but similar in position as well, attention must be paid to the number of the four quantities

$$\frac{ab' - ba'}{b}, \quad \frac{\partial p}{\partial t} \cdot \frac{\partial q}{\partial u} - \frac{\partial p}{\partial u} \cdot \frac{\partial q}{\partial t}, \quad \frac{\partial P}{\partial T} \cdot \frac{\partial Q}{\partial U} - \frac{\partial P}{\partial U} \frac{\partial Q}{\partial T}, \quad \frac{AB' - BA'}{H},$$

which have negative signs. If none or an even number of them have negative signs the first solution must be chosen; if one or three of them have negative signs, the second solution must be chosen. For any other choice the similarity is always reversed.

Moreover it can be shown that, if the above four quantities are designated by r, s, S, R respectively, the equations

$$\frac{r\sqrt{e^2 + g^2 + b^2}}{s} = \pm n, \quad \frac{R\sqrt{E^2 + G^2 + H^2}}{S} = \pm N$$

always hold, where n and N have the same significance as in §5; we omit the easily found proof of this theorem here, however, since this, for our purpose, is not necessary.

STEINER

On Birational Transformations between Two Spaces

(Translated from the German by Professor Arnold Emch, University of Illinois, Urbana, Ill.)

Jakob Steiner (1796–1863) was born in humble circumstances and could not write before he reached the age of fourteen. Pestalozzi (1746–1827) took him into his school at Yverdon, Switzerland, at the age of seventeen and inspired in him a love for mathematics. He went to the University of Heidelberg in 1818 and in 1834 became a professor in the University of Berlin. He was a prolific writer on geometry. In his classic *Systematische Entwickelung der Abhängikeit geometrischer Gestalten von einander* (Berlin, 1832) he established and discussed (pp. 251–270) the so-called skew projection (Schiefe Projektion) and its applications. This projection is based upon two fixed planes, (x) and (x'), and two fixed axes, l and y in space. From every point x in (x) there is, in general, one transversal through l and y which cuts (x') in a point x'. Thus to every point in (x) there corresponds a point in (x'), and conversely. To lines correspond conics, etc. By this construction there is established a general quadratic transformation between two planes, with distinct real fundamental points and lines in both planes. On page 295, Steiner indicates the quadratic transformation between two spaces, and in a footnote he makes the significant statement quoted below, thus clearly realizing the possibility of transformations of higher order, including Cremona transformations beyond the quadratic. For a further discussion see "Selected Topics in Algebraic Geometry," *Bulletin of the National Research Council* (Washington, 1928, Chap. I).

How in this manner other more complex systems of this kind may be established is easily seen. Namely, by every porism in which, for example, the relation between two points is such that, while one of the points describes a line (or a curve), the other describes a definite curve, such a system arises...

CREMONA

On the Geometric Transformations of Plane Figures

(Translated from the Italian by E. Amelotti, M.S. University of Illinois, Urbana, Ill.)

Luigi Cremona was born in Pavia Dec. 7, 1830, and died in Rome June 10, 1903. In 1860 he became professor of higher geometry in Bologna, in 1866 professor of geometry and graphical statistics at Milan, and in 1873 professor of higher mathematics and director of engineering schools in Rome.

Synthetic geometry was studied by him with great success. A memoir in 1866 on cubic surfaces secured half of the Steiner prize from Berlin. He wrote on plane curves, on surfaces, and on birational transformations of plane and solid space. A Cremona transformation is equivalent to a succession of quadratic transformations of Magnus's type. Cremona's theory of transformation of curves was extended by him to three dimensions. For further information concerning the life and works of Cremona the reader is referred to the *Periodico di Matematica per l'Insegnamento Secondario*, Ser. 1, Vol. 5–6, 1890–1891; *Supplemento*, 1901–1902, pp. 113–114; and the *History of Mathematics*, by Florian Cajori, New York, 1926 ed.

In modern algebraic geometry such properties of figures are studied as are invariant under (a) the projective transformation, (b) the Cremona transformation, or (c) the birational transformation. The first clear survey of the aggregate of Cremona transformations in the plane is contained in the article here reported. In this and in later memoirs of Cremona the fundamental properties of such transformations in plane and space are established.

The article here translated is taken from the *Giornale di Matematiche*, of Battaglini, vol. I, ser. 1 (1863), pp. 305–311.

Messrs. Magnus and Schiaparelli, the one in Tome 8 of Crelle's *Journal*, the other in a very recent volume of the memoirs of the Accademia Scientifica di Torino, were seeking for the analytic expression for the geometric transformation of a plane figure into another plane figure under the condition that to any point of one there corresponds only one point of the other, and conversely to each point of the other only one point of the first (*transformation of the first order*). And from the above cited authors it seems that one should conclude that, in the most general situation to the lines of one figure there corresponds, in the other, conics circumscribed about a fixed triangle (real or imaginary), *i. e.*, that the most general transformation of the first order is that which Schiaparelli calls *conical transformation*.

477

But it is evident that by applying to a figure a succession of conical transformations there will result from this composition a transformation which is still of the first order, even though in it, to the right lines of the given figure there would correspond in the transformed plane not conics, but curves of higher order.

. . . .

Upon the Transformation of Plane Figures

I will consider two figures, one located in a plane P, the other in a plane P', and will suppose that the second was deduced from the first by means of any law of transformation although in such manner that to each point of the first figure there correspond only one of the second, and conversely.

The geometric transformations subject to the conditions above mentioned are the only ones which I will examine in this account: and they shall be called "transformations of the first order"[1] to distinguish them from others which are determined by different conditions.

Assuming that the transformation by means of which the proposed figures are deduced, one from the other, are among those of the first order the most general, I then ask the question: What curve of a figure corresponds to right lines of the other?

Let n be the order of the curve which in the plane P' (or P) corresponds to any line whatsoever of the plane P (or P'). Since a line of the plane P is determined by two points a, b, then the two corresponding points a', b', of the plane P' are sufficient to determine the curve which corresponds to the given line. Therefore the curves of the one figure corresponding to the lines of the other form a system such that through two arbitrarily given points only one line passes through them; i.e., those curves form a geometric net of order n (II).

A curve of order n is determined by $\frac{1}{2} n(n + 3)$ conditions; therefore the curve of a figure corresponding to right lines of another are subjected to $\frac{1}{2}n(n + 3) - 2 = \frac{1}{2}(n - 1)(n + 4)$ common conditions.

Two right lines of the one figure have only one point in common, a, determined by them. The point a' corresponding to a will

---

[1] Schiaparelli: "Sulla Trasformazione Geometrica delle Figure ed in Particolare sulla Trasformazione iperbolica" (*Memorie della R. Accademia delle Scienze di Torino*, serie, 2ª, tomo XXI, Torino 1862).

belong to the two curves of order n to which the two lines correspond. And since these two curves must determine the point a', the remaining $n^2 - 1$ intersections must be common to all the curves of the geometric net above mentioned.

Let x_r be the number of r-ple (multiple points of order r) points common to these curves; since an r-ple point common to two curves is equivalent to r^2 intersections of the same, then we will have evidently:

(1) $\qquad x_1 + 4x_2 + 9x_3 + \ldots + (n-1)^2 x_{n-1} = n^2 - 1.$

The $x_1 + x_2 + x_3 + \ldots + x_{n-1}$ points common to the curves of the net constitute the $\frac{1}{2}(n-1)(n+4)$ conditions which determine it. If a curve must pass r times through a given point, that is equivalent to $\frac{1}{2}r(r+1)$ conditions;

(2) $x_1 + 3x_2 + 6x_3 + \ldots + \frac{1}{2}n(n-1)x_{n-1} = \frac{1}{2}(n-1)(n+4).$

Equations (1) and (2) are evidently the only conditions which the integral positive numbers $x_1, x_2, \ldots, x_{n-1}$ must satisfy[1] (or P) corresponds to any line whatsoever of the plane P (or P'). Since a line of the plane P is determined by two points a, b, then the two corresponding points a', b' of the plane P' are sufficient to determine the curve which corresponds to the given line. Therefore the curves of the one figure corresponding to the lines of the other form a system such that through two arbitrarily given points only one line passes through them; i. e., those curves form a geometric net of order n.[2]

A curve of order n is determined by $\frac{1}{2}n(n+3)$ conditions; therefore the curve of a figure corresponding to right lines of another are subjected to $\frac{1}{2}n(n+3) - 2 = \frac{1}{2}(n-1)(n+4)$ common conditions.

Two right lines of the one figure have only one point in common a, determined by them. The point a' corresponding to a will belong to the two curves of order n to which the two lines correspond. And since these two curves must determine the point a', the remaining $n^2 - 1$ intersections must be common to all the curves of the geometric net above mentioned.

---

[1] [Cremona then inserts a footnote explaining how it is that one does not obtain new equations when one considers the curves which in the plane P' correspond to curves of a given order μ in the plane P.]

[2] See my "Introduzione ad una teoria geometrica delle curve piane, Page 71."

Let x_r be the number of r-ple (multiple points of order r) points common to these curves; since an r-ple point common to two curves is equivalent to r^2 intersections of the same then we will have evidently:

(1) $\qquad x_1 + 4x_2 + 9x_3 + \ldots + (n - 1)^2 x_{n-1} = n^2 - 1$

The $x_1 + x_2 + x_3 + \ldots + x_{n-1}$ points common to the curves of the net constitute the $\dfrac{(n - 1)(n + 4)}{2}$ conditions which determine it. If a curve must pass r times through a given point, that is equivalent to $\dfrac{r(r + 1)}{2}$ conditions; therefore

Examples.—For $n = 2$, the equations (1) and (2) reduce to the single equation $x_1 = 3$; *i. e.*, to the lines of a figure there will correspond in the other curves of second order circumscribed about a fixed triangle.

This is the aforesaid "Conical Transformation" considered by Steiner,[1] by Magnus,[2] and by Schiaparelli.[3]

For $n = 3$, one has, from (1) and (2),

$$x_1 = 4, \; x_2 = 1;$$

i. e., to the right lines of the one figure there correspond in the other curves of third order all having a double and four simple points in common.

For $n = 4$, (1) and (2) become

$$x_1 + 4x_2 + 9x_3 = 15,$$
$$x_1 + 3x_2 + 6x_3 = 12,$$

which admit the two solutions

First: $\;\; x_1 = 3, \; x_2 = 3, \; x_3 = 0;$
Second: $x_1 = 6, \; x_2 = 0, \; x_3 = 1;$

etc.

On eliminating x_1 from the equations (1) and (2) one obtains the following:

(3) $\quad x_2 + 3x_3 + \ldots + \dfrac{(n - 1)(n - 2)}{2} x_{n-1} = \dfrac{(n - 1)(n - 2)}{2}$

from which one sees that x_{n-1} can not have other than one of these two values:

$$x_{n-1} = 1, \; x_{n-1} = 0$$

---

[1] *Systematiscbe Entwickelung,* u.s.w., Berlin, 1832, page 251. [See page 476.]
[2] *Crelle's Journal,* t. 8, page 51.
[3] *Loco citato.*

and that in the case $x_{n-1} = 1$ one necessarily has:

$$x_2 = 0, \; x_3 = 0, \ldots, \; x_{n-2} = 0$$

and by virtue of (1) $x_1 = 2(n - 1)$

I propose to prove that the transformation corresponding to these values of $x_1, x_2, \ldots, x_{n-1}$ is, for an arbitrary value of n, geometrically possible.

Let it be supposed that the two figures be located in two distinct planes P, P', in such way that to each point of the first plane there corresponds a unique point of the second, and conversely. I will imagine two directrix curves such that through an arbitrary point of space it will be possible to pass only one line to meet both, and I will consider as correspondents the points in which this line meets the planes P, P'.

Let p and q be the orders of the two directrix curves and r the number of their common points. Assuming an arbitrary point 0 of the space as the vertex of two cones, the directrices of which are the above given curves the orders of these two cones will be p, q and therefore they will have $p.q$ common generators. Included among these are the lines which unite O with the r points common to the two directrix curves, and the remaining $pq - r$ generators common to the two cones will be, consequently, the right lines that from O can be drawn to meet both the one and the other directrix curve. But the lines endowed with such property we wish reduced to only one; therefore it must be true that

$$(4) \qquad\qquad pq - r = 1$$

Furthermore to any line R situated in one of the planes P, P', there will correspond in the other a curve of order n; *i. e.*, a variable line which meets constantly the line R and the two directrix curves of order p, q must generate a warped surface of order n. One seeks therefore the order of the surface generated by a variable line which cuts three given directrices, the first of which is a line R, and the other two of order p, q have r points in common. The number of the lines which meet three given lines and a curve of order p is $2p$: this being the number of points common to the curve of order p and to the hyperboloid which has as directrices the three given lines. This amounts to saying that $2p$ is the order of a warped surface the directrices of which are two curves and a curve of order p. This surface is met by the curve of order q in $2.p.q - r$ points not situated on the curve of order p.

Therefore the order of the warped surface which has for directrices a line and curves of order p, q, having r common points, is $2pq - r$.   Therefore we must have:

(5) $$2pq - r = n$$

From the equations (4) and (5) one gets

(6) $$p.q = n - 1, \quad r = n - 2$$

Let it be supposed that the line R is in plane P, and consider the corresponding curve of order n in plane P', i. e., the intersection of this plane with the warped surface of order $2p.q - r$ previously mentioned.   The curve of which one is dealing will have:

p multiple points of order q; they are the intersections of the plane P' with the directrix curve of order p (in fact from each point of this curve it is possible to draw q lines to meet the other directrix curves and the line R, or in other words the directrix curve of order p is multiple of order q on the warped surface);

q multiple points of order p, and they are the intersections of the plane P' with the directrix curve of order q (because analogously this one is multiple of order p on the warped surface);

$p.q$ simple points of intersection of the right line common to the planes P', P, with the lines which from the points where the directrix of order p cuts the plane P, go to the points where the other directrix cuts the same plane.

These $p + q + pq$ points do not vary, as R varies, i. e., they are points common to all the curves of order n, of plane P', corresponding to the lines of plane P.   Therefore we will have:

$$x_1 = p.q, \ x_p = q, \ x_q = p.$$

and the other x's will be equal to zero; thus the equations (1) and (2) give, having regard to the first of (6):

$$p + q = n$$

And this one combined with the first of (6) gives as a result

$$p = n - 1, q = 1$$

This signifies that of the two directrices, one will be a curve of order $n - 1$ and the other a line which will have $n - 2$ points in common.   This condition can be verified by a line and a plane curve of order $n - 1$ (not situated in the same plane) provided that the latter have a multiple point of order $n - 2$, and the directrix passes through this multiple point.

Also, the directrix of order $n - 1$ can be a twisted curve; because, for example, on the surface of an hyperboloid one can describe[1] a twisted curve K of order $n - 1$ which will be met by cach of the generators of same system in $n - 2$ points (and in consequence by each generator of the other system in only one point). We can therefore assume such twisted curve and a generator D of the first system as directrices of the transformation.

In this transformation, to each point a of the plane P there corresponds one and only one point a' of the plane P', and conversely, which point a' one determines thus. The plane drawn through the point a and through the line D meets the curve K in only one point outside of the line D. This point joined to a gives a line which meets the plane P' in the required point a'.

If R is any line in plane P, the warped surface (of order n) which has as directrices the lines K, D, R, cuts the plane P' in the curve (of order n) corresponding to R. All the curves which analogously correspond to lines have in common a multiple point of order $n - 1$ and $2(n - 1)$ simple points, i. e., First, the point in which D meets the plane P'; Second, the $n - 1$ points in which the plane P' is met by the directrix K; Third, the $n - 1$ points in which the line of intersection of P, P' is met by the lines which unite the point common to the line D and the plane P with points common to the curve K and the same plane P.

In other words: The warped surface analogous to that one the directrices of which are K, D, R, all have in common: First, The directrix D (multiple of order $n - 1$, and thus equivalent to $(n - 1)^2$ common lines); Second, The curvilinear (simple) directrix K; Third, $n - 1$ generators (simple) situated in the plane P. All these curves taken together are equivalent to a curve of order $(n + 1)^2 + 2(n - 1)$. Therefore two warped surfaces (of order n) determined by two lines R, S, in the plane P, will have also in common a line; which evidently unites the point a, of intersection of R, S with the corresponding point a', common to the two curves which in the plane P' correspond to the lines R, S.

If the line R goes through the point d in which point D meets the plane P it is evident that the relative ruled surface decomposes into a cone which has its vertex at d and as directrix the curve K, and into the plane which contains the lines D, R.

If the line passes through one of the points k common to the plane P and curve K, the relative ruled surface decomposes into

---

[1] Comptes rendus de l' Académie de France, 24 Juin, 1861.

the plane which contains the point k and the line D, and into the warped surface of order $n - 2$, having directrices K, D, R.

If the line R passes through two of the points k, the relative ruled surface will decompose into two planes and into a warped surface of order $n - 2$.

And it is also very easy to see that any curve C, of order μ, given in the plane P, gives rise to a warped surface of order μn, for which D is multiple of order $\mu(n - 1)$ and K is multiple of order μ. Therefore to the curve C there will correspond in the plane P' a curve of order μn, having: First; A multiple point of order $\mu(n - 1)$ upon D; Second; $n - 1$ multiple points of order μ, upon K; Third; $n - 1$ multiple points of order μ, upon the line which is a common intersection of P, P'.

Applying to the aforesaid things the principle of duality we will obtain two figures: one composed of lines and planes passing through the point 0; the other of lines and planes passing through another point $0'$; and the two figures will have such relations to each other, that to each plane of one there will correspond only one plane of the other and conversely; and to the lines of any one of the figures there will correspond in the other a conical surface of class n, having in common x_1, x_2, ..., x_{n-1} tangent planes simple and multiple. The numbers x_1, x_2, ..., x_{n-1} will be connected by the same equations (1) and (2).

In particular then, to deduce one figure from the other we can assume as directrices a fixed line D and a developable surface K of class $n - 1$, which has $n - 2$ tangent planes passing through D. Then, given any plane π through 0 which cuts D in a point a; through this point there passes (other than the $n - 2$ planes through D) only one tangent plane which will cut π along a certain line. The plane π' determined by it and the point $0'$ is the correspondent of π.

Cutting then the two figures with two planes P and P' respectively, we will obtain in these, two figures such that to each right line of one there will correspond a single right line in the other and conversely; but to a point of the one of the two planes there will correspond in the other a curve of class n, having a certain number of fixed, simple and multiple tangent lines.

LIE

On a Class of Geometric Transformations

(Translated from the Norwegian by Professor Martin A. Nordgaard, St. Olaf
College, Northfield, Minn.)

Marius Sophus Lie (Dec. 17, 1842–Feb. 18, 1899) was the most prominent
Scandinavian mathematician of his time.   He lived for a time in France, but
at the age of thirty became professor of mathematics at Christiania (Oslo)
and from 1886 to 1898 he held a similar position at Leipzig.   Of his style of
discourse Klein has this to say:

"To fully understand the mathematical genius of Sophus Lie, one must not
turn to the books recently published by him in collaboration with Dr. Engel,
but to his earlier memoirs, written during the first years of his scientific career.
There Lie shows himself the true geometer that he is, while in his later publications, finding that he was but imperfectly understood by the mathematicians
accustomed to the analytical point of view, he adopted a very general analytical
form of treatment that is not always easy to follow."[1]

Lie's earliest writings, when his ideas, as Klein says, were still in their
"nascent" stage, possess a vividness and a happy directness of expression that
is not always noticeable in his later exposition.

It was in 1869–1870, while still a young man, that he made the remarkable
discovery of a contact transformation by which a sphere can be made to correspond to a right line.   He communicated the results of his discovery to the
Christiania Academy of Sciences in July and October, 1870, in a memoir
entitled "Over en Classe geometriske Transformationer."   The memoir is
published in the society's *Proceedings* for 1871, pp. 67–109; the translation of
which is here presented.   It is because of a general impression that the German
version was lacking in the force of the original that it was decided to present
the memoir through a direct translation from the Norwegian instead of relying
upon the one later published in Berlin.

INTRODUCTION[2]

The rapid development of geometry in the present century has
been closely related to and dependent on the philosophic views

---

[1] Felix Klein, in his lecture on Mathematics, at the Evanston Colloquium, 1893.   Macmillan, New York.

[2] The most important points of view in this memoir were communicated to
the Christiania Academy of Sciences in July and October, 1870.   Compare a
note by Mr. Klein and myself in the Berlin Academy's *Monatsbericht* for Dec.
15, 1870.

of the nature of Cartesian geometry,—views which have been set forth in their most general form by Plücker in his earlier works.

Those who have penetrated into the spirit of Plücker's works find nothing essentially new in the idea that one may employ as element in the geometry of space any curve involving three parameters. Since no one, as far as I know, has put this suggestion into effect, the reason is probably that no resultant advantages have seemed likely.

I was led to a general study of this theory by discovering that through a very remarkable representation[1] the theory of principal tangent curves can be led back to the theory of curves of curvature.

Following Plücker's plan I shall discuss the system of equations

$$F_1(x, y, z, X, Y, Z) = 0, F_2(x, y, z, X, Y, Z) = 0$$

which, in a sense that will be explained later, defines a general reciprocal relation between two spaces. If, as a special case, the two equations are linear with respect to each system of variables, we obtain a representation in which to the points of one space there correspond in the other the lines of a Plücker complex of lines. The simplest one of the class of transformations derived in this manner is the well-known Ampère transformation, which by this method appears in a new light. I am now making a special study of the method of this representation; for on this I base *a fundamental relation between the Plücker line geometry and a space geometry in which the element is the sphere,*—a very important relation, it seems to me.

While occupied with this paper I have been continually exchanging opinions and views with Plücker's pupil, Dr. Felix Klein. To him I am indebted for many of the ideas here expressed; for some of them I may not even be able to give the reference.

Let me also remark that this paper has several points of contact with my works on the imaginaries of plane geometry. The reason for my not bringing out this dependence in the present discussion is partly that this relation is to some extent fortuitous, and partly

---

[1] [Lie uses the word "afbildning" which literallym eans picturing or imaging. Since these are uncommon forms in English, we shall use the word "representation" which has been used by later students of the theory. Its graphical connotation is inadequate, however, and we shall use the forms "image" and "imaged" for the words "billede" and "afbildes," consistently used by Lie in his earliest memoirs.

The excessive use of italics is as in the original. It has been thought best to follow Lie's usage as serving to show his points of emphasis.]

that I do not wish to deviate from the customary language of mathematics.[1]

PART I

CONCERNING A NEW SPACE RECIPROCITY

§1

Reciprocity between Two Planes or between Two Spaces

1. The Poncelet-Gergonne theory of reciprocity can be derived for the field of plane geometry from the equation

$$X(a_1x + b_1y + c_1) + Y(a_2x + b_2y + c_2) + (a_3x + b_3y + c_3) = 0$$
(1)

or from the equivalent equation

$$x(a_1X + a_2Y + a_3) + y(b_1X + b_2Y + b_3) + (c_1X + c_2Y + c_3) = 0,$$

provided that (x, y) and (X, Y) are interpreted as Cartesian point coordinates for two planes.

For if we apply the expression *conjugate* to two points (x, y) and (X, Y) whose coordinate values satisfy equation (1), we may say that the points (X, Y) conjugate to a given point (x, y) form a right line which we may interpret as *corresponding* to the given point.

Since all points of a given right line have a common conjugate point in the other plane, their corresponding right lines pass through this common point.

Thus the two planes are imaged, the one on the other, by equation (1) in such a way that, mutually, to the points of one plane correspond the right lines of the other To the points of a given right line λ correspond the right lines that pass through λ's image point.

But this is exactly what constitutes the principle of the Poncelet-Gergonne theory of reciprocity.

Now consider in one plane a multilateral whose vertices are p_1, $p_2, \ldots p_n$, and in the other plane the polygon whose sides S_1, $S_2, \ldots S_n$ correspond to these points. From what has been said it follows also that the vertices S_1S_2, $S_2S_3, \ldots S_{n-1}S_n$, of the latter

---

[1] The theories set forth in this memoir have induced Mr. Klein, in a note just made public (Gesellschaft d. Wissensch. zu Göttingen, March 4, 1870), to carry Plücker's ideas one step forward; for he has demonstrated that the Plücker line geometry (or, in my representation, the corresponding sphere geometry) illustrates the metric geometry of four variables.

multilateral are image points of the sides $p_1p_2, p_2p_3, \ldots p_{n-1}p_n$ of the given polygon, and that the two polygons are thus in a reciprocal relation to one another.

By the process of limits we may pass to considering two curves c and C which correspond in such a way that the tangents of one are imaged as the points of the other. Two such curves are said to be reciprocal to one another in respect to equation (1).

2. Plücker[1] has based a generalization of this theory on the interpretation of the general equation

$$F(x, y, X, Y) = 0. \tag{2}$$

The points (X, Y) [or (x, y)] conjugate to a given point (x, y) [or (X, Y)] now form a curve C [or c] which is represented by equation (2), provided (x, y) [or (X, Y)] be taken as parameters while (X, Y) [or (x, y)] be taken as current coordinates.

Thus, by means of equation (2) the two planes are imaged, the one on the other, in such a way that to the two points in one correspond one-to-one the curves of a certain curve net in the other.

Reasoning as before, we see that to the points of a given curve c [or C] there correspond curves C [or c] which pass through the image point of the given curve.

To a polygon of curves $c(c_1, c_2, \ldots c_n)$ correspond n points P_1, $P_2, \ldots P_n$, which lie in pairs on the curves $C(P_1P_2, P_2P_3, \ldots P_{n-1}P_n)$, whose image points are vertices of the given curvilinear polygon. Here also we come at length to the consideration of curves σ and Σ in the two planes, which are so related that to the points of the one correspond the curves c[or C] that envelope the other. This reciprocal relation, however, is generally not complete, for adjoined forms appear, as a rule.

3. Plücker[2] bases the general reciprocity between two spaces on the interpretation of the general equation

$$F(x, y, z, X, Y, Z) = 0.$$

If F is linear with respect to each system of variables, the Poncelet-Gergonne reciprocity between the two spaces is obtained.

In this memoir, especially in Part One of the same, I aim to make a study of a new space reciprocity to be thought of as coordinate with the Plückerian, and defined by the equations

$$F_1(x, y, z, X, Y, Z) = 0,$$
$$F_2(x, y, z, X, Y, Z) = 0,$$

---

[1] *Analytisch-geometrische Entwickelungen.* T. I. Zweite Abth.

[2] Though I cannot give any reference, I think I am correct in ascribing this reciprocity to Plücker.

where (x, y, z) *and* (X, Y, Z) *are to be interpreted as point coordinates for the two spaces r and R.*

§2

A Space Curve Involving Three Parameters May Be Selected as Element for the Geometry of Space

4. A transformation of geometric propositions which is based on the Poncelet-Gergonne or the Plücker reciprocity may be studied from a higher point of view, as was stressed by Gergonne and Plücker. This view-point will be described here, as it applies also to our new reciprocity.

Cartesian analytic geometry translates any geometric theorem into an algebraic one and effects that the geometry of the plane becomes a physical[1] representation of the algebra of two variables and likewise that the geometry of space becomes an interpretation of the algebra of three variable quantities.

Plücker has called our attention to the fact that the Cartesian analytic geometry is encumbered by a two-fold arbitrariness.

Descartes represents a system of values for the variables x and y by a *point* in the plane; as ordinarily expressed, he has *chosen the point as element for the geometry of the plane*, whereas one could with equal validity employ for this purpose the right line or any curve whatsoever depending on two parameters. In respect to the plane we may therefore look upon the transformation based on the Poncelet-Gergonne reciprocity as consisting of changing from the point to the right line as element, and in the same sense the Plücker reciprocity of the plane consists in introducing a curve involving two parameters as element for the geometry of the plane.

Furthermore, Descartes represents a system of quantities (x, y) by *that* point in the plane whose distances from the given axes are equal to x and y; *from an infinite number of possible coordinate systems he has chosen a particular one.*

The progress made by geometry in the 19th century has been made possible largely because this two-fold arbitrariness in the Cartesian analytic geometry has been clearly recognized as such; the next step should be an effort to utilize these truths still further.

5. The new theories advanced in the following pages are based on the fact that *any space curve involving three parameters may be*

---

[1] [Lie uses the word "sanselig," affecting the senses, material. We could have translated the word with "visual," but that word often refers to graphical representation in analytic geometry.]

selected as element for the geometry of space. If we recall, for example, that the equations for a right line in space contain four essential constants, we readily see that the right lines satisfying one given condition can be employed as elements for a geometry of space which, like our conventional geometry, gives a physical representation of the algebra of three variables.

This, however, causes a certain system of lines,—a *Plücker complex of lines*—to be singled out, and for this reason it is evident that a particular representation of this kind can have only a limited applicability. However, if it is a question of a *study of space relative to a given complex of lines* it may prove very advantageous to choose the right lines of this complex as space element. In metrical geometry the infinitely distant imaginary circle and, hence, the right lines intersecting it are singled out, and *we might therefore have some reason a priori to suppose that in dealing with certain metrical problems, it would be advantageous to introduce these right lines as elements.*

It should be noted that when we as an illustration stated that it is possible to choose the right lines of a line complex as space element, then this is something different—something more particular, if you please—from the ideas that are the basis of Plücker's last work, *Neue Geometrie des Raumes, gegründet auf die Betrachtung der geraden Linie als Raum-Element.* Early in his studies Plücker had observed that it is possible to set up a representation for an algebra which comprises any number of variables by introducing as element a figure depending on the necessary number of parameters. He emphasized[1] particuarly that since the space line has four coordinates, one may, by choosing it as space element, obtain a geometry for which space has *four* dimensions.

§3

The Curve Complex. A New Geometric Interpretation of Partial Differential Equations of the First Order. The Principal Tangent Curves of a Line Complex.

6. Plücker employs the expression *line complex* to designate the assemblage of right lines which satisfy one given condition and which therefore depend on *three* undetermined parameters.

---

[1] *Geometrie des Raumes.* Art. 258. (1846.)

Analogously I shall define a *curve complex* to mean any system of space curves *c*, whose equations

$$f_1(x, y, z, a, b, c) = 0, f_2(x, y, z, a, b, c) = 0 \qquad (3)$$

contain *three essential constants.*

By differentiating (3) with respect to *x, y, z* and eliminating *a, b, c* between the two new and the original equations, we obtain a result in the form

$$f(x, y, z, dx, dy, dz) = 0 \qquad (4)$$

If we interpret *x, y, z* as parameters and *dx, dy, dz* as direction cosines, then by equation (4) every point in space will be associated with a cone, namely the assemblage of the tangents to the complex curves *c* which pass through the point in question. These cones I shall call *elementary complex cones.* I shall also use the expression *elementary complex directions* to indicate any line element (*dx, dy, dz*) belonging to a complex curve *c*. *The assemblage of the elementary complex directions corresponding to a point form the elementary complex cone associated with the point.*

To a given system (3), or, if we choose, to a given complex of curves there corresponds a definite equation *f* = 0; *but an equation f* = 0 *may, on the other hand, be derived from an infinite number of systems* (3).

For, if we choose any relation of the form

$$\psi(x, y, z, dx, dy, dz, \alpha) = 0,$$

where α denotes a constant, and represent by

$$\varphi_1(x, y, z, \alpha, \beta, \gamma) = 0, \varphi_2(X, Y, Z, \alpha, \beta, \gamma) = 0$$

the integral of the simultaneous system

$$f = 0, \psi = 0,$$

then it is clear that if we differentiate $\varphi_1 = 0$, $\varphi_2 = 0$ with respect to *x, y, z* and eliminate α, β, γ we obtain the result *f* = 0.

Every curve of this new complex

$$\varphi_1 = 0, \varphi_2 = 0$$

is enveloped by the curves *c*, inasmuch as its elements are severally complex-directions.

7. According to Monge a partial differential equation of the first order in *x, y, z* is equivalent to the following problem: To find the most general surface which at every one of its points

touches a cone associated with that point, the general equation of the cone in plane coordinates being represented by the given partial differential equation.

Lagrange and Monge have reduced this problem to the determination of a certain complex of curves, the so-called *characteristic curves*, by proving that if we unite into one surface a family of characteristic curves, each of which intersects the curve immediately preceding, an integral surface is always formed.

Note that the equation

$$f(x, y, z, dx, dy, dz) = 0,$$

determined by the characteristic curves as stated above, is of equal value with the partial differential equation itself, for both of these equations are the analytical definition of the same triple infinity of cones.

8. *A more general geometric interpretation of partial differential equations of the first order in x, y, z may be obtained by showing that the problem of finding the most general surface which at every one of its points has a three-point contact with a curve of a given curve complex finds its analytical expression in a partial differential equation of the first order; granted, that the curve in question does not lie wholly on the surface. Furthermore, if $f(x, y, z, dx, dy, dz) = 0$ is the equation determined by the characteristic curves, then will every curve complex whose equations satisfy $f = 0$ stand in the given geometric relation to the given partial differential equation.*

Consider that we have given a complex of curves c which satisfy the equation $f = 0$ and express analytically the requirement that the surface $z = F(x, y)$ have a three-point contact with a curve c at every one of its points, *without excluding the possibility of even a closer contact.* This gives us for the determination of z a partial differential equation of the second order $(\delta_2 = 0)$.[1] But every surface generated by an infinity of c's obviously satisfies the equation $\delta_2 = 0$ and therefore its general integral includes two arbitrary functions. By means of analytical considerations that are in essence very simple, though formally somewhat extensive, I wish to prove that the first order differential equation $\delta_1 = 0$, which corresponds to $f = 0$, satisfies $\delta_2 = 0$. Obviously $\delta_1 = 0$ is not, in general, contained in the general integral mentioned; consequently $\delta_1 = 0$ is a *singular* integral of $\delta_2 = 0$.

---

[1] $\delta_2 = 0$ has the form $A(rt - s^2) + Br + Cs + Dt + E = 0$. Compare a paper by Boole in Crelle's *Journal.* Vol. 61.

The equation $f(x, y, z, dx, dy, dz) = 0$ gives by differentiation,

$$f'_x dx + f'_y dy + f'_z dz + f'_{dx} d^2x + f'_{dy} d^2y + f'_{dz} d^2z = 0, \quad (6)$$

in which dx, dy, dz, d^2x, d^2y, d^2z are considered as belonging to any curve that satisfies $f = 0$. Equation (6) holds, specifically, for the characteristic curves of $\delta_1 = 0$, and if we distinguish these by a subscript, we obtain:

$$f'_{x_1} dx_1 + \ldots f'_{dx_1} d^2x_1 + \ldots = 0$$

Here I shall remark that every curve which touches any of the integral surfaces $U = 0$ of $\delta_1 = 0$, satisfies the equation

$$\frac{dU}{dx} dx + \frac{dU}{dy} dy + \frac{dU}{dz} = 0; \quad (7)$$

and furthermore that every curve which has a three-point contact with $U = 0$ also satisfies the relation:

$$\frac{d^2U}{dx^2}(dx^2) + \ldots \left(\frac{dU}{dx}\right) d^2x \ldots = 0. \quad (8)$$

From this it is seen that every characteristic curve which lies in $U = 0$ satisfies both (7) and (8).

But $U = 0$ touches at every one of its points the associated cone of the system $f = 0$, and therefore the following equations hold:

$$d'_{dx} = \rho\frac{dU}{dx}, f'_{dy} = \rho\frac{dU}{dy}, f'_{dz} = \rho\frac{dU}{dz},$$

where ρ indicates an unknown proportionality factor. Thus the subscripted equation (8) is transformed into the following:

$$\rho\left[\frac{d^2U}{dx_1{}^2}(dx_1)^2 + \ldots\right] + [f'_{dx_1} d^2x_1 + \ldots] = 0.$$

Now we know that

$$f'_{x_1} dx_1 + \ldots + f'_{dx} . d^2x_1 + \ldots = 0$$

Consequently

$$\rho\left[\frac{d^2U}{dx_1{}^2} + \ldots\right] = f'_{x_1}[dx_1 + \ldots]$$

or, by omitting the now unnecessary subscripts:

$$\rho\left[\frac{d^2U}{dx^2}dx^2 + \ldots\right] = f'_x dx + \ldots$$

Since

$$\rho\left[\frac{dU}{dx}d^2x + \frac{dU}{dy}d^2y + \frac{dU}{dz}d^2z\right] = [f'_{dx}d^2x + \ldots]$$

the following equation holds:

$$\rho\left[\frac{dU}{dx}d^2x + \frac{dU}{dy}d^2y + \frac{dU}{dz}d^2z + \frac{d^2U}{dx^2}(dx)^2 + \ldots\right]$$
$$= f'_x dx + f'_y dy + f'_z dz + f'_{dx}d^2x + f'_{dy}d^2y + f'_{dz}d^2z,$$

whose left and right-hand members vanish simultaneously.

Our exposition shows that every curve which satisfies f = 0 and which touches a characteristic curve lying on U = 0 has a three-point contact with this surface; consequently, $\delta_1 = 0$ is a singular integral of $\delta_2 = 0$.

Now we shall prove that $\delta_2 = 0$ has no other singular integral.

For, let every point on an integral surface I of $\delta_2 = 0$ have associated with it a direction, namely, the tangent of the corresponding c of three-point contact. Assuming that I is not generated by a family of c's, there will pass through every point of I two coincident curves c, both tangent to the surface at the point in question. But I is consequently touched at each of its points by the corresponding elementary complex cone; I satisfies the equation $\delta_1 = 0$.

9. Corollary. The determination of the most general surface which at each of its points has a principal tangent not lying on the surface belonging to a given line complex depends on the solution of a first-order partial differential equation whose characteristic curves are enveloped by the lines of the complex. In this case these curves appear as principal tangent curves on the integral surfaces.

We shall give an independent geometric proof for this corollary.

The partial differential equation whose characteristic curves are enveloped by the lines of a given line complex is, according to Monge's theory, the analytical expression of the following problem: To find the most general surface which at every one of its points touches the complex cone corresponding to the point. But if the tangents to a curve belong to a line complex, then is the osculation plane of the same the tangent plane of the corresponding complex cone. Thus the osculation planes of our characteristic curves are tangent planes for all integral surfaces that contain these curves. This might require a few additional words of explanation, but it would be largely a repetition of what has been said before.

Accordingly, every complex of lines determines a complex of curves which are enveloped by the lines of the line complex and which possesses this property: they are principal tangent curves on every surface generated by a system of these curves, every curve intersecting the one immediately preceding. *This complex of curves we shall call the principal tangent curves of the line complex.*

I am indebted to Mr. Klein for the statement that the congruence of right lines which Plücker calls the *singular lines* of a line complex belongs to the aforementioned complex of curves. If the given complex is formed by the tangents of a surface [or by the right lines which cut a curve], then are severally the lines of this line complex singular lines and, hence, principal tangent curves.

§4

The Equations $F_1(x, y, z, X, Y, Z) = 0$, $F_2(x, y, z, X, Y, Z) = 0$, Determine a Reciprocity Between Two Spaces.[1]

10. We shall now begin the study of the space reciprocity determined by the equations

$$F_1(x, y, z, X, Y, Z) = 0$$
$$F_2(x, y, z, X, Y, Z) = 0 \qquad (9)$$

where (x, y, z) and (X, Y, Z) are considered point coordinates in two spaces r and R.[2]

If we use the expression *conjugate* about two points the values of whose coordinates (x, y, z) and (X, Y, Z) fulfill the relations (9), we may say that the points (X, Y, Z) conjugate to a given point (x, y, z) form a curve C which is represented by (9), provided x, y, z are interpreted as parameters and X, Y, Z as current coordinates.

To the points of the space r, therefore, correspond one-to-one the curves C of a certain curve complex in R, and there is likewise in r a complex of curves c holding a similar relation to the points of R.

Thus, by equations (9) the two spaces are imaged, the one in the other, in such a way that to the points in each of the two spaces there correspond one-to-one the curves of a certain complex in the other. As a point describes a complex curve, the complex curve corresponding

---

[1] Compare this article with §1.

[2] Whatever pertains to space r we generally designate by small letters, and whatever refers to space R by capital letters.

to the point will turn[1] *about the image point of the described complex curve.*

11. We may now prove that the equations (9) determine a general reciprocity between figures in the two spaces and, specifically, between curves that are enveloped by the complex curves c and C.

When two curves of one complex have a common point (which obviously is not the case in general), their image points lie on a complex curve. Note, specifically, that two infinitely near complex curves which intersect are imaged as two points whose infinitely small connecting line is an elementary complex direction.

Consider in r a curve σ, enveloped by curves c, and all the curves C which correspond to the points of σ. According to our analysis above, two consecutive C's will intersect, and therefore their aggregate will determine an envelope curve Σ.

It is also evident that as a point moves along Σ, the corresponding c will *envelope* a curve σ^* and it can be shown that σ^* is precisely the original given curve σ.

For, consider on the one hand a curvilinear polygon formed by the complex curves c_1, c_2, c_3, ...c_n, whose vertices are c_1c_2, c_2c_3, ... $c_{n-1}c_n$, and on the other hand the image points P_1, P_2, ...P_n of the curves c. Manifestly these lie in pairs P_1P_2, P_2P_3, ...$P_{n-1}P_n$ on the complex curves C, namely, on those curves which correspond to the vertices of the given polygon. The new polygon in R and the given polygon are therefore in complete reciprocal relation to one another.

By passing to the limit we obtain in the two spaces curves that are enveloped by the complex curves c and C, and which are so reciprocally related that to the points of one correspond the complex curves which envelope the other.

Therefore a curve enveloped by complex curves is imaged in a two-fold sense as another curve likewise enveloped by complex curves. We say the latter is *reciprocal* to the given curve relative to the system of equations (9). Notice also that the elementary complex directions (dx, dy, dz), (dX, dY, dZ) arrange themselves in pairs as reciprocals, and thus that two curved lines, tangent to one another and enveloped by complex curves, are imaged in the other space as curves bearing the same relation to one another.

---

[1] The expression "turn" is in-so-far unfortunate as we, of course, mean a turning accompanied with a change of form.

12. There are other space forms between which equations (9) determine a correspondence, which, however, is not generally a complete reciprocity.

Thus, the points of a given surface f are imaged in R as a double infinity of curves C, that is, a congruence of curves whose focal surface[1] is F. Similarly, there corresponds to the points of F a congruence of curves c, whose focal surface, as we shall see later, contains f as a reducible part.

The elementary complex cones whose vertices lie in the surface f intersect the corresponding tangent planes of the surface in n right lines (n designating the order of the complex cones) and determine n elementary complex directions at every point of f. The continuous succession of these directions form a family of curves n-ply covering the plane f. The curves are one and all enveloped by complex curves c. *The geometric locus of the reciprocal curves of this family of curves, or, if we choose, the assemblage of the image points of the c's that are tangent to f, form the focal surface F.*

To prove this we recall that two infinitely near and intersecting curves C are imaged as two points whose infinitely small connecting line is an elementary complex direction. From a point p_0 on the surface f proceeds n complex directions. Hence C_0, the image curve of p_0, is intersected in n points by neighboring C's belonging to the curve congruence discussed above. The intersection points are the n points that correspond to the n complex curves c which touch the surface f at the point p_0. Thus the points of F are the image of the c's that touch f.

Since the position of f in space r is general, a curve c which touches f at some one point will in general not have any other points of contact with the surface. But all these c's form a congruence in which every c touches the focal system in N points,—N indicating the order of the elementary complex cones in R. Therefore, as was stated above, the focal system of the congruence is broken up into f and a surface φ, to which every c is tangent in $(N - 1)$ points.

Accordingly, in order that the correspondence between surfaces in r and R determined by equations (9) shall be a complete reciproc-

---

[1] In analogy with the terminology applied to congruencies of lines I shall take the focal surface of this congruence of curves to mean the geometrical locus of the intersection points of the infinitely near curves C. If we think of a curve congruence as defined by a linear partial differential equation, then its focal surface is what we ordinarily call the singular integral of the differential equation.

ity, it is necessary and sufficient that both n and N be equal to unity. *The reciprocal relation is generally incomplete, inasmuch as analogous operations on the one hand carry f into F, and on the other hand F into the sum of f and φ.*

The above observations are also valid if f, and consequently F, are surface elements; if f is infinitely small in one direction, the same holds for F.

Finally, consider a curve k not enveloped by complex curves c, together with the surface F, formed by all the C's that correspond to the points of k. The points of a C change into the curves c that pass through the image point of C. Hence, to the points of F correspond the assemblage of curves c intersecting k. *Thus there is a two-fold relation of dependence between k and F.*

The equations (9), which picture the two spaces in one another mutually, accordingly carry given space forms into new ones which hold a reciprocal relation to the given forms and therefore serve to transform geometrical theorems and problems. We shall later make important applications of this principle of transformation to a special form of equations (9).

§5

The Transformation of Partial Differential Equations

13. Legendre[1] was the first to give a general method for transforming, in the language of modern geometry, a partial differential equation in point coordinates x, y, z into a differential equation in plane coordinates t, u, v, or (we might also say) in point coordinates t, u, v for a space related reciprocally to the given space.

In a similar manner, if we introduce the curves c as element for the space r it is possible to transform a partial differential equation in x, y, z into a differential equation in the coordinates X, Y, Z of the new space element. In this we may interpret X, Y, Z as point coordinates for the space R,—an interpretation which will be prominent in our presentation.

Let there be given any partial differential equation of the first order in x, y, z, and all the surfaces ψ which represent its so-called "integral complet," bearing in mind that every other integral surface f may be represented as an envelope of a singly infinite set of ψ's.

---

[1] Compare Plücker, *Geometrie des Raumes.* (1846.) §2.

Consider, in addition, all surfaces Ψ and Φ in space R that correspond to the surfaces ψ and f. We shall prove that every F is the envelope surface of a singly infinite set of Ψ's, that accordingly the surfaces F satisfy a partial differential equation of the first order for which all Ψ's form an "integral complet."

For, if in r there be given two surfaces possessing a common surface element, they will be imaged in R as surfaces that touch one another; and surfaces possessing infinitely many surface elements in common are changed into surfaces that are tangent along a curve in the manner of the given surface.

Assuming this, let us consider an integral surface f_0 and the singly infinite set of Ψ_0's tangent to f_0 along a characteristic curve; and, finally, the corresponding surfaces F_0 and Ψ. It is clear that F_0 has contact with every Ψ along a curve and, hence, that F_0 is the enveloping surface of all the Ψ_0's.

14. Of special interest is the case there the *partial differential equation* that is transformed is precisely the one *determined by the complex curves c* (cf. §3). In this case it may be shown that the corresponding differential equation in X, Y, Z is broken up into two equations, of which one *is precisely the one that corresponds to the complex curves C.*

For, let there be given an integral surface f of the given differential equation in x, y, z, and all the elementary complex cones corresponding to the points of f. By §4, these cones determine, at every point of f, n complex directions, of which in this case two are coincident; hence the family of curves that are enveloped by the complex curves c and lie on the surface f, which we discussed in §4, is broken up into the characteristic curves of f and a curve system that covers f $(n-2)$-fold.

Thus the curve congruence in R corresponding to the points of f has a focal system which is separated into two surfaces, of which one, which we shall call Φ, is tangent to every c at two coincident points, while to the other there are $(n-2)$ points of contact. *Thus the surfaces Φ satisfy the partial differential equation which, according to the theorem in §3, is determined by the complex curves C.*

Noting that Φ is the geometric locus of the reciprocal curves of the characteristic curves of f, we see that two integral surfaces f_1 and f_2, tangent to one another along a characteristic curve k, are transformed into two surfaces Φ_1 and Φ_2 that are tangent to one another along the *reciprocal* curve of k; for k is enveloped by complex curves c.

The characteristic curves of the two partial differential equations which, according to §3, are determined by the curve complexes c and C, are reciprocal curves relative to the system of equations (9).

15. The proposition just stated gives the following general method for transforming partial differential equations of the first order.

Determine by the usual methods the equation

$$f(x, y, z, dx, dy, dz) = 0$$

which the characteristic curves of the given partial differential equation satisfy. Then select a relation of the form

$$\Psi(x, y, z, dx, dy, dz, X) = 0,$$

where X denotes a constant. Let the simultaneous system

$$f = 0, \Psi = 0,$$

be integrated in the form

$$F_1(x, y, z, X, Y, Z) = 0, F_2(x, y, z, X, Y, Z) = 0,$$

where Y and Z are the constants of integration. By differentiation and elimination we obtain a relation of the form

$$F_3(X, Y, Z, dZ, dY, dZ) = 0,$$

which we interpret to be the equation of the characteristic curves of a partial differential equation

$$F_4\left(X, Y, Z, \frac{dZ}{dX}, \frac{dZ}{dY}\right) = 0.$$

Our former discussions show that $F_4 = 0$, derived from $F_3 = 0$ by the usual processes, and the given partial differential equation are mutually dependent in such a manner that if one is integrable, so is the other.

From this we may draw general conclusions concerning the reducing to lower degree of first order partial differential equations defined by a complex of curves of a given order. Every first-order partial differential equation defined by a line complex (§3), for example, may be transformed into a partial differential equation of the second degree.[1]

We may likewise transform every partial differential equation defined by a complex of conics into a differential equation of degree 30.[2]

----

[1] This reduction depends on the fact that every line of a line congruence touches the focal system in two points.   (§4, 12.)

[2] The number 30 results from the product of 6 by $(6 - 1)$; 6 is the number of points in which the focal system of a congruence of conics has contact with each conic.

§6

Concerning the Most General Transformation Which Change Surfaces Mutually Tangent into Similarly Situated Surfaces

16. In the study of partial differential equations an important role is played by transformations expressible in the form

$$X = F_1(x, y, z, p, q), \quad Y = F_2(x, y, z, p, q),$$
$$Z = F_3(x, y, z, p, q),$$

As usual, p and q indicate the partial derivatives $\dfrac{dz}{dx}, \dfrac{dz}{dy}$; P and Q likewise stand for $\dfrac{dZ}{dX}$ and $\dfrac{dZ}{dY}$.

In the following we shall consider the case[1] where the functions F_1, F_2, and F_3 are chosen such that P and Q also depend only on x, y, z, p, q. Thus:

$$P = F_4(x, y, z, p, q); \quad Q = F_5(x, y, z, p, q).$$

Assuming that no relation between X, Y, Z, P, Q can be derived from the above five equations, we shall show that each of the quantities x, y, z, p, q are also expressible as functions of X, Y, Z, P, Q.

If we think of x, y, z and X, Y, Z as point coordinates for r and R, we may say that by a transformation of this kind there is defined a *correspondence between the surface elements of the two spaces,—in fact, the most general correspondence.* We shall show that *these transformations divide into two distinct, coordinate classes, of which one corresponds to the Plücker reciprocity, while the other corresponds to the reciprocity which I have set up in this memoir.*

Eliminating p, q, P, and Q in the five equations

$$X = F_1, \quad Y = F_2, \quad Z = F_3, \quad P = F_4, \quad Q = F_5$$

two essentially different results may come about. We shall either obtain only one equation in x, y, z, X, Y, Z, or there will be two relations obtaining among the quantities. (The existence of *three* mutually independent equations involving the point coordinates of the two spaces assumes that the transformation in question is a *point* transformation.)

But we know that the equation $F(x, y, z, X, Y, Z) = 0$ *always* defines a reciprocal correspondence between the surface elements

---

[1] Cf. Du Bois-Reymond, *Partielle Differential Gleichungen.* §§75–81.

of the two spaces.   I have likewise shown in the preceding pages that the system

$$F_1(x, y, z, X, Y, Z) = 0, \quad F_2(x, y, z, X, Y, Z) = 0$$

always determines a transformation which changes mutually tangent surfaces into similarly situated surfaces.

My statement is therefore proved.

Let me at this time call attention to a remarkable property of these transformations: they change any differential equation of the form

$$A(rt - s^2) + Br + Cs + Dt + E = 0,$$

in which A, B, C, D are dependent only on x, y, z, p, q, into an equation of the same form.   Consequently, if the given equation has a general first integral, so does the resulting equation (Cf. Boole's paper in Crelle's *Journal*, Vol. 61).

PART II

THE PLÜCKER LINE GEOMETRY MAY BE TRANSFORMED INTO A SPHERE GEOMETRY

§7

The Two Curve Complexes are Line Complexes

17. Let us assume that these equations, which image the two spaces in one another, are linear in each system of variables:

$$(10) \begin{cases} 0 = X(a_1x + b_1y + c_1z + d_1) + Y(a_2x + b_2y + c_2z + d_2) \\ \qquad + Z(a_3x + b_3y + c_3z + d_3) + (a_4 + \ldots) \\ 0 = X(\alpha_1x + \beta_1y + \gamma_1z + \delta_1) + Y(\alpha_2x + \beta_2y + \gamma_2z + \delta_2) + \\ \qquad Z(\alpha_3x + \beta_3y + \gamma_3z + \delta_3) + (\alpha_4x + \beta_4y + \gamma_4z + \delta_4). \end{cases}$$

Then clearly the points of the other space which are conjugate to a given point will form a right line.   The two curve complexes are Plücker line complexes.[1]   It follows that the equations (10) define a correspondence between r and R which possesses the following characteristic properties:

(a) *To the points in each space correspond one-to-one the lines of a line complex in the other.*

---

[1] Regarding the theory of line complexes I assume the reader's acquaintance with these two works: Plücker, *Neue Geometrie des Raumes, gegründet auf*, etc. ...(1868–69); Klein, "Zur Theorie der Complexe," *Math. Annalen*, Vol. II.

(*b*) *As a point describes a complex line, the corresponding line in the other space turns about the image point of the described line.*

(*c*) *Curves enveloped by the lines of the two complexes arrange themselves in pairs, as reciprocals, in such a way that the tangents of each one correspond to the points of the other.*

(*d*) *With a surface f in space r there is associated in a two-fold sense a surface F in R. On the one hand F is the focal surface of the line congruence of which f is the image; on the other, the points of F correspond to those tangents of f which belong to the line complex in r.*

(*e*) *On f and F all curves arrange themselves as pairs of conjugates in such a way that to the points of a curve on f [or F] corresponds in the other space a line surface which contains the conjugate curve and is tangent to F or f along the curve.*

(*f*) *To a curve on f enveloped by the lines of the line complex there corresponds conjugately a curve on F also enveloped by complex lines, and these curves are reciprocal in the sense defined in (c).*

Each of the equations (10) determine an anharmonic correspondence between points and planes in the two spaces. Consequently each of our line complexes may be defined as the aggregate of the lines of intersection of planes in anharmonic relation, or as the connecting lines of points in anharmonic relation. But according to Reye the second-degree complex thus defined is identical with a certain line system discussed by Binet. Binet was the first to look upon this system as the aggregate of the stationary axes of revolution of a material body. It has since been studied by several mathematicians, notably Chasles and Reye.

If we particularize the constants in equations (10), we either give the two complexes a special position or we particularize the complexes themselves. As to the special positions complexes assume, they may, for example, coincide; and Mr. Reye has discussed this case in his *Geometrie der Lage* (1868), Part Second, where he also gives the propositions stated in (*a*) and (*b*). As regards the particularized complexes, I shall not enter into a discussion of all the possible special cases, but will emphasize two of the most important degenerations:[1]

---

[1] Lie, "Repräsentation der Imaginaeren," in the proceedings of the Christiania Academy of Sciences (Christiania Videnskabs-Selskab) for February and August, 1869. The space representation there discussed in §§17 and 27–29 is identical with the one discussed here. In §25 I emphasize expressly the first of the two degenerations discussed here.

(1) Both complexes may be *special* and *linear*. This case gives us the well-known transformation of Ampère. We may therefore consider this transformation as based on our introducing as space element the assemblage of right lines intersecting a given line, instead of the point.

(2) One complex may degenerate into the assemblage of right lines that intersect a given conic. In that case the other complex will be a general linear complex. I may mention here that Mr. Noether (*Götting. Nachr.*, 1869) has, on occasions, given a representation of the linear complex in a point space which is identical with the one under discussion. But the conception that *every* space contains a complex whose lines are imaged as the points of the other space, which is fundamental for our purpose, is not touched upon in Mr. Noether's brief presentation.—This is the degeneration of which we shall make a study in the following article. We assume that the fundamental conic is the infinitely distant imaginary circle.

18. We have seen that the two curve complexes are line complexes if the equations of representation are linear in each system of variables. This leads us to investigate whether this sufficient condition is necessary.

If one complex is a general line complex, the elementary complex cones of the corresponding curve complex must be resolved into cones of the second degree. The proof (cf. §4, 12) of this lies in the fact that the lines of a *line* congruence touch the focal surface in *two* points. If one complex is a special line complex, then the elementary complex cones of the corresponding curve complex in the other space will resolve into plane sheaves.

Thus, if both complexes are to be line complexes the elementary complex cones of both spaces must be resolved into second and first degree cones. But if the cones of a line complex may be continually resolved, the complex is itself reducible.[1] We have therefore proved that if two line complexes are imaged in one another as described in the previous article, it follows that either both are of the second degree, or one is a special complex of the second degree and the other linear, or they are both special linear complexes. All three cases are represented by equations (10), and we shall indicate how one may know *that equations* (10) *define the most general representation of two line complexes upon one another.*

---

[1] I know of no proof for this assertion, but I have been told that it is reliable. However, the conclusions based on it are not essential for what follows.

If both complexes are of the second degree it can be shown that the surface of singularity can not be a *curved* surface.

For through each point of this surface there pass two plane sheaves whose lines are imaged in the other space as the points of one right line. From this follows that all of the lines of one sheaf correspond to one and the same point in the other space.

But the assemblage of lines which have not independent images cannot form a complex; they can only, at best, form a congruence or a number of congruences. Since, however, the assemblage of plane sheaves of rays which proceed from each of all the points of a *curved* surface of necessity forms a complex, our assertion that the surface of singularity cannot be a *curved* surface is proved.

If two complexes of the second degree are imaged upon one another—in which case none of them may be a special complex—, the surface of singularity for each will consist of planes, and consequently both line systems are of the kind first studied by Binet.

If a second-degree complex and a linear complex are imaged upon one another, one might in advance conceive of two possible cases: (1) the second degree complex might be formed by all the lines intersecting a conic,—and such a case does exist, according to the above discussion; (2) the second degree complex might consist of all the tangents to a second degree surface. Through considerations having something in common with those I shall use in §12 I have shown that this case does not exist. For if it did, I might deduce, from the fact that a linear complex can be changed into itself by a triple infinity of linear, inter-permutable transformations, that the same would hold for the second degree surface. Which, however, is not so.

§8

Reciprocity between a Linear Complex and the Assemblage of Right Lines Which Intersect the Infinitely Distant Imaginary Circle

19. In what follows we shall make a closer study of the system of equations:

$$
\left.
\begin{aligned}
-\frac{\lambda}{2B}Zz &= x - \frac{1}{2A}(X + iY) \\[2ex]
\frac{1}{2B}(X - iY)z &= y - \frac{1}{2\lambda A}\,Z,
\end{aligned}
\right| i = \sqrt{-1} \qquad (11)
$$

This is linear in respect to both systems of variables and therefore, according to §7, it determines a correspondence between two line complexes. We shall first derive the equations of these complexes in Plücker line coordinates.

Plücker gives the equations of the right line in the form

$$rz = x - \rho, \qquad sz = y - \sigma,$$

where the five quantities r, ρ, s, σ, $(r\sigma - s\rho)$ are considered line coordinates. Therefore, if we regard X, Y, Z as parameters, equations (11) represent the system of right lines whose coordinates satisfy these relations:

$$r = -\frac{\lambda}{2B}Z, \qquad \rho = \frac{1}{2A}(X + iY),$$

$$s = \frac{1}{2B}(X - iY), \qquad \sigma = \frac{1}{2\lambda A}Z.$$

These by the elimination of X, Y, Z, give as the equation of our complex

$$\lambda^2 A\sigma + Br = 0. \tag{12}$$

Thus the line complex in the space r is a linear complex. It is, furthermore, a general linear complex and contains, as we notice, the infinitely distant right line of the xy-plane.

To determine the line complex in R we replace the system (11) by the equivalent equations

$$\left(\frac{\lambda A}{2B}z - \frac{B}{2\lambda Az}\right)Z = X - \left(Ax + B\frac{y}{z}\right),$$

$$\frac{1}{i}\left(\frac{\lambda A}{2B}z + \frac{B}{2\lambda Az}\right)Z = Y - \frac{1}{i}\left(Ax - B\frac{y}{z}\right).$$

Comparing these with the equations of the right line in R,

$$RZ = X - P, \quad SZ = Y - \Sigma, \tag{13}$$

we have

$$R = \frac{\lambda A}{2B}z - \frac{B}{2\lambda Az}, \qquad P = Ax + B\frac{y}{z},$$

$$S = \frac{1}{i}\left(\frac{\lambda A}{2B}z + \frac{B}{2\lambda Az}\right), \qquad \Sigma = \frac{1}{i}\left(Ax - B\frac{y}{z}\right).$$

The equation of the line complex in R is then found to be

$$R^2 + S^2 + 1 = 0. \tag{14}$$

But by (13),

$$R = \frac{dX}{dZ}, \qquad S = \frac{dY}{dZ},$$

and consequently we may write (14) in the form

$$dX^2 + dY^2 + dZ^2 = 0. \tag{15}$$

From which we see that the line complex in R is formed by the imaginary right lines whose length equals zero, or, if we choose, by the lines which intersect the infinitely distant imaginary circle.

By equations (11) *the two spaces are imaged, the one in the other, in such a way that to the points of r there correspond in R the imaginary right lines whose length is zero, while the points of R are imaged as the lines of the linear complex* (12).

It should be noted that as a point moves along a line of this linear complex, the corresponding right line in R describes an infinitesimal sphere,—a point sphere.

20. According to the general theory of reciprocal curves developed in §4, if we know a curve whose tangents belong to one of our line complexes, it is possible to find by simple operations the image curve that is enveloped by the lines of the other complex. Lagrange made a study of the general determination of space curves whose length is equal to zero, whose tangents therefore possess the same property. He found the general equation of these curves. Therefore, according to the analysis above, *it is also possible to set up general formulas for the curves whose tangents belong to a linear complex.*

So as not to digress from our aim we shall refrain from taking up in detail the simple geometric relations that exist between the reciprocal curves of the two spaces.[1]

We must now somewhat modify our previous observations concerning the correspondence between surfaces in the two spaces, inasmuch as all the congruencies of right lines which intersect the infinitely distant circle possess a common focal curve—namely, the circle itself—, and inasmuch as the right lines of a line congruence touch the focal surface at only two points.

For let there be given a surface F in R and let f be the geometric locus of the points in r that correspond to the tangents to F of

---

[1] If the given curve of length zero has a cusp, the corresponding curve in the linear complex has a stationary tangent. In general *stationary tangents* appear as *ordinary singularities*, if curves are regarded as formed by lines, that is, as enveloped by lines of a *given* line complex.

length zero. Then, conversely, F is also the *complete* geometric locus of the image points of the right lines in the linear complex (12) which are tangent to f.

On the other hand, if we have given a surface φ in a general position in r, the instance is like the ordinary case; for then the right lines of the linear complex (12) which touch φ also envelope another surface ψ, the so-called reciprocal polar of φ relative to (12).

This system of lines is imaged in R as a surface Φ, which clearly is the focal surface for two congruences,—one being the assemblage of right lines of length zero which correspond to the points of φ, and the other, the assemblage of the lines having the same relation to the points of ψ.

The tangents of length zero of Φ consequently resolve into two systems; or, we may say, the geodetic curves of length zero of Φ form two distinct families.

In passing I wish to remark that the determination of the *curves which are enveloped by the right lines of a congruence belonging to a linear complex may be reduced, according to our general theory, to finding on the image surface F the geodetic curves whose length is zero.* For these curves are mutually reciprocal (17, f) relative to the system (11).

21. Later we shall find use one or two times for the following two propositions:

a. A surface F of the nth order, which includes the infinitely distant imaginary circle as a p-fold line, is the image of a congruence whose order and, consequently, whose class is $(n - p)$.[1]

For, an imaginary line of zero length intersects F in $(n - p)$ points of the finite space; hence there are always $(n - p)$ lines in the image congruence which pass through a given point, or which lie in a given plane in space r.

b. A curve C of the nth order which intersects the infinitely distant circle in p points is imaged in r as a line surface of order $(2n - p)$.

For, a right line of the linear complex (12) intersects this line surface in as many points, numerically, as there are common points (not infinitely distant) between the curve C and an infinitesimal sphere.

___

[1] Let me here state a proposition which is well-known to every mathematician who works with line geometry, but which is not stated explicitly anywhere, as far as I know: *For a congruence belonging to a linear complex, the order is always numerically equal to the class.*

§9

The Plücker Line Geometry May Be Transformed into a Sphere Geometry

22. In this section we shall give the basis for a *fundamental relation that exists between the Plücker line geometry and a geometry whose element is the sphere.*

For equations (11) transform the right lines of space r into the spheres of space R, and in a two-fold sense (12).

On the one hand the right lines of the complex (12) which intersect a given line l_1, and hence also its reciprocal polar l_2 relative to (12), are imaged as the points of a sphere, according to the proposition in (21, b); on the other hand the points of l_1 and l_2 are changed into the right line generatrices of this sphere.

We arrive at the relations that obtain between the line coordinates of l_1 and l_2 and the coordinate of the center X', Y', Z', and the radius H' of the image sphere, by the following analytic observations:

Let the equations of the line l_1 [or l_2] be

$$rz = x - \rho, \qquad sz = y - \sigma.$$

Also recall that the right lines of the linear complex (12) may be represented by the equations

$$-\frac{\lambda}{2B}Zz = x - \frac{1}{2A}(X + iY)$$

$$\frac{1}{2B}(X - iY)z = y - \frac{1}{2\lambda A}Z.$$

It is clear that if we eliminate x, y, z between these four equations we have the relation which expresses the condition that the right lines intersect l_1. By so doing we arrive at the following relation between the parameters X, Y, Z of these lines, or, if we choose, between the coordinates of the image points:

$$\left[Z - \left(A\sigma\lambda - \frac{Br}{\lambda}\right)\right]^2 + [X - (A\rho + Bs)]^2 + [Y - i(Bs - A\rho)]^2$$

$$= \left[A\lambda\sigma + \frac{B}{\lambda}r\right]^2. \quad (16)$$

The immediate interpretation of this equation confirms the above statements and gives, in addition, the following formulas:

$$X' = A\rho + Bs, \qquad iY' = A\rho - Bs,$$
$$Z' = \lambda A\sigma - \frac{B}{\lambda}r, \quad \pm H' = \lambda A\sigma + \frac{B}{\lambda}r, \qquad (17)$$

or the equivalent formulas:

$$\rho = \frac{1}{2A}(X' + iY'), \qquad s = \frac{1}{2B}(X' - iY'),$$

$$\sigma = \frac{1}{2\lambda A}(Z' \pm H'), \qquad r = -\frac{\lambda}{2B}(Z' \mp H'). \tag{18}$$

(We may without loss omit the primes on the *sphere coordinates* X', Y', Z', H', since, in our conception, the points of space R are spheres of radius zero.)

Formulas (17) and (18) show that a right line in r is imaged as a uniquely defined sphere in R, while to a given sphere there correspond in r two lines

$$(X, Y, Z, + H) \text{ and } (X, Y, Z, - H),$$

which are reciprocal polars relative to the linear complex

$$H = 0 = \lambda A\sigma + \frac{B}{\lambda}r. \tag{12}$$

If H is set equal to zero, formulas (17) and (18) express clearly that the right lines of the complex (12) and the point spheres of space R are of one set in a one-to-one relation.

A plane—that is, a sphere with infinitely large radius—is imaged as two right lines (l_1 and l_2) which intersect the infinitely distant right line of the xy-plane. It follows that the points of l_1 and l_2 are the images of the imaginary lines in the given plane which pass through its infinitely distant circle points.

As a particular case we note that to a plane which touches the infinitely distant imaginary circle there corresponds a line of the complex $H = 0$ parallel to the xy-plane.

23. *Two intersecting right lines l_1 and λ_1, are imaged as spheres in a position of tangency.*

For the polars of l_1 and λ_1 relative to $H = 0$ also intersect one another and consequently the spheres have two common generatrices. But second-degree surfaces whose curves of intersection consist of a conic and two right lines touch one another in three points, the double points of the curve of the section. The image spheres of l_1 and λ_1, therefore, have three points of contact of which two, imaginary and infinitely distant, in common parlance, do not enter into our discussion.

The analytic proof of our theorem follows.

The condition that the two right lines

$$r_1z = x - \rho_1, \qquad r_2z = x - \rho_2$$
$$s_1z = y - \sigma_1, \qquad s_2z = y - \sigma_2$$

intersect is expressed by the equation

$$(r_1 - r_2)(\sigma_1 - \sigma_2) - (\rho_1 - \rho_2)(s_1 - s_2) = 0.$$

This, by aid of (18), gives

$$(X_1 - X_2)^2 + (Y_1 - Y_2)^2 + (Z_1 - Z_2)^2 + (iH_1 - iH_2)^2 = 0,$$

which proves our proposition.

Our theorem shows that the assemblage of right lines which intersect a given line is imaged as the totality of all the spheres which touch a given sphere. Consequently *the image of the special linear complex is known.*

Conversely, to two spheres that are tangent to one another there correspond two pairs of lines so situated that every line in one pair intersects a line in the other.

24. *The representation*[1] *of the general linear complex.* The general linear complex is represented by the equation

$$(r\sigma - \rho s) + mr + n\sigma + p\rho + qs + t = 0. \tag{19}$$

Equations (18) and (19) give us, as the equation of the corresponding "linear complex of spheres"

$$(X^2 + Y^2 + Z^2 - H^2) + MX + NY + PZ + QH + T = 0.[2]$$

In this equation M, N, P, Q, T denote constants that depend upon m, n, p, q, t, *while* X, Y, Z, H *are understood to be (nonhomogeneous) sphere coordinates.*

It is easy to see that the last equation determines all the spheres that intersect at a constant the image sphere of the linear congruence common to the complexes (19) and $H = 0$.

If the simultaneous invariant of these complexes is equal to zero, or if, to use Klein's expression, the two complexes are in involution, then the constant angle is a right angle.

To spheres that intersect a given sphere at a constant angle there correspond in r the right lines of two linear complexes which are reciprocal polars relative to $H = 0$.

We note particularly that the spheres which intersect a given sphere orthogonally are imaged as the right lines of a linear complex in involution with $H = 0$.

---

[1] [Lie uses the word "afbildning," meaning, literally, picture or image.]

[2] This equation may be put in the form

$$(X - X_0)^2 + (Y - Y_0)^2 + (Z - Z_0)^2 + (iH - iH_0)^2 = C^2,$$

where X_0, Y_0, Z_0, H_0, C_0 are understood to be non-homogeneous coordinates of the linear complex. Mr. Klein has called to my attention the fact that the sphere (X_0, Y_0, Z_0, H_0) is the image of the axis of this linear complex.

Let there be given a linear complex whose equation is of the form

$$ar + bs + c\rho + d\sigma + e = 0. \tag{20}$$

The corresponding relation between X, Y, Z, H is also linear, and hence the linear sphere complex is formed by all the *spheres which intersect a given plane at a given constant angle.*

This might also have been deduced from the fact that the complex (20) contains the infinitely distant right line of the xy-plane, and that therefore the congruence common to it and $H = 0$ possesses directrices that intersect this line.

If the complexes (20) and $H = 0$ are in involution, then the lines of (20) are imaged as the totality of spheres that intersect a given plane orthogonally, or, what amounts to the same thing, as the spheres whose centers lie in a given plane.

The four complexes

$$X = 0 = A\rho + Bs, \qquad Z = 0 = \lambda A\sigma - \frac{B}{\lambda}r,$$

$$iY = 0 = A\rho - Bs, \qquad H = 0 = \lambda A\sigma + \frac{B}{\lambda}r,$$

are obviously in involution by pairs. They also contain as a common line the infinitely distant line of the xy-plane.

Thus, the special linear complex (Constant = 0), formed by all the lines parallel to the xy-plane, in conjunction with the four general linear complexes $X = 0$, $Y = 0$, $Z = 0$, $H = 0$, forms a system which we may regard as a degeneration of Mr. Klein's six fundamental complexes. In analogy with our use of X, Y, Z, H as non-homogeneous coordinates for a geometry of four dimensions, with the sphere as element introduced above, we may also use these quantities as non-homogeneous line coordinates.

It is interesting to notice that the linear complexes whose equation is

$$H = \lambda A\sigma + \frac{B}{\lambda}r = \text{constant,}$$

and which, according to the form of the equation, are tangent to one another in a special linear congruence, the directrices of which unite in the infinitely distant line of the xy-plane, are imaged as a family of sphere complexes characterized by the property that all their spheres have equal radii.

25. *Various Representations.* A surface f and all its tangent lines at a given point are imaged as a surface F and all the spheres that are tangent to it at a given point.

A line on f is imaged as a sphere which is tangent to F along a curve

If f is a line surface, then F is a sphere envelope,—a tubular surface.

If, particularly, f is a second degree surface and, hence, contains two systems of right line generatrices, then we may interpret F as a sphere envelope in two ways. It is clear that in this manner we *obtain the most general surface possessing this property (the cyclide)*.

A developable surface changes into the envelope surface of a family of spheres in which two consecutive spheres are tangent to one another throughout,—that is, into an imaginary line surface whose generatrices intersect the infinitely distant imaginary circle. These line surfaces, we know, are precisely the ones characterized by Monge as possessing only one system of curves of curvature.

26. An immediate consequence of Plücker's conception is that if $l_1 = 0$ and $l_2 = 0$ are the equations of two linear complexes, then the equation $l_1 + \mu l_2 = 0$, where μ is a parameter, represents a family of linear complexes that include a common linear congruence. The principle of representation which we employ transforms this theorem into the following:

The spheres K which intersect two given spheres S_1 and S_2 at given angles V_1 and V_2 hold the same relation to infinitely many spheres S. Corresponding to the two directrices of the line congruence are two spheres S, to which all the spheres K are tangent.

The variable line complex $l_1 + \mu l_2 = 0$ intersects the complex $H = 0$ in a linear congruence whose directrices describe a second-degree surface, namely, the section of the three complexes $l_1 = 0$, $l_2 = 0$, $H = 0$. Consequently the spheres S envelope a cyclide. In this instance the cyclide degenerates into a circle along which the different spheres S intersect.

We wish to call attention to the fact that our sphere representation enables us to derive from intersecting discontinuous groups of lines corresponding groups of spheres, and conversely. As an instance, we may apply the well-known theory concerning the twenty-seven right lines of a third-degree surface to prove the existence of groups of twenty-seven spheres, of which each one is tangent to ten of the others.

Conversely, piles of spheres present lines of a linear complex arranged in peculiar, discontinuous arrays.

§10

Transforming Problems Concerning Spheres into Problems of Lines

27. In this section we shall solve a few simple and familiar problems concerning spheres by considering the corresponding line problems that result from our principle of transformation.

Problem I.—*How many spheres are tangent to four given spheres?*

The four spheres are transformed into four pairs of lines (l_1, λ_1), (l_2, λ_2), (l_3, λ_3), (l_4, λ_4). The corresponding problem of lines is, therefore, to find the lines that intersect four lines selected from the eight in such a way that each pair furnishes one line.

Lines l and λ may be arranged in sixteen different groups of four, in such a way that each group contains only one line from each pair; thus:

$$l_1 l_2 l_3 l_4, \ \lambda_1 \lambda_2 \lambda_3 \lambda_4$$
$$l_1 l_2 l_3 \lambda_4, \ \lambda_1 \lambda_2 \lambda_3 l_4$$

$$\cdots \cdots \cdots$$

$$\cdots \cdots \cdots$$

But these sixteen groups are also formed in pairs by lines that are reciprocal polars in respect to $H = 0$. Consequently the pairs of transversals (t_1, t_2) (τ_1, τ_2) of two related groups are also one another's polars in respect to $H = 0$. The last-mentioned four lines are therefore imaged as *two* spheres, and consequently there exist sixteen spheres arranged in eight pairs, which are tangent to four given spheres.

Problem II.—*How many spheres intersect four given spheres at four given angles?*

The spheres which intersect a given sphere at a fixed angle are imaged as those right lines of two linear complexes which are mutually reciprocal polars in respect to $H = 0$. We must therefore observe four pairs of complexes, (l_1, λ_1), (l_2, λ_2), (l_3, λ_3), (l_4, λ_4), and the problem now is, to find those lines which belong to four of these complexes and are selected in such a way that one is taken from each pair.

Four linear complexes have two common lines. Therefore, if we follow the same procedure as was used in the preceding problem, we shall obtain as the solution sixteen spheres arranged in eight pairs.

Our problem is simplified if one or more of the given angles are right angles; for then the spheres orthogonal to a given sphere are imaged as the lines of *one* complex, which is in involution with $H = 0$ (Article 24). If all angles are right angles, the question is, how many lines are common to four linear complexes in involution with $H = 0$. Two such lines are mutually reciprocal polars in respect to $H = 0$, and *consequently there is only one sphere which intersects four given spheres orthogonally.*

Problem III.—To construct the spheres which intersect five given spheres at a fixed angle.

Our principle of transformation changes this problem into the following: To find the linear complexes which contain one line from each of five given pairs $(l_1, \lambda_1) \ldots (l_5, \lambda_5)$.

These ten lines may be arranged in thirty-two different groups of five in a way such that every group contains *one* line of each, thus:

$$(l_1 l_2 l_3 l_4 l_5), \quad (\lambda_1 \lambda_2 \lambda_3 \lambda_4 \lambda_5)$$

. . . .

. . . .

Note that these groups are mutually reciprocal polars by pairs in respect to $H = 0$. Every group gives a line complex and in all we obtain thirty-two linear complexes conjugate in pairs. These are imaged as sixteen linear sphere complexes. The sixteen spheres which are severally intersected at a constant angle by the spheres of these systems are the solutions of our problem.

Two groups of lines, as $(l_1, l_2, \lambda_3, \lambda_4, l_5)$ and $(\lambda_1, l_2, \lambda_3, \lambda_4, l_5)$ contain four common lines. It follows that the two corresponding linear complexes intersect in a linear congruence whose directrices d_1 and d_2 are the transversals of these four lines.

But the complex $H = 0$ intersects this congruence along a second degree surface which is the image of a circle, namely, the section circle of two of the spheres wanted, as also of the image spheres of d_1 and d_2. The latter spheres may be defined by saying they are tangent to four out of five given spheres; hence, by the aid of the construction just described, we may determine a number of circles on any of the spheres wanted.

On each of the sixteen spheres which intersect five given spheres at a constant angle we may construct five circles, provided we can construct the spheres that are tangent to four given spheres.

§11

The Relation between the Theory of Curves of Curvature and the Theory of Principal Tangent Curves

28. The transformation discussed in the previous sections acquires a peculiar interest on account of the following, in my opinion, very important theorem:

To the curves of curvature of a given surface F in space R there correspond in space r line surfaces which touch the imaged surface f along principal tangent curves.

The tangents of the surface f are transformed into spheres that touch F, and the thought lies near that *to the principal tangents of f there correspond the principal spheres of F.* This also proves to be the case.

For f is cut by a principal tangent in three coincident points, and this shows that three consecutive generatrices of the image sphere of the principal tangent touch F. But such a sphere cuts F along a curve which has a cusp in the contact point of the two, and this is precisely a characteristic of principal spheres.

Note, furthermore, that the direction of this cusp is tangent to a curve of curvature. It is then seen that two consecutive points of a principal tangent curve on f are imaged as two lines which touch F at consecutive points of the same curve of curvature. Therefore, *to the principal tangent curves of f, considered as formed by points, there correspond imaginary line surfaces that touch F along curves of curvature.*

But curves on f and F arrange themselves in pairs of conjugate curves in such a way (Article 17, *e*) that the points of one form the image of lines that touch the other surface at points of the conjugate curve. *This proves our theorem.*

The following two illustrations may be regarded as verifications of this proposition.

A sphere in R is the image of a linear congruence, of which the two directrices are to be considered the focal surface. We know that every curve on a sphere is a curve of curvature. Moreover, the directrices appear as principal tangent curves on every line surface belonging to a linear congruence.—An hyperboloid f in space r presents in R a surface which in two ways may be regarded as a sphere envelope. But the line surfaces in the complex $H = 0$ which touch f along its principal tangent curves, that is, along its right line generatrices, are themselves surfaces of the second degree. Consequently *the curves of curvature of the cyclide F are circles.*

An interesting corollary resulting from our theorem is the following:

Kummer's surface of order and class four has algebraic principal tangent curves of order sixteen, and these form the complete contact section between this surface and line surfaces of order eight.

For, Kummer's surface is the focal surface of the general line congruence of order and class two which is imaged (provided it belongs to $H = 0$) as a fourth degree surface containing the infinitely distant circle twice (Article 21, *b*).

Now, Darboux and Moutard[1] have demonstrated that the lines of curvature of the last-mentioned surface are curves of order eight, cutting the infinitely distant imaginary circle in eight points. Hence, these lines are imaged as line surfaces of order eight (Article 21, *b*).

If we recall that the generatrices of these line surfaces are double tangents to the Kummer surface, we shall perceive the correctness of the proposition.[2]

It is clear that the degenerations of the Kummer surface, as, for example, *the wave surface, the Plücker complex plane, the Steiner surface of order four and class three,*[3] *a line surface of the fourth degree, the line surface of the third degree,* also have algebraic principal tangent curves.

29. Mr. Darboux has proved that we can generally determine a line of curvature in finite space on any surface, the curve of contact with the imaginary developable, which simultaneously is circumscribed about the given surface and the infinitely distant imaginary circle.

In consequence of which we can generally point out one principal tangent curve on the focal plane of a congruence of a linear complex, this curve being the geometric locus of points for which the tangent plane is also the plane associated with the linear complex.

For the infinitely small spheres which are tangent to F consist of the points of F and of the above-described imaginary developable. Consequently the right lines of the complex $H = 0$ that are tangent to the image surface f divide into two systems,—one, a system of double tangents, and the other, the assemblage of lines which are tangent to f in the points of a certain curve. But this curve, being the image of an imaginary line surface which touches F along a curve of curvature, is one of the principal tangent curves of f.

This determination of a principal tangent curve becomes illusory, however, if the focal plane—or, more correctly, a reducible part of it—, and not the congruence, is given arbitrarily. For on

---

[1] *Comptes rendus* (1864).

[2] Klein and Lie, in *Berliner Monatsbericht*, Dec. 15, 1870.

[3] Clebsch has determined the principal tangent curves of the Steiner surface.

a surface there is ordinarily only a finite number of points at which the tangent plane is also the plane associated, through a given linear complex, with that point.

It is of interest to note that a line surface whose genetrices belong to a linear complex contains a singly infinite set of points for each of which the tangent plane is also the plane associated, through the linear complex, with that point. The assemblage of these points forms a principal tangent curve, determinable by simple operations,— differentiation and elimination.

Now, Mr. Clebsch has demonstrated that if one principal tangent curve is known on a line surface, the others may be found by quadrature.

The determination of the principal tangent curves on a line surface belonging to a linear complex depends only on quadrature.

Applying our principle of transformation to the statement quoted from Clebsch as well as to its corollary proposition we arrive at the following theorems:

If on a tubular surface (sphere envelope) a non-circular curve of curvature is known, the others may be found by quadrature.

A singly infinite set of spheres which intersect a given sphere S at a constant angle envelope a tubular surface on which one curve of curvature can be given and the others obtained by quadrature.

That a curve of curvature can be found on the tubular surface is apparent also from the fact that the tubular surface intersects S at a constant angle. This curve of intersection must be one of the curves of curvature of the tubular surface, according to a certain well-known proposition. This proposition states: If two surfaces intersect at a constant angle, and the intersection curve is a line of curvature on one surface it is also such a line on the other. But on a sphere every curve is a line of curvature.

§12

The Correspondence between the Transformations of the Two Spaces

30. We may, as stated in Article 16, express our transformations by means of five equations which in the two groups (x, y, z, p, q) (X, Y, Z, P, Q) determine any quantity in one as a function of quantities in the other. If one of the two spaces, r, for example, undergoes a transformation, in which surfaces that are tangent are changed into similar surfaces, the corresponding transformation

of the other space will possess the same property. For, the transformation of r may be expressed by five equations in x_1, y_1, z_1, p_1, q_1, and x_2, y_2, z_2, p_2, q_2,—the subscripts 1 and 2 refer to the two states of r—and these relations are changed by the aid of the representation equation in x, y, z, p, q and X, Y, Z, P, Q, into relations in X_1, Y_1, Z_1, P_1, Q_1 and X_2, Y_2, Z_2, P_2, Q_2. Ths proves our assertion.

If we limit ourselves to linear transformations of r, we find among the corresponding transformations of R the following: *all movements (translational, rotational, and helicoidal), the transformation of similarity, the transformation by reciprocal radii, the parallel transformation*[1] (*transferring from one surface to a parallel surface*), *a reciprocal transformation studied by Mr. Bonnet.*[2] All of these, since they correspond to linear transformations in r, possess the property that they change curves of curvature into curves of curvature. We shall now prove *that to the general linear transformation of r there corresponds the most general transformation of R in which lines of curvature are covariant curves.*

31. In the first place, consider the linear point transformations of r to which correspond linear point transformations of R. It is clear that here we meet only with those transformations of R in which the infinitely distant imaginary circle remains unchanged; but these we do obtain.

For such a linear point transformation of R carries, on the one hand, right lines intersecting the circle into similar right lines, and, on the other hand, spheres into spheres. Thus the corresponding transformation of r is at one and the same time both a point and a line transformation,—that is, a linear point transformation. Which was to be proved.

The general linear transformation of R which does not displace the infinitely distant circle includes seven constants; and it can be built up by translations and rotations in conjunction with the similarity transformation. The corresponding transformation of r, which obviously also involves seven constants, may be characterized by saying that it carries a linear complex $H = 0$ and a certain one of its lines (the infinitely distant line of the xy-plane) into itself. We could also define this transformation by saying that it carries a special linear congruence into itself.

---

[1] Bonnet's "dilation."
[2] *Comptes rendus.* Several times in the 1850's.

The linear point transformation of r corresponding to a translation of R may be determined analytically. A translation is expressed by these equations:

$$X_1 = X_2 + A; \quad Y_1 = Y_2 + B; \quad Z_1 = Z_2 + C; \quad H_1 = H_2.$$

These equations and formulas (17) give the relations

$$r_1 = r_2 + a; \quad s_1 = s_2 + b; \quad \rho_1 = \rho_2 + c; \quad \sigma_1 = \sigma_2 + d.$$

Substituting these expressions in the equations of a right line,

$$r_1 z_1 = x_1 - \rho_1, \quad s_1 z_1 = y_1 - \sigma_1,$$

we obtain, as the definition of the required transformation of r, the following:

$$z_1 = z_2; \quad x_1 = x_2 + a z_2 + c; \quad y_1 = y_2 + b z_2 + d.$$

It is likewise an easy matter to determine analytically the transformation of r corresponding to a *similarity transformation* of R. For, by applying (17), the equations

$$X_1 = m X_2; \quad Y_1 = m Y_2; \quad Z_1 = m Z_2; \quad H_1 = m H_2$$

give the relations

$$r_1 = m r_2; \quad \rho_1 = m \rho_2; \quad s_1 = m s_2; \quad \sigma_1 = m \sigma_2.$$

These relations define a linear transformation of r which may also be expressed by the equations

$$z_1 = z_2; \quad x_1 = m x_2; \quad y_1 = m y_2.$$

But these last equations define a linear point transformation which may be characterized by saying that in it *the points of two right lines remain stationary.*

By geometric considerations we shall show that rotations of R are also changed into transformations of the kind just described. Let A be the axis of rotation and M and N the two points of the imaginary circle that are not displaced by the rotation. It is clear that all the imaginary lines which intersect A and pass through M and N keep their position during the rotation. It follows that the same obtains for the image points of these lines, which form two right lines parallel to the xy-plane.

32. Transformation of the space R by reciprocal radii carries points into points, spheres into spheres and, finally, right lines of length zero into similar lines. The corresponding transformation of r is therefore a *linear point* transformation which carries the complex $H = 0$ into itself. If we note further that in the

transformation by reciprocal radii the points and right line generatrices of a certain sphere keep their position, it is clear that the corresponding reciprocal point transformation will not displace the points of the two right lines.

Mr. Klein[1] has called attention to the fact that the transformation just mentioned may be thought of as consisting of two transformations relative to two linear complexes in involution. In this case, $H = 0$ is one complex; the other is the one that corresponds to the assemblage of spheres which intersect orthogonally the fundamental sphere of the given reciprocal radii transformation.

From which it is clear that to a surface F which is carried into itself by a reciprocal radii transformation, there corresponds in space r a congruence belonging to $H = 0$, which is its own reciprocal polar in respect to a linear complex in involution with $H = 0$. The focal surface (f) of the congruence in question is thus its own reciprocal polar in respect to both the linear complexes. Consequently the totality of the double tangents of f is generally broken up into three congruences, two of which belong to $H = 0$ and to the complex in involution with $H = 0$.

33. Now consider, on the one hand, all line transformations of r by which right lines that intersect one another are changed into similar lines[2] and, on the other, the corresponding transformations of R which possess the property that they change spheres into spheres and spheres that are tangent into similarly places spheres.

This line transformation changes the assemblage of tangents to a surface f_1 into the totality of tangents to another surface f_2. Particularly, the principal tangents of f_1 change into the principal tangents of f_2,—this irrespective of whether the line transformation considered is a point transformation or a point-plane transformation.

By the corresponding transformation of R the triple infinity of spheres that touch a surface F_1 is changed into the totality of spheres which have a similar relation to F_2; and, specifically, the principal spheres of F_2. From this it follows that there is a correspondence between the lines of curvature of F_1 and F_2, in the sense that if in a relation $\phi(X_1, Y_1, Z_1, P_1, Q_1) = 0$, valid along a line of curvature for F_1, we substitute for X_1, Y_1, Z_1, P_1, Q_1 the values

---

[1] "Zur Theorie ———," in *Math. Annalen*, Vol. II.

[2] We must here consider two essentially different cases; for lines that are concurrent may be changed either into similarly placed lines or into lines that are coplanar.

X_2, Y_2, Z_2, P_2, Q_2, we obtain an equation which is valid for one of the curves of curvature of F_2.

I shall now prove that every transformation of R of the form

$$X_1 = F_1\left(X_2,\ Y_2,\ Z_2,\ \frac{dZ_2}{dX_2},\ \frac{dZ_2}{dY_2},\frac{d^2Z_2}{dX_2}\cdots\frac{d^{m+n}Z_2}{dX_2{}^m dY_2{}^n}\right)$$

$$Y_1 = F_2\left(X_2,\ Y_2,\ Z_2,\dots\dots\dots\dots\cdots\frac{d^{m+n}Z_2}{dX_2{}^m dY_2{}^n}\right)$$

$$Z_1 = F_3\left(X_2,\ Y_2,\ Z_2,\dots\dots\dots\dots\cdot\frac{d^{m+n}Z_2}{dX_2{}^m dY_2{}^n}\right)$$

which changes the lines of curvature of any given surface into lines of curvature of the new surface, corresponds, by my representation, to a linear transformation of r.

The proof of this reduces at once to showing that if a transformation of r changes the principal tangent curves of any surface into principal tangent curves of the transformed surface, then intersecting right lines are changed by the same transformation into similarly situated lines.

To begin with, the transformation in question must change right lines into right lines; because the right line is the only curve which is a principal tangent curve for every surface containing same.

Furthermore, that to right lines that intersect correspond right lines in the same relative position may be deduced from the fact that the developable surface is the only line surface so constituted that through each of its points passes only one principal tangent curve. Our transformation, therefore, changes developable surfaces into developable surfaces.

Hence, our statement is proved.

It may be remarked that, corresponding to the two essentially different kinds of linear *transformations there exist two distinct classes of transformations for which curves of curvature are covariant curves.*

If among the aforementioned transformations of R we choose those that are point transformations, we obtain *the most general point transformation of R in which lines of curvature are covariant curves,* a problem first solved by *Liouville.* That conformity is preserved even in the smallest parts is due to the fact that infinitesimal spheres carry into infinitesimal spheres.

The *parallel transformation* is known to carry lines of curvature into lines of curvature, and it is in reality easy to verify that the corresponding transformation of r is a linear point transformation.

For the equations

$$X_1 = X_2; \qquad Y_1 = Y_2; \qquad Z_1 = Z_2; \qquad H_1 = H_2 + A$$

are transformed (compare with our observations on translation in article 31) into relations of the form

$$z_1 = z_2; \qquad x_1 = x_2 + az_2 + b; \qquad y_1 = y_2 + cz_2 + d.$$

34. *Mr. Bonnet* has frequently discussed a transformation which he defines by the equations

$$Z_2 = iZ_2\sqrt{1 + p_2{}^2 + q_2{}^2}; \qquad x_1 = x_2 + p_2z_2; \qquad y_1 = y_2 + q_2z_2,$$

where the two subscripts refer to the given and to the transformed surface.

He proves that this transformation is a reciprocal one, in the sense that if applied twice it leads back to the given surface; that it transforms lines of curvature into lines of curvature; that, finally, if H_1 and H_2 indicate radii of curvature at corresponding points and if ζ_1 and ζ_2 are z-ordinates of the corresponding centers of curvature, these relations come about:

$$\zeta_1 = iH_2, \; H_1 = -i\zeta_2 \quad (\alpha)$$

Bonnet's transformation is the image of a transformation of r in respect to the linear complex $Z + iH = 0$. This we shall prove. If we recall that $X = 0$, $Y = 0$, $Z = 0$, $H = 0$ are in involution by pairs, we shall find that the coordinates of two right lines which are mutually polars in respect to $Z + iH = 0$ satisfy these relations:

$$X_1 = X_2; \qquad Y_1 = Y_2; \qquad Z_1 = iH_2; \qquad H_1 = -iZ_2. \quad (\beta)$$

But if X, Y, Z, H are interpreted as sphere coordinates, these formulas determine a relation by pairs among all the spheres of the space, precisely the same as the transformation of Bonnet.

For the principal spheres of a surface F_1 are by this changed into the principal spheres of surface F_2, and herein we recognize Bonnet's formulas (α). Moreover, if we think of F_1 as generated by point spheres, then the equations (β) define F_2 as an envelope of spheres whose centers lie in the plane $Z = 0$; for $Z_2 = 0$, since $H_1 = 0$. This leads exactly to the geometric construction given by Mr. Bonnet.

MÖBIUS, CAYLEY, CAUCHY, SYLVESTER, AND CLIFFORD

On Geometry of Four or More Dimensions

(Selections and translations made by Professor Henry P. Manning, Brown University, Providence, R. I.)

All references to a geometry of more than three dimensions before 1827 are in the form of single sentences pointing out that we cannot go beyond a certain point in some process because there is no space of more than three dimensions, or mentioning something that would be true if there were such a space. For the next fifty or sixty years the subject is treated more positively, but still in a fragmentary way, single features being developed to be used in some memoir on a different subject. The following selections are from some of the more interesting of those memoirs which of themselves, and because of the standing of the authors in the mathematical world, were to have apparently the chief influence in the further growth of this subject. The first article, by Möbius, is from *Der barycentrische Calcul* (Leipzig, 1827), a work from which other extracts have been made on pages 670–677 for another purpose. A brief biographical note accompanies that translation.

MÖBIUS

On Higher Space[1]

§139, page 181.   If, given two figures, to each point of one corresponds a point of the other so that the distance between any two points of one is equal to the distance between the corresponding points of the other, then the figures are said to be *equal and similar.*

. . . .

§140, pages 182–183.   *Problem.—To construct a system of n points which is equal and similar to a given system of n points.*

Solution.   Let $A, B, C, D, \ldots$, be the points of the given system, and $A', B', C', D', \ldots$, the corresponding points of the system to be constructed.   We have to distinguish three cases according as the points of the first set lie on a line, or in a plane, or in space.

. . . .

Finally, if the given system lies in space, then A' is entirely arbitrary, B' is an arbitrary point of the spherical surface which has A' for center and AB for radius, C' is an arbitrary point of the circle in which the two spherical surfaces drawn from A' with AC as radius and from B' with BC as radius intersect, and D' is one of the two points in which the three spherical surfaces drawn from A' with AD, from B' with BD, and from C' with CD as radii intersect.   In the same way as D' will also each of the remaining points, for example, E', be found, only that of the two common intersections of the spherical surfaces drawn from A', B', C', with AE, BE, CE as radii, that one is taken which lies on the same side or on the opposite side of the plane $A'B'C'$ as D', according as the one or the other is the case with the corresponding points in the given system.

For the determination of A' therefore no distance is required, for the determination of B' one, for the determination of C' two, and for the determination of each of the remaining $n - 3$ points three. Therefore in all

$$1 + 2 + 3(n - 3) = 3n - 6$$

distances are required.

Remark.—Thus only for the point D', and for none of the remaining points, are we free to choose between the two intersections

---

[1] (From *Der barycentrische Calcul*, Leipzig, 1827, part 2, Chapter 1.)

on the three spherical surfaces falling on opposite sides of the plane $A'B'C'$. These two intersections are distinguished from each other in this way, that looking from one the order of the points A', B', C' is from right to left, but from the other from left to right, or, as also we can express it, that the former point lies on the left, the latter on the right of the plane $A'B'C'$. Now according as we choose for D' the one or the other of these two points, so also will the order formed be the same or different from that in which the point D appears from the points A, B, C. In both cases are the systems A, B, C, D,..., and A', B', C', D',... indeed equal and similar, but only in the first case can they be brought into coincidence.

It seems remarkable that solid figures can have equality and similarity without having coincidence, while always, on the contrary, with figures in a plane or systems of points on a line equality and similarity are bound with coincidence. The reason may be looked for in this, that beyond the solid space of three dimensions there is no other, none of four dimensions. If there were no solid space, but all space relations were contained in a single plane, then would it be even as little possible to bring into coincidence two equal and similar triangles in which corresponding vertices lie in opposite orders. Only in this way can we accomplish this, namely by letting one triangle make a half revolution around one of its sides or some other line in its plane, until it comes into the plane again. Then with it and the other triangle will the order of the corresponding vertices be the same, and it can be made to coincide with the other by a movement in the plane without any further assistance from solid space.

The same is true of two systems of points A, B,...and A', B',... on one and the same straight line. If the directions of AB and $A'B'$ are opposite, then in no way can a coincidence of corresponding points be brought about by a movement of one system along the line, but only through a half revolution of one system in a plane going through the line.

For the coincidence of two equal and similar systems, A, B, C, D,...and A', B', C', D',...in space of three dimensions, in which the points D, E,...and D', E',...lie on opposite sides of the planes ABC and $A'B'C'$, it will be necessary, we must conclude from analogy, that we should be able to let one system make a half revolution in a space of four dimensions. But since such a space cannot be thought, so is also coincidence in this case impossible.

CAYLEY

On Higher Space

Arthur Cayley (1821–1895) was Sadlerian professor of mathematics at Cambridge. He wrote memoirs on nearly all branches of mathematics and, in particular created the theory of invariants. The extract is from his memoir, written in French, "On some Theorems of Geometry of Position," Crelle's *Journal*, vol. 31, 1846, pp. 213–227; *Mathematical Papers*, vol. I, Number 50, pp. 317–328.

In taking for what is given any system of points and lines we can draw through pairs of given points new lines, or find new points, namely, the points of intersection of pairs of given lines, and so on. We obtain in this way a new system of points and lines, which can have the property that several of the points are situated on the same line or several of the lines pass through the same point, which gives rise to so many theorems of the geometry of position. We have already studied the theory of several of these systems; for example, that of four points, of six points situated by twos on three lines which meet in a point, of six points three by three on two lines, or, more generally, of six points on a conic (this last case that of the mystic hexagram of Pascal, is not yet exhausted, we shall return to it in what follows), and also some systems in space. However, there exist systems more general than those which have been examined, and whose properties can be perceived in a manner almost intuitive, and which, I believe, are new.

Commence with the case most simple. Imagine a number n of points situated in any manner in space, which we will designate by 1, 2, 3,...n. Let us pass lines through all the combinations of two points, and planes through all the combinations of three points. Then cut these lines and planes by any plane, the lines in points and the planes in lines. Let $\alpha\beta$ be the point which corresponds to the line drawn through the two points α and β, let $\beta\gamma$ be the point which corresponds to that drawn through β and γ and so on. Further let $\alpha\beta\gamma$ be the line which corresponds to the plane passed through the three points α, β, and γ, etc. It is clear that the three points $\alpha\beta$, $\alpha\gamma$, and $\beta\gamma$ will be situated on the line

$\alpha\beta\gamma$. Then, representing by N_2, N_3,...the numbers of the combinations of n letters 2 at a time, 3 at a time, etc., we have the following theorem.

THEOREM I.—*We can form a system of N_2 points situated 3 at a time on N_3 lines, to wit, representing the points by* 12, 13, 23, *etc., and the lines by* 123, *etc., the points* 12, 13, 23, *will be situated on the line* 123, *and so on.*

For $n = 3$ and $n = 4$ this is all very simple; we have three points on a line, or 6 points, 3 at a time, on 4 lines. There results no geometrical property. For $n = 5$ we have 10 points, 3 at a time on as many lines, to wit the points

$$12 \quad 13 \quad 14 \quad 15 \quad 23 \quad 24 \quad 25 \quad 34 \quad 35 \quad 45,$$

and the lines

$$123 \quad 124 \quad 125 \quad 134 \quad 135 \quad 145 \quad 234 \quad 235 \quad 245 \quad 345.$$

The points 12, 13, 14, 23, 24, 34 are the angles of an arbitrary quadrilateral,[1] the point 15 is entirely arbitrary, the point 25 is situated on the line passing through the points 12 and 15, but its position on this line is arbitrary. We will determine then the points 35 and 45, 35 as the point of intersection of the line passing through 13 and 15 and the line passing through 23 and 25, that is, of the lines 135 and 235, and the point 45 as the point of intersection of the lines 145 and 245. The points 35 and 45 will have the geometrical property of being in a line with 34, or all three will be in the same line 345.

. . . .

Page 217.

The general theorem, Theorem I, can be considered as the expression of an analytical fact, which ought equally well to hold in considering four coordinates instead of three. Here a geometrical interpretation holds which is applied to the points in space. We can, in fact, without having recourse to any metaphysical notion in regard to the possibility of a space of four dimensions, reason as follows (all of this can also be translated into language purely analytical): In supposing four dimensions of space it is necessary to consider *lines* determined by two points, *half-planes* determined by three points, and *planes*[2] determined by four points

---

[1] It is necessary always to have regard to the difference between quadrilateral and quadrangle. Each quadrilateral has four sides and six angles; each quadrangle has four angles and six sides.

[2] [His plane is what we call a hyperplane and his half-plane is an ordinary plane, and so he has to distinguish between a plane and an ordinary plane.]

(two planes intersect in a half-plane, etc.). Ordinary space can be considered as a plane, and it will cut a plane in an ordinary plane, a half-plane in an ordinary line and a line in an ordinary point. All this being granted, let us consider a number, n, of points, combining them by two, three, and four, in lines, half-planes, and planes, and then cut the system by space considered as a plane. We obtain the following theorem of geometry of three dimensions:

THEOREM VII.—*We can form a system of N_2 points, situated 3 by 3 in N_3 lines which themselves are situated 4 by 4 in N_4 planes. Representing the points by 12, 13, etc., the points situated in the same line are 12, 13, 23, and lines being represented by 123, etc., as before, the lines 123, 124, 134, 234 are situated in the same plane, 1234.*

In cutting this figure by a plane we obtain the following theorem of plane geometry:

THEOREM VIII.—*We can form a system of N_3 points situated 4 by 4 on N_4 lines. The points ought to be represented by the notation 123, etc., and the lines by 1234, etc. Then 123, 124, 134, 234 are in the same line designated by 1234.*

. . . .

CAUCHY

On Higher Space

When Louis Phillippe came to the throne Augustin Louis Cauchy (1789–1857), was unwilling to take the oath required by the government and for a while was in exile in Switzerland and Italy, but he returned to Paris in 1838 and finally became professor at the École Polytechnique. For a further biographical note see page 635. The article here translated is his "Memoir on Analytic Loci, *Comptes Rendus*, vol. XXIV, p. 885 (May 24, 1847); *Complete Works*, first series, vol. X, Paris, 1897, p. 292. It is one of a collection of memoirs on radical polynomials, a radical polynomial being a polynomial

$$\alpha + \beta\rho + \gamma\rho^2 + \ldots \eta\rho^{n-1},$$

where ρ is a primitive root of the equation

$$x^n = 1.$$

Consider several variables, x, y, z, ... and various explicit functions, u, v, w, ... of these variables. To each system of values of the variables x, y, z, ... will generally correspond determined values of the functions u, v, w, ... Moreover, if the variables are in number only two or three they can be thought of as representing the rectangular coordinates of a point situated in a plane or in space, and therefore each system of values of the variables can be thought of as corresponding to a determined point. Finally, if the variables x, y or x, y, z are subject to certain conditions represented by certain inequalities, the different systems of values of x, y, z for which the conditions are satisfied will correspond to different points of a certain locus, and the lines or surfaces which limit this locus in the plane in question or in space will be represented by the equations into which the given inequalities are transformed when in them we replace the sign $<$ or $>$ by the sign $=$.

Conceive now that the number of variables x, y, z, ... becomes greater than three. Then each system of values of x, y, z, ... will determine what we shall call an *analytical point* of which these variables are the coordinates, and to this point will correspond a certain value of each function of x, y, z, ... Further, if the variables are subject to conditions represented by inequalities, the systems of values of x, y, z, ... for which these conditions are satisfied will

530

correspond to analytical points, which together will form what we shall call an *analytical locus*. Moreover, this locus will be limited by analytical envelopes whose equations will be those to which the given inequalities are reduced when in them we replace the sign $<$ or $>$ by the sign $=$.

We shall also call *analytical line* a system of analytical points whose coordinates are expressed by aid of given linear functions of one of them. Finally, the *distance* of two analytical points will be the square root of the sum of the squares of the differences between the corresponding coordinates of these two points.

The consideration of analytical points and loci furnishes the means of clearing up a great many delicate questions, and especially those which refer to the theory of radical polynomials.

. . . .

SYLVESTER

On Higher Space

James Joseph Sylvester (1814–1897) was barred from certain honors in England because he was a Jew. He was professor at the University College, London, and at the Royal Military Academy in Woolwich. For a short time he taught at the University of Virginia. When the Johns Hopkins University was started he went there to take the lead in the advance of higher mathematics in this country. In 1883 he returned to England and became Savilian professor of geometry at Oxford. The article quoted is "On the Center of Gravity of a Truncated Triangular pyramid, and on the Principles of Barycentric Perspective." It appeared in the *Philosophical Magazine*, vol. XXVI, 1863, pp. 167–183; *Collected Mathematical Papers*, vol. II, Cambridge, 1908, pp. 342–357.

There is a well-known geometrical construction for finding the center of gravity of a plane quadrilateral which may be described as follows.

Let the intersection of the two diagonals (say Q) be called the *cross-center*, and the intersection of the lines bisecting opposite sides (say O) the *mid-center* (which, it may be observed, is the center of gravity of the four angles viewed as equal weights), then the center of gravity is in the line joining these two centers produced past the latter (the mid-center), and at a distance from it equal to one-third of the distance between the two centers. In a word, if G be the center of gravity of the quadrilateral, QOG will be in a right line and $OG = \frac{1}{3}QO$.

The frustum of a pyramid is the nearest analogue in space to a quadrilateral in a plane since the latter may be regarded as the frustum of a triangle. The analogy, however, is not perfect, inasmuch as a quadrilateral may be regarded as a frustum of either of two triangles, but the pyramid to which a given frustum belongs is determinate. Hence à *priori* reasonable doubts might have been entertained as to the possibility of extending to the pyramidal frustum the geometrical method of centering the plane quadrilateral. The investigation subjoined dispels this doubt, and will be found to lead to the perfect satisfaction, under a somewhat unexpected form, of the hoped-for analogy.

Let abc and $\alpha\beta\gamma$ be the two triangular faces, and $a\alpha$, $b\beta$, and $c\gamma$ the edges of the quadrilateral faces of a pyramidal frustum. Then this frustum may be resolved in six different ways into three different pyramids as shown in the annexed double triad of schemes.

a	b	c	α		b	c	a	β		c	a	b	γ
b	c	α	β		c	a	β	γ		a	b	γ	α
c	α	β	γ		a	β	γ	α		b	γ	α	β

b	a	c	β		a	c	b	α		c	b	a	γ
a	c	β	α		c	b	α	γ		b	a	γ	β
c	β	α	γ		b	α	γ	β		a	γ	β	α

If then, taking any one of the above schemes, we draw a plane through the centers[1] of the three pyramids of which it is composed, the six planes thus drawn will meet in a point, which will be the center of the frustum.[2]

Let the point in which αa, βb, and γc meet when produced be the origin of coordinates, and $bc\beta\gamma$, $ca\gamma\alpha$, and $ab\alpha\beta$ be taken as the planes of x, y, z and let $4a, 0, 0; 0, 4b, 0; 0, 0, 4c$ be the coordinates of a, b, c, and $4\alpha, 0, 0; 0, 4\beta, 0; 0, 0, 4\gamma$ those of α, β, γ. Consider the first of the schemes above written.

$a + \alpha, b,\quad c$ will be the coordinates of the center of $abc\alpha$,

$\alpha,\quad b + \beta, c$ will be the coordinates of the center of $bc\alpha\beta$,

$\alpha,\quad \beta,\quad c + \gamma$ will be the coordinates of the center of $c\alpha\beta\gamma$,

because, as everyone knows, the center of a triangular pyramid is the same as that of its angles regarded as of equal weight. But again, if we define as the *mid-center* the center of the six angles of the frustum regarded as of equal weight, its coordinates will be

$$\frac{2a + 2\alpha}{3}, \frac{2b + 2\beta}{3}, \frac{2c + 2\gamma}{3};$$

and if we substitute for each of the three centers last named, points lying, respectively, in a right line with them and the mid-center, on the opposite side of the mid-center and at distances from it double

---

[1] I shall throughout in future for greater brevity hold myself at liberty to use the word center to mean center of gravity.

[2] I shall hereafter show that these six planes all touch the same cone, of which, as also of its polar reciprocal, I have succeeded in obtaining the equation.

those of these centers themselves, these quasi-images of the centers in question will have for their coordinates

$$0, \ 2\beta, \ 2\gamma,$$
$$2a, \ 0, \ 2\gamma,$$
$$2a, \ 2b, \ 0.$$

These points are, accordingly, the centers of the lines $\beta\gamma$, γa, and ab, respectively.

And a similar conclusion will apply to each of the six schemes. Hence, using in general (p, q) to mean the middle of the line pq, and by the collocation of the symbols for three points understanding the plane passing through them, it is clear

1. That the six planes

$$(\beta, \gamma) \ (\gamma, a) \ (a, b) \quad (\gamma, \alpha) \ (\alpha, b) \ (b, c) \quad (\alpha, \beta) \ (\beta, c) \ (c, a)$$
$$(\gamma, \beta) \ (\beta, a) \ (a, c) \quad (\alpha, \gamma) \ (\gamma, b) \ (b, a) \quad (\beta, \alpha) \ (\alpha, c) \ (c, b)$$

will meet in a single point which may be called the *cross-center*, being the true analogue of the intersection of the two diagonals of a quadrilateral figure in the plane.

2. That if we join this cross-center (say Q) with O the mid-center, and produce QO to G, making $OG = \frac{1}{2}QO$, G will be the center of the frustum $abc\alpha\beta\gamma$.[1]

It may be satisfactory to some of my readers to have a direct verification of the above.

Let then

$$A = \frac{a^2bc - \alpha^2\beta\gamma}{abc - \alpha\beta\gamma}, \ B = \frac{ab^2c - \alpha\beta^2\gamma}{abc - \alpha\beta\gamma}, \ C = \frac{abc^2 - \alpha\beta\gamma^2}{abc - \alpha\beta\gamma}.$$

A moment's reflection will serve to show that A, B, C are the coordinates of the center of the frustum.[2]

---

[1] [The three centers of the three tetrahedrons lie in a plane through the center G. Drawing lines from these three points and G to the mid-center O, and laying off on these lines produced beyond O any given multiples of these lines, we shall have three points corresponding to the three given points and a fourth point Q corresponding to G, all lying in a plane parallel to the first plane. We can do this for any three of the six planes through G and get planes whose intersection will be Q, and then from Q and O we can get G by reversing the process. If we take the multiplier to be 2, the three new points will be very simple as pointed out in the footnote.]

[2] [These expressions can be obtained, for example, by considering the frustum as the difference of two pyramids with a common vertex at the origin.]

Again, the first three of the six planes last referred to will be found to have for their equations, respectively,

$$\beta\gamma x + \gamma ay + abz = 2a\gamma(b + \beta),$$
$$bcx + \gamma\alpha y + \alpha bz = 2b\alpha(c + \gamma),$$
$$\beta cx + cay + \alpha\beta z = 2c\beta(a + \alpha).$$

The determinant

$$\begin{vmatrix} \beta\gamma & \gamma a & ab \\ bc & \gamma\alpha & \alpha b \\ \beta c & ca & \alpha\beta \end{vmatrix} = (abc - \alpha\beta\gamma)^2.$$

The determinant

$$\begin{vmatrix} \gamma a & ab & 2a\gamma(b + \beta) \\ \gamma\alpha & \alpha b & 2b\alpha(x + \gamma) \\ ca & \alpha\beta & 2c\beta(a + \alpha) \end{vmatrix}$$
$$= 2\alpha a(bc - \beta\gamma)(abc - \alpha\beta\gamma) = 2[(\alpha^2\beta\gamma - a^2bc)(abc - \alpha\beta\gamma) \\ + (a + \alpha)(abc - \alpha\beta\gamma)^2].$$

Hence if x, y, and z be the coordinates of the intersection of the above-mentioned three planes,

$$x = -2A + 2(a + \alpha),$$
$$y = -2B + 2(b + \beta),$$
$$z = -2C + 2(c + \gamma),$$

and the same will evidently be true of the other ternary system of planes, so that all six planes intersect in a single point Q, of which x, y, and z above written are the coordinates. And the coordinates of O being

$$\frac{2a + 2\alpha}{3}, \frac{2b + 2\beta}{3}, \frac{2c + 2\gamma}{3},$$

and those of G being A, B, C, it is obvious that QOG is a right line, and $OG = \frac{1}{2}QO$, as was to be shown.

The analogy with the quadrilateral does not end here. There is a construction[1] for the center of a quadrilateral still easier than

---

[1] This is the mode of statement (except that the important notion of opposite points was not explicitly contained in it) which, accidently meeting my eye in a proof sheet of some geometrical notes (by an anonymous author) intended for further insertion in the forthcoming (if not forthcome) number of the *Quarterly Journal of Mathematics*, led to the long train of reflections embodied in this paper, which but for that casual glance would never have seen the light. The same construction, under another and somewhat less eligible form, is given in the *Mathematician* (a periodical now extinct, edited by Dr. Rutherford and Mr. Fenwick, both of the Royal Military Academy), 1847, volume II, page 292, and is therein stated by the latter gentleman to have, "as he believes, first appeared in the *Mechanics Magazine*, and subsequently in the *Lady's Diary* for 1830."

that above cited, which may be expressed in general terms by aid of a simple definition. Agreè to understand by the *opposite* to a point L on a limited line AB a point M such that L and M are at equal distances from the center of AB but on opposite sides of it. Then we may affirm that the center of a quadrilateral is the center of the triangle whose apices are the intersection of its two diagonals (that is, the cross-center) and the opposites of that intersection on these two diagonals, respectively. So now, if we agree to understand by opposite points on a limited triangle two points on a line with the center of the triangle and at equal distances from it on opposite sides, and bear in mind that the cross-center of a pyramidal frustum is the intersection of either of two distinct ternary systems of triangles which may be called the two systems of cross-triangles,[1] we may affirm that the center of a pyramidal frustum is the center of a pyramid whose apices are its cross-center, and the opposites of that center on the three components of either of its systems of cross-planes. This is easily seen, for if we take the first of the two systems, their respective centers will be

$$\frac{4a}{3}, \frac{2b+2\beta}{3}, \frac{4\gamma}{3}; \frac{4a}{3}, \frac{4b}{3}, \frac{2c+2\gamma}{3}; \frac{2a+2\alpha}{3}, \frac{4\beta}{3}, \frac{4c}{3}.$$

Thus the three opposites to the cross-center whose coordinates are

$$-2A + 2(a + \alpha), \quad -2B + 2(b + \beta), \quad -2C + 2(c + \gamma),$$

will have for their x coordinates[2]

$$\frac{2a}{3} - 2\alpha + 2A; \quad -2a + \frac{2\alpha}{3} + 2A; \quad -\frac{2a}{3} - \frac{2\alpha}{3} + 2A;$$

for their y coordinates

$$\frac{2b}{3} - 2\beta + 2B; \quad -2b + \frac{2\beta}{3} + 2B; \quad -\frac{2b}{3} - \frac{2\beta}{3} + 2B;$$

and for their z coordinates

$$\frac{2c}{3} - 2\gamma + 2C; \quad -2c + \frac{2\gamma}{3} + 2C; \quad -\frac{2c}{3} - \frac{2\gamma}{3} + 2C;$$

---

[1] From the description given previously, it will be seen that a cross-triangle of the frustum is one which has its apices at the centers of either diagonal of any quadrilateral face and of the two edges conterminous but not in the same face with that diagonal.

[2] These are not arranged so that the three coordinates of a point are in a column. There is a certain cyclical shifting in the second and third lines. If we think of the nine coordinates in the arrangement here as forming a determinant, we get the coordinates of the three opposites separately by taking the three negative diagonals.

and consequently the center of the pyramid whose apices are the cross-center and its three opposites will be *A, B, C,* that is, will be the center of gravity of the frustum, as was to be shown.[1]

It is clear that these results may be extended to space of higher dimensions. Thus in the corresponding figure in space of four dimensions bounded by the hyperplanar quadrilaterals[2] *abcd* and $\alpha\beta\gamma\delta$, which will admit of being divided into four hyperpyramids in 24 different ways, all corresponding to the type

$$a \ b \ c \ d \ \alpha$$
$$b \ c \ d \ \alpha \ \beta$$
$$c \ d \ \alpha \ \beta \ \gamma$$
$$d \ \alpha \ \beta \ \gamma \ \delta,$$

---

[1] I at one time supposed that *a, b, c, α, β, γ,* formed two systems of diagonal planes, and that there were thus two cross-centers, and dreamed a dream of the construction for the center of gravity of the pyramidal frustum based upon this analogy, inserted (it is true as a conjecture only) in the *Quarterly Journal of Mathematics,* but the nature of things is ever more wonderful than the imagination of men's minds, and her secrets may be won, but cannot be snatched from her. Who could have imagined *à priori* that for the purposes of this theory a diagonal of a quadrilateral was to be viewed as a line drawn through two opposite angles of the figure regarded, not as themselves, but as their own center of gravity. Some of my readers may remember a single case of a similar autometamorphism which occurred to myself in an algebraical inquiry, in which I was enabled to construct the canonical form of a six-degreed binary quantic from an analogy based on the same for a four-degreed one, by considering the square of a certain function which occurs in the known form as consisting of two factors, one the function itself, the other a function morphologically derived from, but happening for that particular case to coincide with the function. The parallelism is rendered more striking from the fact of 4 and 6 being the numbers concerned in each system of analogies, those numbers referring to degrees in one theory and to angular points in the other. It is far from improbable that they have their origin in some common principle, and that so in like manner the parallelism will be found to extend in general to any quantic of degree 2n, and the corresponding barycentric theory of the figure with 2n apices (n of them in one hyperplane and n in another), which is the problem of a hyperpyramid in space of n dimensions. The probability of this being so is heightened by the fact of the barycentric theory admitting, as is hereafter shown, of a descriptive generalization, descriptive properties being (as is well known) in the closest connection with the theory of invariants. Much remains to be done in fixing the canonic forms of the higher even-degreed quantics, and this part of their theory may hereafter be found to draw important suggestions from the hypergeometry above referred to, if the supposed alliance have a foundation in fact.

[2] [This word should be *tetrahedrons.*]

there will be a *cross-center* given by the intersection of any four
out of 24 hyperplanes resoluble into six sets of four each,[1]—
one such set of four being given in the scheme subjoined, where in
general *pqr* means the point which is the center of (p, q, r) and the
collocation of four points means the hyperplane passing through
them, namely,

$$\beta\gamma\delta \quad \gamma\delta a \quad \delta ab \quad abc,^2$$
$$\gamma\delta\alpha \quad \delta\alpha b \quad \alpha bc \quad bca,$$
$$\delta\alpha\beta \quad \alpha\beta c \quad \beta cd \quad cdb,$$
$$\alpha\beta\gamma \quad \beta\gamma d \quad \gamma da \quad dac.$$

The *mid-center* will mean the center of the eight angles a, b, c, d,
α, β, γ, δ, regarded as of equal weight, and to find the center of
the hyper-pyramidal frustum we may either produce the line
joining the cross-center with the mid-center through the latter
and measure off three-fifths of the distance of the joining line on
the part produced (as in the preceding cases we measured off two-
fourths and one-third of the analogous distance) or we may take
the four opposites of the cross-center on the four components of
any one of the six systems of hyperplanar tetrahedrons of which
it is the intersection, and find the center of the hyperpyramid so
formed. The point determined by either construction will be
the center of gravity of the hyperpyramidal frustum in question.
And so for space of any number of dimensions.   It will of course be
seen that a general theorem of determinants[3] is contained in

---

[1] [The second, third, and fourth of a set may be obtained from the first by
taking the cylical permutations of the Roman letters with the same permutations of the Greek letters.]

[2] [The last letters of these four lines should be c, d, a, b.]

[3] We learn indirectly from this how to represent under the form of determinants of the ith order, and that in a certain number of ways, the general
expressions

$$(l_1 l_2 \ldots l_i - \lambda_1 \lambda_2 \ldots \lambda_i)^{i-1}$$

and

$$l_1 \lambda_1 (l_2 l_3 \ldots l_i - \lambda_2 \lambda_3 \ldots \lambda_i)(l_1 l_2 \ldots l_i - \lambda_1 \lambda_2 \ldots \lambda_i)^{i-2}$$

a strange conclusion to be able to draw incidentally from a hyper-theory of
center of gravity!   Thus, for example, on taking $i = 4$ we shall find

$$\begin{vmatrix} bcd & cd\alpha & d\alpha\beta & \alpha\beta\gamma \\ \beta\gamma\delta & cda & da\beta & a\beta\gamma \\ b\gamma\delta & \gamma\delta\alpha & dab & ab\gamma \\ bc\delta & c\delta\alpha & \delta\alpha\beta & abc \end{vmatrix} = (abcd - \alpha\beta\gamma\delta)^3.$$

the assertion that for space of n dimensions there will be $n!$ quasi-planes all intersecting in the same point, as also in the general relation connecting this point (the cross-center) with the mid-center and center of gravity, of each of which it is easy to assign the value of the coordinates in the general case.

. . . .

And again

$$\begin{vmatrix} \alpha d(bc + c\beta + \beta\gamma) & cd\alpha & da\beta & \alpha\beta\gamma \\ \beta a(cd + d\gamma + \gamma\delta) & cda & da\beta & a\beta\gamma \\ \gamma b(da + a\delta + \delta\alpha) & \gamma\delta\alpha & dab & ab\gamma \\ \delta c(ab + b\alpha + \alpha\beta) & c\delta\alpha & \delta a\beta & abc \end{vmatrix} = a\alpha(bcd - \beta\gamma\delta)(abcd - \alpha\beta\gamma\delta)^2$$

The number of these representations will not be 24, that is 4!, but only 12, the half of that number, because it will be easily seen that the cycle $abcd$, $\alpha\beta\gamma\delta$ will lead to the same determinants, only differently arranged, as the cycles $bcda$, $\beta\gamma\delta\alpha$. I believe the law is that the number of varieties of such representations is $i!$ or $\frac{1}{2}i!$ according as i is odd or even. The expression $ab - \alpha\beta$ at once conjures up the idea of a determinant. We now see that there is an equally natural determinantive representation, or system of representations, of $(abc - \alpha\beta\gamma)^2$, $(abcd - \alpha\beta\gamma\delta)^3$, etc.

CLIFFORD

On Higher Space

William Kingdon Clifford (1848–1879), was professor of mathematics and mechanics in University College, London from 1871 to the time of his death. The following article is his solution of a "Problem in Probability," in the *Educational Times*, January, 1866, Problem 1878, proposed by himself. *A line of length a is broken up into pieces at random; prove that* (1) *the chance that they cannot be made into a polygon of n sides is* $n2^{1-n}$; *and* (2) *the chance that the sum of the squares described on them does not exceed* $\dfrac{a^2}{(n-1)}$ *is*

$$\left(\frac{\pi}{n^2 - n}\right)^{\frac{1}{2}n - 1} \frac{\Gamma(n)}{\Gamma\{\frac{1}{2}(n+1)\}}, \frac{1}{n^{\frac{1}{2}}}.$$

Solution by the proposer. November, 1866; reprinted in *Mathematical Questions with Solutions*, vol. VI, London, 1866, pp. 83–87; also in *Mathematical Papers*, London, 1882, pp. 601–607.

1. Let us define as follows. A point is taken *at random* on a (finite or infinite) straight line when the chance that the point lies on a finite portion of the line varies as the length of that portion. And a line is broken up at random when the points of division are taken at random.

Now the n pieces will always be capable of forming a polygon except when one of them is greater than the sum of all the rest, that is, greater than half the line. The first part of the question may therefore be stated thus: $n - 1$ *points are taken at random on a finite line; to find the chance that some one of the intervals shall be greater than half the line.*

. . . .

4. *Third Solution.*—To make this clear I will state first the previously known analogous solutions in the cases where $n = 3$ and $n = 4$. When the line is divided into three pieces, call them x, y, and z, and take their lengths for the coordinates of a point P in geometry of three dimensions. Then, since

$$x + y + z = a \qquad (1)$$

and x, y, and z are all positive, the point P must be somewhere on the surface of the equilateral triangle determined on the plane (1) by the coordinate planes. Now consider those points on the

540

triangle for which $x > \frac{1}{2}a$. These are cut off by the plane $x = \frac{1}{2}a$, and it is easy to see that this plane cuts off from one corner of the triangle a similar triangle of half the linear dimensions, and therefore of one-fourth the area. Now there are three corners cut off. Their joint area is therefore three-fourths of the area of the triangle, and the chance required is accordingly $\frac{3}{4}$.

When the line is divided into four pieces, take the first three pieces as the coordinates of a point in space. Then we have $x + y + z < a$ and x, y, and z all positive. So the point must lie within the content of the tetrahedron bounded by the plane $x + y + z = a$ and the coordinate planes. Now if $x + y + z < \frac{1}{2}a$ the fourth piece must be greater than $\frac{1}{2}a$. The points for which this is the case are cut off by the plane $x + y + z = \frac{1}{2}a$ and it is easily seen as before that this plane cuts off from one corner of the tetrahedron a similar tetrahedron of half the linear dimensions, and therefore of one eighth the volume. So also the plane $x = \frac{1}{2}a$ cuts off from another corner a similar tetrahedron of half the linear dimensions. Since therefore there are four corners cut off, their joint volume is four-eighths or one-half of the volume of the tetrahedron, and the chance required is accordingly $\frac{1}{2}$.

5. Now consider the analogous case in geometry of n dimensions. Corresponding to a closed area and a closed volume we have something which I shall call a *confine*. Corresponding to a triangle and to a tetrahedron there is a confine with $n + 1$ corners or vertices which I shall call a *prime confine*[1] as being the simplest form of confine. A prime confine has also $n + 1$ *faces*, each of which is, not a plane, but a prime confine of $n - 1$ dimensions. Any two vertices may be joined by a straight line, which is an *edge* of the confine. Through each vertex pass n edges. A prime confine may be *regular*, which it is when any three vertices form an equilateral triangle; or *rectangular*, which it is when the edges through some one vertex are all equal and at right angles to one another.

To solve the question for general values of n we may adopt as a type either of the geometrical solutions given for the cases $n = 3$ and $n = 4$. First take the lengths of the n pieces for the coordinates of a point in geometry of n dimensions. Then, since their sum is a and they are all positive, the point must lie within a

---

[1] [The term now commonly used is *simplex*. In space of four dimensions this is a *pentabedroid*.]

certain regular prime confine of $n - 1$ dimensions. The supposition that a certain piece is greater than $\frac{1}{2}a$ cuts off from one corner of the confine a similar confine of half the linear dimensions and therefore of 2^{1-n} times the content. As there are n corners, their joint content is $n2^{1-n}$ times the content of the confine. The chance required is consequently $n2^{1-n}$. Or take the lengths of the first $n - 1$ pieces as the coordinates of a point in geometry of $n - 1$ dimensions. The point will then lie within a certain rectangular confine of $n - 1$ dimensions, and the investigation proceeds as before, the n corners being cut off in the same manner.

6. It will be seen that this third solution involves in a geometrical form the assumption of which some sort of proof was given in the first solution. Let us make this extension of our fundamental definition:—A point is taken at random in a (finite or infinite) space of n dimensions when the chance that the point lies in a finite portion of the space varies as the content of that portion. The assumption is that when the lengths of the pieces into which a line is broken up are taken as coordinates of a point, then if the line is broken up at random the point is taken at random and *vice-versa*. The proof of this assumption may be shown to involve a geometrical proposition equivalent to the integration by parts of the differential in Art. 3.[1]

Making this assumption, we may solve the second part of the question by the method of the third solution of the first part. I will first state the previously known analogous solution of the case where $n = 3$. The question in this case is, *If a line of length a be broken into three pieces at random find the chance that the sum of the squares of these pieces shall be less than $\frac{1}{2}a^2$.* Take the lengths of the three pieces for coordinates x, y, and z of a point P in geometry of three dimensions. Then, as before, the point must lie somewhere in the area of the equilateral triangle determined on the plane $x + y + z = a$ by the coordinate planes. But if also the sum of the squares of the pieces is less than a certain quantity m^2, then the point P must lie within a certain circle

---

[1] [The assumption of which "some sort of a proof was given in the first solution" is that the chance that the rth piece reckoning from one end of the line shall be greater than $\frac{1}{2}a$ is equal to the chance that the $(r + 1)$th piece sha l be greater than $\frac{1}{2}a$. In the second solution (Art. 3) the chance that the rth piece shall be greater than $\frac{1}{2}a$ is proved equal to the integral

$$\frac{(n-1)!}{(n-r)!(r-2)!} \int_0^{\frac{1}{2}a} \left(\frac{x}{a}\right)^{r-2} \left(\frac{1}{2} - \frac{x}{a}\right)^{n-r+1} \frac{dx}{a}.]$$

determined on the plane $x + y + z = a$ by the sphere $x^2 + y^2 + z^2 = m^2$. Now in the case where $m^2 = \frac{1}{2}a^2$ this circle is the circle inscribed in the equilateral triangle, so that the question reduces itself to this one:—

To find in terms of an equilateral triangle the area of its inscribed circle.

Now let us go a little further and consider the case in which $n = 4$. Here we shall have to take a point P in geometry of four dimensions. The point must lie somewhere in the regular tetrahedron determined on the hyperplane $x + y + z + w = a$ by the coordinate hyperplanes. If also the sum of the squares of the pieces is less than a certain quantity m^2, then the point P must lie within a certain sphere determined on the hyperplane $x + y + z + w = a$ by the quasi-sphere $x^2 + y^2 + z^2 + w^2 = m^2$. In the particular case where m is the perpendicular from the vertex on the base of a rectangular tetrahedron each of whose equal edges is of length a, or $m^2 = \frac{1}{3}a^2$ this sphere is the sphere inscribed in the regular tetrahedron.[1] The question is therefore reduced to this one:—

To find in terms of a regular tetrahedron the volume of its inscribed sphere.

Now a similar reduction holds in the general case; namely, the question can always be reduced to this one:—

To find in terms of a regular prime confine of $n - 1$ dimensions the content of its inscribed quasi-sphere.

This question I proceed to solve.

7. Let $n - 1 = p$. *The perpendicular from any vertex on the opposite face of a regular prime confine in p dimensions*

$$= \left(\frac{p+1}{2p}\right)^{\frac{1}{2}} \cdot \text{(edge)}.$$

For let O be the vertex in question, OA, OB,...the p edges through O. Draw through each vertex A a space of $p - 1$ dimensions parallel to the face opposite to A. The p spaces thus drawn will intersect in a point P such that OP is the diagonal of a confine analogous to a parallelogram and to a parallelepiped. Then OP is p times the perpendicular from O on the opposite

---

[1] [Each face of the regular tetrahedron is the base of a rectangular tetrahedron formed with its vertices and the origin, and the perpendicular from the origin upon this face will be the radius of a hypersphere which is tangent to the face, and so intersects the hyperplane of the tetrahedron in the inscribed sphere.]

face of the regular confine; for the perpendicular is the projection of one edge at a certain angle, while OP is the projection at the same angle of a broken line consisting of p edges.[1]

We have also[2]

$$OP^2 = OA^2 + OB^2 + OC^2 + \ldots + 2OA.OB \cos AOB + \ldots$$
$$= \Sigma.OA^2 + \Sigma.OA.OB \text{ [since } \cos AOB = \tfrac{1}{2}, \text{ etc.]}$$
$$= [p + \tfrac{1}{2}p(p - 1)].OA^2 = \tfrac{1}{2}p(p + 1).OA^2,$$

therefore (perpendicular)$^2 = \dfrac{OP^2}{p^2} = \dfrac{p + 1}{2p}.(\text{edge})^2.$

If the confine were rectangular, or all the angles at O right angles, we should have $\cos AOB = 0$, etc., and so

$$(\text{perpendicular})^2 = \frac{1}{p}(\text{edge})^2 = \frac{a^2}{n - 1},$$

which proves that the question does always reduce itself to the one now under consideration.

The content of a regular prime confine in p dimensions whose edge is a is[3]

$$= \frac{a^p}{p!}\left(\frac{p + 1}{2^p}\right)^{\frac{1}{2}}$$

Suppose this formula true for $p - 1$ dimensions; that is, let

$$V_{p-1} = \frac{a^{p-1}}{(p - 1)!}\left(\frac{p}{2^{p-1}}\right)^{\frac{1}{2}}.$$

Now, content of confine $= \dfrac{1}{p} \times$ perpendicular $\times$ content of face, or

$$V_p = \frac{a}{p}\left(\frac{p + 1}{2p}\right)^{\frac{1}{2}}.V_{p-1} = \frac{a^p}{p!}\left(\frac{p + 1}{2^p}\right)^{\frac{1}{2}}.$$

Hence the formula, if true for one value of p, is true for the next. It can be immediately verified in the case of $p = 1$. Therefore it is generally true.

The radius of the inscribed quasi-sphere $\rho = \dfrac{a}{\{2p(p + 1)\}^{\frac{1}{2}}}.$

---

[1] [The perpendicular produced to p times its length will give us OP. In fact, if we take the edges through O for a system of oblique axes the coordinates of P will all be equal and the line OP must make equal angles with the axes.]

[2] [See Salmon's *Geometry of Three Dimensions,* fourth edition, Dublin, 1882, p. 11.]

[3] [The edge in our case is $a\sqrt{2}$, but the ratio of the inscribed quasisphere does not depend on the length of the edge.]

We can divide the regular confine into $p + 1$ equal confines, each having the center of the inscribed quasi-sphere for vertex, and the content of one of these $= \dfrac{\rho}{p} \times$ content of face. But the sum of them all is equal to the content of the whole confine. Hence $(p + 1)\rho =$ perpendicular of confine[1]

$$= a\left(\frac{p+1}{2p}\right)^{\frac{1}{2}}, \text{ or, } \rho = \frac{a}{\{2p(p+1)\}^{\frac{1}{2}}}$$

The content of the quasi-sphere $= \rho^p \dfrac{\{\Gamma(\frac{1}{2})\}^p}{\Gamma(\frac{1}{2}p + 1)}$.

For it is the value of $\int\int\int \ldots dx\, dy\, dz \ldots$, the integral being so taken as to give to the variables all values consistent with the condition that $x^2 + y^2 + z^2 + \ldots$ is not greater than ρ^2 (see Todhunter's *Integral Calculus*, Art.[2] 271). Let C_p denote this content. Then

$$C_p = \rho^p \frac{\{\Gamma(\frac{1}{2})\}^p}{\Gamma(\frac{1}{2}p + 1)} = \frac{a^p}{(2p^2 + 2p)^{\frac{1}{2}p}} \cdot \frac{\{\Gamma(\frac{1}{2})\}^p}{\Gamma(\frac{1}{2}p + 1)}.$$

Therefore

$$\frac{C_p}{V_p} = \left(\frac{\pi}{p^2 + p}\right)^{\frac{1}{2}p} \cdot \frac{\Gamma(p + 1)}{\Gamma(\frac{1}{2}p + 1)} \cdot \frac{1}{(p + 1)^{\frac{1}{2}}}.$$

Restore $n - 1$ for p and we get the answer to the question, namely,

$$\left(\frac{\pi}{n^2 - n}\right)^{\frac{1}{2}(n-1)} \cdot \frac{\Gamma(n)}{\Gamma\{\frac{1}{2}(n + 1)\}} \cdot \frac{1}{n^{\frac{1}{2}}}.$$

8. The following are applications of the same method.

If a line be broken up at random into n pieces, the chance of an assigned two of them (the pth and qth from one end) being together greater than half the line is $n2^{1-n}$.

If n pieces be cut off at random, one from each of n equal lines, the chance that the pieces cannot be made into a polygon is $\dfrac{1}{(n - 1)!}$.

---

[1] [Two confines having the same base are to each other as their altitudes.]
[2] [In some editions at least (4th and 7th) this is Art. 275.]

IV. FIELD OF PROBABILITY

FERMAT AND PASCAL ON PROBABILITY

(Translated from the French by Professor Vera Sanford, Western Reserve University, Cleveland, Ohio.)

Italian writers of the fifteenth and sixteenth centuries, notably Pacioli (1494), Tartaglia (1556), and Cardan (1545), had discussed the problem of the division of a stake between two players whose game was interrupted before its close. The problem was proposed to Pascal and Fermat, probably in 1654, by the Chevalier de Méré, a gambler who is said to have had unusual ability "even for the mathematics." The correspondence which ensued between Fermat and Pascal, was fundamental in the development of modern concepts of probability, and it is unfortunate that the introductory letter from Pascal to Fermat is no longer extant. The one here translated, written in 1654, appears in the *Œuvres de Fermat* (ed. Tannery and Henry, Vol. II, pp. 288–314, Paris, 1894) and serves to show the nature of the problem. For a biographical sketch of Fermat, see page 213; of Pascal, page 67. See also pages 165, 213, 214, and 326.

Monsieur,

If I undertake to make a point with a single die in eight throws, and if we agree after the money is put at stake, that I shall not cast the first throw, it is necessary by my theory that I take $\frac{1}{6}$ of the total sum to be impartial because of the aforesaid first throw.

And if we agree after that that I shall not play the second throw, I should, for my share, take the sixth of the remainder that is $\frac{5}{36}$ of the total.

If, after that, we agree that I shall not play the third throw, I should to recoup myself, take $\frac{1}{6}$ of the remainder which is $\frac{25}{216}$ of the total.

And if subsequently, we agree again that I shall not cast the fourth throw, I should take $\frac{1}{6}$ of the remainder or $\frac{125}{1296}$ of the total, and I agree with you that that is the value of the fourth throw supposing that one has already made the preceding plays.

But you proposed in the last example in your letter (I quote your very terms) that if I undertake to find the six in eight throws and if I have thrown three times without getting it, and if my opponent

546

(Facing page 546.)

proposes that I should not play the fourth time, and if he wishes me to be justly treated, it is proper that I have $^{125}/_{1296}$ of the entire sum of our wagers.

This, however, is not true by my theory. For in this case, the three first throws having gained nothing for the player who holds the die, the total sum thus remaining at stake, he who holds the die and who agrees to not play his fourth throw should take $\frac{1}{6}$ as his reward.

And if he has played four throws without finding the desired point and if they agree that he shall not play the fifth time, he will, nevertheless, have $\frac{1}{6}$ of the total for his share. Since the whole sum stays in play it not only follows from the theory, but it is indeed common sense that each throw should be of equal value.

I urge you therefore (to write me) that I may know whether we agree in the theory, as I believe (we do), or whether we differ only in its application.

I am, most heartily, etc.,

Fermat.

Pascal to Fermat
Wednesday, July 29, 1654

Monsieur,—

1. Impatience has seized me as well as it has you, and although I am still abed, I cannot refrain from telling you that I received your letter in regard to the problem of the points[1] yesterday evening from the hands of M. Carcavi, and that I admire it more than I can tell you. I do not have the leisure to write at length, but, in a word, you have found the two divisions of the points and of the dice with perfect justice. I am thoroughly satisfied as I can no longer doubt that I was wrong, seeing the admirable accord in which I find myself with you.

I admire your method for the problem of the points even more than that of the dice. I have seen solutions of the problem of the dice by several persons, as M. le chevalier de Méré, who proposed the question to me, and by M. Roberval also. M. de Méré has

---

[1] [The editors of these letters note that the word *parti* means the division of the stake between the players in the case when the game is abandoned before its completion. *Parti des dés* means that the man who holds the die agrees to throw a certain number in a given number of trials. For clarity, in this translation, the first of these cases will be called the problem of the points, a term which has had a certain acceptance in the histories of mathematics, while the second may by analogy be called the problem of the dice.]

never been able to find the just value of the problem of the points nor has he been able to find a method of deriving it, so that I found myself the only one who knew this proportion.

2. Your method is very sound and it is the first one that came to my mind in these researches, but because the trouble of these combinations was excessive, I found an abridgment and indeed another method that is much shorter and more neat, which I should like to tell you here in a few words; for I should like to open my heart to you henceforth if I may, so great is the pleasure I have had in our agreement. I plainly see that the truth is the same at Toulouse and at Paris.

This is the way I go about it to know the value of each of the shares when two gamblers play, for example, in three throws, and when each has put 32 pistoles at stake:

Let us suppose that the first of them has *two* (points) and the other *one*. They now play one throw of which the chances are such that if the first wins, he will win the entire wager that is at stake, that is to say 64 pistoles. If the other wins, they will be *two* to *two* and in consequence, if they wish to separate, it follows that each will take back his wager that is to say 32 pistoles.

Consider then, Monsieur, that if the first wins, 64 will belong to him. If he loses, 32 will belong to him. Then if they do not wish to play this point, and separate without doing it, the first should say "I am sure of 32 pistoles, for even a loss gives them to me. As for the 32 others, perhaps I will have them and perhaps you will have them, the risk is equal. Therefore let us divide the 32 pistoles in half, and give me the 32 of which I am certain besides." He will then have 48 pistoles and the other will have 16.

Now let us suppose that the first has *two* points and the other *none*, and that they are beginning to play for a point. The chances are such that if the first wins, he will win all of the wager, 64 pistoles. If the other wins, behold they have come back to the preceding case in which the first has *two* points and the other *one*.

But we have already shown that in this case 48 pistoles will belong to the one who has *two* points. Therefore if they do not wish to play this point, he should say, "If I win, I shall gain all, that is 64. If I lose, 48 will legitimately belong to me. Therefore give me the 48 that are certain to be mine, even if I lose, and let us divide the other 16 in half because there is as much chance that you will gain them as that I will." Thus he will have 48 and 8, which is 56 pistoles.

Let us now suppose that the first has but *one* point and the other *none*. You see, Monsieur, that if they begin a new throw, the chances are such that if the first wins, he will have *two* points to *none*, and dividing by the preceding case, 56 will belong to him. If he loses, they will be point for point, and 32 pistoles will belong to him. He should therefore say, "If you do not wish to play, give me the 32 pistoles of which I am certain, and let us divide the rest of the 56 in half. From 56 take 32, and 24 remains. Then divide 24 in half, you take 12 and I take 12 which with 32 will make 44.

By these means, you see, by simple subtractions that for the first throw, he will have 12 pistoles from the other; for the second, 12 more; and for the last 8.

But not to make this more mysterious, inasmuch as you wish to see everything in the open, and as I have no other object than to see whether I am wrong, the value (I mean the value of the stake of the other player only) of the last play of *two* is double that of the last play of *three* and four times that of the last play of *four* and eight times that of the last play of *five*, etc.

3. But the ratio of the first plays is not so simple to find. This therefore is the method, for I wish to disguise nothing, and here is the problem of which I have considered so many cases, as indeed I was pleased to do: *Being given any number of throws that one wishes, to find the value of the first.*

For example, let the given number of throws be 8. Take the first eight even numbers and the first eight uneven numbers as:

$$2, 4, 6, 8, 10, 12, 14, 16$$

and

$$1, 3, 5, 7, \quad 9, 11, 13, 15.$$

Multiply the even numbers in this way: the first by the second, their product by the third, their product by the fourth, their product by the fifth, etc.; multiply the odd numbers in the same way: the first by the second, their product by the third, etc.

The last product of the even numbers is the *denominator* and the last product of the odd numbers is the numerator of the fraction that expresses the value of the first throw of *eight*. That is to say that if each one plays the number of pistoles expressed by the product of the even numbers, there will belong to him [who forfeits the throw] the amount of the other's wager expressed by the product of the odd numbers. This may be proved, but with much difficulty by combinations such as you have imagined, and I have not been able to prove it by this other method which I am about to tell you, but only by that of combinations. Here are the theorems which lead up to this which are properly arithmetic propositions regarding combinations, of which I have found so many beautiful properties:

4. If from any number of letters, as 8 for example,

$$A, B, C, D, E, F, G, H,$$

you take all the possible combinations of 4 letters and then all possible combinations of 5 letters, and then of 6, and then of 7, of 8, etc., and thus you would take all possible combinations, I say that if you add together half the combinations of 4 with each of the higher combinations, the sum will be the number equal to the number of the quaternary progression beginning with 2 which is half of the entire number.

For example, and I shall say it in Latin for the French is good for nothing:

If any number whatever of letters, for example 8,

$$A, B, C, D, E, F, G, H,$$

be summed in all possible combinations, by fours, fives, sixes, up to eights, I say, if you add half of the combinations by fours, that is 35 (half of 70) to all the combinations by fives, that is 56, and all the combinations by sixes, namely 28, and all the combinations by sevens, namely 8, and all the combinations by eights namely 1, the sum is the fourth number of the quaternary progression whose first term is 2. I say the fourth number for 4 is half of 8.

The numbers of the quaternary progressions whose first term is 2 are

$$2, 8, 32, 128, 512, \text{etc.},$$

of which 2 is the first, 8 the second, 32 the third, and 128 the fourth. Of these, the 128 equals:

+ 35 half of the combinations of 4 letters
+ 56 the combinations of 5 letters
+ 28 the combinations of 6 letters
+ 8 the combinations of 7 letters
+ 1 the combinations of 8 letters.

5. That is the first theorem, which is purely arithmetic. The other concerns the theory of the points and is as follows:

It is necessary to say first: if one (player) has *one* point out of 5, for example, and if he thus lacks 4, the game will infallibly be decided in 8 throws, which is double 4.

The value of the first throw of 5 in the wager of the other is the fraction which has for its numerator the half of the combinations of 4 things out of 8 (I take 4 because it is equal to the number of points that he lacks, and 8 because it is double the 4) and for the denominator this same numerator plus all the higher combinations.

Thus if I have one point out of 5, $35/128$ of the wager of my opponent belongs to me. That is to say, if he had wagered 128 pistoles, I would take 35 of them and leave him the rest, 93.

But this fraction $35/128$ is the same as $105/384$, which is made by the multiplication of the even numbers for the denominator and the multiplication of the odd numbers for the numerator.

You will see all of this without a doubt, if you will give yourself a little trouble, and for that reason I have found it unnecessary to discuss it further with you.

6. I shall send you, nevertheless, one of my old Tables; I have not the leisure to copy it, and I shall refer to it.

You will see here as always, that the value of the first throw is equal to that of the second, a thing which may easily be proved by combinations.

You will see likewise that the numbers of the first line are always increasing; those of the second do the same; those of the third the same.

But after that, those of the fourth line diminish; those of the fifth etc. This is odd.

If each wagers 256 on

	6 throws	5 throws	4 throws	3 throws	2 throws	1 throw
First throw	63	70	80	96	128	256
Second	63	70	80	96	128	
Third	56	60	64	64		
Fourth	42	40	32			
Fifth	24	16				
Sixth	8					

There belongs to me of the 256 pistoles of my opponent for the

		If each wagers 256 on					
		6 throws	5 throws	4 throws	3 throws	2 throws	1 throw
Of the 256 pistoles of my opponent, there belongs to me for the	First throw	63	70	80	96	128	256
	First two throws	126	140	160	192	256	
	First three throws	182	200	224	256		
	First four throws	224	240	256			
	First five throws	248	256				
	First six throws	256					

7. I have no time to send you the proof of a difficult point which astonished M. (de Méré) so greatly, for he has ability but he is not a geometer (which is, as you know, a great defect) and he does not even comprehend that a mathematical line is infinitely divisible and he is firmly convinced that it is composed of a finite number of points. I have never been able to get him out of it. If you could do so, it would make him perfect.

He tells me then that he has found an error in the numbers for this reason:

If one undertakes to throw a *six* with a die, the advantage of undertaking to do it in 4 is as 671 is to 625.

If one undertakes to throw double sixes with two dice the disadvantage of the undertaking is 24.

But nonetheless, 24 is to 36 (which is the number of faces of two dice)[1] as 4 is to 6 (which is the number of faces of one die).

This is what was his great scandal which made him say haughtily that the theorems were not consistent and that arithmetic was demented. But you will easily see the reason by the principles which you have.

I shall put all that I have done with this in order when I shall have finished the treatise on geometry[2] on which I have already been working for some time.

8. I have also done something with arithmetic on which subject, I beg you to give me your advice.

---

[1] [Clearly, the number of possible ways in which two dice can fall.]

[2] [Perhaps the manuscript which Leibniz saw, but which is not now extant.]

I proposed the lemma which every one accepts, that the sum of as many numbers as one wishes of the continuous progression from unity as

$$1, 2, 3, 4,$$

being taken by twos is equal to the last term 4 multiplied into the next greater, 5. That is to say that the sum of the integers[1] in A being taken by twos is equal to the product

$$A \times (A + 1).$$

I now come to my theorem:

If one be subtracted from the difference of the cubes of any two consecutive numbers, the result is six times all the numbers contained in the root of the lesser number.

Let the two roots R and S differ by unity. I say that $R^3 - S^3 - 1$ is equal to six times the sum of the numbers contained in S.

Let S be called A, then R is $A + 1$. Therefore the cube of the root R or $A + 1$ is

$$A^3 + 3A^2 + 3A + 1^3.$$

The cube of S, or A, is A^3, and the difference of these is $R^3 - S^3$; therefore, if unity be subtracted, $3A^2 + 3A$ is equal to $R^3 - S^3 - 1$. But by the lemma, double the sum of the numbers contained in A or S is equal to $A \times (A + 1)$; that is, to $A^2 + A$. Therefore, six times the sum of the numbers in A is equal to $3A^2 + 3A$. But $3A^2 + 3A$ is equal to $R^3 - S^3 - 1$. Therefore $R^3 - S^3 - 1$ is equal to six times the sum of the numbers contained in A or S. *Quod erat demonstrandum.* No one has caused me any difficulty in regard to the above, but they have told me that they did not do so for the reason that everyone is accustomed to this method today. As for myself, I mean that without doing me a favor, people should admit this to be an excellent type of proof. I await your comment, however, with all deference. All that I have proved in arithmetic is of this nature.

9. Here are two further difficulties: I have proved a plane theorem making use of the cube of one line compared with the cube of another. I mean that this is purely geometric and in the greatest rigor. By these means I solved the problem: "Any four planes, any four points, and any four spheres being given, to find a sphere which, touching the given spheres, passes through

---

[1] ["...des nombres contenus dans A."]

the given points, and leaves on the planes segments in which given angles may be inscribed;"[1] and this one: "Any three circles, any three points, and any three lines being given, to find a circle which touches the circles and the points and leaves on the lines an arc in which a given angle may be inscribed."

I solved these problems in a plane, using nothing in the construction but circles and straight lines, but in the proof I made use of solid loci,[2]—of parabolas, or hyperbolas. Nevertheless, inasmuch as the construction is in a plane, I maintain that my solution is plane, and that it should pass as such.

This is a poor recognition of the honor which you have done me in putting up with my discourse which has been plaguing you so long. I never thought I should say two words to you and if I were to tell you what I have uppermost in my heart,—which is that the better I know you the more I honor and admire you,— and if you were to see to what degree that is, you would allot a place in your friendship for him who is, Monsieur, your etc.

Pascal to Fermat
Monday, August 24, 1654

Monsieur,

1. I was not able to tell you my entire thoughts regarding the problem of the points by the last post,[3] and at the same time, I have a certain reluctance at doing it for fear lest this admirable harmony which obtains between us and which is so dear to me should begin to flag, for I am afraid that we may have different opinions on this subject. I wish to lay my whole reasoning before you, and to have you do me the favor to set me straight if I am in error or to indorse me if I am correct. I ask you this in all faith and sincerity for I am not certain even that you will be on my side.

When there are but *two* players, your theory which proceeds by combinations is very just. But when there are three, I believe I have a proof that it is unjust that you should proceed in any other manner than the one I have. But the method which I have disclosed to you and which I have used universally is common to all imaginable conditions of all distributions of points, in the place of that of combinations (which I do not use except in partic-

---

[1] ["...capable d'angles donnés."]

[2] [A common name for conics.]

[3] ["...par l'ordinaire passé." Cf. the English expression, by the "last ordinary."]

ular cases when it is shorter than the general method), a method which is good only in isolated cases and not good for others.

I am sure that I can make it understood, but it requires a few words from me and a little patience from you.

2. This is the method of procedure when there are *two* players: If two players, playing in several throws, find themselves in such a state that the first lacks *two* points and the second *three* of gaining the stake, you say it is necessary to see in how many points the game will be absolutely decided.

It is convenient to suppose that this will be in *four* points, from which you conclude that it is necessary to see how many ways the four points may be distributed between the two players and to see how many combinations there are to make the first win and how many to make the second win, and to divide the stake according to that proportion. I could scarcely understand this reasoning if I had not known it myself before; but you also have written it in your discussion. Then to see how many ways four points may be distributed between two players, it is necessary to imagine that they play with dice with two faces (since there are but two players), as heads and tails, and that they throw four of these dice (because they play in four throws). Now it is necessary to see how many ways these dice may fall. That is easy to calculate. There can be *sixteen*, which is the second power of *four*; that is to say, the square. Now imagine that one of the faces is marked *a*, favorable to the first player. And suppose the other is marked *b*, favorable to the second. Then these four dice can fall according to one of these sixteen arrangements:

a	a	a	a	a	a	a	a	b	b	b	b	b	b	b	b
a	a	a	a	b	b	b	b	a	a	a	a	b	b	b	b
a	a	b	b	a	a	b	b	a	a	b	b	a	a	b	b
a	b	a	b	a	b	a	b	a	b	a	b	a	b	a	b
1	1	1	1	1	1	1	2	1	1	1	2	1	2	2	2

and, because the first player lacks two points, all the arrangements that have two *a*'s make him win. There are therefore 11 of these for him. And because the second lacks three points, all the arrangements that have three *b*'s make him win. There are 5 of these. Therefore it is necessary that they divide the wager as 11 is to 5.

There is your method, when there are *two* players, whereupon you say that if there are more players, it will not be difficult to make the division by this method.

3. On this point, Monsieur, I tell you that this division for the two players founded on combinations is very equitable and good, but that if there are more than two players, it is not always just and I shall tell you the reason for this difference.   I communicated your method to [some of] our gentlemen, on which M. de Roberval made me this objection:

That it is wrong to base the method of division on the supposition that they are playing in *four* throws seeing that when one lacks *two* points and the other *three*, there is no necessity that they play *four* throws since it may happen that they play but *two* or *three*, or in truth perhaps *four*.

Since he does not see why one should pretend to make a just division on the assumed condition that one plays *four* throws, in view of the fact that the natural terms of the game are that they do not throw the dice after one of the players has won; and that at least if this is not false, it should be proved.   Consequently he suspects that we have committed a paralogism.

I replied to him that I did not found my reasoning so much on this method of combinations, which in truth is not in place on this occasion, as on my universal method from which nothing escapes and which carries its proof with itself.   This finds precisely the same division as does the method of combinations.   Furthermore, I showed him the truth of the divisions between two players by combinations in this way: Is it not true that if two gamblers finding according to the conditions of the hypothesis that one lacks *two* points and the other *three*, mutually agree that they shall play four complete plays, that is to say, that they shall throw four two-faced dice all at once,—is it not true, I say, that if they are prevented from playing the four throws, the division should be as we have said according to the combinations favorable to each?   He agreed with this and this is indeed proved.   But he denied that the same thing follows when they are not obliged to play the four throws.   I therefore replied as follows:

It is not clear that the same gamblers, not being constrained to play the four throws, but wishing to quit the game before one of them has attained his score, can without loss or gain be obliged to play the whole four plays, and that this agreement in no way changes their condition?   For if the first gains the two first points of four, will he who has won refuse to play two throws more, seeing that if he wins he will not win more and if he loses he will not win less?   For the two points which the other wins are not sufficient for him since he lacks three, and there are not enough [points] in four throws for each to make the number which he lacks.

It certainly is convenient to consider that it is absolutely equal and indifferent to each whether they play in the natural way of the game, which is to finish as soon as one has his score, or whether they play the entire four throws. Therefore, since these two conditions are equal and indifferent, the division should be alike for each. But since it is just when they are obliged to play.the four throws as I have shown, it is therefore just also in the other case.

That is the way I prove it, and, as you recollect, this proof is based on the equality of the two conditions true and assumed in regard to the two gamblers, the division is the same in each of the methods, and if one gains or loses by one method, he will gain or lose by the other, and the two will always have the same accounting.

4. Let us follow the same argument for *three* players and let us assume that the first lacks *one* point, the second *two*, and the third *two*. To make the division, following the same method of combinations, it is necessary to first discover in how many points the game may be decided as we did when there were two players. This will be in three points for they cannot play three throws without necessarily arriving at a decision.

It is now necessary to see how many ways three throws may be combined among three players and how many are favorable to the first, how many to the second, and how many to the third, and to follow this proportion in distributing the wager as we did in the hypothesis of the two gamblers.

It is easy to see how many combinations there are in all. This is the third power of 3; that is to say, its cube, or 27. For if one throws three dice at a time (for it is necessary to throw three times), these dice having three faces each (since there are three players), one marked *a* favorable to the first, one marked *b* favorable to the second, and one marked *c* favorable to the third,—it is evident that these three dice thrown together can fall in 27 different ways as:

a	a	a	a	a	a	a	a	a	b	b	b	b	b	b	b	b	b	c	c	c	c	c	c	c	c	c
a	a	a	b	b	b	c	c	c	a	a	a	b	b	b	c	c	c	a	a	a	b	b	b	c	c	c
a	b	c	a	b	c	a	b	c	a	b	c	a	b	c	a	b	c	a	b	c	a	b	c	a	b	c
1	1	1	1	1	1	1	1	1	1	1	1	1			1			1	1	1	1			1		
													2	2		2						2				
																	3						3		3	3

Since the first lacks but *one* point, then all the ways in which there is one *a* are favorable to him. There are 19 of these. The second lacks *two* points. Thus all the arrangements in which there are two *b*'s are in his favor. There are 7 of them. The third lacks *two* points. Thus all the arrangements in which there are two *c*'s are favorable to him. There are 7 of these.

If we conclude from this that it is necessary to give each according to the proportion 19, 7, 7, we are making a serious mistake and I would hesitate to believe that you would do this. There are several cases favorable to both the first and the second, as *abb* has the *a* which the first needs, and the two *b*'s which the second needs. So too, the *acc* is favorable to the first and third.

It therefore is not desirable to count the arrangements which are common to the two as being worth the whole wager to each, but only as being half a point. For if the arrangement *acc* occurs, the first and third will have the same right to the wager, each making their score. They should therefore divide the wager in half. If the arrangement *aab* occurs, the first alone wins. It is necessary to make this assumption:

There are 13 arrangements which give the entire wager to the first, and 6 which give him half and 8 which are worth nothing to him. Therefore if the entire sum is one pistole, there are 13 arrangements which are each worth one pistole to him, there are 6 that are each worth ½ a pistole, and 8 that are worth nothing.

Then in this case of division, it is necessary to multiply

13 by one pistole which makes 13
6 by one half which makes   3
8 by zero which makes   0
Total 27   Total 16

and to divide the sum of the values 16 by the sum of the arrangements 27, which makes the fraction $\frac{16}{27}$ and it is this amount which belongs to the first gambler in the event of a division; that is to say, 16 pistoles out of 27.

The shares of the second and the third gamblers will be the same:

There are 4 arrangements which are worth 1 pistole; multiplying, 4
There are 3 arrangements which are worth ½ pistole; multiplying, 1½
And   20 arrangements which are worth nothing   0
Total   27   Total 5½

Therefore 5½ pistoles belong to the second player out of 27, and the same to the third. The sum of the 5½, 5½, and 16 makes 27.

5. It seems to me that this is the way in which it is necessary to make the division by combinations according to your method, unless you have something else on the subject which I do not know. But if I am not mistaken, this division is unjust.

The reason is that we are making a false supposition,—that is, that they are playing three throws without exception, instead of the natural condition of this game which is that they shall not play except up to the time when one of the players has attained the number of points which he lacks, in which case the game ceases.

It is not that it may not happen that they will play three times, but it may happen that they will play once or twice and not need to play again.

But, you will say, why is it possible to make the same assumption in this case as was made in the case of the two players? Here is the reason: In the true condition [of the game] between three players, only one can win, for by the terms of the game it will terminate when one [of the players] has won. But under the assumed conditions, two may attain the number of their points, since the first may gain the one point he lacks and one of the others may gain the two points which he lacks, since they will have played only three throws. When there are only two players, the assumed conditions and the true conditions concur to the advantage of both. It is this that makes the greatest difference between the assumed conditions and the true ones.

If the players, finding themselves in the state given in the hypothesis,—that is to say, if the first lacks *one* point, the second *two*, and the third *two;* and if they now mutually agree and concur in the stipulation that they will play *three* complete throws; and if he who makes the points which he lacks will take the entire sum if he is the only one who attains the points; or if two should attain them that they shall share equally,—*in this case*, the division should be made as I give it here: the first shall have 16, the second 5½, and the third 5½ out of 27 pistoles, and this carries with it its own proof on the assumption of the above condition.

But if they play simply on the condition that they will not necessarily play three throws, but that they will only play until one of them shall have attained his points, and that then the play shall cease without giving another the opportunity of reaching his score, then 17 pistoles should belong to the first, 5 to the second, and 5 to the third, out of 27. And this is found by my general method which also determines that, under the preceeding condition, the first should have 16, the second $5\frac{1}{2}$, and the third $5\frac{1}{2}$, without making use of combinations,—for this works in all cases and without any obstacle.

6. These, Monsieur, are my reflections on this topic on which I have no advantage over you except that of having meditated on it longer, but this is of little [advantage to me] from your point of view since your first glance is more penetrating than are my prolonged endeavors.

I shall not allow myself to disclose to you my reasons for looking forward to your opinions. I believe you have recognized from this that the theory of combinations is good for the case of two players by accident, as it is also sometimes good in the case of three gamblers, as when one lacks *one* point, another *one*, and the other *two*[1] because, in this case, the number of points in which the game is finished is not enough to allow two to win, but it is not a general method and it is good only in the case where it is necessary to play exactly a certain number of times.

Consequently, as you did not have my method when you sent me the division among several gamblers, but [since you had] only that of combinations, I fear that we hold different views on the subject.

I beg you to inform me how you would proceed in your research on this problem. I shall receive your reply with respect and joy, even if your opinions should be contrary to mine. I am etc.

Fermat to Pascal
Saturday, August 29, 1654

Monsieur,

1. Our interchange of blows still continues, and I am well pleased that our thoughts are in such complete adjustment as it seems since they have taken the same direction and followed the same road. Your recent *Traité du triangle aritbmétique*[2] and its applications are an authentic proof and if my computations do

---

[1] [Evidently a misprint, since two throws may be needed.]
[2] [See p. 67.]

me no wrong, your eleventh consequence[1] went by post from Paris to Toulouse while my theorem on figurate numbers,[2] which is virtually the same, was going from Toulouse to Paris.

I have not been on watch for failure while I have been at work on the problem and I am persuaded that the true way to escape failure is by concurring with you. But if I should say more, it would be of the nature of a compliment and we have banished that enemy of sweet and easy conversation.

It is now my turn to give you some of my numerical discoveries, but the end of the parliament augments my duties and I hope that out of your goodness you will allow me due and almost necessary respite.

2. I will reply however to your question of the three players who play in two throws. When the first has one [point] and the others none, your first solution is the true one and the division of the wager should be 17, 5, and 5. The reason for this is self-evident and it always takes the same principle, the combinations making it clear that the first has 17 changes while each of the others has but five.

3. For the rest, there is nothing that I will not write you in the future with all frankness. Meditate however, if you find it convenient, on this theorem: The squared powers of 2 augmented by unity[3] are always prime numbers. [That is,]

The square of 2 augmented by unity makes 5 which is a prime number;

The square of the square makes 16 which, when unity is added, makes 17, a prime number;

The square of 16 makes 256 which, when unity is added, makes 257, a prime number;

The square of 256 makes 65536 which, when unity is added, makes 65537, a prime number;

and so to infinity.

This is a property whose truth I will answer to you. The proof of it is very difficult and I assure you that I have not yet been able to find it fully. I shall not set it for you to find unless I come to the end of it.

---

[1] [From the *Traité du triangle arithmétique*,—"Each cell on the diagonal is double that which preceded it in the parallel or perpendicular rank."]

[2] [I. e., the theorem that $A(A + 1)$ is double the triangular number $1 + 2 + 3 + \ldots A$, See p. 553.]

[3] [I. e. $2^{2^n} + 1$. Euler (1732) showed the falsity of the statement.]

This theorem serves in the discovery of numbers which are in a given ratio to their aliquot parts, concerning which I have made many discoveries. We will talk of that another time.

I am Monsieur, yours etc.

Fermat.

At Toulouse, the twenty ninth of August, 1654.

Fermat to Pascal
Friday, September 25, 1654

Monsieur,

1. Do not be apprehensive that our argument is coming to an end. You have strengthened it yourself in thinking to destroy it and it seems to me that in replying to M. de Roberval for yourself you have also replied for me.

In taking the example of the three gamblers of whom the first lacks one point, and each of the others lack two, which is the case in which you oppose, I find here only 17 combinations for the first and 5 for each of the others; for when you say that the combination *acc* is good for the first, recollect that everything that is done after one of the players has won is worth nothing. But this combination having made the first win on the first die, what does it matter that the third gains two afterwards, since even when he gains thirty all this is superfluous? The consequence, as you have well called it "this fiction," of extending the game to a certain number of plays serves only to make the rule easy and (according to my opinion) to make all the chances equal; or better, more intelligibly to reduce all the fractions to the same denomination.

So that you may have no doubt, if instead of *three* parties you extend the assumption to *four*, there will not be 27 combinations only, but 81; and it will be necessary to see how many combinations make the first gain his point later than each of the others gains two, and how many combinations make each of the others win two later than the first wins one. You will find that the combinations that make the first win are 51 and those for each of the other two are 15, which reduces to the same proportion. So that if you take five throws or any other number you please, you will always find three numbers in the proportion of 17, 5, 5. And accordingly I am right in saying that the combination *acc* is [favorable] for the first only and not for the third, and that *cca* is only for the third and not for the first, and consequently my law of combinations is the same for three players as for two, and in general for all numbers.

2. You have already seen from my previous letter that I did not demur at the true solution of the question of the three gamblers for which I sent you the three definite numbers, 17, 5, 5. But because M. de Roberval will perhaps be better satisfied to see a solution without any dissimulation and because it may perhaps yield to abbreviations in many cases, here is an example:

The first may win in a single play, or in two or in three.

If he wins in a single throw, it is necessary that he makes the favorable throw with a three-faced die at the first trial. A single die will yield three chances. The gambler then has ⅓ of the wager because he plays only one third.

If he plays twice, he can gain in two ways,—either when the second gambler wins the first and he the second, or when the third wins the throw and when he wins the second. But two dice produce 9 chances. The player than has ⅔ of the wager when they play twice.

But if he plays three times, he can win only in two ways, either the second wins on the first throw and the third wins the second, and he the third; or when the third wins the first throw, the second the second, and he the third; for if the second or the third player wins the two first, he will win the wager and the first player will not. But three dice give 27 chances of which the first player has ²/₂₇ of the chances when they play three rounds.

The sum of the chances which makes the first gambler win is consequently ⅓, ⅔, and ²/₂₇, which makes 17/27.

This rule is good and general in all cases of the type where, without recurring to assumed conditions, the true combinations of each number of throws give the solution and make plain what I said at the outset that the extension to a certain number of points is nothing else than the reduction of divers fractions to the same denomination. Here in a few words is the whole of the mystery, which reconciles us without doubt although each of us sought only reason and truth.

3. I hope to send you at Martinmas an abridgment of all that I have discovered of note regarding numbers. You allow me to be concise [since this suffices] to make myself understood to a man [like yourself] who comprehends the whole from half a word. What you will find most important is in regard to the theorem that every number is composed of one, two, or three triangles;[1] of

---

[1] [*I. e.*, triangular numbers.]

one, two, three, or four squares;[1] of one, two, three, four, or five pentagons; of one, two, three, four, five, or six hexagons, and thus to infinity.

To derive this, it is necessary to show that every prime number which is greater by unity than a multiple of 4 is composed of two squares, as 5, 13, 17, 29, 37, etc.

Having given a prime number of this type, as 53, to find by a general rule the two squares which compose it.

Every prime number which is greater by unity than a multiple of 3, is composed of a square and of the triple of another square, as 7, 13, 19, 31, 37, etc.

Every prime number which is greater by 1 or by 3 than a multiple of 8, is composed of a square and of the double of another square, as 11, 17, 19, 41, 43, etc.

There is no triangle of numbers whose area is equal to a square number.

This follows from the invention of many theorems of which Bachet vows himself ignorant and which are lacking in Diophantus.

I am persuaded that as soon as you will have known my way of proof in this type of theorem, it will seem good to you and that it will give you the opportunity for a multitude of new discoveries, for it follows as you know that *multi pertranseant ut augeatur scientia.*

When I have time, we will talk further of magic numbers and I will summarize my former work on this subject.

I am, Monsieur, most heartily your etc.

Fermat.

The twenty-fifth of September.
I am writing this from the country, and this may perhaps delay my replies during the holidays.

Pascal to Fermat
Tuesday, October 27, 1654

Monsieur,

Your last letter satisfied me perfectly.   I admire your method for the problem of the points, all the more because I understand it well.   It is entirely yours, it has nothing in common with mine, and it reaches the same end easily.   Now our harmony has begun again.

But, Monsieur, I agree with you in this, find someone elsewhere to follow you in your discoveries concerning numbers, the state-

---

[1] [See page 91.]

ments of which you were so good as to send me. For my own part, I confess that this passes me at a great distance; I am competent only to admire it and I beg you most humbly to use your earliest leisure to bring it to a conclusion. All of our gentlemen saw it on Saturday last and appreciate it most heartily. One cannot often hope for things that are so fine and so desirable. Think about it if you will, and rest assured that I am etc.

Pascal.

Paris, October 27, 1654.

DE MOIVRE

ON THE LAW OF NORMAL PROBABILITY

(Edited by Professor Helen M. Walker, Teachers College, Columbia
University, New York City.)

Abraham de Moivre (1667–1754) left France at the revocation of the Edict
of Nantes and spent the rest of his life in London, where he solved problems
for wealthy patrons and did private tutoring in mathematics. He is best
known for his work on trigonometry, probability, and annuities. On November 12, 1733 he presented privately to some friends a brief paper of seven pages
entitled "Approximatio ad Summam Terminorum Binomii $\overline{a + b}^n$ in Seriem
expansi." Only two copies of this are known to be extant. His own translation with some additions, was included in the second edition (1738) of *The
Doctrine of Chances*, pages 235–243.

This paper gave the first statement of the formula for the "normal curve,"
the first method of finding the probability of the occurrence of an error of a
given size when that error is expressed in terms of the variability of the distribution as a unit, and the first recognition of that value later termed the
probable error. It shows, also, that before Stirling, De Moivre had been
approaching a solution of the value of factorial *n*.

*A Method of approximating the Sum of the Terms of the Binomial
$\overline{a+b}^n$ expanded into a Series, from whence are deduced some practical Rules to estimate the Degree of Assent which is to be given to
Experiments.*

Altho' the Solution of Problems of Chance often require that
several Terms of the Binomial $\overline{a+b}^n$ be added together, nevertheless in very high Powers the thing appears so laborious, and of so
great a difficulty, that few people have undertaken that Task;
for besides *James* and *Nicolas Bernoulli*, two great Mathematicians, I know of no body that has attempted it; in which, tho'
they have shewn very great skill, and have the praise which is
due to their Industry, yet some things were farther required;
for what they have done is not so much an Approximation as the
determining very wide limits, within which they demonstrated
that the Sum of the Terms was contained. Now the Method
which they have followed has been briefly described in my *Miscellanea Analytica*, which the Reader may consult if he pleases,

unless they rather chuse, which perhaps would be the best, to consult what they themselves have writ upon that Subject: for my part, what made me apply myself to that Inquiry was not out of opinion that I should excel others, in which however I might have been forgiven; but what I did was in compliance to the desire of a very worthy Gentleman, and good Mathematician, who encouraged me to it: I now add some new thoughts to the former; but in order to make their connexion the clearer, it is necessary for me to resume some few things that have been delivered by me a pretty while ago.

I. It is now a dozen years or more since I had found what follows; If the Binomial $1+1$ be raised to a very high Power denoted by n, the ratio which the middle Term has to the Sum of all the Terms, that is, to 2^n, may be expressed by the Fraction $\dfrac{2A \times \overline{n-1}^n}{n^n \times \sqrt{n-1}}$, wherein A represents the number of which the Hyperbolic Logarithm is $\dfrac{1}{12}-\dfrac{1}{360}+\dfrac{1}{1260}-\dfrac{1}{1680}$, &c. but because the Quantity $\dfrac{\overline{n-1}^n}{n^n}$ or $\overline{1-\dfrac{1}{n}}^n$ is very nearly given when n is a high Power, which is not difficult to prove, it follows that, in an infinite Power, that Quantity will be absolutely given, and represent the number of which the Hyperbolic Logarithm is -1; from whence it follows, that if B denotes the Number of which the Hyperbolic Logarithm is $-1+\dfrac{1}{12}-\dfrac{1}{360}+\dfrac{1}{1260}-\dfrac{1}{1680}$, &c. the Expression above-written will become $\dfrac{2B}{\sqrt{n-1}}$ or barely $\dfrac{2B}{\sqrt{n}}$, and that therefore if we change the Signs of that Series, and now suppose that B represents the Number of which the Hyperbolic Logarithm is $1-\dfrac{1}{12}+\dfrac{1}{360}-\dfrac{1}{1260}+\dfrac{1}{1680}$, &c. that Expression will be changed into $\dfrac{2}{B\sqrt{n}}$.

When I first began that inquiry, I contented myself to determine at large the Value of B, which was done by the addition of some Terms of the above-written Series; but as I perceiv'd that it converged but slowly, and seeing at the same time that what I had done answered my purpose tolerably well, I desisted from proceeding farther, till my worthy and learned Friend Mr. *James Stirling*, who had applied himself after me to that inquiry, found that the Quantity B did denote the Square-root of the Circumference of a Circle whose Radius is Unity, so that if that Circumference be called c, the Ratio of the middle Term to the Sum of all the Terms will be expressed by $\dfrac{2}{\sqrt{nc}}$.[1]

But altho' it be not necessary to know what relation the number B may have to the Circumference of the Circle, provided its value be attained, either by pursuing the Logarithmic Series before mentioned, or any other way; yet I own with pleasure that this discovery, besides that it has saved trouble, has spread a singular Elegancy on the Solution.

II. I also found that the Logarithm of the Ratio which the middle Term of a high Power has to any Term distant from it by an Interval denoted by l, would be denoted by a very near approximation, (supposing $m = \frac{1}{2}n$) by the Quantities $\overline{m+l-\frac{1}{2}}\times$log. $\overline{m+l-1} + \overline{m-l+\frac{1}{2}}\times$log.$\overline{m-l+1} - 2m\times$log. $m + $log.$\dfrac{m+l}{m}$.

COROLLARY I.

This being admitted, I conclude, that if m or $\frac{1}{2}n$ be a Quantity infinitely great, then the Logarithm of the Ratio, which a Term distant from the middle by the Interval l, has to the middle Term, is $-\dfrac{2ll}{n}$.[2]

COROLLARY 2.

The Number, which answers to the Hyperbolic Logarithm $-\dfrac{2ll}{n}$, being

$$1 - \frac{2ll}{n} + \frac{4l^4}{2nn} - \frac{8l^6}{6n^3} + \frac{16l^8}{24n^4} - \frac{32l^{10}}{120n^5} + \frac{64l^{12}}{720n^6}, \&c.$$

---

[1] [Under the circumstances of De Moivre's problem, nc is equivalent to $8\sigma^2\pi$, where σ is the standard deviation of the curve. This statement therefore shows that De Moivre knew the maximum ordinate of the curve to be

$$y_0 = \frac{1}{\sigma\sqrt{2\pi}}.]$$

[2] [Since $n = 4\sigma^2$ under the assumptions made here, this is equivalent to stating the formula for the curve as

$$y = y_0 e^{-\frac{x^2}{2\sigma^2}}.]$$

it follows, that the Sum of the Terms intercepted between the Middle, and that whose distance from it is denoted by l, will be

$$\frac{2}{\sqrt{nc}} \text{ into } l - \frac{2l^3}{1\times3n} + \frac{4l^5}{2\times5nn} - \frac{8l^7}{6\times7n^3} + \frac{16l^9}{24\times9n^4} - \frac{32l^{11}}{120\times11n^5}, \text{ \&c.}$$

Let now l be supposed $= s\sqrt{n}$, then the said Sum will be expressed by the Series

$$\frac{2}{\sqrt{c}} \text{ into } \int - \frac{2\int^3}{3} + \frac{4\int^5}{2\times5} - \frac{8\int^7}{6\times7} + \frac{16\int^9}{24\times9} - \frac{32\int^{11}}{120\times11}, \text{ \&c.}[1]$$

Moreover, if $\int$ be interpreted by $\frac{1}{2}$, then the Series will become

$$\frac{2}{\sqrt{c}} \text{ into } \frac{1}{2} - \frac{1}{3\times4} + \frac{1}{2\times5\times8} - \frac{1}{6\times7\times16} + \frac{1}{24\times9\times32} - \frac{1}{120\times11\times64},$$

&c. which converges so fast, that by help of no more than seven or eight Terms, the Sum required may be carried to six or seven places of Decimals: Now that Sum will be found to be 0.427812, independently from the common Multiplicator $\frac{2}{\sqrt{c}}$, and therefore to the Tabular Logarithm of 0.427812, which is $\overline{9}.6312529$, adding the Logarithm of $\frac{2}{\sqrt{c}}$, viz. $\overline{9}.9019400$, the Sum will be $\overline{19}.5331929$, to which answers the number 0.341344.

Lemma.

If an Event be so dependent on Chance, as that the Probabilities of its happening or failing be equal, and that a certain given number n of Experiments be taken to observe how often it happens and fails, and also that l be another given number, less than $\frac{1}{2}n$, then the Probability of its neither happening more frequently than $\frac{1}{2}n+l$ times, nor more rarely than $\frac{1}{2}n-l$ times, may be found as follows.

Let L and L be two Terms equally distant on both sides of the middle Term of the Binomial $\overline{1+1}^n$ expanded, by an Interval equal to l; let also $\int$ be the Sum of the Terms included between L and L together with the Extreams, then the Probability required will be rightly expressed by the Fraction $\frac{\int}{2^n}$, which being founded on the common Principles of the Doctrine of Chances, requires no Demonstration in this place.

---

[1] [The long s which De Moivre employed in this formula is not to be mistaken for the integral sign.]

COROLLARY 3.

And therefore, if it was possible to take an infinite number of Experiments, the Probability that an Event which has an equal number of Chances to happen or fail, shall neither appear more frequently than $\frac{1}{2}n + \frac{1}{2}\sqrt{n}$ times, nor more rarely than $\frac{1}{2}n - \frac{1}{2}n\sqrt{n}$ times, will be express'd by the double Sum of the number exhibited in the second Corollary, that is, by 0.682688, and consequently the Probability of the contrary, which is that of happening more frequently or more rarely than in the proportion above assigned will be 0.317312, those two Probabilities together compleating Unity, which is the measure of Certainty: Now the Ratio of those Probabilities is in small Terms 28 to 13 very near.

COROLLARY 4.

But altho' the taking an infinite number of Experiments be not practicable, yet the preceding Conclusions may very well be applied to finite numbers, provided they be great, for Instance, if 3600 Experiments be taken, make $n = 3600$, hence $\frac{1}{2}n$ will be $= 1800$, and $\frac{1}{2}\sqrt{n} = 30$, then the Probability of the Event's neither appearing oftner than 1830 times, nor more rarely than 1770, will be 0.682688.

COROLLARY 5.

And therefore we may lay this down for a fundamental Maxim, that in high Powers, the Ratio, which the Sum of the Terms included between two Extreams distant on both sides from the middle Term by an Interval equal to $\frac{1}{2}\sqrt{n}$, bears to the Sum of all the Terms, will be rightly express'd by the Decimal 0.682688, that is $\frac{28}{41}$ nearly.

Still, it is not to be imagin'd that there is any necessity that the number n should be immensely great; for supposing it not to reach beyond the 900th Power, nay not even beyond the 100th, the Rule here given will be tolerably accurate, which I have had confirmed by Trials.

But it is worth while to observe, that such a small part as is $\frac{1}{2}\sqrt{n}$ in respect to n, and so much the less in respect to n as n increases, does very soon give the Probability $\frac{28}{41}$ or the Odds of 28 to 13; from whence we may naturally be led to enquire, what are the

Bounds within which the proportion of Equality is contained; I answer, that these Bounds will be set at such a distance from the middle Term, as will be expressed by $\frac{1}{4}\sqrt{2n}$ very near; so in the case above mentioned, wherein n was supposed $= 3600$, $\frac{1}{4}\sqrt{2n}$ will be about 21.2 nearly, which in respect to 3600, is not above $\frac{1}{169}$th part: so that it is an equal Chance nearly, or rather something more, that in 3600 Experiments, in each of which an Event may as well happen as fail, the Excess of the happenings or failings above 1800 times will be no more than about 21.

COROLLARY 6.

If l be interpreted by $\sqrt{n}$, the Series will not converge so fast as it did in the former case when l was interpreted by $\frac{1}{2}\sqrt{n}$, for here no less than 12 or 13 Terms of the Series will afford a tolerable approximation, and it would still require more Terms, according as l bears a greater proportion to $\sqrt{n}$, for which reason I make use in this case of the Artifice of Mechanic Quadratures, first invented by Sir *Isaac Newton*, and since prosecuted by Mr. *Cotes*, Mr. *James Stirling*, myself, and perhaps others; it consists in determining the Area of a Curve nearly, from knowing a certain number of its Ordinates A, B, C, D, E, F, &c. placed at equal Intervals, the more Ordinates there are, the more exact will the Quadrature be; but here I confine myself to four, as being sufficient for my purpose: let us therefore suppose that the four Ordinates are A, B, C, D, and that the Distance between the first and last is denoted by l, then the Area contained between the first and the last will be $\dfrac{1 \times \overline{A+D} + 3 \times \overline{B+C}}{8} \times l$; now let us take the Distances $0\sqrt{n}$, $\frac{1}{6}\sqrt{n}$, $\frac{2}{6}\sqrt{n}$, $\frac{3}{6}\sqrt{n}$, $\frac{4}{6}\sqrt{n}$, $\frac{5}{6}\sqrt{n}$, $\frac{6}{6}\sqrt{n}$, of which every one exceeds the preceding by $\frac{1}{6}\sqrt{n}$, and of which the last is $\sqrt{n}$; of these let us take the four last, *viz.* $\frac{3}{6}\sqrt{n}$, $\frac{4}{6}\sqrt{n}$, $\frac{5}{6}\sqrt{n}$, $\frac{6}{6}\sqrt{n}$, then taking their Squares, doubling each of them, dividing them all by n, and prefixing to them all the Sign $-$, we shall have $-\frac{1}{2}$, $-\frac{8}{9}$, $-\frac{25}{18}$, $-\frac{2}{1}$, which must be look'd upon as Hyperbolic Logarithms, of which consequently the corresponding numbers, *viz.* 0.60653, 0.41111, 0.24935, 0.13534 will stand for the four Ordinates A, B, C, D.

Now having interpreted l by $\frac{1}{2}\sqrt{n}$, the Area will be found to be $=0.170203 \times \sqrt{n}$, the double of which being multiplied by $\frac{2}{\sqrt{nc}}$, the product will be 0.27160; let therefore this be added to the Area found before, that is, to 0.682688, and the Sum 0.95428 will shew, what after a number of Trials denoted by n, the Probability will be of the Event's neither happening oftner than $\frac{1}{2}n+\sqrt{n}$ times, nor more rarely than $\frac{1}{2}n-\sqrt{n}$, and therefore the Probability of the contrary will be 0.04572, which shews that the Odds of the Event's neither happening oftner nor more rarely than within the Limits assigned are 21 to 1 nearly.

And by the same way of reasoning, it will be found that the Probability of the Event's neither appearing oftner $\frac{1}{2}n+\frac{3}{2}\sqrt{n}$, nor more rarely than $\frac{1}{2}n-\frac{3}{2}\sqrt{n}$ will be 0.99874, which will make it that the Odds in this case will be 369 to 1 nearly.

To apply this to particular Examples, it will be necessary to estimate the frequency of an Event's happening or failing by the Square-root of the number which denotes how many Experiments have been, or are designed to be taken; and this Square-root, according as it has been already hinted at in the fourth Corollary, will be as it were the Modulus by which we are to regulate our Estimation; and therefore suppose the number of Experiments to be taken is 3600, and that it were required to assign the Probability of the Event's neither happening oftner than 2850 times, nor more rarely than 1750, which two numbers may be varied at pleasure, provided they be equally distant from the middle Sum 1800, then make the half difference between the two numbers 1850 and 1750, that is, in this case, $50 = \int \sqrt{n}$; now having supposed $3600 = n$, then $\sqrt{n}$ will be $=60$, which will make it that 50 will be $=60\int$, and consequently $\int = \frac{50}{60} = \frac{5}{6}$; and therefore if we take the proportion, which in an infinite power, the double Sum of the Terms corresponding to the Interval $\frac{5}{6}\sqrt{n}$, bears to the Sum of all the Terms, we shall have the Probability required exceeding near.

Lemma 2.

In any Power $\overline{a+b}^n$ expanded, the greatest Term is that in which the Indices of the Powers of a and b, have the same proportion to one another as the Quantities themselves a and b; thus taking the 10th Power of $a+b$, which is $a^{10}+10a^9b+45a^8b^2+120a^7b^3+210a^6b^4+252a^5b^5+210a^4b^6+120a^3b^7+45a^2b^8+10ab^9+b^{10}$; and supposing that the proportion of a to b is as 3 to 2, then the Term $210a^6b^4$ will be the greatest, by reason that the Indices of the Powers of a and b, which are in that Term, are in the proportion of 3 to 2; but supposing the proportion of a to b had been as 4 to 1, then the Term $45a^8b^2$ had been the greatest.

Lemma 3.

If an Event so depends on Chance, as that the Probabilities of its happening or failing be in any assigned proportion, such as may be supposed of a to b, and a certain number of Experiments be designed to be taken, in order to observe how often the Event will happen or fail; then the Probability that it shall neither happen more frequently than so many times as are denoted by $\frac{an}{a+b}+l$,

nor more rarely than so many times as are denoted by $\frac{an}{a+b}-l$, will be found as follows:

Let L and R be equally distant by the Interval l from the greatest Term; let also S be the Sum of the Terms included between L and R, together with those Extreams, then the Probability required will be rightly expressed by $\frac{S}{\overline{a+b}^n}$.

Corollary 8.[1]

The Ratio which, in an infinite Power denoted by n, the greatest Term bears to the Sum of all the rest, will be rightly expressed by the Fraction $\frac{a+b}{\sqrt{abnc}}$, wherein c denotes, as before, the Circumference of a Circle for a Radius equal to Unity.

Corollary 9.

If, in an infinite Power, any Term be distant from the Greatest by the Interval l, then the Hyperbolic Logarithm of the Ratio which that Term bears to that Greatest will be expressed by the Fraction $-\frac{\overline{a+b}^2}{2abn}\times ll$; provided the Ratio of l to n be not a finite

---

[1] [Numbered as in the original. There is no corollary 7 in the text.]

Ratio, but such a one as may be conceived between any given number p and $\sqrt{n}$, so that l be expressible by $p\sqrt{n}$, in which case the two Terms L and R will be equal.

COROLLARY 10.

If the Probabilities of happening and failing be in any given Ratio of unequality, the Problems relating to the Sum of the Terms of the Binomial $\overline{a+b}^{\,n}$ will be solved with the same facility as those in which the Probabilities of happening and failing are in a Ratio of Equality.

From what has been said, it follows, that Chance very little disturbs the Events which in their natural Institution were designed to happen or fail, according to some determinate Law; for if in order to help our conception, we imagine a round piece of Metal, with two polished opposite faces, differing in nothing but their colour, whereof one may be supposed to be white, and the other black; it is plain that we may say, that this piece may with equal facility exhibit a white or black face, and we may even suppose that it was framed with that particular view of shewing sometimes one face, sometimes the other, and that consequently if it be tossed up Chance shall decide the appearance; but we have seen in our LXXXVIIth Problem, that altho' Chance may produce an inequality of appearance, and still a greater inequality according to the length of time in which it may exert itself, still the appearances, either one way or the other, will perpetually tend to a proportion of Equality; but besides we have seen in the present Problem, that in a great number of Experiments, such as 3600, it would be the Odds of above 2 to 1, that one of the Faces, suppose the white, shall not appear more frequently than 1830 times, nor more rarely than 1770, or in other Terms, that it shall not be above or under the perfect Equality by more than $\frac{1}{120}$ part of the whole number of appearances; and by the same Rule, that if the number of Trials had been 14400 instead of 3600, then still it would be above the Odds of 2 to 1, that the appearances either one way or other would not deviate from perfect Equality by more than $\frac{1}{260}$ part of the whole, but in 1000000 Trials it would be the Odds of above 2 to 1, that the deviation from perfect Equality would not be more than by $\frac{1}{2000}$ part of the whole.   But the

Odds would increase at a prodigious rate, if instead of taking such narrow limits on both sides the Term of Equality, as are represented by $\frac{1}{2}\sqrt{n}$, we double those Limits or triple them; for in the first case the Odds would become 21 to 1, and in the second 369 to 1, and still be vastly greater if we were to quadruple them, and at last be infinitely great; and yet whether we double, triple or quadruple them, &c. the Extension of those Limits will bear but an inconsiderable proportion to the whole, and none at all, if the whole be infinite, of which the reason will easily be perceived by Mathematicians, who know, that the Square-root of any Power bears so much a less proportion to that Power, as the Index of it is great.

And what we have said is also applicable to a Ratio of Inequality, as appears from our 9th Corollary. And thus in all cases it will be found, that altho' Chance produces irregularities, still the Odds will be infinitely great, that in process of Time, those Irregularities will bear no proportion to the recurrency of that Order which naturally results from original Design.

LEGENDRE

On Least Squares

(Translated from the French by Professor Henry A. Ruger and Professor Helen M. Walker, Teachers College, Columbia University, New York City.)

The great advances in mathematical astronomy made during the early year) of the nineteenth century were due in no small part to the development of the method of least squares. The same method is the foundation for the calculus of errors of observation now occupying a place of great importance in the scientific study of social, economic, biological, and psychological problems. Gauss says in his work on the *Theory of the Motions of the Heavenly Bodies* (1809) that he had made use of this principle since 1795 but that it was first published by Legendre. The first statement of the method appeared as an appendix entitled "Sur la Méthode des moindres quarrés" in Legendre's *Nouvelles méthodes pour la détermination des orbites des comètes*, Paris, 1805. The portion of the work translated here is found on pages 72–75.

Adrien-Marie Legendre (1752–1833) was for five years a professor of mathematics in the École Militaire at Paris, and his early studies on the paths of projectiles provided a background for later work on the paths of heavenly bodies. He wrote on astronomy, the theory of numbers, elliptic functions, the calculus, higher geometry, mechanics, and physics. His work on geometry, in which he rearranged the propositions of Euclid, is one of the most successful textbooks ever written.

On the Method of Least Squares

In the majority of investigations in which the problem is to get from measures given by observation the most exact results which they can furnish, there almost always arises a system of equations of the form

$$E = a + bx + cy + fz + \&c.$$

in which a, b, c, f, &c. are the known coefficients which vary from one equation to another, and x, y, z, &c. are the unknowns which must be determined in accordance with the condition that the value of E shall for each equation reduce to a quantity which is either zero or very small.

If there are the same number of equations as unknowns x, y, z, &c., there is no difficulty in determining the unknowns, and the error E can be made absolutely zero. But more often the number

576

of equations is greater than that of the unknowns, and it is impossible to do away with all the errors.

In a situation of this sort, which is the usual thing in physical and astronomical problems, where there is an attempt to determine certain important components, a degree of arbitrariness necessarily enters in the distribution of the errors, and it is not to be expected that all the hypotheses shall lead to exactly the same results; but it is particularly important to proceed in such a way that extreme errors, whether positive or negative, shall be confined within as narrow limits as possible.

Of all the principles which can be proposed for that purpose, I think there is none more general, more exact, and more easy of application, than that of which we have made use in the preceding researches, and which consists of rendering the sum of the squares of the errors a *minimum*. By this means there is established among the errors a sort of equilibrium which, preventing the extremes from exerting an undue influence, is very well fitted to reveal that state of the system which most nearly approaches the truth.

The sum of the squares of the errors $E^2 + E'^2 + E''^2 +$ &c. being

$$(a \ + bx \ + cy \ + fz \ + \text{&c.})^2$$
$$+ \ (a' \ + b'x \ + c'y \ + f'z \ + \text{&c.})^2$$
$$+ \ (a'' + b''x + c''y + f''z + \text{&c.})^2$$
$$+ \ \text{&c.,}$$

if its *minimum* is desired, when x alone varies, the resulting equation will be

$$0 = \int ab + x\int b^2 + y\int bc + z\int bf + \text{&c.,}$$

in which by $\int ab$ we understand the sum of similar products, i.e., $ab + a'b' + a''b'' +$ &c.; by $\int b^2$ the sum of the squares of the coefficients of x, namely $b^2 + b'^2 + b''^2 +$ &c., and similarly for the other terms.

Similarly the *minimum* with respect to y will be

$$0 = \int ac + x\int bc + y\int c^2 + z\int fc + \text{&c.,}$$

and the *minimum* with respect to z,

$$0 = \int af + x\int bf + y\int cf + z\int f^2 + \text{&c.,}$$

in which it is apparent that the same coefficients $\int bc$, $\int bf$, &c. are common to two equations, a fact which facilitates the calculation.

In general, *to form the equation of the* minimum *with respect to one of the unknowns, it is necessary to multiply all the terms of each given equation by the coefficient of the unknown in that equation, taken with regard to its sign, and to find the sum of these products.*

The number of equations of *minimum* derived in this manner will be equal to the number of the unknowns, and these equations are then to be solved by the established methods. But it will be well to reduce the amount of computation both in multiplication and in solution, by retaining in each operation only so many significant figures, integers or decimals, as are demanded by the degree of approximation for which the inquiry calls.

Even if by a rare chance it were possible to satisfy all the equations at once by making all the errors zero, we could obtain the same result from the equations of *minimum;* for if after having found the values of x, y, z, &c. which make E, E', &c. equal to zero, we let x, y, z, &c. vary by $\delta x, \delta y, \delta z$, &c., it is evident that E^2, which was zero, will become by that variation $(a\delta x + b\delta y + c\delta z + \&c.)^2$. The same will be true of E'^2, E''^2, &c. Thus we see that the sum of the squares of the errors will by variation become a quantity of the second order with respect to $\delta x, \delta y$, &c., which is in accord with the nature of a minimum.

If after having determined all the unknowns x, y, z, &c., we substitute their values in the given equations, we will find the value of the different errors E, E', E'', &c., to which the system gives rise, and which cannot be reduced without increasing the sum of their squares. If among these errors are some which appear too large to be admissible, then those equations which produced these errors will be rejected, as coming from too faulty experiments, and the unknowns will be determined by means of the other equations, which will then give much smaller errors. It is further to be noted that one will not then be obliged to begin the calculations anew, for since the equations of minimum are formed by the addition of the products made in each of the given equations, it will suffice to remove from the addition those products furnished by the equations which would have led to errors that were too large.

The rule by which one finds the mean among the results of different observations is only a very simple consequence of our general method, which we will call *the method of least squares.*

Indeed, if experiments have given different values a', a'', a''', &c. for a certain quantity x, the sum of the squares of the errors will be $(a' - x)^2 + (a'' - x)^2 + (a''' - x)^2 +$ &c., and on making that sum a minimum, we have

$$o = (a' - x) + (a'' - x) + (a''' - x) + \&c.,$$

from which it follows that

$$x = \frac{a' + a'' + a''' + \&c.}{n},$$

n being the number of the observations.

In the same way, if to determine the position of a point in space, a first experiment has given the coordinates a', b', c'; a second, the coordinates a'', b'', c''; and so on, and if the true coordinates of the point are denoted by x, y, z; then the error in the first experiment will be the distance from the point (a', b', c') to the point (x, y, z). The square of this distance is

$$(a' - x)^2 + (b' - y)^2 + (c' - z)^2.$$

If we make the sum of the squares of all such distances a minimum, we get three equations which give

$$x = \frac{\int a}{n}, \quad y = \frac{\int b}{n}, \quad z = \frac{\int c}{n},$$

n being the number of points given by the experiments. These formulas are precisely the ones by which one might find the common center of gravity of several equal masses situated at the given points, whence it is evident that the center of gravity of any body possesses this general property.

If we divide the mass of a body into particles which are equal and sufficiently small to be treated as points, the sum of the squares of the distances from the particles to the center of gravity will be a minimum.

We see then that the method of least squares reveals to us, in a fashion, the center about which all the results furnished by experiments tend to distribute themselves, in such a manner as to make their deviations from it as small as possible. The application which we are now about to make of this method to the measurement of the meridian will display most clearly its simplicity and fertility.[1]

---

[1] [An application of the method to an astronomical problem follows.]

CHEBYSHEV (Tchebycheff)

Theorem Concerning Mean Values

(Translated from the French by Professor Helen M. Walker, Teachers College, Columbia University, New York City.)

The inequality which Chebyshev (Tchebycheff) derived in his paper on mean values is an important contribution to the theory of dispersion. In this paper by simple algebra, without approximation or the aid of the calculus, he reached a result from which both "Jacques Bernoulli's Theorem" and Poisson's "Law of Large Numbers" can be derived as special cases. The selection printed here was translated from the Russian into French by M. N. de Khanikof and appeared in Liouville's *Journal de mathématiques pures et appliquées ou recueil mensuel de mémoires sur les diverses parties des mathématiques*, 2nd series, XII (1867), 177–184. The same material is to be found in his *Œuvres*, I (1899).

Pafnutii Lvovitch Chebyshev (Tchebysheff)[1] (1821–1894) was, after Lobachevsky, Russia's most celebrated mathematician. Even as a small boy he was greatly interested in mechanical inventions, and it is said that in his first lesson in geometry he saw the bearing of the subject upon mechanics and therefore resolved to master it. At the age of twenty he received his diploma from the University of Moscow, having already received a medal for a work on the numerical solution of algebraic equations of higher orders.

Chebyshev's father was a Russian nobleman, but after the famine of 1840 the estate was so reduced that for the rest of his life he was forced to practice extreme economy, spending money freely for nothing except the mechanical models of his various inventions. He never married, but devoted himself solely to science.

Chebyshev collaborated with Bouniakovsky in bringing out the two large volumes of the collected works of Euler in 1849 and this seems to have turned his thoughts to the theory of numbers, and particularly to the very difficult problem of the distribution of primes. In 1850 he established the existence of limits within which must be comprised the sum of the logarithms of primes inferior to a given number. In 1860 he was made a correspondent of the Institut de France, and in 1874 an *associé étranger*. He was also a foreign member of the Royal Society of London.

From 1847 to 1882 he was professor of mathematics at the University of St. Petersburg, and at different periods during this time he taught analytic

---

[1] The name is spelled in many ways such as Chebychef, Chebichev, Tchebychev, Tchébycheff, Tschebycheff. For further biographical details the reader is referred to a brochure by A. Vassilief entitled *P. L. Tchébycbef et son Œuvre Scientifique* (Turin, 1898) reprinted from the *Bollettino di bibliografia e storia delle scienze matematicbe pubblicato per cura di Gino Loria*, 1898, or to the sketch by C. A. Possé in the *Dictionnaire des écrivains et savants russes rédigé par M. Vénguerof*, reprinted in Volume II of Markof's edition of Chebyshev's *Œuvres*.

geometry, higher algebra, theory of numbers, integral calculus, theory of probabilities, the calculus of finite differences, the theory of elliptic functions, and the
theory of definite integrals, and his biographers are agreed that the quality of his
teaching was no less remarkable than that of his research. Chebyshev made
important contributions to the theory of numbers, theory of least squares,
interpolation theory, calculus of variations, infinite series, and the theory of
probability, and published nearly a hundred memoirs on these and other
mathematical topics, being best known for his work on primes. The very
day before his death he received his friends as usual and discoursed upon the
subject of a simple rule he had discovered for the rectification of a curve.

On the Mean Values

If we agree to speak of the *mathematical expectation* of any
magnitude as the sum of all the values which it may assume
multiplied by their respective probabilities, it will be easy for us
to establish a very simple theorem concerning the limits between
which shall be contained a sum of any values whatever.

THEOREM. *If we designate by* a, b, c..., *the mathematical
expectations of the quantities* x, y, z..., *and by* a_1, b_1, c_1..., *the
mathematical expectations of their squares* x^2, y^2, z^2..., *the probability that the sum* $x + y + z$... *is included within the limits*

$$a + b + c + \ldots + \alpha\sqrt{a_1 + b_1 + c_1 + \ldots - a^2 - b^2 - c^2 - \ldots},$$

and

$$a + b + c + \ldots - \alpha\sqrt{a_1 + b_1 + c_1 + \ldots - a^2 - b^2 - c^2 - \ldots},$$

will always be larger than $1 - \dfrac{1}{\alpha^2}$, *no matter what the size of* α.

Proof. Let

$$x_1, x_2, x_3, \ldots, x_l,$$
$$y_1, y_2, y_3, \ldots, y_m,$$
$$z_1, z_2, z_3, \ldots z_n,$$
$$\ldots \ldots \ldots \ldots$$

be all conceivable values of the quantities x, y, z,..., and let

$$p_1, p_2, p_3, \ldots, p_l,$$
$$q_1, q_2, q_3, \ldots, q_m,$$
$$r_1, r_2, r_3, \ldots, r_n,$$
$$\ldots \ldots \ldots \ldots$$

be the respective probabilities of these values, or, better, the
probabilities of the hypotheses

$$x = x_1, x_2, x_3, \ldots, x_l,$$
$$y = y_1, y_2, y_3, \ldots, y_m,$$
$$z = z_1, z_2, z_3, \ldots, z_n,$$
$$\ldots \ldots \ldots \ldots$$

In accordance with this notation, the mathematical expectations of the magnitudes $x, y, z, \ldots$, and of $x^2, y^2, z^2, \ldots$ will be expressed as follows:

(1)
$$\begin{cases} a = p_1x_1 + p_2x_2 + p_3x_3 + \ldots + p_lx_l, \\ b = q_1y_1 + q_2y_2 + q_3y_3 + \ldots + q_my_m, \\ c = r_1z_1 + r_2z_2 + r_3z_3 + \ldots + r_nz_n, \end{cases}$$

. . . .

(2)
$$\begin{cases} a_1 = p_1x_1^2 + p_2x_2^2 + p_3x_3^2 + \ldots + p_lx_l^2, \\ b_1 = q_1y_1^2 + q_2y_2^2 + q_3y_3^2 + \ldots + q_my_m^2, \\ c_1 = r_1z_1^2 + r_2z_2^2 + r_3z_3^2 + \ldots + r_nz_n^2, \end{cases}$$

. . . .

Now since the assumptions we have just made concerning the quantities $x, y, z, \ldots$ are the only ones possible, their probabilities will satisfy the following equations:

(3)
$$\begin{cases} p_1 + p_2 + p_3 + \ldots + p_l = 1, \\ q_1 + q_2 + q_3 + \ldots + q_m = 1, \\ r_1 + r_2 + r_3 + \ldots + r_n = 1, \end{cases}$$

. . . .

It will now be easy for us to find by the aid of equations (1), (2), and (3), to what the sum of the values of the expression

$$(x_\lambda + y_\mu + z_\nu + \ldots - a - b - c - \ldots)^2 p_\lambda q_\mu r_\nu \ldots,$$

will reduce if we make successively

$$\lambda = 1, 2, 3, \ldots, l,$$
$$\mu = 1, 2, 3, \ldots, m,$$
$$\nu = 1, 2, 3, \ldots, n^1$$

Indeed when this expression is developed we have

$$p_\lambda q_\mu r_\nu \ldots x_\lambda^2 + p_\lambda q_\mu r_\nu \ldots y_\mu^2 + p_\lambda q_\mu r_\nu \ldots z_\nu^2 + \ldots$$
$$+ 2p_\lambda q_\mu r_\nu \ldots x_\lambda y_\mu + 2p_\lambda q_\mu r_\nu \ldots x_\lambda z_\nu + 2p_\lambda q_\mu r_\nu \ldots y_\mu z_\nu + \ldots$$
$$- 2(a + b + c + \ldots)p_\lambda q_\mu r_\nu \ldots x_\lambda$$
$$- 2(a + b + c + \ldots)p_\lambda q_\mu r_\nu \ldots y_\mu$$
$$- 2(a + b + c + \ldots)p_\lambda q_\mu r_\nu \ldots z_\nu - \ldots$$
$$+ (a + b + c + \ldots)^2 p_\lambda q_\mu r_\nu \ldots$$

Giving to λ in this expression all the values from $\lambda = 1$ to $\lambda = l$, and summing the results of these substitutions, we will obtain the sum as follows:

---

[1] [The original has here $\nu + 1, 2, 3, \ldots, n \ldots$ which is obviously a misprint.]

$$q_\mu r_\nu \ldots (p_1 x_1{}^2 + p_2 x_2{}^2 + p_3 x_3{}^2 + \ldots + p_l x_l{}^2)$$
$$+ (p_1 + p_2 + p_3 + \ldots + p_l) q_\mu r_\nu \ldots y_\mu{}^2$$
$$+ (p_1 + p_2 + p_3 + \ldots + p_l) q_\mu r_\nu \ldots z_\nu{}^2$$
$$+ 2(p_1 x_1 + p_2 x_2 + p_3 x_3 + \ldots + p_l x_l) q_\mu r_\nu \ldots y_\mu$$
$$+ 2(p_1 x_1 + p_2 x_2 + p_3 x_3 + \ldots + p_l x_l) q_\mu r_\nu \ldots z_\nu$$
$$+ 2(p_1 + p_2 + p_3 + \ldots + p_l) q_\mu r_\nu \ldots y_\mu z_\nu$$
$$+ \ldots \ldots \ldots \ldots \ldots \ldots \ldots \ldots \ldots \ldots \ldots \ldots \ldots$$
$$- 2(a + b + c + \ldots)(p_1 x_1 + p_2 x_2 + p_3 x_3 + \ldots + p_l x_l)$$
$$q_\mu r_\nu \ldots$$
$$- 2(a + b + c + \ldots)(p_1 + p_2 + p_3 + \ldots + p_l) q_\mu r_\nu \ldots y_\mu$$
$$- 2(a + b + c + \ldots)(p_1 + p_2 + p_3 + \ldots + p_l) q_\mu r_\nu \ldots z_\nu - \ldots$$
$$+ (a + b + c + \ldots)^2 (p_1 + p_2 + p_3 + \ldots + p_l) q_\mu r_\nu \ldots$$

If by means of equations (1), (2), and (3) we substitute in place of the sums

$$p_1 x_1 + p_2 x_2 + p_3 x_3 + \ldots + p_l x_l,$$
$$p_1 x_1{}^2 + p_2 x_2{}^2 + p_3 x_3{}^2 + \ldots + p_l x_l{}^2$$

and

$$p_1 + p_2 + p_3 + \ldots + p_l$$

their values a, a_1 and 1, we will obtain the following formula:

$$a_1 q_\mu r_\nu \ldots + q_\mu r_\nu \ldots y_\mu{}^2 + q_\mu r_\nu \ldots z_\nu{}^2 + \ldots$$
$$+ 2a q_\mu r_\nu \ldots y_\mu + 2a q_\mu r_\nu \ldots z_\nu + 2 q_\mu r_\nu \ldots y_\mu z_\nu + \ldots$$
$$- 2(a + b + c \ldots) a q_\mu r_\nu \ldots - 2(a + b + c \ldots)$$
$$q_\mu r_\nu \ldots z_\nu - \ldots$$
$$+ (a + b + c \ldots)^2 q_\mu r_\nu \ldots$$

If we give to μ in this formula the values

$$\mu = 1, 2, 3, \ldots m,$$

then sum the expressions which result from these substitutions, and substitute for the sums

$$q_1 y_1 + q_2 y_2 + q_3 y_3 + \ldots + q_m y_m,$$
$$q_1 y_1{}^2 + q_2 y_2{}^2 + q_3 y_3{}^2 + \ldots + q_m y_m{}^2,$$
$$q_1 + q_2 + q_3 + \ldots + q_m,$$

their values b, b_1, and 1 derived from equations (1), (2) and (3) we will obtain the following expression:

$$a_1 r_\nu \ldots + b_1 r_\nu \ldots + r_\nu \ldots z_\nu{}^2 + \ldots$$
$$+ 2ab r_\nu \ldots + 2a r_\nu \ldots z_\nu + 2b r_\nu \ldots z_\nu + \ldots$$
$$- 2(a + b + c + \ldots) a r_\nu \ldots - 2(a + b + c + \ldots) b r_\nu \ldots$$
$$- 2(a + \cdot b + c + \ldots) r_\nu \ldots z_\nu - \ldots + (a + b + c + \ldots)^2 r_\nu$$
$$\ldots$$

By treating ν in the same manner we will see that the sum of all the values of the expression

$$(x_\lambda + y_\mu + z_\nu + \ldots - a - b - c \ldots)^2 p_\lambda q_\mu r_\nu \ldots$$

derived by letting
$$\lambda = 1, 2, 3, \ldots, l,$$
$$\mu = 1, 2, 3, \ldots, m,$$
$$\nu = 1, 2, 3, \ldots, n,$$
$$\ldots$$

will be equal to

$$a_1 + b_1 + c_1 + \ldots + 2ab + 2ac + 2bc + \ldots - 2(a + b + c \ldots)a$$
$$- 2(a + b + c \ldots)b - 2(a + b + c \ldots)c - \ldots$$
$$+ (a + b + c \ldots)^2.$$

Upon developing this expression it reduces to

$$a_1 + b_1 + c_1 + \ldots - a^2 - b^2 - c^2 \ldots$$

Hence we conclude that the sum of the values of the expression

$$\frac{(x_\lambda + y_\mu + z_\nu + \ldots - a - b - c - \ldots)^2}{\alpha^2(a_1 + b_1 + c_1 + \ldots - a^2 - b^2 - c^2 - \ldots)} p_\lambda q_\mu r_\nu \ldots,$$

which we obtain by making

$$\lambda = 1, 2, 3, \ldots, l,$$
$$\mu = 1, 2, 3, \ldots, m,$$
$$\nu = 1, 2, 3, \ldots, n,$$
$$\ldots$$

will be equal to $\dfrac{1}{\alpha^2}$. Now it is evident that by rejecting from that sum all the terms in which the factor

$$\frac{(x_\lambda + y_\mu + z_\nu + \ldots - a - b - c - \ldots)^2}{\alpha^2(a_1 + b_1 + c_1 + \ldots - a^2 - b^2 - c^2 - \ldots)}$$

is less than 1 and by substituting unity for all those larger than 1, we will decrease that sum, and it will be less than $\dfrac{1}{\alpha^2}$. But this reduced sum will be formed of only those products $p_\lambda q_\mu r_\nu \ldots,$ which correspond to the values of $x_\lambda, y_\mu, z_\nu, \ldots$ for which the expression

$$(4) \qquad \frac{(x_\lambda + y_\mu + z_\nu + \ldots - a - b - c - \ldots)^2}{\alpha^2(a_1 + b_1 + c_1 + \ldots - a^2 - b^2 - c^2 - \ldots)} > 1,$$

and it will evidently represent the probability that $x, y, z \ldots$ have values which satisfy condition (4).

This same probability can be replaced by $1 - P$, if we designate by P the probability that the values of $x, y, z, \ldots$ do not satisfy condition (4), or better—and this is the same thing—that the quantities have values for which the ratio

$$\frac{(x + y + z + \ldots - a - b - c - \ldots)^2}{\alpha^2(a_1 + b_1 + c_1 + \ldots - a^2 - b^2 - c^2 - \ldots)}$$

is not >1. Consequently the sum $x + y + z \ldots$ is included within the limits

$$a + b + c + \ldots + \alpha\sqrt{a_1 + b_1 + c_1 + \ldots - a^2 - b^2 - c^2 - \ldots}$$

and

$$a + b + c + \ldots - \alpha\sqrt{a_1 + b_1 + c_1 + \ldots - a^2 - b^2 - c^2 - \ldots}$$

Hence it is evident that the probability P must satisfy the inequality

$$1 - P < \frac{1}{\alpha^2},$$

which gives us

$$P > 1 - \frac{1}{\alpha^2},$$

which was to have been proved.

If N be the number of the quantities $x, y, z, \ldots$, and if in the theorem which we have just demonstrated we set

$$\alpha = \frac{\sqrt{N}}{t},$$

and divide by N both the sum $x + y + z + \ldots$ and its limits

$$a + b + c + \ldots + \alpha\sqrt{a_1 + b_1 + c_1 + \ldots - a^2 - b^2 - c^2 - \ldots}$$

and

$$a + b + c + \ldots - \alpha\sqrt{a_1 + b_1 + c_1 + \ldots - a^2 - b^2 - c^2 - \ldots},$$

we will obtain the following theorem concerning the mean values.

THEOREM. *If the mathematical expectations of the quantities $x, y, z, \ldots$ and $x^2, y^2, z^2, \ldots$ be respectively $a, b, c, \ldots a_1, b_1, c_1, \ldots$, be probability that the difference between the arithmetic mean of the N quantities $x, y, z, \ldots$ and the arithmetic mean of the mathematical expectations of these quantities will not exceed*

$$\frac{1}{t}\sqrt{\frac{a_1 + b_1 + c_1 + \ldots}{N} - \frac{a^2 + b^2 + c^2 + \ldots}{N}}$$

will always be larger than $1 - \dfrac{t^2}{N}$ whatever the value of t.

Since the fractions $\dfrac{a_1 + b_1 + c_1 + \ldots}{N}$ and $\dfrac{a^2 + b^2 + c^2 + \ldots}{N}$ express the mean of the quantities a_1, b_1, c_1, … and $a_1{}^2$, $b_1{}^2$, $c_1{}^2$, …, whenever the mathematical expectations a, b, c, … a_1, b_1, c_1, … do not exceed a given finite limit, the expression

$$\sqrt{\frac{a_1 + b_1 + c_1 + \ldots}{N} - \frac{a^2 + b^2 + c^2 + \ldots}{N}}$$

will have a finite value, no matter how large the number N, and in consequence we may make the value of

$$\frac{1}{t}\sqrt{\frac{a_1 + b_1 + c_1 + \ldots}{N} - \frac{a^2 + b^2 + c^2 + \ldots}{N}}$$

as small as we wish by giving to t a value sufficiently large. Now since, no matter what the value of t, if the number N approaches infinity the fraction $\dfrac{t^2}{N}$ will approach zero, by means of the preceding theorem, we reach the conclusion:

THEOREM. *If the mathematical expectations of the quantities* U_1, U_2, U_3, … *and of their squares* $U_1{}^2$, $U_2{}^2$, $U_3{}^2$, … *do not exceed a given finite limit, the probability that the difference between the arithmetic mean of* N *of these quantities and the arithmetic mean of their mathematical expectations will be less than a given quantity, becomes unity as* N *becomes infinite.*

For the particular hypothesis that the quantities U_1, U_2, U_3, … are either unity or zero, as when an event E either happens or fails on the 1st, 2nd, 3rd, … Nth *trial*, we note that the sum $U_1 + U_2 + U_3 + \ldots + U_N$ will give the number of *repetitions* of the event E in N trials, and that the arithmetic mean

$$\frac{U_1 + U_2 + U_3 + \ldots + U_N}{N}$$

will represent the ratio of the number of *repetitions* of the event E to the number of *trials*. In order to apply to this case our last theorem, let us designate by P_1, P_2, P_3, … P_N the probabilities of the event E in the 1st, 2nd, 3rd … Nth trial. The mathematical expectations of the quantities $U_1 + U_2 + U_3 + \ldots + U_N$ and of their squares $U_1{}^2$, $U_2{}^2$, $U_3{}^2$, …, $U_N{}^2$ will be expressed, in conformity with our notation, as

$$P_1 1 + (1 - P_1)0,\ P_2 1 + (1 - P_2)0,\ P_3 1 + (1 - P_3)0,\ \ldots$$
$$P_1 1^2 + (1 - P_1)0^2,\ P_2 1^2 + (1 - P_2)0^2,\ P_3 1^2 + (1 - P_3)0^2, \ldots$$

Hence we see that the mathematical expectations are P_1, P_2, P_3, ... and that the arithmetic mean of the first N expectations is

$$\frac{P_1 + P_2 + P_3 + \ldots + P_N}{N},$$

that is to say, the arithmetic mean of the probabilities P_1, P_2, P_3, ...P_N.

As a consequence of this, and by virtue of the preceding theorem, we arrive at the following conclusion:

When the number of trials becomes infinite, we obtain a probability —which may even be approximately one if we so wish—that the difference between the arithmetic mean of the probabilities of the event, during the trials, and the ratio of the number of repetitions of that event to the total number of trials, is less than any given quantity.

In the particular case in which the probability remains constant during all the trials, we have the theorem of Bernoulli.

LAPLACE

On the Probability of the Errors in the Mean Results of a Great Number of Observations and on the Most Advantageous Mean Result

(Translated from the French by Dr. Julian L. C. A. Gÿs, Harvard University, Cambridge, Mass.)

Pierre-Simon, Marquis de Laplace (1749–1827), born at Beaumont-en-Auge (Calvados), the son of a farmer, was in his early years professor of mathematics at the military school in his native city. He took part in the founding of the École Polytechnique and the École Normale. He dealt mostly with problems of celestial mechanics, and to the works of Newton, Halley, Clairaut, d'Alembert, and Euler on the consequences of universal gravitation, he added many personal contributions relating to the variations of the motion of the moon, the perturbations of the planets Jupiter and Saturn, the theory of the satellites of Jupiter, the velocity of the rotation of the ring of Saturn, aberration, the motion of the comets, and the tides. He was also famous for the invention of the cosmogonic system which bears his name. His *Théorie analytique des probabilités* ranks among the most important works done in the field of probability theory. In the edition of Mme. Vve. Courcier, Paris, 1820, it is preceded by a most interesting introduction which was first published under the title *Essai philosophique sur les probabilités*.
were published under the auspices of the Academy of Sciences in 1886.

The extract under consideration has been taken from the *Œuvres complètes de Laplace*, Vol. VII, published under the auspices of the Académie des Sciences, Paris, 1886 (Book 2, Chapter IV, pp. 304–327).

The interest in the passage lies in presenting the line of reasoning by which Laplace arrived at what is generally known as the law of errors of Gauss. Laplace certainly discovered the law before Gauss published his way of deriving it from his well-known postulates on errors. The method of Laplace is entirely different from that of Gauss. It should be noted that De Moivre gave a proof of the same law in 1733. (See page 566).

CHAPTER IV

On the probability of the errors in the mean results of a great number of observations and on the most advantageous mean result.

18. Let us now consider the mean results of a great number of observations of which the law of the frequency of the errors is known. Let us first assume that for each observation the errors may equally be

$$-n, \ -n+1, \ -n+2, \ldots -1, 0, 1, \ldots n-2, \ n-1, \ n.$$

The probability of each error will be $\dfrac{1}{2n+1}$. If we call the number of observations s, the coefficient of $c^{l\omega\sqrt{-1}}$ in the development of the polynomial

$$\left\{ \begin{array}{l} c^{-n\omega\sqrt{-1}} + c^{-(n-1)\omega\sqrt{-1}} + c^{-(n-2)\omega\sqrt{-1}}\ldots \\ \ldots + c^{-\omega\sqrt{-1}} + 1 + c^{\omega\sqrt{-1}}\ldots + c^{n\omega\sqrt{-1}} \end{array} \right\}^{s}$$

will be the number of combinations in which the sum of the errors is l.[1] This coefficient is the term independent of $c^{\omega\sqrt{-1}}$ and of its powers in the development of the same polynomial multiplied by $c^{-l\omega\sqrt{-1}}$, and it is clearly equal to the term independent of ω in the same development multiplied by $\dfrac{c^{l\omega\sqrt{-1}} + c^{-l\omega\sqrt{-1}}}{2}$ or by $\cos l\omega$

Thus we have for the expression of this coefficient,

$$\frac{1}{\pi}\int d\omega.\,\cos l\omega.\,(1 + 2\cos\omega + 2\cos 2\,\omega\ldots + 2\cos n\omega)^{s},$$

the integral being taken from $\omega = 0$ to $\omega = \pi$.

We have seen (Book I, art. 36) that this integral is[2]

$$\frac{(2n+1)^{s}\sqrt{3}}{\sqrt{n.(n+1).2s\pi}}\cdot c^{-\frac{3\,l^{2}}{n(n+1)s}};$$

the total number of combinations of the errors is $(2n+1)^{s}$. Dividing the former quantity by the latter, we have for the probability that the sum of the errors of the s observations be l,

$$\frac{\sqrt{3}}{\sqrt{n(n+1).2s\pi}}\cdot c^{-\frac{3\,l^{2}}{n(n+1).s}},$$

If we set

$$l = 2t.\sqrt{\frac{n.(n+1).s}{6}},$$

the probability that the sum of the errors will be within the limits $+2T\sqrt{\dfrac{n.(n+1).s}{6}}$ and $-2T.\sqrt{\dfrac{n.(n+1).s}{6}}$ will be equal to

$$\frac{2}{\sqrt{\pi}}.\int dt.c^{-t^{2}},$$

---

[1] [Here c stands for what we now represent by e.]

[2] [In section 36 of his Book I, Laplace computes the coefficient of $a^{\pm l}$ in the development of the polynomial

$$(a^{-n} + a^{-n+1} + \ldots + a^{-1} + 1 + a + \ldots + a^{n-1} + a^{n})^{s}$$

where $a = c^{\omega\sqrt{-1}}$, in the case of a very large exponent.]

the integral being taken from $t = 0$ to $t = T$. This expression holds for the case of n infinite. Then, calling $2a$ the interval between the limits *of* the errors of each observation, we have $n = a$, and the preceding limits become $\pm \dfrac{2T.a.\sqrt{s}}{\sqrt{6}}$: thus the probability that the sum of the errors will be included within the limits $\pm ar.\sqrt{s}$ is

$$2.\sqrt{\frac{3}{2\pi}}.\int dr.c^{-\frac{3}{2}r^2}.$$

This is also the probability that the mean error will be included within the limits $\pm\dfrac{ar}{\sqrt{s}}$; for the mean error is obtained by dividing the sum of the errors by s.

The probability that the sum of the inclinations of the orbits of s comets will be included within the given limits, assuming that all inclinations are equally possible from zero to a right angle, is evidently the same as the preceding probability. The interval $2a$ of the limits of the errors of each observation is in this case the interval $\pi/2$ of the limits of the possible inclinations. Thus the probability that the sum of the inclinations will be included within the limits $\pm\dfrac{\pi.r\sqrt{s}}{4}$ is $2.\sqrt{\dfrac{3}{2\pi}}.\int dr.c^{-\frac{3}{2}r^2}$ which agrees with what has been found in section 13.[1]

Let us assume in general that the probability of each error positive or negative, may be expressed by $\phi(x/n)$, x and n being infinite numbers. Then, in the function

$$1 + 2\cos\omega + 2\cos 2\omega + 2\cos 3\omega \ldots + 2\cos n\omega,$$

each term such as $2\cos x\omega$ must be multiplied by $\phi(x/n)$. But we have

$$2\phi\left(\frac{x}{n}\right)\cdot\cos x\omega = 2\phi\left(\frac{x}{n}\right) - \frac{x^2}{n^2}\cdot\phi\left(\frac{x}{n}\right)\cdot n^2\omega^2 + \text{etc.}$$

Thus by putting

$$x' = \frac{x}{n}, \qquad dx' = \frac{1}{n},$$

---

[1] [In that section, Laplace finds the same result by considering the problem of the inclinations of the orbits as an application of this problem: given an urn containing $(n + 1)$ balls numbered from 0 to n, to find the probability of attaining the sum s by i drawings if the ball drawn is returned each time.]

the function

$$\phi\left(\frac{0}{n}\right) + 2\phi\left(\frac{1}{n}\right) \cdot \cos \omega + 2\phi\left(\frac{2}{n}\right) \cdot \cos 2\omega \ldots + 2\phi\left(\frac{n}{n}\right) \cdot \cos n\omega,$$

becomes

$$2n. \int dx'.\phi(x') - n^3\omega^2. \int x'^2 dx'.\phi(x') + \text{etc.},$$

the integrals being taken from $x' = 0$ to $x' = 1$. Then let

$$k = 2\int dx'.\phi(x'), \qquad k'' = \int x'^2 dx'.\phi(x'), \text{ etc.}$$

The preceding series becomes

$$nk.\left(1 - \frac{k''}{k} \cdot n^2\omega^2 + \text{etc.}\right).$$

Now the probability that the sum of the errors of s observations will lie within the limits $\pm l$ is, as is easily verified by the preceding reasoning,

$$\frac{2}{\pi} \cdot \int \int d\omega.dl. \cos l\omega \left\{ \begin{array}{l} \phi\left(\frac{0}{n}\right) + 2\phi\left(\frac{1}{n}\right) \cdot \cos \omega + 2\phi\left(\frac{2}{n}\right) \cdot \cos 2\omega \ldots \\[2mm] \qquad\qquad \ldots + 2\phi\left(\frac{n}{n}\right) \cdot \cos n\omega \end{array} \right\}^s$$

the integral being taken from $\omega = 0$ to $\omega = \pi$. This probability is then

$$2.\frac{(nk)^s}{\pi} \cdot \int \int d\omega.dl. \cos l\omega \left(1 - \frac{k''}{k} \cdot n^2\omega^2 - \text{etc.}\right)^s. \qquad (u)$$

Let us assume that

$$\left(1 - \frac{k''}{k} \cdot n^2\omega^2 - \text{etc.}\right)^s = c^{-t^2}.$$

In taking hyperbolic logarithms, when s is a large number, we have very nearly

$$s.\frac{k''}{k} \cdot n^2\omega^2 = t^2;$$

which yields

$$\omega = \frac{t}{n}.\sqrt{\frac{k}{k''s}}.$$

If we then observe that the quantity nk or $2.\int dx.\phi(x/n)$ which expresses the probability that the error of an observation is included within the limits $\pm n$, should be equal to unity, the function (u) becomes

$$\frac{2}{n\pi}\cdot\sqrt{\frac{k}{k''s}}\cdot\int\int dl.dt.c^{-t^2}\cdot\cos\left(\frac{lt}{n}\cdot\sqrt{\frac{k}{k''s}}\right);$$

the integral with respect to t being taken between $t = 0$ and $t = \pi n\sqrt{\dfrac{k''s}{k}}$ or to $t = \infty$, n being assumed to be infinite. But from Book I, section 25[1] we have

$$\int dt.\cos\left(\frac{lt}{n}\cdot\sqrt{\frac{k}{k''s}}\right)\cdot c^{-t^2} = \frac{\sqrt{\pi}}{2}\cdot c^{-\frac{l^2}{4n^2}\cdot\frac{k}{k''s}}.$$

Then setting

$$\frac{l}{n} = 2t'\cdot\sqrt{\frac{k''s}{k}};$$

the function (u) becomes

$$\frac{2}{\sqrt{\pi}}\cdot\int dt'.c^{-t'^2}.$$

Thus, calling the interval included between the limits of errors of each observation $2a$ as above, the probability that the sum of the errors of s observations will be included within the limits $\pm ar.\sqrt{s}$ is

$$\sqrt{\frac{k}{k''\pi}}\cdot\int dr.c^{-\frac{kr^2}{4k''}},$$

if $\phi\left(\dfrac{x}{n}\right)$ is constant.   Then $\dfrac{k}{k''s} = 6$, and this probability becomes

$$2.\sqrt{\frac{3}{2\pi}}\int dr.c^{-\frac{3}{2}r^2},$$

which is conformable to what we found above.

If $\phi\left(\dfrac{x}{n}\right)$ or $\phi(x')$ is a rational and entire function of x', we have by the method of section 15, the probability that the sum of the errors shall be included within the limits $\pm ar.\sqrt{s}$ expressed by a series of powers of s, $2s$, etc., of quantities of the form

$$s - \mu \pm r.\sqrt{s},$$

---

[1] [Result found by Laplace in his chapter on integration by approximation of differentials which contain factors raised to high powers.]

in which μ increases in arithmetic progression, these quantities being continuous until they become negative. By comparing this series with the preceding expression of the same probability, we obtain the value of the series very accurately. And relative to this type of sequence we obtain theorems analogous to those which have been given in section 42, Book I, on the finite differences of powers of one variable.

If the law of frequency of the errors is expressed by a negative exponential that can extend to infinity and in general if the errors can extend to infinity then a becomes infinite and some difficulties may arise with the application of the preceding method. In all cases we shall set

$$\frac{x}{b} = x', \qquad \frac{1}{b} = dx',$$

b being an arbitrary finite quantity. And by following the above analysis exactly, we shall find for the probability that the sum of the errors of the s observations be included between the limits $\pm br.\sqrt{s}$,

$$\sqrt{\frac{k}{k''\pi}} \cdot \int dr.c^{-\frac{kr^2}{4k''}},$$

an expression in which we should observe that $\phi\left(\frac{x}{b}\right)$ or $\phi(x')$ expresses the probability of the error $\pm x$, and that we have

$$k = 2\int dx'.\phi(x'), \qquad k'' = \int x'^2 dx'.\phi(x'),$$

the integrals being taken from $x' = 0$ to $x' = \infty$.

19. Let us now determine the probability that the sum of the errors of a very greater number of observations shall be included within the given limits, disregarding the signs of the errors, i. e., taking them all as positive. To that end, let us consider the series

$$\phi\left(\frac{n}{n}\right).c^{-n\omega\sqrt{-1}} + \phi\left(\frac{n-1}{n}\right).c^{-(n-1)\omega\sqrt{-1}} \ldots + \phi\left(\frac{0}{n}\right)\ldots$$

$$\ldots + \phi\left(\frac{n-1}{n}\right).c^{(n-1)\omega\sqrt{-1}} + \phi\left(\frac{n}{n}\right).c^{n\omega\sqrt{-1}},$$

$\phi\left(\frac{x}{n}\right)$ being the ordinate of the probability curve of the errors, corresponding to the error $\pm x$, x being considered as well as n as formed by an infinite number of unities. By raising this series to the s-th power, after having changed the sign of the negative exponentials, the coefficient of an arbitrary exponential, say $c^{(l+\mu s).\omega\sqrt{-1}}$, is the probability that the sum of the errors disregarding their sign, is $l + \mu s$; hence the probability is equal to

$$\frac{1}{2\pi} \cdot \int d\omega \cdot c^{-(l+\mu s)\omega\sqrt{-1}} \left\{ \begin{array}{l} \phi\left(\frac{0}{n}\right)+2\phi\left(\frac{1}{n}\right)\cdot c^{\omega\sqrt{-1}} \\ +2\phi\left(\frac{2}{n}\right)\cdot c^{2\omega\sqrt{-1}}\ldots+2\phi\left(\frac{n}{n}\right)\cdot c^{n\omega\sqrt{-1}} \end{array} \right\}$$

the integral with respect to ω being taken from $\omega = -\pi$ to $\omega = \pi$. Then, in that interval, the integral

$\int d\omega \cdot c^{-r\omega\sqrt{-1}}$, or $\int d\omega \cdot (\cos r\omega - \sqrt{-1} \sin r\omega)$, vanishes for all values of r that are not zero.

The development with respect to the powers of ω yields

$$\log\left\{ c^{-\mu s\omega\sqrt{-1}} \cdot \left[\phi\left(\frac{0}{n}\right) + 2\phi\left(\frac{1}{n}\right)\cdot c^{\omega\sqrt{-1}}\ldots+2\phi\left(\frac{n}{n}\right)\cdot c^{n\omega\sqrt{-1}} \right]^s \right\}$$

$$= s\cdot\log\left\{ \begin{array}{l} \phi\left(\frac{0}{n}\right) + 2\phi\left(\frac{1}{n}\right) + 2\phi\left(\frac{2}{n}\right)\ldots+ 2\phi\left(\frac{n}{n}\right) \\ +2\omega\sqrt{-1}\cdot\left[\phi\left(\frac{1}{n}\right) + 2\phi\left(\frac{2}{n}\right)\ldots+ n\phi\left(\frac{n}{n}\right) \right] \\ -\omega^2\cdot\left[\phi\left(\frac{1}{n}\right) + 2^2\phi\left(\frac{2}{n}\right)\ldots+n^2\phi\left(\frac{n}{n}\right) \right] \end{array} \right\} -\mu s\omega\sqrt{-1}. \tag{1}$$

Therefore, setting $\frac{x}{n} = x'$, $\quad \frac{1}{n} = dx'$,

$2\int dx'\cdot\phi(x') = k$, $\quad \int x'dx'\cdot\phi(x') = k'$, $\quad \int x'^2 dx'\cdot\phi(x') = k''$, $\int x'^3 dx'\cdot\phi(x') = k'''$, $\quad \int x'^4 dx'\cdot\phi(x') = k^{IV}$, etc.

the integrals being taken from $x' = 0$ to $x' = 1$, the second member of the equation (1) becomes

$$s\cdot\log nk + s\cdot\log\left(1+\frac{2\cdot k'}{k}\cdot n\omega\sqrt{-1} - \frac{k''}{k}\, n^2\omega^2 - \text{etc.}\right) - \mu s\omega\sqrt{-1}.$$

As the error of each observation necessarily falls between the limits $\pm n$, we have $nk = 1$; and thus the preceding quantity becomes

$$s.\left(\frac{2k'}{k} - \frac{\mu}{n}\right).n\omega\sqrt{-1} - \frac{(kk'' - 2k'^2)\cdot s\cdot n^2\omega^2}{k^2} - \text{etc.}$$

By putting

$$\frac{\mu}{n} = \frac{2k'}{k},$$

and neglecting the powers of ω higher than the square, this quantity reduces to its second term and the preceding probability becomes

$$\frac{1}{2\pi}\int d\omega \cdot c^{-l\omega\sqrt{-1}-\frac{(kk''-2k'^2)}{k^2}\cdot s\cdot n^2\omega^2}$$

Let

$$\beta = \frac{k}{\sqrt{kk''-2k'^2}}, \quad \omega = \frac{\beta t}{n.\sqrt{s}}, \quad \frac{l}{n} = r.\sqrt{s}.$$

The preceding integral becomes

$$-\frac{\beta^2 r^2}{4} = \frac{1}{2\pi}\cdot\frac{c}{n.\sqrt{s}}\cdot\int \beta dt.c^{-\left(t+\frac{l\beta\sqrt{-1}}{2n\sqrt{s}}\right)^2}.$$

This integral should be taken from $t = -\infty$ to $t = \infty$; and then the preceding quantity becomes

$$\frac{\beta}{2\sqrt{\pi}.n.\sqrt{s}}\cdot c^{-\frac{\beta^2 r^2}{4}}.$$

On multiplying by dl or by $ndr.\sqrt{s}$ the integral

$$\frac{1}{2\sqrt{\pi}}\int \beta dr.c^{-\frac{\beta^2 r^2}{4}}$$

will be the probability that the value of l and consequently the sum of the errors of the observations is included between the limits $\frac{2k'}{k}\cdot$ as $\pm ar.\sqrt{s}$, $\pm a$ being the limits of the errors of each observation, limits which we designate by $\pm n$ when we imagine them split up into an infinity of parts.

Thus we see that the most probable sum of the errors, disregarding the sign, is that which corresponds to $r = 0$. This sum is $\frac{2k'}{k}\cdot as$. In the case when $\phi(x)$ is constant, $\frac{2k'}{k} = \frac{1}{2}$, the most probable sum of the errors is then half of the largest possible sum, which sum is equal to sa. But if $\phi(x)$ is not a constant and if it decreases when the error x increases, then $\frac{2k'}{k}$ is less than $\frac{1}{2}$ and the sum of the errors disregarding the sign is less than half of the greatest sum possible.

By the same analysis, we can determine the probability that the sum of the squares of the errors will be $1 + \mu s$. It is easily seen that the expression of the probability is the integral

$$\frac{1}{2\pi}\cdot\int d\omega.c^{-(l+\mu s)\omega\sqrt{-1}}\left\{\begin{array}{l}\phi\left(\frac{0}{n}\right)+2\phi\left(\frac{1}{n}\right).c^{\omega\sqrt{-1}}+2\phi\left(\frac{2}{n}\right).c^{2^2\omega\sqrt{-1}}\\ \quad\cdots+2\phi\left(\frac{n}{n}\right).c^{n^2\omega\sqrt{-1}}\end{array}\right\}^s,$$

taken from $\omega = -\pi$ to $\omega = \pi$. Following the preceding analysis precisely, we will have

$$\mu = \frac{2n^2.k''}{k};$$

and putting

$$\beta' = \frac{k}{\sqrt{kk^{IV} - 2k''^2}}$$

the probability that the sum of the squares of the errors of s observations will lie between the limits $\frac{2k''}{k} \cdot a^2 s \pm a^2 r. \sqrt{s}$ will be

$$\frac{1}{2\pi} \int \beta' dr.c^{-\frac{\beta'^2 r^2}{4}} .$$

The most probable sum is that which corresponds to $r = 0$ and therefore it is $\frac{2k''}{k} \cdot a^2 s$. If s is an exceedingly large number, the result of the observations will differ very little from that value and will therefore yield very satisfactorily the factor $\frac{a^2 \cdot k''}{k}$.

20. When it is desired to correct an element already known to a good approximation by the totality of a great number of observations, we form equations of condition as follows: Let z be the correction of an element and let β be the observation, the analytic expression for which is a function of the element. By substituting for this element its approximate value plus the correction z and reducing to a series with respect to z and neglecting the square of z, this function will take the form $b + pz$. Setting it equal to the observed quantity β we obtain

$$\beta = b + pz;$$

z would thus be determined if the observation were exact, but since it is susceptible to error, we have exactly, calling that error ϵ to terms of order z^2

$$\beta + \epsilon = b + pz;$$

and by setting

$$\beta - b = \alpha,$$

we have

$$\epsilon = pz - \alpha.$$

Each observation yields a similar equation that may be written for the $(i + 1)$th observation as follows:

$$\epsilon^{(i)} = p^{(i)} \cdot z - \alpha^{(i)}.$$

Combining these equations, we have

$$S \cdot \epsilon^{(i)} = z \cdot S \cdot p^{(i)} - S \cdot \alpha^{(i)}, \tag{1}$$

where the symbol S holds for all values of i from $i = 0$ to $i = s - 1$, s being the total number of observations. Assuming that the sum of the errors is zero, this yields

$$z = \frac{S \cdot \alpha^{(i)}}{S \cdot p^{(i)}}.$$

This is what we usually call the *mean result of the observations*.

We have seen in section 18 that the probability that the sum of the errors of s observations be included within the limits $\pm ar \cdot \sqrt{s}$ is

$$\sqrt{\frac{k}{k''\pi}} \cdot \int dr \cdot c^{-\frac{kr^2}{4k''}}.$$

Call $\pm u$ the error in the result z. Substituting $\pm ar \cdot \sqrt{s}$ for $S \cdot \epsilon^{(i)}$ in equation (1), and $\frac{S \cdot \alpha^{(i)}}{S \cdot p^{(i)}} \pm u$ for z, this yields

$$r = \frac{u \cdot S \cdot p^{(i)}}{a \cdot \sqrt{s}}.$$

The probability that the error of the result z will be included within the limits $\pm u$ is thus

$$\sqrt{\frac{k}{k''s\pi}} \cdot S \cdot p^{(i)} \cdot \int \frac{du}{a} \cdot c^{-\frac{ku^2 \cdot (S \cdot P^{(i)})^2}{4k''a^2s}}.$$

Instead of supposing that the sum of the errors is zero, we may suppose that an arbitrary linear function of these errors is zero, as

$$m \cdot \epsilon + m^{(1)} \cdot \epsilon^{(1)} + m^{(2)} \cdot \epsilon^{(2)} \ldots + m^{(s-1)} \cdot \epsilon^{(s-1)}, \tag{m}$$

$m, m^{(1)}, m^{(2)}$ being positive or negative integers. By substituting in this function (m) the values given by the equations of condition in the place of ϵ, $\epsilon^{(1)}$ etc., this becomes

$$z \cdot S \cdot m^{(i)} p^{(i)} - S \cdot m^{(i)} \alpha^{(i)}.$$

Setting the function (m) equal to zero yields

$$z = \frac{S \cdot m^{(i)} \alpha^{(i)}}{S \cdot m^{(i)} p^{(i)}}.$$

Let u be the error in this result so that

$$z = \frac{S \cdot m^{(i)} \alpha^{(i)}}{S \cdot m^{(i)} p^{(i)}} + u.$$

The function (m) becomes

$$u.S.m^{(i)}p^{(i)}.$$

Let us determine the probability of the error u, when the number of the observations is large.

For this, let us consider the product

$$\int \phi\left(\frac{x}{a}\right).c^{mx\omega\sqrt{-1}} \times \int \phi\left(\frac{x}{a}\right).c^{m^{(1)}x\omega\sqrt{-}} \cdots \times \int \phi\left(\frac{x}{a}\right).c^{m(s-1)x\omega\sqrt{-1}},$$

the symbol $\int$ extending over all the values of x, from the extreme negative value to the extreme positive value. As above, $\phi\left(\dfrac{x}{a}\right)$ is the probability of an error x in each observation, x being as is a, assumed to be formed of an infinite number of parts taken as unity. It is clear that the coefficient of an arbitrary exponential $c^{l\omega\sqrt{-1}}$ in the development of this product will be the probability that the sum of the errors of the observations, multiplied respectively by m, $m^{(1)}$ etc., in other words, the function (m), shall be equal to l. Then multiplying the latter product by $c^{-l\omega\sqrt{-1}}$, the term independent of $c^{\omega\sqrt{-1}}$ and of its powers in this new product will represent the same probability. If we assume as we did here, that the probability of positive errors is the same as that of negative errors, we may combine the terms multiplied by $c^{mx\omega\sqrt{-1}}$ and by $c^{-mx\omega\sqrt{-1}}$ in the sum $\int \phi\left(\dfrac{x}{a}\right).c^{mx\omega\sqrt{-1}}$. Then this sum will take the form $2\int \phi\left(\dfrac{x}{a}\right).\cos mx\omega$. And so for all similar sums. Hence the probability for the function (m) to be equal to l is

$$\frac{1}{2\pi}.\int d\omega \left| \begin{array}{l} c^{-l\omega\sqrt{-1}} \times 2\int \phi\left(\dfrac{x}{a}\right).\cos mx\omega \\ \\ \times 2\int \phi\left(\dfrac{x}{a}\right).\cos m^{(1)}x\omega \cdots \times 2\int \phi\left(\dfrac{x}{a}\right).\cos m^{(s-1)}x\omega \end{array} \right| ;$$

the integral being taken from $\omega = -\pi$ to $\omega = \pi$. Reducing the cosines to a series yields

$$\int \phi\left(\frac{x}{a}\right).\cos mx\omega = \int \phi\left(\frac{x}{a}\right) - \frac{1}{2}.m^2a^2.\omega^2.\int \frac{x^2}{a^2}.\phi\left(\frac{x}{a}\right) + \text{etc.}$$

Letting $x/a = x'$ and observing that since the variation of x is unity, $dx' = 1/a$; we obtain

$$\int \phi\left(\frac{x}{a}\right) = a.\int dx'.\phi(x').$$

As above, let us call k the integral $2\int dx'.\phi(x')$ taken from $x' = 0$ to its extreme positive value, and k'' the integral $\int x'^2 dx'$ taken over the same limits, and so on, thus we will have

$$2\int \phi\left(\frac{x}{a}\right).\cos mx\omega = ak.\left(1 - \frac{k''}{k}.m^2a^2\omega^2 + \frac{k^{IV}}{12k}.m^4a^4\omega^4 - \text{etc.}\right).$$

The logarithm of the second member of this equation is

$$-\frac{k''}{k}.m^2a^2\omega^2 + \frac{kk^{IV} - 6k''^2}{12k^2}.m^4a^4\omega^4 - \text{etc.} + \log ak,$$

ak or $2a\int dx'.\phi(x')$ expresses the probability that the error of each observation shall be included within the limits, a thing which is certain. We then have $ak = 1$. This reduces the preceding logarithm to

$$-\frac{k''}{k}.m^2a^2\omega^2 + \frac{kk^{IV} - 6k''^2}{12k^2}.m^4a^4\omega^4 - \text{etc.}$$

From this it is easy to conclude that the product

$$2\int \phi\left(\frac{x}{a}\right).\cos mx\omega \times 2\int \phi\left(\frac{x}{a}\right).\cos m^{(1)}x\omega \ldots \times 2\int \phi\left(\frac{x}{a}\right)\cos m^{(s-1)}x\omega,$$

is

$$\left(1 + \frac{kk^{IV} - 6k''^2}{12k^2}.a^4\omega^4.S.m^{(i)4} + \text{etc.}\right).c^{-\frac{k''}{k}a^2\omega^2.S.m^{(i)2}}.$$

The preceding integral (i) reduces then to

$$\frac{1}{2\pi}.\int d\omega.\left\{1 + \frac{kk^{IV} - 6k''^2}{12k^2}.a^4\omega^4.S.m^{(i)4} + \text{etc.}\right\}$$
$$\times c^{-l\omega\sqrt{-1} - \frac{k''}{k}.a^2\omega^2.S.m^{(i)2}}.$$

Setting $sa^2\omega^2 = t^2$, this integral becomes

$$\frac{1}{2a\pi\sqrt{s}}\int dt.\left\{1 + \frac{kk^{IV} - 6k''^2}{12k^2}\cdot\frac{S.m^{(i)4}}{s^2}.t^4 + \text{etc.}\right\}$$
$$\times c^{-\frac{lt\sqrt{-1}}{a\sqrt{s}} - \frac{k''}{k}\cdot\frac{S.m^{(i)2}}{s}t^2};$$

$S.m^{(i)2}$, $S.m^{(i)4}$ are evidently quantities of order s. Thus $\dfrac{S.m^{(i)4}}{s^2}$ is of order $1/s$. Then neglecting the terms of the latter order with respect to unity, the above integral reduces to

$$\frac{1}{2a\pi.\sqrt{s}}\int dt.c^{-\frac{lt\sqrt{-1}}{a\sqrt{s}} - \frac{k''}{k}\cdot\frac{Sm^{(i)2}}{s}.t^2}.$$

The integral with respect to ω must be taken from $\omega = -\pi$ to $\omega = \pi$, the integral with respect to t must be taken from $t = -a\pi.\sqrt{s}$ to $t = a\pi.\sqrt{s}$, and in such cases the exponential under the radical sign is negligible at the two limits, either because s is a large number or because a is supposed to be divided up into an infinity of parts taken as unity. It is therefore permissible to take the integral from $t = -\infty$ to $t = \infty$. Letting

$$t' = \sqrt{\frac{k''.S.m^{(i)2}}{ks}}.\left\{t + \frac{l.\sqrt{-1}.k.\sqrt{s}}{2a.k''.S.m^{(i)2}}\right\},$$

the preceding integral function becomes

$$\frac{c^{-\frac{kl^2}{4k''.a^2.S.m^{(i)2}}}}{2a\pi\sqrt{\frac{k''}{k}.S.m^{(i)2}}}.\int dt'.c^{-t'^2}.$$

The integral with respect to t' should be taken, just as the integral with respect to t, from $t' = -\infty$ to $t' = \infty$, so that the above quantity reduces to the following one

$$\frac{c^{-\frac{kl^2}{4k''.a^2.S.m^{(i)2}}}}{2a.\sqrt{\pi}\sqrt{\frac{k''}{k}.S.m^{(i)2}}}$$

Setting $l = ar.\sqrt{s}$ and observing that, since the variation of l is unity, $adr = 1$, we will have

$$\frac{\sqrt{s}}{2.\sqrt{\frac{k''\pi}{k}.S.m^{(i)2}}}.\int dr.c^{-\frac{kr^2.s}{4k''.S.m^{(i)2}}}$$

for the probability that the function (m) be included within the limits zero and $ar.\sqrt{S}$, the integral being taken from r equal to zero.

Here we need to know the probability of the error u in the element as determined by setting the function (m) equal to zero. This function being assumed to be equal to l or to $ar.\sqrt{s}$, we have, according to previous relations

$$u.S.m^{(i)}p^{(i)} = ar.\sqrt{s}.$$

Substituting this value in the preceding integral function, this one becomes

$$\frac{S.m^{(i)}p^{(i)}}{2a\sqrt{\frac{k''\pi}{k}.S.m^{(i)2}}}.\int du.c^{-\frac{ku^2.(S.m^{(i)}p^{(i)})^2}{4k''.a^2.S.m^{(i)2}}}$$

This is the expression for the probability that the value of u be included between the limits zero and u. It is also the expression for the probability that u will be included between the limits zero and $-u$. Setting

$$u = 2at.\sqrt{\frac{k''}{k}}.\frac{\sqrt{S.m^{(i)2}}}{S.m^{(i)}p^{(i)}},$$

the preceding integral function becomes

$$\frac{1}{\sqrt{\pi}}.\int dt.c^{-t^2}.$$

Now as the probability remains the same, t remains the same, and the interval of the two limits of u becomes smaller and smaller, the smaller $a.\sqrt{\frac{k''}{k}}.\frac{\sqrt{S.m^{(i)2}}}{S.m^{(i)}p^{(i)}}$ becomes. This interval remaining the same, the value of t and consequently the probability that the error of the element will fall within this interval, is the larger as the same quantity $a.\sqrt{\frac{k''}{k}}\frac{\sqrt{S.m^{(i)2}}}{S.m^{(i)}p^{(i)}}$ is smaller. It is then necessary to chose a system of factors $m^{(i)}$ which will make this quantity a *minimum*. And as a, k, k'' are the same in all these systems, we must chose the system that will make $\frac{\sqrt{S.m^{(i)2}}}{S.m^{(i)}p^{(i)}}$ a minimum.

It is possible to arrive at the same result in the following way. Let us consider again the expression for the probability that u will be within the limits zero and u. The coefficient of du in the differential of that expression is the ordinate of the probability curve of the errors u in the element, errors which are represented by the abscissas of that curve which can be extended to infinity on both sides of the ordinate corresponding to $u = 0$. This being said, all errors whether positive or negative must be looked on as either a disadvantage or a real loss, in some game. Now by means of the probability theory, which has been expounded at some length in the beginning of this book, that disadvantage is computed by adding the products of each disadvantage by its corresponding probability.   The mean value of the error to fear in excess is thus equal to the integral

$$\frac{\int u du \cdot S \cdot m^{(i)} p^{(i)} \cdot c^{-\frac{ku^2 \cdot (S.m^{(i)}p^{(i)})^2}{4k''.a^2.S.m^{(i)2}}}}{2a\sqrt{\frac{k''\pi}{k}}.S.m^{(i)2}}$$

the integral being taken from $u = 0$ to u infinite; thus the error is

$$a\sqrt{\frac{k''}{k\pi}} \cdot \frac{\sqrt{S.m^{(i)2}}}{S.m^{(i)}p^{(i)}}.$$

The same quantity taken with the $-$ sign gives the mean error to fear in deficiency.   It is evident that the system of the factors $m^{(i)}$ which must be chosen is such that these errors are *minima* and therefore such that

$$\frac{\sqrt{S.m^{(i)2}}}{S.m^{(i)}p^{(i)}}$$

is a *minimum*.

If we differentiate this function with respect to $m^{(i)}$ we will have, equating this derivative to zero, by the condition for a *minimum*,

$$\frac{m^{(i)}}{S.m^{(i)2}} = \frac{p^{(i)}}{S.m^{(i)}p^{(i)}}$$

This equation holds whatever i may be, and as the variation of i cannot affect the fraction $\frac{S.m^{(i)2}}{S.m^{(i)}p^{(i)}}$, setting this fraction equal to μ, we have

$$m = \mu \cdot p, \qquad m^{(1)} = \mu.p^{(1)}, \dots \qquad m^{(s-1)} = \mu.p^{(s-1)};$$

and whatever $p, p^{(1)}$, etc. may be, we may take μ so that the numbers m, $m^{(1)}$, etc., are integers as the above analysis assumes.   Then we have

$$z = \frac{S.p^{(i)}\alpha^{(i)}}{S.p^{(i)2}},$$

and the mean error to fear becomes $\pm \dfrac{a.\sqrt{\frac{k''}{k\pi}}}{\sqrt{S.p^{(i)2}}}$.   Under every hypothesis that can be made about the factors m, $m^{(i)}$, etc., this is the least mean error possible.

If we set the values of m, $m^{(1)}$, etc. equal to ± 1, the mean error to fear will be smaller when the sign $\pm$ is determined so that $m^{(i)}p^{(i)}$ will be positive.   This amounts to supposing that

$1 = m = m^{(1)} =$ etc., and to preparing the equations of condition in such a way that the coefficient of z in each of them be positive. This is done by the ordinary method. The mean result of the observations is then

$$z = \frac{S.\alpha^{(i)}}{S.p^{(i)}},$$

and the mean error to fear whether it be in excess or in deficiency is equal to

$$\pm \frac{a.\sqrt{\dfrac{k''.s}{k\pi}}}{S.p^{(i)}}.$$

But this error is greater than the former which as has been seen is the smallest possible. Moreover this can be shown as follows. It suffices to prove the inequality

$$\frac{\sqrt{s}}{S.p^{(i)}} > \frac{1}{\sqrt{S.p^{(i)2}}},$$

or

$$s.S.p^{(i)2} > (S.p^{(i)})^2.$$

Indeed, $2pp^{(1)}$ is less than $p^2 + p^{(1)2}$ since $(p^{(1)} - p)^2$ is positive. Hence it is permissible to substitute for $2pp^{(1)}$ in the second member of the above inequality the quantity $p^2 + p^{(1)2} - f$, f being positive. Making similar substitutions for all similar products, that second member will be equal to the first one minus a positive quantity. The result

$$z = \frac{S.p^{(i)}\alpha^{(i)}}{S.p^{(i)2}}$$

to which corresponds the minimum of the mean error to fear, is the same as that given by the method of least squares of the errors of observations; for, as the sum of these squares is

$$(p \cdot z - \alpha)^2 + (p^{(1)} \cdot z - \alpha^{(1)})^2 \ldots + (p^{(s-1)} \cdot z - \alpha^{(s-1)})^2;$$

the *minimum* condition of this function yields when z varies, the preceding expression. Preference should thus be given to this method, for all laws of frequency of the errors whatsoever they may be are the laws on which the ratio $\dfrac{k''}{k}$ depends.

If $\phi(x)$ is a constant, this ratio is equal to $\frac{1}{6}$. It is less than $\frac{1}{6}$ if $\phi(x)$ varies in such a way that it decreases when x increases as is natural to suppose. Adopting the mean law of errors given in section 15, according to which $\phi(x)$ is equal to $\frac{1}{2a} \cdot \log \frac{a}{x}$, we have $\frac{k''}{k} = \frac{1}{18}$. As to the limits $\pm a$, we may take for those limits, the deviations from the mean result which would cause the rejection of an observation.

But, by means of the observations themselves, it is possible to determine the factor $a \cdot \sqrt{\frac{k''}{k}}$ in the expression for the mean error. Indeed it has been seen in the preceding section that the sum of the squares of the errors in the observations is very nearly equal to $2s \cdot \frac{a^2 k''}{k}$ and that it becomes extremely probable when there is a great number of observations for the observed sum not to differ from that value by an appreciable amount. We may set them equal to each other. Now the observed sum is equal to $S \cdot \epsilon^{(i)2}$ or to $S \cdot (p^{(i)} \cdot z - \alpha^{(i)})^2$. Substituting for z its value $\frac{S \cdot p^{(i)} \alpha^{(i)}}{S \cdot p^{(i)2}}$; it is found that

$$2s \cdot \frac{a^2 k''}{k} = \frac{S \cdot p^{(i)2} \cdot S \cdot \alpha^{(i)2} - (S \cdot p^{(i)} \alpha^{(i)})^2}{S \cdot p^{(i)2}}.$$

The above expression for the mean error to fear in the result z then becomes

$$\pm \frac{\sqrt{S \cdot p^{(i)2} \cdot S \cdot \alpha^{(i)2} - (S \cdot p^{(i)} \alpha^{(i)})^2}}{S \cdot p^{(i)2} \cdot \sqrt{2s\pi}}$$

an expression in which nothing appears that is not given by the observations or by the coefficients of the equations of condition.

V. FIELD OF THE CALCULUS, FUNCTIONS, QUATERNIONS

CAVALIERI'S APPROACH TO THE CALCULUS

(Translated from the Latin by Professor Evelyn Walker, Hunter College, New York City.)

Bonaventura Francesco Cavalieri (Milan, 1598-Bologna, 1647), a Jesuit, was a pupil of Galileo. In order to prove his fitness for the Chair of Mathematics at the University of Bologna, he submitted, in 1629, the manuscript of his famous work, *Geometria Indivisibilibus Continuorum Nova quadam ratione promota*, which he published in 1635. This publication exerted an enormous influence upon the development of the calculus. Cavalieri was the author of a number of less important works, among them his *Exercitationes Geometricæ Sex*, which is still sometimes mentioned.

The following extract, known as Cavalieri's theorem, is from the *Geometria Indivisibilibus*, Book VII, Theorem 1, Proposition 1.[1]

Any plane figures, constructed between the same parallels, in which [plane figures] any straight lines whatever having been drawn equidistant from the same parallels, the included portions of any straight line are equal, will also be equal to one another; and any solid figures, constructed between the same parallel planes, in which [solid figures] any planes whatever having been drawn equidistant from the same parallel planes, the plane figures of any plane so drawn included within these solids, are equal, the [solid figures] will be equal to one another.

Now let the figures compared with one another, the plane as well as the solid, be called analogues, in fact even up[2] to the ruled lines or parallel planes between which they are assumed to lie, as it will be necessary to explain.

---

[1] This translation has been compared with and checked by that of G. W. Evans in *The American Mathematical Monthly*, XXIV, 10 (December 1917), pp. 447–451. The diagram and lettering used by Evans have been adopted as being more convenient than those of Cavalieri, whose diagram is not only poorly printed, but has its points designated by numbers as well as by both Roman and Greek letters.

[2] The expression used is "juxta regulas lineas," literally "next to the ruled lines."]

Let there be any plane figures, *ABC, XYZ*, constructed between the same parallels, *PQ, RS*; but *DN, OU*, any parallels to *PQ, RS*, having been drawn, the portion [*s*], for example of the *DN*, included within the figures, namely *JK, LM*, are equal to each other, and besides, the portions *EF, GH*, of the *OU* taken together (for the figure *ABC*, for example, may be hollow within following the contour *FgG*), are likewise equal to the *TV*; and let this happen in any other lines equidistant from the *PQ*.   I say that the figures *ABC, XYZ*, are equal to each other

For either of the figures *ABC, XYZ*, as the *ABC*, having been taken with the portions of the parallels *PQ, RS*, coterminous with it, namely with *PA, RB*, let it be superimposed upon the remaining figure *XYZ*, but so that the [lines] *PA, RB*, may fall upon *AQ* and *CS*; then either the whole [figure] *ABC* coincides with the whole [figure] *XYZ*, and so, since they coincide with each other, they are equal, or not; yet there may be some part which coincides with another part, as *XMC'YTbL*, a part of the figure *ABC*, with *XMC'YTbL*, a part of the figure *XYZ*.

By the superposition of the figures effected in such a way that portions of the parallels *PQ, RS*, coterminous with the two figures, are superposed in turn, it is evident that whatever straight lines included within the figures were in line with each other, they still remain in line with each other, as, for example, since *EF, GH*, were in the same line *TV*, the said superposition having been made, they will still remain in line with themselves, obviously *E'F'*, *TH'*, in line with the *TV*, for the distance of the *EF, GH*, from *PQ* is equal to the distance [of] *TV* from the same *PQ*.   Whence, however many times *PA* is laid upon *AQ*, wherever it may be done, *EF, GH*, will always remain in line with the *TV*; which is clearly apparent also for any other lines whatsoever parallel to *PQ* in each figure.

But when a part of one figure, as *ABC*, necessarily coincides with a part of the figure *XYZ* and not with the whole, while the superposition is made according to such a rule as has been stated, it will be demonstrated thus.   For since, any parallels whatever having been drawn to the [line] *AD*, the portions of them included within the figures, which were in line with one another still remain in line with one another after the superposition, they, of course, being equal by hypothesis before superposition, then after superposition the portions of the [lines] parallel to the *AD*, included within the superposed figures, will likewise be equal; as, for example, $E'F'$, TH', taken together, will be equal to the TV; therefore if the $E'F'$, TH', together do not coincide with the whole TV, then a part coinciding with some part, as TH' with TH' itself, $E'F'$ will be equal to the $H'V$, and in fact $E'F'$ will be in the residuum of the superposed figure ABC, [and] $H'V$ indeed in the residuum of the figure XYZ upon which the superposition was made. In the same way we shall show [that] to any [line] whatsoever parallel to the PQ, included within the residuum of the superposed figure ABC, as it were $LB'YTF'$, there corresponds an equal straight line, in line [with it], which will be in the residuum of the figure XYZ upon which the superposition was made; therefore, the superposition having been carried out in accordance with this rule, when there is left over any [part] of the superposed figure

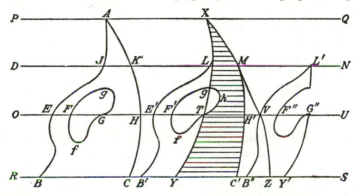

which does not fall on the figure upon which the superposition was made, it must be that some [part] of the remaining figure also is left over, upon which nothing has been superposed.

Since, moreover, to each one of the straight lines parallel to PQ included within the residuum or residuua (because there may be several residual figures) of the superposed figure ABC or XYZ, there corresponds, in line [with it], another straight line in the residuum or residua of the figure XYZ, it is manifest that these residual figures, or aggregates of residua, are between the same parallels. Therefore since the residual figure $LB'YTF'$ is between the parallels DN, RS, likewise the residual figure or aggregate of the residual figures of the XYZ, because it has within it the frusta Tbg, $MC'Z$, will be between the same parallels DN, RS. For if it did not extend both ways to the parallels DN, RS, as, for example, if it extended indeed all the way to DN, but not all the way to RS, but only as far as OU, to the straight lines included within the frustum $E'B'YfF'$, parallel to the [line] PQ, there would not correspond in the residuum of the figure XYZ, or [in] the aggregate of the residua, other straight lines, as, it was proved above, is necessary. Therefore these residua or the aggregates of residua are between the same parallels, and the portions of the [lines] included therein parallel to the PQ, RS, are equal to one another, as we have shown above. Therefore the remainders or the aggregates of the remainders, are in that condition in which, it was assumed just now, were the figures ABC and XYZ; that is, [they are] likewise analogues.

Then again let the superposition of the residua be made, but so that the parallels KL, CY, may be placed upon the parallels LN, YS, and the part $VB''Z$ of the frustum $LB'YTF'$ may coincide with the part $VB''Z$ of the frustum $MC'Z$. Then we shall show, as above, [that] while there is a residuum of one there is also a residuum of the other, and these residua, or aggregates of residua, are between the same parallels. Now let $L'VZY'G''F''$ be the residuum with respect to the figure ABC, but let the residua $MC'B''V$, Tbg, whose aggregate is between the same parallels as the residuum $L'VZY'G''F''$, belong to the figure XYZ, of course between the parallels DN, RS; then if again a superposition of these residua is made, but so that the parallels between which they lie may always be superposed in turn, and it may be understood that this is always to be done, until the whole figure ABC will have been superposed, I say [that] the whole [of it] must coincide with the XYZ; otherwise, if there is any residuum, as of the figure XYZ, upon which nothing had been superposed, then there would be some residuum of the figure ABC which would not have been superposed, as we have shown above to be necessary. But it has been stated that the whole ABC was superposed upon the XYZ; therefore they are so superposed one [part] after another, that there is a residuum of neither; therefore they are so superposed that they coincide with each other; therefore the figures ABC, XYZ, are equal to each other.

Now let there be constructed in the same diagram any two solid figures whatever, ABC, XYZ, between the same parallel planes PQ, RS; then any planes DN, OU, having been drawn equidistant from the aforesaid [planes], let the figures which are included within the solids, and which lie in the same plane, always be equal to each other, as JK equal to LM, and EF, GH, taken together (for a solid figure, for example ABC, may be hollow in any way within, following the surface *FfGg*) equal to the [figure] *TV*. I say that these solid figures are equal.

For if we superpose the solid *ABC*, together with the portions *PA*, *RC* of the planes *PQ*, *RS*, coterminous with it, upon the solid *XYZ*, so that the plane *PA* may be upon [the plane] *PQ*, and [the plane] *RC* on the plane *RS*, we shall show (as we did above with respect to the portions of the lines parallel to the *PQ* included within the plane figures *ABC*, *XYZ*) that the figures included within the solids *ABC*, *XYZ*, which were in the same plane, will also remain in the same plane after superposition, and therefore thus far the figures included within the superposed solids and parallel to the *PQ*, *RS*, are equal.

Then unless the whole solid coincides with the [other] whole [solid] in the first superposition, there will remain residual solids, or solids composed of residua in either solid, which will not be superposed upon one another; for when, for example, the figures *E'F'*, *TH'*, are equal to the figure *TV*, the common figure *TH'* having been subtracted, the remainder *E'F'* will be equal to the remainder *H'V*; and this will happen in any plane whatsoever parallel to the plane *PQ*, meeting the solids *ABC*, *XYZ*. Therefore having a residuum of one solid, we shall always have a residuum of the other also. And it will be evident, according to the method applied in the first part of this proposition concerning plane figures, that the residual solids or the aggregates of the residua will always be between the same parallel planes, as the residua *LB'YTF'*, *MC'Z*, *Tbg*, are between the parallel planes *DN*, *RS*, and likewise [that they are] analogues.

If therefore these residua also are superposed so that the plane *DL* is placed upon the plane *LN*, and *RY* upon *YS*, and this is understood to be done continually until [the one] which is superposed, as *ABC*, taken all together, as the whole *ABC*, will be coincident with the whole *XYZ*. For the whole solid *ABC* having been superposed upon the *XYZ*, unless they coincide with one another, there will be some residuum of one, as of the solid *XYZ*, and therefore there would be some residuum of the solid *XB'C'* or *ABC*, and this would not have been superposed, which is absurd; for it has already been stated that the whole solid *ABC* was superposed upon the *XYZ*. Therefore there will not be any residuum in these solids. Therefore they will coincide. Therefore the said solid figures *ABC*, *XYZ*, will be equal to each other. Which [things] were to be demonstrated.

FERMAT

On Maxima and Minima

(Translated from the French by Dr. Vera Sanford, Teachers College, Columbia University, New York City.)

Supplementing the communication from Fermat to Pascal (see page 289), giving some of his ideas on analytic geometry, the following letter to Roberval shows how his mind was working toward one of the combinations of the calculus with the Cartesian system. The letter was written on Monday, September 22, 1636, a year before Descartes published *La Geometrie*. See the *Œuvres de Fermat* (ed. Tannery and Henry, Vol. II, pp. 71-74, Paris, 1894). For a biographical sketch of Fermat, see page 397; for the introductory pages of Descartes, see pages 213, 214.

Monsieur,

1. With your permission, I shall postpone writing you on the subject of the propositions of mechanics until you shall do me the favor of sending me the demonstration of your theorems which I trust to see as soon as possible according to the promise you made me.

2. On the subject of the method of *maxima* and *minima,* you know that as you have seen the work which M. Despagnet gave you, you have seen mine which I sent him about seven years since at Bordeaux.

At that time, I recollect that M. Philon received a letter from you in which you proposed that he find the greatest cone of all those whose conical surface is equal to a given circle. He forwarded it to me and I gave the solution to M. Prades to return to you. If you search your memory, you will perhaps recall it and also the fact that you set this question as a difficult one that had not then been solved. If I discover your letter, which I kept at the time, among my papers, I will send it to you.

3. If M. Despargnet laid my method before you as I then sent it to him, you have not seen its most beautiful applications, for I have made use of it by amplifying it a little:

(1) For the solution of problems such as that of the conoid which I sent you in my last letter.

(2) For the construction of tangents to curved lines on which subject I set you the problem: To draw a tangent at a given point on the conchoid of Nicomedes.

(3) For the discovery of the centers of gravity of figures of every type, even of figures that differ from the ordinary ones such as my conoid and other infinite figures of which I shall show examples if you desire them.

(4) For numerical problems in which there is a question of aliquot parts and which are all very difficult.

4. It is by this method that I discovered 672 [the sum of] whose factors are twice the number itself, just as the factors of 120 are twice 120.

It is by the same method also that I discovered the infinite numbers that make the same thing[1] as 220 and 284, that is to say, the [sum of the] factors of the first are equal to the second, and those of the second are equal to the first. If you wish to see an example of this to test the question, these two numbers 17296 and 18416 satisfy the conditions.

I am certain that you told me that this question and others of its type are very difficult. I sent the solution to M. de Beaugrand some time ago.

I have also found numbers which exceeded the aliquot parts of a given number in a given ratio, and several others.

5. These are the four types of problems included in my method which perhaps you did not know.

With reference to the first, I have squared[2] infinite figures bounded by curved lines; as for example, if you imagine a figure such as a parabola, of such a type that the cubes of the ordinates[3] are in proportion to the segments which they cut from the diameter. This figure is something like a parabola and it differs from one only in the fact that in a parabola we take the ratio of the squares while in this figure I take that of the cubes. It is for this reason that M. de Beaugrand, to whom I put this problem calls it a *cubical parabola*.[4]

I have also proved that this figure is in the sesquialter[5] ratio to the triangle of the same base and height. You will find on investi-

---

[1] [The "amicable numbers" 220 and 284 were known at an early date, possibly as early as the Pythagoreans. The second pair were discovered by Fermat whose expression "the infinite numbers" meant that the author had discovered a general rule for the formation of these numbers.]

[2] [Found the area of.]

[3] [...que les cubes des appliquées soient en proportion des lignes qu'elles coupent du diamètre."]

[4] ["parabole solide."]

[5] [*I. e.*, the ratio of 3:2.]

gation that it was necessary for me to follow another method than that of Archimedes for the quadrature of the parabola, and that I should never have found it by his method.

6. Since you found my theorem on the conoid excellent, here

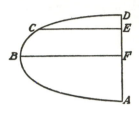

is its most general case: If a parabola with the vertex B and axis BF and ordinate AF be revolved about the straight line AD, a new type of conoid will be produced in which a section cut by a plane perpendicular to the axis will have the ratio to the cone on the same base and with the same altitude that 8 has to 5.

If, indeed, the plane cuts the axis in unequal segments, as at E, the segment of the conoid $ABCE$ is to the cone of the same base and altitude as five times the square ED with twice the rectangle AED and the rectangle of DF and AE is to five times the square on ED. And likewise the segment of the conoid DCE is to the cone with the same base and altitude as five times the square AE added to twice the rectangle AED and the rectangle of DF and DE is to five times the square on AE.

For the proof, besides the aid which I have from my method, I make use of inscribed and circumscribed cylindars.

7. I have omitted the principal use of my method which is in the discovery of plane and solid loci. It had been of particular service to me in finding the plane loci which I found so difficult before:

If from any number of given points, straight lines are drawn to a single point, and if these lines are such that they are equally spaced by a given amount from each other, then the point lies on a circle which is given in position.

All that I shall tell you are but examples, for I can assure you that for each of the preceding points I have found a very great number of exceedingly beautiful theorems. I shall send you their proof if you wish. May I however beg you to try them soon and to give me your solutions.

8. Finally, since the last letter which I wrote you, I have discovered the theorem which I set you. It caused me the greatest difficulty and it did not occur to me at an earlier date.

I beg you to share some of your reflections with me and to believe me etc.

ISAAC NEWTON

(*Facing page* 613.)

NEWTON

On Fluxions

(Translated from the Latin by Professor Evelyn Walker, Hunter College, New York City.)

Sir Isaac Newton (1642–1727) was the son of a Lincolnshire farmer. In 1660 he entered Trinity College, Cambridge, where he became the pupil of Isaac Barrow by whom his future work was strongly influenced. His discoveries in mathematics and physics began as early as 1664, although he did not publish any of his work until many years later. In 1669 he succeeded Barrow as Lucasian professor of mathematics at Trinity. Later he became warden of the mint and member of parliament, and was knighted by Queen Anne. He was elected fellow of the Royal Society in 1672, and from 1703 until his death was its president. In 1699 he was made foreign associate of the Académie des Sciences. He is buried in Westminister Abbey.[1]

His best known work is the *Principia*, or to give it its full title, *Philosophiæ Naturalis Principia Mathematica*, published in 1687, containing his theory of the universe based on his law of gravitation. Every high-school boy knows his name in connection with the binomial theorem, and more advanced students in connection with infinite series and the theory of equations. But his mathematical fame is due most of all to his invention of the calculus. His first development of the subject proceeded by means of infinite series as told by Wallis in his *Algebra*, 1685, Later he used the method that is most commonly associated with his name, that of fluxions, as exemplified in the *Quadratura Curvarum*, 1704.[2] Both of these naturally entail the use of infinitely small quantities. Finally in his *Principia* he explains the use of prime and ultimate ratios. The following quotations from the sources specified show the three points of view.

[Integration by Means of Infinite Series[3]]

He doth therein, not only give us many such Approximations... but he lays down general Rules and Methods... And gives

---

[1] For a brief summary of the life of Newton and a good bibliography for the same, see David Eugene Smith, *History of Mathematics*, Vol. I, p. 398.

[2] Charles Hayes published this method of Newton in a work of his own, *A Treatise of Fluxions: or, An Introduction to Mathematical Philosophy*, London, 1704. Nine years after Newton's death John Colson published *The Method of Fluxions and Infinite Series...from the Author's Latin Original not yet made publick...*, London, 1736.

[3] [John Wallis (see page 46) says in his *Algebra* (1685, p. 330) that he had seen the two letters written by Newton to Oldenburg, June 13 and October 24, 1676, containing Newton's discoveries in the realm of infinite series. The quotation is from Wallis.]

instances, how those Infinite or Interminate Progressions may be accommodated, to the Rectifying of Curve Lines...; Squaring of Curve-lined Figures; finding the Length of Archs,....

. . . .

[Newton's Method of Fluxions[1]]

Therefore, considering that quantities, which increase in equal times, and by increasing are generated, become greater or less according to the greater or less velocity with which they increase and are generated; I sought a method of determining quantities from the velocities of the motions, or [of the] increments, with which they are generated; and calling these velocities of the motions, or [of the] increments, *fluxions*, and the generated quantities *fluents*, I fell by degrees, in the years 1665 and 1666, upon the method of fluxions, which I have made use of here in the quadrature of curves.

Fluxions are very nearly as the augments of the fluents generated in equal, but very small, particles of time; and, to speak accurately, they are in the *first ratio* of the nascent augments; but they may be expounded by any lines which are proportional to them.

. . . .

It amounts to the same thing if the fluxions are taken in the ultimate ratio of the evanescent parts.[2]

. . . .

Let the quantity x flow uniformly, and let it be proposed to find the fluxion of x^n.

In the time that the quantity x, by flowing, becomes $x + o$, the quantity x^n will become $\overline{x + o}|^n$, that is, by the method of infinite series,

$$x^n + nox^{n-1} + \frac{n^2 - n}{2}oox^{n-2} + \text{etc.}$$

---

[1] [The quotation that follows is from *Quadratura Curvarum*, published with Newton's *Opticks: or, a Treatise of the Reflexions, Refractions, Inflexions and Colours of Light. Also Two Treatises of the Species and Magnitude of Curvilinear Figures*, London, 1704. The second of the treatises mentioned is the *Quadratura Curvarum*, pp. 165–211. We quote from the Introduction and give one proposition from the work itself. The translation as given here does not differ, except in a few unimportant details, from that of John Stewart, published in London, 1745.]

[2] [Newton here gives some examples. He makes the tangent coincide with the limiting position of the secant by making the ordinate of the second point of intersection of the secant with the curve move up into coincidence with that of the first. See page 617.]

And the augments o and $nox^{n-1} + \dfrac{n^2 - n}{2}oox^{n-2} +$ etc. are to one another as

$$1 \text{ and } nx^{n-1} + \frac{n^2 - n}{2}ox^{n-2} + \text{ etc.}$$

Now let these augments vanish, and their ultimate ratio will be as

$$1 \text{ to } nx^{n-1}.$$

From the fluxions to find the fluents is a much more difficult problem, and the first step of the solution is to find the quadrature of curves; concerning which I wrote the following some time ago.[1]

In what follows I consider indeterminate quantities as increasing or decreasing by a continued motion, that is, by flowing or ebbing, and I designate them by the letters z, y, x, v, and their fluxions or velocities of increasing I denote by the same letters pointed $\dot{z}$, $\dot{y}$, $\dot{x}$, $\dot{v}$. There are likewise fluxions or mutations, more or less swift, of these fluxions, which may be called the second fluxions of the same quantities z, y, x, v, and may be thus designated: $\ddot{z}$, $\ddot{y}$, $\ddot{x}$, $\ddot{v}$; and the first fluxions of these last, or the third fluxions of z, y, x, v, thus: $\dddot{z}$, $\dddot{y}$, $\dddot{x}$, $\dddot{v}$; and the fourth fluxions thus: $\ddddot{z}$, $\ddddot{y}$, $\ddddot{x}$, $\ddddot{v}$.

And after the same manner that $\dddot{z}$, $\dddot{y}$, $\dddot{x}$, $\dddot{v}$, are the fluxions of the quantities $\ddot{z}$, $\ddot{y}$, $\ddot{x}$, $\ddot{v}$, and these the fluxions of the quantities $\dot{z}$, $\dot{y}$, $\dot{x}$, $\dot{v}$, and these last the fluxions of the quantities z, y, x, v; so the quantities z, y, x, v, may be considered as the fluxions of others which I shall designate thus: $\acute{z}$, $\acute{y}$, $\acute{x}$, $\acute{v}$; and these as fluxions of others $\acute{\acute{z}}$, $\acute{\acute{y}}$, $\acute{\acute{x}}$, $\acute{\acute{v}}$; and these last as the fluxions of still others $\acute{\acute{\acute{z}}}$, $\acute{\acute{\acute{y}}}$, $\acute{\acute{\acute{x}}}$, $\acute{\acute{\acute{v}}}$. Therefore $\acute{\acute{\acute{z}}}$, $\acute{\acute{z}}$, $\acute{z}$, z, $\dot{z}$, $\ddot{z}$, $\dddot{z}$, $\ddddot{z}$, etc., designate a series of quantities whereof every one that follows is the fluxion of the preceding, and every one that goes before is a flowing quantity having the succeeding one as its fluxion.

· · · · · · · · · ·

And it is to be remembered that any preceding quantity in this series is as the area of a curvilinear figure of which the succeeding quantity is the rectangular ordinate, and [of which] the abscissa is z; . . .

---

[1] [This is the end of the introduction.]

Prop. 1. *Prob.* 1. *An equation being given involving any number of flowing quantities, to find the fluxions.*

Solution. Let every term of the equation be multiplied by the index of the power of every flowing quantity that it involves, and in every multiplication let a side [or root] of the power be changed into its fluxion, and the aggregate of all the products, with their proper signs, will be the new equation.

Explication. Let a, b, c, d, etc., be determinate and invariable quantities, and let any equation be proposed involving the flowing quantities z, y, x, v, etc., as

$$x^3 - xy^2 + a^2z - b^3 = 0.$$

Let the terms be first multiplied by the indices of the powers of x, and in every multiplication, for the root or x of one dimension, write $\dot{x}$, and the sum of the terms will be

$$3\dot{x}x^2 - \dot{x}y^2.$$

Let the same be done in y, and it will produce

$$-2xy\dot{y}.$$

Let the same be done in z, and it will produce

$$aa\dot{z}.$$

Let the sum of these results be placed equal to nothing, and the equation will be obtained

$$3\dot{x}x^2 - \dot{x}y^2 - 2xy\dot{y} + aa\dot{z} = 0.$$

I say that the relation of the fluxions is defined by this equation.

Demonstration.—For let o be a very small quantity, and let $o\dot{z}$, $o\dot{y}$, $o\dot{x}$, be the moments, that is the momentaneous synchronal increments, of the quantities z, y, x. And if the flowing quantities are just now z, y, x, these having been increased after a moment of time by their increments $o\dot{z}$, $o\dot{y}$, $o\dot{x}$, these quantities will become $z + \dot{z}o$, $y + o\dot{y}$, $x + o\dot{x}$; which being written in the first equation for z, y and x, give this equation:

$$x^3 + 3x^2o\dot{x} + 3xoo\dot{x}\dot{x} + o^3\dot{x}^3 - xy^2 - o\dot{x}y^2 - 2xo\dot{y}y - 2xo^2\dot{y}y$$
$$\cdot\; xo^2\dot{y}\dot{y} - \dot{x}o^3\dot{y}\dot{y} + a^2z + a^2o\dot{z} - b^3 = 0.$$

Let the former equation be subtracted [from the latter] and the remainder be divided by 0, and it will be

$$3\dot{x}x^2 + 3\dot{x}\dot{x}ox + \dot{x}^3o^2 - xy^2 - 2x\dot{y}y - 2\dot{x}o\dot{y}y - xo\dot{y}\dot{y} - \dot{x}o^2\dot{y}\dot{y}$$
$$+ a^2z = 0.$$

Let the quantity o be diminished infinitely, and, neglecting the terms which vanish, there will remain

$$3\dot{x}x^2 - \dot{x}y^2 - 2x\dot{y}y + a^2\dot{z} = 0.$$

Q. E. D.[1]

If the points are distant from each other by an interval, however small, the secant will be distant from the tangent by a small interval. That it may coincide with the tangent and the last ratio be found, the two points must unite and coincide altogether. In mathematics errors, however small, must not be neglected.

. . . .

[*The Method of Prime and Ultimate Ratios*[2]]

Quantities, as also ratios of quantities, which constantly tend toward equality in any finite time, and before the end of that time approach each other more nearly than [with] any given difference whatever, become ultimately equal . . .

The objection is that there is no ratio[3] of evanescent quantities, which obviously, before they have vanished, is not ultimate; when they have vanished, there is none. But also by the same like argument it may be contended that there is no ultimate velocity of a body arriving at a certain position; for before the body attains the position, this is not ultimate; when it has attained [it], there is none. And the answer is easy: By ultimate velocity I understand that with which the body is moved, neither before it arrives at the ultimate position and the motion ceases, nor thereafter, but just when it arrives; that is, that very velocity with which the body arrives at the ultimate position and with which the motion ceases. And similarly for the motion of evanescent quantities is to be understood the ratio of the quantities, not before they vanish, nor thereafter, but [that] with which they vanish. And likewise the first nascent ratio is the ratio with which they begin. And the prime and ultimate amount is to be [that] with which they begin and cease (if you will, to increase and diminish). There

---

[1] [The *Quadratura Curvarum*, 1704, besides explaining the method of fluxions, also anticipates the method of prime and ultimate ratios, which is practically the modern method of limits. The paragraph which follows occurs earlier in the text.]

[2] [This translation has been made from *Philosophiæ Naturalis Principia Mathematica Auctore Isaaco Newtono*. Amsterdam, 1714. The first edition was published in 1687. The selections are from pages 24 and 33.]

[3] [Newton's word is "proportio."]

exists a limit which the velocity may attain at the end of the motion, but [which it may] not pass.   This is the ultimate velocity.   And the ratio of the limit of all quantities and proportions, beginning and ceasing, is equal...

The ultimate ratios in which quantities vanish, are not really the ratios of ultimate quantities, but the limits toward which the ratios of quantities, decreasing without limit, always approach; and to which they can come nearer than any given difference, but which they can never pass nor attain before the quantities are diminished indefinitely.[1]

---

[1] Acknowledgment is hereby made to G. H. Graves, whose article, "Development of the Fundamental Ideas of the Differential Calculus," in *The Mathematics Teacher*, Vol. III (1910–1911), pp. 82–89, has been freely used.

Leibniz.

(*Facing page 619.*)

LEIBNIZ

On the Calculus

(Translated from the Latin by Professor Evelyn Walker, Hunter College, New York City.)

Gottfried Wilhelm, Freiherr von Leibniz (Leipzig, 1646– Hannover, 1716), ranks with Newton as one of the inventors of the calculus. He was an infant prodigy, teaching himself Latin at the age of eight, and taking his degree in law before the age of twenty-one. In the service of the Elector of Mainz, and later in that of three successive dukes of Braunschweig-Lüneburg, he travelled extensively through England, France, Germany, Holland, Italy, everywhere seeking the acquaintance of prominent scholars. He finally settled at Hannover as librarian to the duke. In 1709 he was made a Baron of the Empire. When, in 1714, the Duke of Hannover crossed to England to become George I., he refused to allow Leibniz to accompany him. This embittered the last years of Leibniz's life.

His was a most versatile genius. He wrote on mathematics, natural science, history, politics, jurisprudence, economics, philosophy, theology, and philology. He invented a calculating machine that would add, subtract, multiply, divide, and even extract roots.

He was elected to membership in the Royal Society of London (1673), and to foreign membership in the Académie des Sciences (1700). He founded the Akademie der Wissenschaften (1700), and became its president for life. Many of his articles appear in the *Acta Eruditorum*, the organ of the last named society.[1]

His interest in the calculus must have been aroused while he was visiting England in 1672, where he probably heard from Oldenburg that Newton had some such method. His own development of the subject seems, however, to have been independent of that of Newton, while it shows the influence of both Barrow and Pascal.[2] He never published a work on the calculus, but confined himself to short articles in the *Acta Eruditorum*, and to piecemeal explanations of his discoveries in letters which he wrote to other mathematicians.

Clearly we are indebted to him for the following contributions to the development of the calculus:

1. He invented a convenient symbolism.
2. He enunciated definite rules of procedure which he called algorithms.
3. He realized and taught that quadratures constitute only a special case of integration; or, as he then called it, the inverse method of tangents.
4. He represented transcendental lines by means of differential equations.

---

[1] For a brief sketch of the life of Leibniz see Smith, *History of Mathematics*, Vol. I., p. 417; also other histories of mathematics and the various encyclopedias.

[2] For example, his use of a characteristic triangle.

These points are illustrated in the following selections from two articles that were published in the *Acta Eruditorum*.[1]

The following extract is from "A new method for maxima and minima..." by Gottfried Wilhelm von Leibniz.[2]

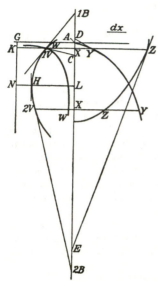

Let there be an axis AX and several curves, as VV, WW, YY, ZZ, whose ordinates VX, WX, YX, ZX, normal to the axis, are called respectively, v, w, y, z; and the AX, cut off from the axis, is called x. The tangents are VB, WC, YD, ZE, meeting the axis in the points B, C, D, E, respectively. Now some straight line chosen arbitrarily is called dx, and the straight [line] which is to dx as v (or w, or y, or z) is to VB (or WC, or YD, or ZE), is called dv (or dw, or dy, or dz) or the difference of the v's (or the w's, or the y's, or the z's). These things assumed, the rules of the calculus are as follows:

If a is a given constant,
$$da = 0,$$
and
$$d\overline{ax} = a\,dx;$$

if
$$y = v,$$

(or [if] any ordinate whatsoever of the curve YY [is] equal to any corresponding ordinate of the curve VV),
$$dy = dv.$$

Now, addition and subtraction:
if
$$z - y + w + x = v,$$
$$d\overline{z - y + w + x} = dv,$$
or
$$= dz - dy + dw + dx.$$

---

[1] The Latin is frequently bad.   The translator wishes to acknowledge her indebtedness to Professors Carter and Hahn, both of Hunter College, who kindly made a number of corrections.

[2] "Nova methodus pro maximis & minimis, itemque tangentibus, qua nec irrationales quantitates moratur, & singulare pro illis calculi genus, per G.G.L." (his Latin initials) *Acta Eruditorum*, October, 1684.

Multiplication:

$$\overline{dvx} = xdv + vdx,$$

or by placing

$$y = xv,$$
$$dy = xdv + vdx.$$

. . . .

Yet it must be noticed that the converse is not always given by a differential equation, except with a certain caution, of which [I shall speak] elsewhere.

Next, division:

$$d\frac{v}{y} = \frac{\pm\, vdy \mp ydv}{yy}$$

$\left(\text{or } z \text{ being placed equal to } \dfrac{v}{y}\right)$

$$dz = \frac{\pm\, vdy \mp ydv}{yy}$$

Until this sign may be correctly written, whenever in the calculus its differential is simply substituted for the letter, the same sign is of course to be used, and $+dx$ [is] to be written for $+x$, and $-dx$ [is] to written for $-x$, as is apparent from the addition and subtraction done just above; but when an exact value is sought, or when the relation of the z to x is considered, then [it is necessary] to show whether the value of the dz is a positive quantity, or less than nothing, or as I should say, negative; as will happen later, when the tangent ZE is drawn from the point Z, not toward A, but in the opposite direction or below X, that is, when the ordinates z decrease with the increasing abscissas x. And because the ordinates v sometimes increase, sometimes decrease, dv will be sometimes a positive, sometimes a negative quantity; and in the former case the tangent $1V1B$ is drawn toward A, in the latter $2V2B$ is drawn in the opposite direction. Yet neither happens in the intermediate [position] at M, at which moment the v's neither increase nor decrease, but are at rest; and therefore dv becomes equal to 0, where nothing represents a quantity [which] may be either positive or negative, for $+0$ equals -0; and at that place the v, obviously the ordinate LM, is maximum (or if the convexity turns toward the axis, minimum) and the tangent to the curve at M is drawn neither above X, where it approaches the axis in the direction of A, nor below X in the contrary direction, but is parallel to the axis. If dv is infinite with respect to the dx, then the tangent is perpendicular to the axis, or it is the ordinate itself. If dv and dx [are] equal, the tangent makes half a right angle with the axis. If, with increasing ordinates v, their increments or differences dv also increase (or if, the dv's being positive, the ddv's, the differences of the differences are also positive, or [the dv's being] negative, [the ddv's are also] negative), the curve turns [its] convexity toward the axis; otherwise [its] concavity.[1] Where indeed the increment is maximum or minimum, or where the increments from decreasing become increasing, or the contrary, there is a point of opposite flexion, and the concavity and convexity are interchanged, provided that the ordinates too do not become decreasing from increasing or the contrary, for then the concavity or convexity would remain; but it is impossible that the amounts of change[2] should continue to increase or decrease while the ordinates become decreasing from increasing or the contrary. And so a point of flexion occurs when, neither v nor dv being 0, yet ddv is 0. Whence, furthermore, problems of opposite flexion have, not two equal roots, like problems of maximum, but three.

Powers:[3]

$$\cdots\cdots\cdots$$

$$dx^a = ax^{a-1}dx,$$

for example,

$$dx^3 = 3x^2dx;$$

$$d\frac{1}{x^a} = -\frac{adx}{x^{a-1}},$$

for example, if

$$w = \frac{1}{x^3},$$

$$dw = -\frac{3dx}{x^4}.$$

Roots:

$$d\sqrt[b]{x^a} = \frac{a}{b}\sqrt[b]{x^{a-b}}dx.$$

---

[1] [In the original the words concavity and convexity are interchanged, but farther on in the article the statement is made correctly.]

[2] [The word that Leibniz uses is "crementa."]

[3] [There are several mistakes in this paragraph and the following one but, as they are obviously printer's errors, they have been corrected by the translator.]

(Hence

$$d\sqrt[3]{y} = \frac{dy}{2\sqrt[3]{y}},$$

for in this case a is 1, and b is 2; therefore $\frac{a}{b}\sqrt[b]{x^{a-b}}$ is $\frac{1}{2}\sqrt[3]{y^{-1}}$; now y^{-1} is the same as $\frac{1}{y}$, from the nature of the exponents of a geometric progression, and $\frac{1}{\sqrt[3]{y}}$ is $\sqrt[3]{y^{-1}}$.)

$$d\frac{1}{\sqrt[b]{x^a}} = \frac{-adx}{b\sqrt[b]{x^{a+b}}}.$$

Again the rule for an integral power would suffice for determining fractions as well as roots, for the power may be a fraction while the exponent is negative, and it is changed into a root when the exponent is a fraction; but I have preferred to deduce these consequences myself rather than to leave them to be deduced by others, since they are completely general and of frequent occurrence; and in a matter which is itself involved it is preferable to take ease[1] [of operation] into account.

From this rule, known as an algorithm, so to speak, of this calculus, which I call differential, all other differential equations may be found by means of a general calculus, and maxima and minima, as well as tangents [may be] obtained, so that there may be no need of removing fractions, nor irrationals, nor other aggregates, which nevertheless formerly had to be done in accordance with the methods published up to the present. The demonstration of all [these things] will be easy for one versed in these matters, who also takes into consideration this one point which has not received sufficient attention heretofore, that dx, dy, dv, dw, dz, can be treated as proportional to the momentaneous differences, whether increments or decrements, of x, y, v, w, z (each in its order).[2]

. . . .

---

[1] [The word used is "facilitati."]

[2] [Leibniz now proceeds to illustrate his rules by means of a number of examples. These are followed by selections from the article "On abstruse geometry...," "De geometria recondita et analysi indivisibilium atque infinitorum, addenda bis qua sunt in *Actis* a. 1684, Maji, p. 233; Octob. p. 467; Decem. p. 585." G. G. L. *Acta Eruditorum*, June, 1686.

The errors in the pages to which reference is made have been corrected by the translator.]

Since, furthermore the method of investigating indefinite quadratures or their impossibilities is with me only a special case (and indeed an easier one) of the far greater problem which I call the *inverse method of tangents*, in which is included the greatest part of all transcendental geometry; and because it could always be solved algebraically, all things were looked upon as discovered;[1] and nevertheless up to the present time I see no satisfactory result from it; therefore I shall show how it can be solved no less than the indefinite quadrature itself. Therefore, inasmuch as algebraists formerly assumed letters or general numbers for the quantities sought, in such transcendental problems I have assumed general or indefinite equations for the lines sought, for example, the abscissa and the ordinate [being represented] by the usual x and y, my equation for the line sought is,

$$0 = a + bx + cy + exy + fx^2 + gy^2, \text{ etc.;}$$

by the use of this indefinitely stated equation, I seek the tangent to a really definite line (for it can always be determined, as far as need be),[2] and comparing what I find with the given property of the tangents, I obtain the value[s] of the assumed letters, a, b, c, etc., and even establish the equation of the line sought, wherein occasionally certain [things] still remain arbitrary; in which case innumerable lines may be found satisfying the question, which was so involved that many, considering the problem as not sufficiently defined at last, believed it impossible. The same things are also established by means of series. But, according to the calculation to be effected, I use many things, concerning which [I shall speak] elsewhere. And if the comparison does not succeed, I decide that the line sought is not algebraic but transcendental.

Which being done, in order that I may discover the species of the transcendence (for some transcendentals depend upon the general section of a ratio, or upon logarithms, others upon the general section of an angle, or upon the arcs of a circle, others upon other more complex indefinite questions); therefore, besides the letters x and y, I assume still a third, as v, which signifies a transcendental quantity, and from these three I form the general

---

[1] [The Latin is, ...& quod algebraice semper posset solvi, omnia reperta haberentur, & vero nihil adhuc de eo extare video satisfaciens, ...]

[2] [The Latin of this parenthesis is: (semper enim determinari potest, quousque assurgi opus sit). This is not good Latin, but the meaning is probably as we have given it.]

equation for the line sought, from which I look for the tangent to the line according to my method of tangents published in the *Acta*, October, 1684, which does not preclude[1] transcendentals. Thence, comparing what I discover with the given property of the tangents to the curve, I find, not only the assumptions, the letters *a*, *b*, *c*, etc., but also the special nature of the transcendental.

. . . .

Let the ordinate be *x*, the abscissa *y*, let the interval between the perpendicular and the ordinate. . . be *p*; it is manifest at once by my method that

$$pdy = xdx,$$

. . . .

which differential equation being turned into a summation becomes

$$\int pdy = \int xdx.$$

But from what I have set forth in the method of tangents, it is manifest that

$$d\overline{\tfrac{1}{2}xx} = xdx;$$

therefore, conversely,

$$\tfrac{1}{2}xx = \int xdx$$

(for as powers and roots in common calculation, so with us sums and differences or $\int$ and d, are reciprocals). Therefore we have

$$\int pdx = \tfrac{1}{2}xx. \qquad\qquad \text{Q. E. D.}$$

Now I prefer to employ *dx* and similar [symbols], rather than letters for them, because the *dx* is a certain modification of the *x*, and so by the aid of this it turns out that, since the work must be done through the letter *x* alone, the calculus obviously proceeds with its own[2] powers and differentials, and the transcendental relations are expressed between *x* and another [quantity]. For which reason, likewise, it is permissible to express transcendental

---

[1] [The Latin word is "moratur" which means, literally, "it does not linger," or "it does not take into consideration." But as the method given really does include the case of transcendentals, accuracy of translation must be sacrificed in the interest of truth.]

[2] [That is the powers and differentials of the *x*.]

lines by an equation; for example, if the arc is a, the versine x, then we shall have

$$a = \int dx : \overline{\sqrt{2x - xx}},$$

and if y is the ordinate of the cycloid, then

$$y = \sqrt{2x - xx} + \int dx : \overline{\sqrt{2x - xx}},$$

which equation perfectly expresses the relation between the ordinate y and the abscissa x. and from it all the properties of the cycloid can be demonstrated; and the analytic calculus is extended in this way to those lines which hitherto have been excluded for no greater cause than that they were believed unsuited[1] to it; also the Wallisian interpolations and innumerable other things are derived from this source.

. . . .

It befell me, up to the present a tyro in these matters, that, from a single aspect of a certain demonstration concerning the magnitude of a spherical surface, a great light suddenly appeared. For I saw that in general a figure formed by perpendiculars to a curve, and the lines applied ordinatewise to the axis (in the circle, the radii), is proportional to the surface of that solid which is generated by the rotation of the figure about the axis. Wonderfully delighted by which theorem, since I did not know that such a thing was known to others, I straightway devised the triangle which in all curves I call the characteristic [triangle], the sides of which would be indivisible (or, to speak more accurately, infinitely small) or differential quantities; whence immediately, with no trouble, I established countless theorems, some of which I afterward observed in the works of Gregory and Barrow.

. . . .

Finally I discovered the supplement of algebra for transcendental quantities, of course, my calculus of indefinitely small quantities, which, the differential as well as [that] of either summations or quadratures, I call, and aptly enough if I am not mistaken, the *analysis of indivisibles and infinites*, which having been once revealed, whatever of this kind I had formerly wondered about seems only child's play and a jest.

---

[1] [The Latin is "incapaces," literally, "incapable of."]

BERKELEY

Analyst and Its Effect upon the Calculus

(Selected and Edited by Professor Florian Cajori, University of California, Berkeley, Calif.)

By the publication of the *Analyst* in 1734, Dean (afterward Bishop) Berkeley profoundly influenced mathematical thought in England for more than half a century. His brilliant defence of his views against the criticisms of James Jurin of Cambridge and John Walton of Dublin, and the controversies among the mathematicians themselves which were started by Berkeley's *Analyst* (London, 1734), led to a clarifying of mathematical ideas as found in two important English books:—Benjamin Robin's *Discourse Concerning the Nature and Certainty of Sir Isaac Newtons Method of Fluxions and of Prime and Ultimate Ratios*, 1735, and Colin Maclaurin's *Treatise of Fluxions*, 1742. Robins greatly improved the theory of limits. Both Robins and Maclaurin banished from their works the fixed infinitesimal. In some recent books, one meets with the statement that it was Weierstrass who first banished the fixed infinitesimal from the calculus. This claim needs to be qualified by the historical fact that already in the eighteenth century the fixed infinitesimal was excluded from the works on the calculus written by Robins, Maclaurin, and also by Simon Lhuilier on the Continent.

The treatment of the calculus as initiated by Leibniz, became known in Great Britain earlier than the theory of fluxions. The Scotsman, John Craig, used the Leibnizian notation, in print, in 1685. The Newtonian fluxional notation was first printed in John Wallis's *Algebra* of 1693. Harris, Hayes, and Stone, though using the term "fluxion" and the notation of Newton, nevertheless drew their inspiration, on matters relating to mathematical concepts, from continental writers who followed Leibniz. These facts explain the reason why Berkeley devoted considerable attention to the calculus of Leibniz.

The *Analyst* is a book of 104 pages. It is addressed to "an infidel mathematician," generally supposed to have referred to the astronomer, Edmund Halley. There is no evidence of religious skepticism in Halley's published writings; his alleged "infidelity" rests only upon common repute. In the selections here given no effort has been made to preserve the capitalization of many of the words as in the original edition.

The Analyst:

A Discourse Addressed to an Infidel Mathematician

Though I am a stranger to your person, yet I am not, Sir, a stranger to the reputation you have acquired in that branch of learning which hath been your peculiar study; nor to the authority that you therefore assume in things foreign to your profession; nor to the abuse that you, and too many more of the like character, are known to make of such undue authority, to the misleading of unwary persons in matters of the highest concernment, and whereof your mathematical knowledge can by no means qualify you to be a competent judge...

Whereas then it is supposed that you apprehend more distinctly, consider more closely, infer more justly, and conclude more accurately than other men, and that you are therefore less religious because more judicious, I shall claim the privilege of a Freethinker; and take the liberty to inquire into the object, principles, and method of demonstration admitted by the mathematicians of the present age, with the same freedom that you presume to treat the principles and mysteries of Religion; to the end that all men may see what right you have to lead, or what encouragement others have to follow you...

The Method of Fluxions is the general key by help whereof the modern mathematicians unlock the secrets of Geometry, and consequently of Nature. And, as it is that which hath enabled them so remarkably to outgo the ancients in discovering theorems and solving problems, the exercise and application thereof is become the main if not the sole employment of all those who in this age pass for profound geometers. But whether this method be clear or obscure, consistent or repugnant, demonstrative or precarious, as I shall inquire with the utmost impartiality, so I submit my inquiry to your own judgment, and that of every candid reader.—Lines are supposed to be generated[1] by the motion of points, planes by the motion of lines, and solids by the motion of planes. And whereas quantities generated in equal times are greater or lesser according to the greater or lesser velocity wherewith they increase and are generated, a method hath been found to determine quantities from the velocities of their generating motions. And such velocities are called fluxions: and the quantities generated are called flowing quantities. These fluxions are said to be nearly as the increments of the flowing quantities, generated in the least equal particles of time; and to be accurately in the first proportion of the nascent, or in the last of the evanescent increments. Sometimes, instead of velocities, the momentaneous

---

[1] *Introd. ad Quadraturam Curvarum.*

increments or decrements of undetermined flowing quantities are considered, under the appellation of moments.

By moments we are not to understand finite particles. These are said not to be moments, but quantities generated from moments, which last are only the nascent principles of finite quantities. It is said that the minutest errors are not to be neglected in mathematics: that the fluxions are celerities, not proportional to the finite increments, though ever so small; but only to the moments or nascent increments, whereof the proportion alone, and not the magnitude, is considered. And of the aforesaid fluxions there be other fluxions, which fluxions of fluxions are called second fluxions. And the fluxions of these second fluxions are called third fluxions: and so on, fourth, fifth, sixth, etc., *ad infinitum*. Now, as our Sense is strained and puzzled with the perception of objects extremely minute, even so the Imagination, which faculty derives from sense, is very much strained and puzzled to frame clear ideas of the least particles of time, or the least increments generated therein: and much more so to comprehend the moments, or those increments of the flowing quantities in *statu nascenti,* in their very first origin or beginning to exist, before they become finite particles. And it seems still more difficult to conceive the abstracted velocities of such nascent imperfect entities. But the velocities of the velocities—the second, third, fourth, and fifth velocities, etc.—exceed, if I mistake not, all human understanding. The further the mind analyseth and pursueth these fugitive ideas the more it is lost and bewildered; the objects, at first fleeting and minute, soon vanishing out of sight. Certainly, in any sense, a second or third fluxion seems an obscure Mystery. The incipient celerity of an incipient celerity, the nascent augment of a nascent augment, *i. e.,* of a thing which hath no magnitude—take it in what light you please, the clear conception of it will, if I mistake not, be found impossible; whether it be so or no I appeal to the trial of every thinking reader. And if a second fluxion be inconceivable, what are we to think of third, fourth, fifth fluxions, and so on without end?...

All these points, I say, are supposed and believed by certain rigorous exactors of evidence in religion, men who pretend to believe no further than they can see. That men who have been conversant only about clear points should with difficulty admit obscure ones might not seem altogether unaccountable. But he who can digest a second or third fluxion, a second or third differ-

ence, need not, methinks, be squeamish about any point in divinity
. . .

Nothing is easier than to devise expressions or notations for fluxions and infinitesimals of the first, second, third, fourth, and subsequent orders, proceeding in the same regular form without end or limit $\dot{x}.\ \ddot{x}.\ \dddot{x}.\ \ddddot{x}$ etc. or $dx.\ ddx.\ dddx.\ ddddx.$ etc. These expressions, indeed, are clear and distinct, and the mind finds no difficulty in conceiving them to be continued beyond any assignable bounds. But if we remove the veil and look underneath, if, laying aside the expressions, we set ourselves attentively to consider the things themselves which are supposed to be expressed or marked thereby, we shall discover much emptiness, darkness, and confusion; nay, if I mistake not, direct impossibilities and contradictions. Whether this be the case or no, every thinking reader is entreated to examine and judge for himself... .

This is given for demonstration.[1] Suppose the product or rectangle AB increased by continual motion: and that the momentaneous increments of the sides A and B are a and b. When the sides A and B were deficient, or lesser by one half of their moments, the rectangle was $\overline{A - \frac{1}{2}a} \times \overline{B - \frac{1}{2}b}$, i. e., $AB - \frac{1}{2}aB - \frac{1}{2}bA + \frac{1}{4}ab$. And as soon as the sides A and B are increased by the other two halves of their moments, the rectangle becomes $\overline{A + \frac{1}{2}a} \times \overline{B + \frac{1}{2}b}$ or $AB + \frac{1}{2}aB + \frac{1}{2}bA + \frac{1}{4}ab$. From the latter rectangle subduct the former, and the remaining difference will be $aB + bA$. Therefore the increment of the rectangle generated by the entire increments a and b is $aB + bA$. Q. E. D. But it is plain that the direct and true method to obtain the moment or increment of the rectangle AB, is to take the sides as increased by their whole increments, and so multiply them together, $A + a$ by $B + b$, the product whereof $AB + aB + bA + ab$ is the augmented rectangle; whence, if we subduct AB the remainder $aB + bA + ab$ will be the true increment of the rectangle, exceeding that which was obtained by the former illegitimate and indirect method by the quantity ab. And this holds universally by the quantities a and b be what they will, big or little, finite or infinitesimal, increments, moments, or velocities. Nor will it avail to say that

---

[1] *Philosophiae Naturalis Principia Mathematica*, Lib. II, lem. 2.

ab is a quantity exceedingly small: since we are told that *in rebus mathematicis errores quam minimi non sunt contemnendi*[1]...

But, as there seems to have been some inward scruple or consciousness of defect in the foregoing demonstration, and as this finding the fluxion of a given power is a point of primary importance, it hath therefore been judged proper to demonstrate the same in a different manner, independent of the foregoing demonstration. But whether this method be more legitimate and conclusive than the former, I proceed now to examine; and in order thereto shall premise the following lemma:—"If, with a view to demonstrate any proposition, a certain point is supposed, by virtue of which certain other points are attained; and such supposed point be itself afterwards destroyed or rejected by a contrary

---

[1] *Introd. ad Quadraturam Curvarum.*

[Of interest are the remarks on Newton's reasoning, made in 1862 by Sir William Rowan Hamilton in a letter to Augustus De Morgan: "It is very difficult to understand the *logic* by which Newton proposes to prove, that the *momentum* (as he calls it) of the *rectangle* (or product) *AB* is equal to *aB* + *bA*, if the *momenta* of the sides (or factors) *A* and *B* be denoted by *a* and *b*. His mode of getting rid of *ab* appeared to me long ago (I must confess it) to involve so much of *artifice*, as to deserve to be called *sophistical;* although I should not like to say so publicly. He subtracts, you know $\left(A - \frac{1}{2}a\right)(B - \frac{1}{2}b)$ from $\left(A + \frac{1}{2}a\right)\left(B + \frac{1}{2}b\right)$; whereby, of course, *ab* disappears in the result. But by *what right*, or *what reason* other than to give an unreal air of *simplicity* to the calculation, does he *prepare* the *products* thus? Might it not be argued similarly that the difference,

$$\left(A + \frac{1}{2}a\right)^3 - \left(A - \frac{1}{2}a\right)^3 = 3aA^2 + \frac{1}{4}a^3$$

was the moment of A^3; and is it not a sufficient *indication* that the mode of procedure adopted is not the fit one for the subject, that it quite *masks* the notion of a *limit;* or rather has the appearance of treating that notion as foreign and irrelevant, notwithstanding all that had been said so well before, in the First Section of the First Book? Newton does not seem to have cared for being very consistent in his *philosophy*, if he could anyway get hold of *truth* or what he considered to be such..." From *Life of Sir William Rowan Hamilton* by R. P. Graves, Vol. 3, p. 569.

We give also Weissenborn's objection to Newton's procedure of taking half of the increments *a* and *b*; with equal justice one might take, says he,

$$\left(A + \frac{2}{3}a\right)\left(B + \frac{2}{3}b\right) - \left(A - \frac{1}{3}a\right)\left(B - \frac{1}{3}b\right),$$

and the result would then be $Ab + Ba + \frac{1}{3}ab$. From H. Weissenborn's *Principien der höheren Analysis in ihrer Entwickelung von Leibniz bis auf Lagrange*, Halle, 1856, p. 42.]

supposition; in that case, all the other points attained thereby, and consequent thereupon, must also be destroyed and rejected, so as from thenceforward to be no more supposed or applied in the demonstration."[1]   This is so plain as to need no proof.

Now, the other method of obtaining a rule to find the fluxion of any power is as follows.   Let the quantity x flow uniformly, and be it proposed to find the fluxion of x^n.   In the same time that x by flowing becomes $x + o$, the power x^n becomes $\overline{x + o}|^n$, i. e., by the method of infinite series

$$x^n + nox^{n-1} + \frac{nn - n}{2}\, oox^{n-2} + \&c.,$$

and the increments

$$o \text{ and } nox^{n-1} + \frac{nn - n}{2}\, oox^{n-2} + \&c.$$

are one to another as

$$1 \text{ to } nx^{n-1} + \frac{nn - n}{2}\, ox^{n-2} + \&c.$$

Let now the increments vanish, and their last proportion will be 1 to nx^{n-1}.   But it should seem that this reasoning is not fair or conclusive.   For when it is said, let the increments vanish, i. e., let the increments be nothing, or let there be no increments, the former supposition that the increments were something, or that there were increments, is destroyed, and yet a consequence of that supposition, i. e., an expression got by virtue thereof, is retained. Which, by the foregoing lemma, is a false way of reasoning.   Certainly when we suppose the increments to vanish, we must

---

[1] [Berkeley's *lemma* was rejected as invalid by James Jurin and some other mathematical writers.   The first mathematician to acknowledge openly the validity of Berkeley's *lemma* was Robert Woodhouse in his *Principles of Analytical Calculation*, Cambridge, 1803, p. XII.   Instructive, in this connection, is a passage in A. N. Whitehead's *Introduction to Mathematics*, New York and London, 1911, p. 227.   Whitehead does not mention Berkeley's *lemma* and probably did not have it in mind.   Nevertheless, Whitehead advances an argument which is essentially the equivalent of Berkeley's, though expressed in different terms.   When discussing the difference-quotient $\dfrac{(x + h)^2 - x^2}{h}$, Whitehead says: "In reading over the Newtonian method of statement, it is tempting to seek simplicity by saying that $2x + b$ is $2x$, when b is zero. *But this will not do; for it thereby abolishes the interval from x to $x + b$*, over which the average increase was calculated.   The problem is, how to keep an interval of length b over which to calculate the average increase, and at the same time to treat b as if it were zero.   Newton did this by the conception of a limit, and we now proceed to give Weierstrass's explanation of its real meaning."]

suppose their proportions, their expressions, and everything else derived from the supposition of their existence, to vanish with them...

I have no controversy about your conclusions, but only about your logic and method: how you demonstrate? what objects you are conversant with, and whether you conceive them clearly? what principles you proceed upon; how sound they may be; and how you apply them?...

Now, I observe, in the first place, that the conclusion comes out right, not because the rejected square of *dy* was infinitely small, but because this error was compensated by another contrary and equal error[1]...

The great author of the method of fluxions felt this difficulty, and therefore he gave in to those nice abstractions and geometrical metaphysics without which he saw nothing could be done on the received principles: and what in the way of demonstration he hath done with them the reader will judge. It must, indeed, be acknowledged that he used fluxions, like the scaffold of a building, as things to be laid aside or got rid of as soon as finite lines were found proportional to them. But then these finite exponents are found by the help of fluxions. Whatever therefore is got by such exponents and proportions is to be ascribed to fluxions: which must therefore be previously understood. And what are these fluxions? The velocities of evanescent increments. And what are these same evanescent increments? They are neither finite quantities, nor quantities infinitely small, nor yet nothing. May we not call them the ghosts of departed quantities...?

You may possibly hope to evade the force of all that hath been said, and to screen false principles and inconsistent reasonings, by a general pretence that these objections and remarks are *metaphysical*. But this is a vain pretence. For the plain sense and truth of what is advanced in the foregoing remarks, I appeal to the understanding of every unprejudiced intelligent reader...

And, to the end that you may more clearly comprehend the force and design of the foregoing remarks, and pursue them still farther in your own meditations, I shall subjoin the following Queries:—

Query 1. Whether the object of geometry be not the proportions of assignable extensions? And whether there be any need

---

[1] [Berkeley explains that the calculus of Leibniz leads from false principles to correct results by a "Compensation of errors." The same explanation was advanced again later by Maclaurin, Lagrange, and, independently, by L. N. M. Carnot in his *Réflexions sur la métaphysique du calcul infinitésimal*, 1797.]

of considering quantities either infinitely great or infinitely small?
. . .

Qu. 4. Whether men may properly be said to proceed in a scientific method, without clearly conceiving the object they are conversant about, the end proposed, and the method by which it is pursued?. . .

Qu. 8. Whether the notions of absolute time, absolute place, and absolute motion be not most abstractely metaphysical? Whether it be possible for us to measure, compute, or know them?
. . .

Qu. 16. Whether certain maxims do not pass current among analysts which are shocking to good sense? And whether the common assumption, that a finite quantity divided by nothing is infinite, be not of this number?[1]. . .

Qu. 31. Where there are no increments, whether there can be any *ratio* of increments? Whether nothings can be considered as proportional to real quantities? Or whether to talk of their proportions be not to talk nonsense? Also in what sense we are to understand the proportion of a surface to a line, of an area to an ordinate? And whether species or numbers, though properly expressing quantities which are not homogeneous, may yet be said to express their proportion to each other?. . .

Qu. 54. Whether the same things which are now done by infinites may not be done by finite quantities? And whether this would not be a great relief to the imaginations and understandings of mathematical men?. . .

Qu. 63. Whether such mathematicians as cry out against mysteries have ever examined their own principles?

Qu. 64. Whether mathematicians, who are so delicate in religious points, are strictly scrupulous in their own science? Whether they do not submit to authority, take things upon trust, and believe points inconceivable? Whether they have not *their* mysteries, and what is more, their repugnances and contradictions?". . .

---

[1] [The earliest exclusion of division by zero in ordinary elementary algebra, on the ground of its being a procedure that is inadmissible according to reasoning based on the fundamental assumptions of this algebra, was made in 1828, by Martin Ohm, in his *Versuch eines vollkommen consequenten Systems der Mathematik*, Vol. I, p. 112. In 1872, Robert Grassmann took the same position. But not until about 1881 was the necessity of excluding division by zero explained in elementary school books on algebra.

CAUCHY

On the Derivatives and Differentials of Functions of a Single Variable

(Translated from the French by Professor Evelyn Walker, Hunter College, New York City.)

Augustin-Louis Cauchy[1] (1789–1857), the well-known French mathematician and physicist, at the age of twenty-four gave up his chosen career as an engineer in order to devote himself to the study of pure mathematics. Soon afterward he became a teacher at the École Polytechnique. In 1816 he won the *Grand Prix* of the Institut for his mémoire on wave propagation. His greatest contributions to mathematics are embodied in the rigorous methods which he introduced. Of treatises and articles in scientific journals he published in all seven hundred and eighty-nine. His greatest achievement in the domain of the calculus was his scientifically correct derivation of the differential of a function, which he accomplished by means of the device that has come to be known as Cauchy's fraction. His treatment of the matter is as follows.[2]

Third Lesson

Derivatives[3] of Functions of a Single Variable

When the function $y = f(x)$ lies continuously between two given limits of x, and there is assigned to the variable a value included between these two limits, an infinitely small increment given to the variable produces an infinitely small increment of the function itself. Consequently, if we then place $\Delta x = i$, the two terms of the *ratio of the differences*

$$(1) \qquad \frac{\Delta y}{\Delta x} = \frac{f(x + i) - f(x)}{i}$$

will be infinitely small quantities. But while these terms indefinitely and simultaneously approach the limit zero, the ratio itself may converge toward another limit, either positive or nega-

---

[1] See C. A. Valson, *La Vie et les Traveaux du Baron Caucby*, Paris, 1868.

[2] The two extracts quoted are from *Résumé des Leçons données à l'École Royale Polytechnique sur le Calcul Infinitésimal*, Paris, 1823. The present translation was made from the same work as republished in *Œuvres Complètes d'Augustin Caucby*, Sér. II, Tome IV, Paris, 1889.

[3] [Cauchy's word is "derivées"]

tive. This limit, when it exists, has a fixed value for each particular value of x; but it varies with x. Thus, for example, if we take $f(x) = x^m$, m designating a whole number, the ratio between the infinitely small differences will be

$$\frac{(x + i)^m - x^m}{i} = mx^{m-1} + \frac{m(m - 1)}{1.2}x^{m-2}i + \ldots + i^{m-1},$$

and it will have for [its] limit the quantity mx^{m-1}, that is to say, a new function of the variable x. It will be the same in general, only the form of the new function which serves as the limit of the ratio $\frac{f(x + i) - f(x)}{i}$ will depend upon the form of the given function $y = f(x)$. In order to indicate this dependence, we give to the new function the name *derived function*, and we designate it, with the help of an accent, by the notation

$$y' \text{ or } f'(x).$$
$$\ldots\ldots\ldots\ldots^1$$

Fourth Lesson

Differentials of Functions of a Single Variable

Let $y = f(x)$ always be a function of the independent variable x; i an infinitely small quantity, and b a finite quantity. If we place $i = \alpha b$, α also will be an infinitely small quantity, and we shall have identically

$$\frac{f(x + i) - f(x)}{i} = \frac{f(x + \alpha b) - f(x)}{\alpha b},$$

whence there will result

(1) $$\frac{f(x + \alpha b) - f(x)}{\alpha} = \frac{f(x + i) - f(x)}{i}b.$$

The limit toward which the first member of equation (1) converges, while the variable α approaches zero indefinitely, the quantity b remaining constant, is what is called the *differential* of the function $y = f(x)$. We indicate this differential by the characteristic d, as follows:

$$dy \text{ or } df(x).$$

---

[1] [Cauchy then differentiates various functions using the above definition.]

It is easy to obtain its value when we know that of the derived function y' or $f'(x)$. Indeed, taking the limits of the two members of equation (1), we shall find generally

(2) $$df(x) = bf'(x).$$

In the special case where $f(x) = x$, equation (2) reduces to

(3) $$dx = b.$$

Therefore the differential of the independent variable x is nothing else than the finite constant b. That granted, equation (2) will become

(4) $$df(x) = f'(x)dx$$

or, what amounts to the same thing,

(5) $$dy = y'dx.$$

It follows from these last [equations] that the derived function $y' = f'(x)dx$ of any function $y = f(x)$ is precisely equal to dy/dx, that is to say, to the ratio between the differential of the function and that of the variable, or, if we wish, to the coefficient by which it is necessary to multiply the second differential in order to obtain the first. It is for this reason that we sometimes give to the derived function the name of *differential coefficient*.[1]

---

[1] [After this Cauchy gives the rules for differentiating various elementary functions, algebraic, exponential, trigonometric and antitrigonometric.]

EULER

On Differential Equations of the Second Order

(Translated from the Latin by Professor Florian Cajori, University of
California, Berkeley, Calif.)

Euler's article from which we here quote represents the earliest attempt to
introduce general methods in the treatment of differential equations of the
second order. It was written when Leonhard (Léonard) Euler was in his
twenty-first year and was residing in St. Petersburg, now Leningrad.

The title of the article is "A New Method of reducing innumerable differential equations of the second degree to differential equations of the first degree"
(Nova methodvs innvmerabiles aeqvationes differentiales secvndi gradvs
redvcendi ad aequationes differentiales primi gradvs). It was published in
the *Commentarii academiae scientiarvm imperialis Petropolitanae*, Tomvs III
ad annvm 1728, Petropoli, 1732, pp. 124–137.

When Euler in this article speaks of the "degree" of a differential equation,
he means what we now call the "order" of such an equation. Observe also
that he uses the letter c to designate 2.718..., the base of the natural system
of logarithms. The first appearance (see page 95) of the letter e, in print, as
the symbol for 2.178..., is in Euler's *Mechanica* (1733). We quote from
Euler's article of 1728:

1. When analysts come upon differential equations of the second
or any higher degree [order], they resort to two modes of solution.
In the first mode they inquire whether it is easy to integrate
them; if it is, they attain what they seek. When, however, an
integration is either utterly impossible or at least more difficult,
they endeavor to reduce them to differentials of the first degree,
concerning which it is certainly easier to tell whether they can be
resolved. Thus far no differential equations, save only those of
the first degree, can be resolved by known [general] methods...

3. However, if in a differentio-differential equation one or the
other of the indeterminates [variables] is absent, it is easy to reduce
it to a simple differential, by substituting in the place of the
differential of the missing quantity an expression composed of a
new indeterminate multiplied by the other differential...As
in the equation $Pdy^n = Qdv^n + dv^{n-2}ddv$, where P and Q signify
any functions of y, and dy is taken to be constant. Since v does

not appear in the equation, let $dv = zdy$, then $ddv = dzdy$. Substituting these, yields the equation $Pdy^n = Qz^n dy^n + z^{n-2} dy^{n-1} dz$, and dividing by dy^{n-1}, gives the equation $Pdy = Qz^n dy + z^{n-2} dz$; this is a simple differential.

4. Except in this manner, no one, as far as I know, has thus far reduced other differentio-differential equations to differentials of the first degree, unless, perhaps, they admitted of being easily integrated directly. It is here that I advance a method by which to be sure not all, but numberless differentio-differential equations in whatever manner affected by any one variable, may be reduced to a simpler differential. Thus I am brought around to those reductions in which, by a certain substitution, I transform them [the differential equations] into others in which one of the indeterminates is wanting. This done, by the aid of the preceding section, the equations thus treated are reduced finally to differential equations of the first degree.

5. In this connection I observe this property of the exponential quantities, or rather powers, the exponent of which is variable, the quantity thus raised remaining constant, that if they are differentiated and differentiated again, the variable itself is restricted, so that it always affects only the exponent, and the differentials are composed of the integral itself multiplied by the differentials of the exponent. A quantity of this kind is c^x where c denotes the number, the logarithm of which is unity; its differential is $c^x dx$, its differentio-differential $c^x(ddx + dx^2)$, where x does not enter, except in the exponent. Considering these things, I observed that if in a differentio-differential equation, exponentials are thus substituted in place of the indeterminates, these variables remain only in the exponents. This being understood, these quantities must be so adapted, when substituted in place of the indeterminates that, after the substitution is made, they do not resist being removed by division; in this manner one indeterminate or the other is eliminated and only its differentials remain.

6. This process is not applicable in all cases. But I have noticed that it holds for three types of differential equations of the second degree. The first type embraces all those equations which have only two terms...

7. All equations of the first type are embraced under this general formula: $ax^m dx^p = y^n dy^{p-2} ddy$, where dx is taken to be constant... To reduce that equation I place $x = c^{av}$, and $y = c^v t$. There result $dx = \alpha c^{av} dv$, and $dy = c^v(dt + tdv)$. And from this,

$ddx = \alpha c^{\alpha v}(ddv + \alpha dv^2)$ and $ddy = c^v(ddt + 2dtdv + tddv + tdv^2)$. But since dx is taken constant, one obtains $ddx = 0$, and $ddv = -\alpha dv^2$. Writing this in place of ddv, there follows $ddy = c^v \cdot (ddt + 2dtdv + (1 - \alpha)tdv^2)$. Substituting these values in place of x and y in the given equation, it is transformed into the following, $ac^{\alpha v(m+p)}\alpha^p dv^p = c^{(n+p-1)v}t^n(dt + tdv)^{p-2}(ddt + 2dtdv + (1 - \alpha) \cdot tdv^2)$.

8. Now α should be so determined that the exponentials may be eliminated by division. To do this it is necessary that $\alpha v(m+p) = (n + p - 1)v$, whence one deduces $\alpha = \dfrac{n + p - 1}{m + p}$. Thus, α being determined, the above equation is changed to the following

$$a\left(\frac{n + p - 1}{m + p}\right)^p dv^p = t^n(dt + tdv)^{p-2}.$$
$$\left(ddt + 2dtdv + \frac{m - n + 1}{m + p}tdv^2\right).$$

This may be deduced from the given equation directly, if I place $x = c^{(n+p-1)v:(m+p)}$, and $y = c^v t$. But $n + p - 1$ is the number of the dimensions which y determines; and $m + p$ which x determines. It is easy, therefore, in any special case to find α and to substitute the result. In the derived equation, since v is absent, place $dv = zdt$, then $ddv = zddt + dzdt$, but

$$ddv = -\alpha dv^2 = \frac{1 - n - p}{m + p}z^2dt^2.$$

From this follows $ddt = \dfrac{-dzdt}{z} + \dfrac{1 - n - p}{m + p}zdt^2$. After substituting these, there emerges

$$a\left(\frac{n + p - 1}{m + p}\right)^p z^p dt^p = t^n(dt + tzdt)^{p-2}\left(\frac{1 - n - p}{m + p}zdt^2\right.$$
$$\left. - \frac{dzdt}{z} + 2zdt^2 + \frac{m - n + 1}{m + p}tzzdt^2\right).$$

This divided by dt^{p-1} gives

$$a\left(\frac{n + p - 1}{m + p}\right)^p z^p dt = t^n(1 + tz)^{p-2}\left(\frac{1 + 2m - n + p}{m + p}zdt\right.$$
$$\left. - \frac{dz}{z} + \frac{m - n + 1}{m + p}tz^2dt\right).$$

9. The proposed general equation $ax^m dx^p = y^n dy^{p-2} ddy$ is thus reduced to this differential of the first degree

$$a\left(\frac{n+p-1}{m+p}\right)^p z^{p+1} dt = t^n (1+tz)^{p-2}\left(\frac{1+2m-n+p}{m+p} z^2 dt\right.$$
$$\left. + \frac{m-n+1}{m+p} tz^3 dt - dz\right),$$

the derived equation being multiplied by z. This equation may be obtained in one step from the one given, by placing in the first substitution $\int z\,dt$ in place of v. One should therefore take

$x = c^{(n+p-1)\int z\,dt:(m+p)}$ and in place of y take $c^{\int z\,dt} t$; or what amounts to the same thing, place $x = c^{(n+p-1)\int z\,dt}$ and $y = c^{(m+p)\int z\,dt} t$. . .

10. We illustrate what we have derived in general terms by particular examples. Let $xdxdy = yddy$, which by division by dy, is reduced to $xdx = ydy^{-1}ddy$. Comparing this with the general equation one obtains $a = 1$, $m = 1$, $p = 1$, $n = 1$. Substituting these in the differential equation of the first degree, the given equation reduces to

$$\frac{1}{2}z^2 dt = t(1+tz)^{-1}\left(\frac{3}{2}z^2 dt + \frac{1}{2}tz^3 dt - dz\right),$$

which becomes $z^2 dt + tz^3 dt = 3tz^2 dt + t^2 z^3 dt - 2tdz$. The given equation, $xdxdy = yddy$, may be [directly] reduced to this by taking $x = c^{\int z\,dt}$ and $y = c^{2\int z\,dt} t$. Therefore, the construction [i. e., resolution] of the proposed equation depends upon the construction of the derived differential equation. . .

11. The second type of differential equations which by my method I can reduce to differentials of the first degree, encompasses those which in the separate terms hold the same number of dimensions which the indeterminates and their differentials establish.[1] A general equation of this kind is the following: $ax^m y^{-m-1} dx^p dy^{2-p} + bx^n y^{-n-1} dx^q dy^{2-q} = ddy$. In its separate terms the dimensions of the indeterminates [and their differentials] is unity. Also, dx is taken constant. This assumed equation is composed of only three terms, but as many as desired may be added to the above, the procedure remaining the same. There may be added $ex^r y^{-r-1} dx^q dy^{2-q}$ and as many of this kind as may be desired. . .

---

[1] [That is, the differential equations are homogeneous in x, y, dx, dy, and d^2y.]

12. I reduce the given equation by substituting c^v for x, and $c^v t$ for y. Since therefore $x = c^v$ and $y = c^v t$, there follows $dx = c^v dv$ and $dy = c^v (dt + t dv)$; and from these, $ddx = c^v (ddv + dv^2)$ and $ddy = c^v (ddt + 2 dt dv + t dv^2 + t ddv)$. Since dx is taken to be constant, $ddx = 0$ and therefore $ddv = -dv^2$, and from this there results $ddy = c^v (ddt + 2 dt dv)$. These values of x, y, dx, dy, and ddy, when placed in the equation, transform it into the following:

$$ac^v t^{-m-1} dv^p (dt + t dv)^{2-p} + bc^v t^{-n-1} dv^q (dt + t dv)^{2-q} =$$
$$c^v (ddt + 2 dt dv).$$

Dividing by c^v, this becomes

$$at^{-m-1} dv^p (dt + t dv)^{2-p} + bt^{-n-1} dv^q (dt + t dv)^{2-q} = ddt + 2 dt dv.$$

Since v is absent from the equation, I place $dv = z dt$, and there will be $ddv = z ddt + dz dt$, but $ddv = -dv^2 = -z^2 dt^2$; therefore,

$$ddt = -z dt^2 - \frac{dz dt}{z}.$$

From this results the equation,

$$at^{-m-1} z^p dt^p (dt + z t dt)^{2-p} + bt^{-n-1} z^q dt^q (dt + z t dt)^{2-q} =$$
$$- z dt^2 - \frac{dz dt}{z} + 2 z dt^2,$$

or in better arrangement,

$$at^{-m-1} z^p dt (1 + zt)^{2-p} + bt^{-n-1} z^q dt (1 + zt)^{2-q} = z dt - \frac{dz}{z}.$$

13. This differential equation of the first degree may be derived from the given one by a single step, namely the straightway assumption that $x = c^{\int z dt}$ and $y = c^{\int z dt} t \ldots$

18. The third type of equations, which I treat by this method of reducing, comprises those in which one or the other of the indeterminates in the separate terms hold the same number of dimensions. Here two cases are to be distinguished, according as the differential of the variable having everywhere the same dimension, is to be taken constant or not. To the first case[1] belongs the following general equation

$$Px^m dy^{m+2} + Qx^{m-b} dx^b dy^{m+2-b} = dx^m ddy.$$

In this, x has the dimension m in each term, and dx is taken constant. Here P and Q signify any functions of y. For reducing this there

---

[1] [That is, the case in which the differential equation is homogeneous in x and dx.]

is need of only one substitution to wit, $x = c^v$, so that $dx = c^v dv$ and $ddx = c^v(ddv + dv^2) = 0$, and consequently, $ddv = -dv^2$. There results from this substitution

$$Pdy^{m+2} + Qdv^b dy^{m+2-b} = dv^m ddy,$$

of course, after dividing by c^{mv}.

19. Because of the absence of v in the derived equation, it can be reduced by substituting zdt for dv...

20. The other case of equations of the third type relates to the following general equation,

$$Px^m dy^{m+1} + Qx^{m-b} dx^b dy^{m-b+1} = dx^{m-1} ddx.$$

In this equation dy is taken constant, P and Q denoting any functions of y. And as one sees, x has the same dimension m in each term.[1] Take as before $x = c^v$; then $dx = c^v dv$, and $ddx = c^v(ddv + dv^2)$. When these are substituted in the equation, there results after division by c^{mv},

$$Pdy^{m+1} + Qdv^b dy^{m-b+1} = dv^{m+1} + dv^{m-1} ddv.$$

This equation is reduced as follows: Since v is absent, take $dv = zdy$, and, dy being constant, $ddv = dzdy$. Consequently, the last equation is changed to

$$Pdy^{m+1} + Qz^b dy^{m+1} = z^{m+1} dy^{m+1} + z^{m-1} dy^m dz.$$

But this, divided by dy^m, gives $Pdy + Qz^b dy = z^{m+1} dy + z^{m-1} dz$. Upon the reduction of this derived equation depends, therefore, the reduction of the given equation.

21. From this it will be understood, I trust, how differential equations of the second degree relating to one or another of the three types may be treated....

---

[1] [That is, the differential equation is homogenous in x, dx, and ddx.]

BERNOULLI

On the Brachistochrone Problem

(Translated from the Latin by Dr. Lincoln La Paz, National Research Fellow in Mathematics, The University of Chicago, Chicago, Ill.)

Jean (Johann, John) Bernoulli was born in Basel, Switzerland in 1667. He was professor of physics and mathematics at Groningen from 1695 until the chair of mathematics at Basel was vacated by the death of his elder brother, Jacques (Jakob, James) in 1705. Thereafter he was professor of mathematics at Basel until his own death in 1748. For further biographical details consult Merian, *Die Mathematiker Bernoulli* (Basel, 1860) or *Allgemeine Deutsche Biographie*, II, pp. 473–76.

The material translated in the following pages is collected in convenient form in Johannis Bernoulli, *Opera Omnia*, Lausanne and Geneva, 1742, vol. I, p. 161, pp. 166–169, pp. 187–193. The original sources are cited below in connection with the translations.

The calculus of variations is generally regarded as originating with the papers of Jean Bernoulli on the problem of the brachistochrone. It is true that Galileo in 1630–38[1] and Newton in 1686[2] had considered questions later recognized as belonging to the field of the calculus of variations. Their inquiries, however, are not looked upon as constituting the origin of this subject; since generality escaped them not only in the conception and formulation of their problems but also in the methods of attack which they devised.

On the contrary the writings of Jean Bernoulli show that he was not only fully aware of the difference between the ordinary problems of maxima and minima and the more difficult question he proposed, but also that he attained a fairly complete if not precise idea of the simpler problems of the calculus of variations *in general*. The terms in which he stated the problem of the brachistochrone may be readily extended to cover the formulation of the general case of the simplest class of variation problems in the plane. The curves he introduced under the name of *synchrones* for this problem furnish the first illustration of that important family of curves, now known as *transversals*, which is associated with the extremals of a problem in the calculus of variations; and in the fact noted by him, that the times of fall are equal along arcs intercepted by a synchrone on the cycloidal extremals of the brachistochrone problem which pass through a fixed point, we have the first instance of the beautiful *transversal theorem* of Kneser.[3]

---

[1] Galileo, *Dialog über die beiden hauptsächlichsten Weltsysteme* (1630) translation by Strauss, pp. 471–72; *Dialogues concerning Two New Sciences* (1638), translation by Crew and De Salvio, p. 239.

[2] Newton, *Principia*, Book II, Section VII, Scholium to Proposition XXXIV.

[3] Kneser, *Lehrbuch der Variationsrechnung*, 1900, p. 48;
Bolza, *Lectures on the Calculus of Variations*, University of Chicago Press, 1904, §33.

The reader of the translation which follows will note Jean Bernoulli's statement that he found a second or direct solution of the problem he proposed. In fact such a direct solution is mentioned in several of the letters which passed between Leibniz and Jean in 1696 as well as in the remarks which the former made on the subject of the brachistochrone problem in the *Acta Eruditorum* for May, 1697. However this direct demonstration which rests on the fundamental idea of general applicability employed by Jacques Bernoulli in obtaining his solution of the problem (namely that if a curve as a whole furnishes a minimum then the same property appertains to every portion of it) was not published until 1718 when both Jacques and Leibniz were dead. This fact is apparently regarded by those who believe Jean plagiarized from his brother Jacques as invalidating the former's claim of having secured a second solution. Jean for his part asserted that he delayed the publication of his second method in deference to counsel given by Leibniz in 1696.[1]

In any event it is regrettable that estimates of the relative value of the more mature methods of the two brothers often seem to be influenced by opinions which have been expressed with regard to the relative generality of their early solutions of the original brachistochrone problem, opinions which have in many cases been unfavorable to Jean Bernoulli. It is interesting to note in this connection that it was the opinion of as well qualified a student as Lagrange, if we may judge by statements made in his famous paper of 1762,[2] that all of the early solutions of the brachistochrone problem were found by special processes. In fact Lagrange emphasizes the part of Jean no less than that of Jacques in pioneering work on a general method in the calculus of variations.

NEW PROBLEM

Which Mathematicians Are Invited to Solve[3]

If two points A and B are given in a vertical plane, to assign to a mobile particle M the path AMB along which, descending under its own weight, it passes from the point A to the point B in the briefest time.

To arouse in lovers of such things the desire to undertake the solution of this problem, it may be pointed out that the question proposed does not, as might appear, consist of mere speculation having therefore no use. On the contrary, as no one would readily believe, it has great usefulness in other branches of science such as mechanics. Meanwhile (to forestall hasty judgment) [it may be remarked that] although the straight line *AB* is indeed the shortest between the points *A* and *B*, it nevertheless is not the

---

[1] Consult in regard to this matter: Cantor, *Geschichte der Mathematik*, Vol. III, chap. 96, especially p. 226, p. 430, p. 439; Leibniz and Jean Bernoulli, *Commercium Philosophicum et Mathematicum*, Lausanne and Geneva, 1745, vol. I, p. 167, p. 178, especially p. 183 pp. 253–4, p. 266; Jean Bernoulli, *Opera Omnia*, vol. II, pp. 266–7.

[2] Lagrange, *Miscellanea Taurinensia*, vol. II, p. 173.

[3] [From the *Acta Eruditorum*, Leipzig, June, 1696, p. 269.]

path traversed in the shortest time.   However the curve AMB, whose name I shall give if no one else has discovered it before the end of this year, is one well known to geometers.

. . . .

PROCLAMATION

Made Public at Groningen, [Jan.], 1697

Jean Bernoulli public professor of mathematics pays his best respects to *the most acute mathematicians of the entire world.*

Since it is known with certainty that there is scarcely anything which more greatly excites noble and ingenious spirits to labors which lead to the increase of knowledge than to propose difficult and at the same time useful problems through the solution of which, as by no other means, they may attain to fame and build for themselves eternal monuments among posterity; so I should expect to deserve the thanks of the mathematical world if, imitating the example of such men as Mersenne, Pascal, Fermat, above all that recent anonymous Florentine enigmatist,[1] and others, who have done the same before me, I should bring before the leading analysts of this age some problem upon which as upon a touchstone they could test their methods, exert their powers, and, in case they brought anything to light, could communicate with us in order that everyone might publicly receive his deserved praise from us.

The fact is that half a year ago in the June number of the *Leipzig Acta* I proposed such a problem whose usefulness linked with beauty will be seen by all who successfully apply themselves to it. [An interval of] six months from the day of publication was granted to geometers, at the end of which, if no one had brought a solution to light, I promised to exhibit my own. This interval of time has passed and no trace of a solution has appeared. Only the celebrated Leibniz, who is so justly famed in the higher geometry has written[2] me that he has by good fortune solved this,

---

[1] Vincentius Viviani, A°. 1692. Aenigma Geometricum proposuit, d° miro opificio Testudinis quadrabilis Hemisphaericae; see *Acta Eruditorum* of this year, June, p. 274, or Vita Viviani in *Hist. Acad. Reg. Scient.*, Paris, A.e 1703. [The problems proposed by the other mathematicians referred to are well known.]

[2] [Leibniz and Jean Bernoulli, *Commercium Philosophicum et Mathematicum*, vol. I, p. 172. Leibniz in *Acta Erud.*, May, 1697, p. 202 credits Galileo with originally proposing the brachistochrone problem.]

as he himself expresses it, very beautiful and hitherto unheard of problem; and he has courteously asked me to extend the time limit to next Easter in order that in the interim the problem might be made public in France and Italy and that no one might have cause to complain of the shortness of time allotted. I have not only agreed to this commendable request but I have decided to announce myself the prolongation [of the time interval] and shall now see who attacks this excellent and difficult question and after so long a time finally masters it. For the benefit of those to whom the *Leipzig Acta* is not available, I here repeat the problem.

Mechanical—Geometrical Problem on the Curve of Quickest Descent.

To determine the curve joining two given points, at different distances from the horizontal and not on the same vertical line, along which a mobile particle acted upon by its own weight and starting its motion from the upper point, descends most rapidly to the lower point.

The meaning of the problem is this: Among the infinitely many curves which join the two given points or which can be drawn from one to the other, to choose the one such that, if the curve is replaced by a thin tube or groove, and a small sphere placed in it and released, then this [sphere] will pass from one point to the other in the shortest time.

In order to exclude all ambiguity let it be expressly understood that we here accept the hypothesis of Galileo, of whose truth, when friction is neglected, there is now no reasonable geometer who has doubt: *The velocities actually acquired by a heavy falling body are proportional to the square roots of the heights fallen through.* However our method of solution is entirely general and could be used under any other hypotheses whatever.

Since nothing obscure remains we earnestly request all the geometers of this age to prepare, to attack, to bring to bear everything which they hold concealed in the final hiding places of their methods. Let who can seize quickly the prize which we have promised to the solver. Admittedly this prize is neither of gold nor silver, for these appeal only to base and venal souls from which we may hope for nothing laudable, nothing useful for science. Rather, since virtue itself is its own most desirable reward and fame is a powerful incentive, we offer the prize, fitting for the man of noble blood, compounded of honor, praise, and approbation; thus we shall crown, honor, and extol, publicly and privately, in letter and by word of mouth the perspicacity of our great Apollo.

If, however, Easter passes and no one is discovered who has solved our problem, then we shall withhold our solution from the world no longer; then, so we hope, the incomparable Leibniz will permit to see the light his own solution and the one obtained by us which we confided to him long ago. If geometers will study these solutions which are drawn from deep lying sources, we have no doubt they will appreciate the narrow bounds of the ordinary geometry and will value our discovery so much the more, as so few have appeared to solve our extraordinary problem, even among those who boast that through special methods, which they commend so highly, they have not only penetrated the deepest secrets of geometry but also extended its boundaries in marvellous fashion; although their golden theorems, which they imagine known to no one, have been published by others long before.[1]

. . . .

The curvature of a beam of light in a non-uniform medium, and the solution of the problem proposed in the Acta 1696, p. 269, of finding the brachistochrone, *i. e., the curve along which a heavy particle slides down from given point to given point in the shortest time; and of the construction of the* synchrone, *or the wave-front of the beam.*[2]

Up to this time so many methods which deal with *maxima* and *minima* have appeared that there seems to remain nothing so subtle in connection with this subject that it cannot be penetrated by their discernment—so they think who pride themselves either as the originators of these methods or as their followers. Now the students may swear by the word of their master as much as they please, and still, if they will only make the effort, they will see that our problem cannot in any way be forced into the narrow confines imposed by their methods, which extend only so far as to determine a *maximum* or *minimum* among given quantities finite or infinite in number. Truly where the very quantities which are involved in our problem, from among which the maximum or minimum is to be found, are no more determinate than the very thing one is seeking—this is a task, this is difficult labor! Even those distinguished men, Descartes, Fermat, and others, who once contended as vigorously for the superiority of their

---

[1] [This remark is to be regarded as a covert thrust at Newton. As a matter of fact Newton, when the problem finally came to his attention, solved it immediately.]

[2] [From the *Acta Eruditorum*, Leipzig, May, 1697, p. 206.]

methods as if they fought for God and country[1] or in their place now their disciples, must frankly confess that the methods handed down from these same authorities are here entirely inadequate. It is neither my nature nor my purpose to ridicule the discoveries of others. These men certainly accomplished much and attained in admirable fashion the goal they had set for themselves. For just as in their writings we find no consideration whatever of this type of maxima and minima, so indeed they have not recommended their methods for any but common problems.

I do not propose to give a universal method, [a thing] that one might search for in vain; but instead particular methods of procedure by means of which I have happily unraveled this problem—methods which, indeed, are successful not only in this problem but also in many others. I decided to submit my solution immediately to the celebrated Leibniz, while others sought other solutions, in order that he might publish it together with his own in case he found one. That he would indeed find a solution I had no doubt, for I am sufficiently well acquainted with the genius of this most sagacious man. In fact, while I write this, I learn from one of the letters with which he frequently honors me that my problem had pleased him beyond [my] expectation, and (since it attracted him by its beauty, so he says, as the apple attracted Eve) he was immediately in possession of the solution. The future will show what others will have accomplished. In any case the problem deserves that geometers devote some time to its solution since such a man as Leibniz, so busy with many affairs, thought it not useless to devote his time to it. And it is reward enough for them that, if they solve it, they obtain access to hidden truths which they would otherwise hardly perceive.

With justice we admire Huygens because he first discovered that a heavy particle falls down along a *common cycloid* in the same time no matter from what point on the *cycloid* it begins its motion. But you will be petrified with astonishment when I say that precisely this *cycloid*, the *tautochrone of Huygens* is our required *brachistochrone*.[2] I arrived at this result along two different

---

[1] [For an interesting first hand account of this dispute between Fermat and Descartes on the subject of maxima and minima see the sequence of letters collected in *Œuvres de Fermat*, Paris, 1894, vol. II, pp. 126–168. See also page 610 of this Source Book.]

[2] [For a description of the cycloid and its properties see Teixeira, *Traité des Courbes Spéciales Remarquables*, 1909, vol. II, pp. 133–149, especially §540. See also R. C. Archibald, "Curves, Special, in the *Encyclopaedia Britannica*, 14th edition.]

paths, one indirect and one direct. When I followed the first [path] I discovered a wonderful accordance between the curved orbit of a ray of light in a continuously varying medium and our *brachistochrone curve.* I also observed other things in which I do not know what is concealed which will be of use in dioptrics. Consequently what I asserted when I proposed the problem is true, namely that it was *not mere speculation but would prove to be very useful in other branches of science,* as for example in dioptrics. But as what we say is confirmed by the thing itself, here is the first method of solution!

Fermat has shown in a letter to de la Chambre (see *Epist. Cartesii Lat.,* Tome III, p. 147, and Fermatii *Opera Mathem.* p. 156 et. seq.) that a ray of light which passes from a rare into a dense medium is bent toward the normal in such a manner that the ray (which by hypothesis proceeds successively from the source of light to the point illuminated) traverses the path which is shortest in time. From this principle he shows that the sine of the angle of incidence and the sine of the angle of refraction are directly proportional to the rarities of the media, or to the reciprocals of the densities; that is, in the same ratio as the velocities with which the ray traverses the media. Later the most acute Leibniz in *Act. Erud.,* 1682, p. 185 et. seq., and soon thereafter the celebrated Huygens in his treatise *de Lumine,* p. 40, proved in detail and justified by the most cogent arguments this same physical or rather metaphysical principle, which Fermat, contented with his geometric proof and all too ready to renounce the validity of his [least time] principle, seems to have abandoned under the pressure of Clerselier.[1]

---

[1] [The following remarks supply the historical background necessary for an appreciation of these statements of Bernoulli:

Fermat in a letter to de la Chambre in 1657 (see *Oeuvres de Fermat,* II, p. 354) emphasized his belief that Descartes had given no valid proof of his law of refraction (the law now credited to Snell). Fermat formulated his Least Time Principle in this letter and guaranteed that he could deduce from it all of the experimentally known properties of refraction by use of his method for solving problems of maxima and minima.

In 1662 Fermat, in compliance with a request made by de la Chambre, actually applied his Principle to the determination of the law of refraction. (*Oeuvres,* II, p. 457.) Since the Cartesian (Snell) law of refraction had been deduced by Descartes on the hypothesis that the velocity of light in a rare medium is less than in a dense medium, an assumption that Fermat regarded as obviously false; Fermat, employing the contrary hypothesis, looked forward with certainty to the discovery of a different law of refraction. To his amaze-

If we now consider a medium which is not uniformly dense but [is] as if separated by an infinite number of sheets lying horizontally one beneath another, whose interstices are filled with transparent material of rarity increasing or decreasing according to a certain law; then it is clear that a ray which may be considered as a tiny sphere travels not in a straight but instead in a certain curved path. (The above-mentioned Huygens notes this in his treatise *de Lumine* but did not determine the nature of the curve itself.) This path is such that a particle traversing it with velocity continuously increasing or diminishing in proportion to the rarity, passes from point to point in the shortest time. Since the sines of [the angles of] refraction in every point are respectively as the rarities of the media or the velocities of the particle it is evident also that the curve will have this property, that the sines of its [angles of] inclination to the vertical are everywhere proportional to the velocities. In view of this one sees without difficulty that the *brachistochrone* is the curve which would be traced by a ray of

---

ment he found on carrying through all the details of his minimizing process that the application of his Principle led to precisely the same law of refraction as that established by Descartes. Fermat was so confounded by this unexpected result that he agreed to cede the victory to the Cartesians; although his distrust of Descartes' mode of proof was manifest.

The Cartesian Clerselier impressed by the fact that if the Least Time Principle were true Descartes' hypothesis with regard to the velocity of light must be false (for he could find no error in Fermat's geometrical proof) applied himself zealously to overthrowing this principle. (*Œuvres*, II, letter CXIII, p. 464; letter CXIV, p. 472.)

Fermat (*Œuvres*, II, p. 483) apparently disgusted with the matter wrote in his answer to Clerselier:

"As to the principal question, it seems to me that I have often said not only to M. de la Chambre but to you that I do not pretend and I have never pretended to be in the secret confidence of nature. She moves by paths obscure and hidden which I have never made the attempt to penetrate. I have merely offered her a little aid from geometry in connection with the subject of refraction, in case this aid would be of use to her. But since you assure me that she can take care of her own affairs without this assistance, and that she is content to follow the path prescribed to her by M. Descartes, I abandon to you with all my heart my supposed conquest of physics [*i. e.* the Least Time Principle]; and I shall be content if you will leave me in possession of my problem of pure geometry *taken in the abstract*, by means of which we can find the path of a moving particle which passes through two different media and which seeks to achieve its motion in the shortest time." Fermat's renunciation of his Principle seems, however, to have been a transitory one; for, in 1664 we find him again attacking on the basis of this Principle Descartes's deduction of the law of refraction. (*Œuvres*, II, letter CXVI, p. 485.)]

light in its passage through a medium whose rarity is proportional to the velocity which a heavy particle attains in falling vertically. For whether the increase in the velocity depends on the nature of the medium, more or less resistant, as in the case of the ray of light, or whether one removes the medium, and supposes that the acceleration is produced by means of another agency but according to the same law, as in the case of gravity; since in both cases the curve is in the end supposed to be traversed in the shortest time, what hinders us from substituting the one in place of the other?

In this way we can solve our problem generally, whatever we assume to be the law of acceleration. For it is reduced to finding the curved path of a ray of light in a medium varying in rarity

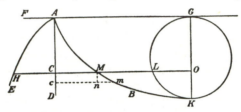

arbitrarily. Let therefore *FGD* be the medium, bounded by the horizontal *FG* in which the radiating point *A* [is situated]. Let the vertical *AD* be the axis of the given curve *AHE*, whose associated *HC* determine the rarities of the medium at the heights *AC*, or the velocities of the ray, or corpuscle, at the points *M*. Let the curved ray itself which is sought be *AMB*. Call *AC*, *x*; *CH*, *t*; *CM*, *y*; the differential *Cc*, *dx*; diff. *nm*, *dy*; diff. *Mm*, *dz*; and let *a* be an arbitrary constant. Take *Mm* for the whole sine,[1] *mn* for the sine of the angle of refraction or of inclination of the curve to the vertical, and then by what we have just said, *mn* is to *HC* in constant ratio, that is $dy : t = dz : a$. This gives the equation $ady = tdz$, or $aady^2 = ttdz^2 = ttdx^2 + ttdy^2$; which when reduced gives the general differential equation $dy = tdx : \sqrt{(aa - tt)}$ for the required curve *AMB*. Thus I have with one stroke solved two remarkable problems, one optical the other mechanical, and [have accomplished] more than I required of others; I have shown that the two problems which are taken from entirely distinct fields of mathematics are nevertheless of the same nature.

Let us now consider a special case, namely that arising on the customary hypothesis first introduced and proved by Galileo,

---

[1] [The Latin reading was changed from "*pro radio*" in the *Acta Erud.*, May 1697 to "*pro sinu toto*" in the *Opera Omnia*, 1742.]

according to which the velocity of heavy falling bodies varies as the square root of the distance fallen through; for this indeed is properly the problem.   Under this assumption the given curve *AHE* will be a parabola, that is, $tt = ax$ and $t = \sqrt{ax}$.   If this is substituted in the general equation we find $dy = dx\sqrt{\dfrac{x}{a-x}}$ from which I conclude that the *brachistochrone* curve is the *ordinary cycloid*.   In fact if one rolls the circle *GLK*, whose diameter is a, on *AG*, and if the beginning of rotation is in *A* itself; then the point *K* describes a cycloid, which is found to have the same differential equation $dy = dx\sqrt{\dfrac{x}{a-x}}$, calling *AC*, x, and *CM*, y.   Also this can be shown analytically from the preceding as follows:

$$dx\sqrt{\frac{x}{a-x}} = xdx : \sqrt{(ax-xx)}$$
$$= adx : 2\sqrt{(ax-xx)} - (adx - 2xdx) : 2\sqrt{(ax\text{ - }xx)};$$

also $(adx - 2xdx) : 2\sqrt{(ax - xx)}$ is the differential quantity whose sum[1] is $\sqrt{(ax - xx)}$ or *LO*; and $adx : 2\sqrt{(ax - xx)}$ is the differential of the arc *GL* itself; and therefore, summing the equation $dy = dx\sqrt{\dfrac{x}{a-x}}$, we have y or *CM* = *GL* − *LO*, hence *MO* = *CO* − *GL* + *LO*. Since indeed (assuming *CO* = semiperiphery *GLK*) *CO* − *GL* = *LK*, we will have *MO* = *LK* + *LO*, and, cancelling *LO*, *ML* = *LK*; which shows the curve *KMA* to be the cycloid.

   In order to completely solve the problem we have yet to show how from a given point, as vertex, we can draw the *brachistochrone*, or cycloid, which passes through a second given point. This is easily accomplished as follows: Join the two given points

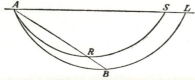

A, *B*, by the straight line *AB*, and describe an arbitrary cycloid on the horizontal *AL*, having its initial point in *A*, and cutting the line *AB* in *R*; then the diameter of the circle which traces the required cycloid *ABL* passing through *B* is to the

---

[1] [Bernoulli uses sum for integral.]

diameter of the circle which traces the cycloid *ARS* as *AB* to *AR*.[1]

Before I conclude, I cannot refrain from again expressing the amazement which I experienced over the unexpected identity of *Huygens's tautochrone* and our *brachistochrone.* Furthermore I think it is noteworthy that this identity is found only under the hypothesis of Galileo so that even from this we may conjecture that nature wanted it to be thus.   For, as nature is accustomed to proceed always in the simplest fashion, so here she accomplishes two different services through one and the same curve, while under every other hypothesis two curves would be necessary the one for oscillations of equal duration the other for quickest descent. If, for example, the velocity of a falling body varied not as the square root but as the cube root of the height [fallen through], then the *brachistochrone* would be algebraic, the *tautochrone* on the other hand transcendental; but if the velocity varied as the height [fallen through] then the curves would be algebraic, the one a circle, the other a straight line.[2]

---

[1] [Compare Bliss, *Calculus of Variations*, 1925, pp. 55–57.]

[2] [Denote by B_1 and B_2 the hypotheses which Bernoulli here suggests, and observe that he is concerned with a particle falling from rest in the velocity fields specified.

Under both B_1 and B_2, in case the initial velocity of the falling particle is different from zero, the integrand functions in the corresponding brachistochrone integrals, $T = \int_0^l \frac{ds}{v}$, are regular along the entire arc joining the given points A and B.   On the contrary when the initial velocity is zero the integrand functions are singular at the initial point A and further investigation is necessary.   For the corresponding case under the ordinary Galilean hypothesis consult Bliss, *Calculus of Variations*, 1925, p. 68; Kneser, *Lehrbuch der Variationsrechnung*, 2nd. Edt., 1925, p. 63.

The hypothesis B_2 is inadmissible in the case of the tautochrone.   This may be shown by applying the method of Puiseux (see *Jour. de Math.*, [Liouville's], Ser. 1, vol. 9, p. 410; compare Appell, *Traité de Mécanique Rationelle*, vol. I, p. 351, and MacMillan, *Statics and the Dynamics of a Particle*, p. 225) to the integral $T = C \int_0^b \frac{\sqrt{1 + [f'(x)]^2}\,dx}{[b - x]^k}$ which represents the time of fall from the height $x = b$ to the height $x = 0$ along the curve whose equation is $y = f(x)$ when the velocity of fall is proportional to the kth power of the distance fallen through.   We find in fact that for this integral to be independent of the value assigned to b it is necessary that $1 + [f'(x)]^2 = \delta^2 x^{2k-2}$.   Consequently the integral is not well defined when $k = 1$.

This objection to the hypothesis B_2 was first raised and justified, essentially as above, by P. Stäckel in Oswald's *Klassiker der exakten Wissenschaften*, No. 46, 1894, Anmerkungen, p. 137.   The formula for $f'(x)$ obtained in the

Geometers, I believe, will not be ungrateful if in conclusion I give the solution of a problem, just as worthy of consideration, which occurred to me while I was writing out what has gone before. *We require in the ver-* 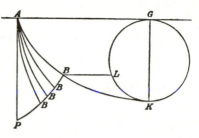 *tical plane the curve* PB, *which may be called the* Synchrone, *to every point* B *of which a heavy particle, descending from* A *along the cycloids* AB *with common vertex, arrives in the same time.* Let *AG* be horizontal and *AP* vertical. The meaning of the problem is as follows, that on each cycloid described on *AG*, a portion *AB* should be marked off, such that a heavy particle descending from *A* requires the same time to traverse it as it would require in falling from a given vertical height *AP*; when this is done, the point *B* will be in the *synchrone curve* PB which we seek.

If what we said above concerning light rays is considered attentively, it will be very evident that this curve is the same one which Huygens represents by the line *BC* in the figure on page 44 of his treatise *de Lumine*, and calls a *wave-front;* and just as the wave-front is cut orthogonally by all rays emanating from the light source *A*, as Huygens notes most opportunely; so our [curve] *PB* cuts all the cycloids with the point *A* as common vertex, at right angles. But if one had chosen to state the problem in this purely geometrical fashion: *to find the curve which cuts at right angles all the cycloids with common vertex;* then the problem would have been very difficult for geometers. However from the other point of view, regarding it as a falling body [problem], I construct [the curve] easily as follows. Let *GLK* be the generator circle of the cycloid *ABK*, and *GK* its diameter. Mark off the arc *GL* equal to the mean proportional between the given segment *AP* and the diameter *GK*. I say that *LB* drawn parallel to the horizontal *AG* cuts the cycloid *ABK* in the point *B*. If anyone wishes to try out his method on other [problems], let him seek the curve which cuts at right angles curves given successively in position (not indeed algebraic curves, for that would be by no means difficult, but transcendental curves), e. g., logarithmic curves on a common axis and passing through the same point.

last paragraph shows that the equation of the tautochrone found by Stäckel is wrong. The error remains uncorrected in the revised (1914) edition of his article.]

ABEL

On Integral Equations

(Translated from the German by Professor J. D. Tamarkin, Brown University, Providence, R. I.)

The name of Niels Henrik Abel (b. August 5, 1802, d. April 6, 1829) deserves a place among those of the creators of our science, such as Newton, Euler, Gauss, Cauchy, and Riemann. During his short life Abel made numerous contributions to mathematics of the utmost importance and significance. Although his work was concentrated primarily on algebra and the integral calculus, his name will always be remembered in connection with many other branches of analysis, particularly the theory of integral equations, whose systematic development by Volterra, Fredholm and Hilbert began some 70 years after Abel's work.

We give here the translation of a short article under the title "Auflösung einer mechanischen Aufgabe," *Journal für die reine und angewandte Mathematik* (Crelle), Vol. I, 1826, pp. 153⁻157; *Œuvres Complètes*, Nouvelle édition par L. Sylow et S. Lie, Vol, I, 1881, pp. 97⁻101. This is a revised and improved version of an earlier paper: "Solution de quelques problèmes a l'aide d'intégrales définies" (in Norwegian), *Magazin for Naturviden-skaberne*, Aargang I, Bind 2, Christiania, 1823; *Œuvres Complètes*. Vol. I, pp. 11⁻18.

Abel solves here the famous problem of tautochrone curves by reducing it to an integral equation which now bears his name. His solution is very elegant and needs but slight modification to be presented in modern form. This solution and the formulas

$$\phi(x) = \int_0^\infty dq f(q) \cos qx, \qquad f(q) = \frac{2}{\pi} \int_0^\infty dx \phi(x) \cos qx$$

given by Fourier[1] are perhaps the first examples of an explicit

---

[1] "Théorie de mouvement de la chaleur dans les corps solides" *Mémoires de l'Académie royale des sciences de l'Institut de France*, Vol. 4, 1819–1820 (published in 1824) pp. 185–555 (489). This memoir was presented by Fourier in 1811 and was awarded a prize in 1812.

determination of an unknown function from an equation in which this function appears under the integral sign. An equation which can be reduced to that of Abel was given almost simultaneously by Poisson,[1] without solution. There exists now an extended literature devoted to Abel's equation and to analogous integral equations. The whole question is closely related to the notion of integrals and derivatives of non-integral order. The possibility of such operations was first suggested by Leibniz (1695) and Euler;[2] the notion was considerably developed by Liouville and Riemann,[3] and at present it has many important applications to various problems of pure and applied analysis.

Solution of a Mechanical Problem

Let $BDMA$ be any curve. Let BC be a horizontal and CA a vertical line. Let a particle move along the curve under the action of gravity, starting from the point D. Let τ be the time elapsed when the particle arrives at a given point A, and let a be the distance EA. The quantity τ is a function of a which depends upon the form of the curve and con-versely, the form of the curve will depend upon this function. We shall investigate how is it possible to find the equation of the curve by means of a definite integral, if τ is a given continuous function of a.

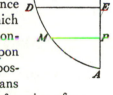

Let $AM = s$, $AP = x$ and let t be the time in which the particle describes the arc DM. By the rules of mechanics we have

[1] "Second Mémoire sur la distribution de la chaleur dans les corps solides," *Journal de l'École Polytechnique*, Cahier 19, Vol. 12, 1823, pp. 249–403 (299).

[2] Leibniz, *Mathematische Schriften*, herausgegeben by C. I. Gerhardt, Halle Vol. 3, 1855 (letters to Johann Bernoulli), Vol. 4, 1859 (letters to Wallis).

L. Euler, "De progressionibus transcendentibus seu quarum termini generales algebraice dari nequeunt," *Commentarii Academiae Scientiarum Petropolitanae*, Vol. 5, 1730–1731, pp. 36–57 (55–57); *Opera Omnia* (1)14, pp. 1–25 (23–25).

[3] J. Liouville, "Mémoire sur quelques questions de géométrie et de mécanique, et sur un nouveau genre de calcul pour résoudre ces questions," *Journal de l'École Polytechnique*, Cahier 21, Vol. 13, 1832, pp. 1–69; "Mémoire sur le calcul des différentielles a l'indices quelconques," *ibidem*, pp. 71–162.

B. Riemann, "Versuch einer allgemeinen Auffassung der Integration und Differentiation," *Werke*, 2nd edition, 1892, pp. 353–366.

$-\dfrac{ds}{dt} = \sqrt{a - x},$[1] whence $dt = -\dfrac{ds}{\sqrt{a - x}}.$ Consequently, on integrating from $x = a$ to $x = 0$,

$$\tau = -\int_a^0 \frac{ds}{\sqrt{a - x}} = \int_0^a \frac{ds}{\sqrt{a - x}},$$

where $\displaystyle\int_\alpha^\beta$ means that the limits of integration correspond to $x = \alpha$ and $x = \beta$ respectively. Let now $\tau = \phi(a)$ be the given function; then

$$\phi(a) = \int_0^a \frac{ds}{\sqrt{a - x}}$$

will be the equation from which s is to be determined as a function of x. Instead of this equation we shall consider another, more general one,

$$\phi(a) = \int_0^a \frac{ds}{(a - x)^n}, \ [0 < n < 1]$$

from which we shall try to derive the expression for s in terms of x.

If $\Gamma(\alpha)$ designates the function

$$\Gamma(\alpha) = \int_0^1 dx \left(\log \frac{1}{x} \right)^{\alpha - 1}, \left[= \int_0^\infty e^{-z} z^{\alpha - 1} dz \right]$$

t is known that

$$\int_0^1 y^{\alpha - 1}(1 - y)^{\beta - 1} dy = \frac{\Gamma(\alpha)\Gamma(\beta)}{\Gamma(\alpha + \beta)}$$

where α and β must be greater than zero. On setting $\beta = 1 - n$ we find

$$\int_0^1 \frac{y^{\alpha - 1} dy}{(1 - y)^n} = \frac{\Gamma(\alpha)\Gamma(1 - n)}{\Gamma(\alpha + 1 - n)},$$

whence, on putting $z = ay$,

$$\int_0^a \frac{z^{\alpha - 1} dz}{(a - z)^n} = \frac{\Gamma(\alpha)\Gamma(1 - n)}{\Gamma(\alpha + 1 - n)} a^{\alpha - n}.$$

---

[1] [If we designate by $v_0 = 0$ and v the velocities of the particle at the points D and M respectively and by g the acceleration due to gravity, then the equation of energy gives $v^2 - v_0^2 = 2g(a - x)$, whence

$$v = \frac{ds}{dt} = -\sqrt{2g}\sqrt{a - x}.$$

Thus the equation in the text corresponds to a choice of units such that $2g = 1$.]

Multiply by $da/(x-a)^{1-n}$ and integrate from $a=0$ to $a=x$:

$$\int_0^x \frac{da}{(x-a)^{1-n}} \int_0^a \frac{z^{\alpha-1}dz}{(a-z)^n} = \frac{\Gamma(\alpha)\Gamma(1-n)}{\Gamma(\alpha+1-n)} \int_0^x \frac{a^{\alpha-n}da}{(x-a)^{1-n}}.$$

Setting $a=xy$ we have

$$\int_0^a \frac{a^{\alpha-n}da}{(x-a)^{1-n}} = x^\alpha \int_0^1 \frac{y^{\alpha-n}dy}{(1-y)^{1-n}} = x^\alpha \frac{\Gamma(\alpha-n+1)\Gamma(n)}{\Gamma(\alpha+1)};$$

hence

$$\int_0^x \frac{da}{(x-a)^{1-n}} \int_0^a \frac{z^{\alpha-1}dz}{(a-z)^n} = \frac{x^\alpha \Gamma(n)\Gamma(1-n)\Gamma(\alpha)}{\Gamma(\alpha+1)}.$$

But, by a known property of the Γ-function,

$$\Gamma(\alpha+1) = \alpha\Gamma(\alpha),$$

whence, by substitution,

$$\int_0^x \frac{da}{(x-a)^{1-n}} \int_0^a \frac{z^{\alpha-1}dz}{(a-z)^n} = \frac{x^\alpha}{\alpha}\Gamma(n)\Gamma(1-n).$$

Multiplying this by $\alpha\phi(\alpha)d\alpha$ and integrating with respect to α [between any constant limits], we have

$$\int_0^x \frac{da}{(x-a)^{1-n}} \int_0^a \frac{(\int\phi(\alpha)\alpha z^{\alpha-1}d\alpha)dz}{(a-z)^n} = \Gamma(n)\Gamma(1-n)\int\phi(\alpha)x^\alpha d\alpha.$$

Setting

$$\int\phi(\alpha)x^\alpha d\alpha = f(x)$$

and differentiating, we have

$$\int\phi(\alpha)\alpha x^{\alpha-1}d\alpha = f'(x), \quad \int\phi(\alpha)\alpha z^{\alpha-1}d\alpha = f'(z).$$

Then

$$\int_0^x \frac{da}{(x-a)^{1-n}} \int_0^a \frac{f'(z)dz}{(a-z)^n} = \Gamma(n)\Gamma(1-n)f(x)[1]$$

[1] [This identity follows immediately from the Dirichlet's formula

$$(*) \qquad \int_0^x da \int_0^a F(a,z)dz = \int_0^x dz \int_z^x F(a,z)da \qquad (1)$$

(Bôcher, *An introduction to the study of integral equations*, 1909, p. 4) under certain restrictive assumptions as to $f(z)$. For instance, it suffices to assume that $f'(z)$ is continuous and $f(0)=0$. Setting in $(*)$

$$F(a,z) = (x-a)^{n-1}(a-z)^{-n}f'(z),$$

we see at once that the left-hand member of the equation in the text reduces to

$$\int_0^x f'(z)dz \int_z^x (x-a)^{n-1}(a-z)^{-n}da.$$

or, since

$$\Gamma(n)\Gamma(1 - n) = \frac{\pi}{\sin n\pi},$$

(1) $$f(x) = \frac{\sin n\pi}{\pi} \int_0^x \frac{da}{(x - a)^{1-n}} \int_0^a \frac{f'(z)dz}{(a - z)^n}.$$

By means of this formula it is easy to find s from the equation

$$\phi(a) = \int_0^a \frac{ds}{(a - x)^n}.$$

Multiplying this equation by $\dfrac{\sin n\pi}{\pi} \dfrac{da}{(x - \alpha)^{1-n}}$ and integrating from $a = 0$ to $a = x$ we have

$$\frac{\sin n\pi}{\pi} \int_0^x \frac{\phi(a)da}{(x - a)^{1-n}} = \frac{\sin n\pi}{\pi} \int_0^x \frac{da}{(x - a)^{1-n}} \int \frac{ds}{(a - x)^n};$$

hence, by (1),

$$s = \frac{\sin n\pi}{\pi} \int_0^x \frac{\phi(a)da}{(x - a)^{1-n}}.^1$$

Substituting $a = z + t(x - z)$ in the interior integral, we reduce it to

$$\int_0^1 t^{-n}(1 - t)^{n-1}dt = \Gamma(n)\Gamma(1 - n)$$

with the final result

$$\int_0^x (x - \alpha)^{n-1}da \int_0^a (\alpha - z)^{-n}f'(z)dz = \Gamma(n)\Gamma(1 - n)\int_0^x f'(z)dz =$$
$$\Gamma(n)\Gamma(1 - n)f(x).$$

Strictly speaking, the method used in the text establishes the identity in question only for the functions $f(x)$ which can be represented by definite integrals of the form

$$\int \phi(\alpha)x^\alpha d\alpha$$

but the investigation of the possibility of such a representation requires the solution of an integral equation of more complicated form than the given one.]

[1] [Two observations should be made concerning the solution obtained.

1. Since the function s replaces $f(x)$ of (1), it must satisfy the restrictions imposed upon $f(x)$, for instance $s'(x)$ must be continuous and $s(0) = 0$, which is natural in view of the physical interpretation of $s(x)$. This imposes certain restrictions upon the given function $\phi(a)$; it is easily verified that the conditions above are satisfied provided $\phi'(a)$ is continuous and $\phi(0) = 0$. The last condition again follows quite naturally from the integral equation of the problem.

2. If all the conditions above are satisfied, identity (1) shows immediately that the solution of the problem is unique, for, if $f(z)$ is a solution, then the interior integral in (1) reduces to $\phi(a)$, which yields the solution obtained in the text.]

Now let $n = \frac{1}{2}$; then

$$\phi(a) = \int_0^a \frac{ds}{\sqrt{a - x}}$$

and

$$s = \frac{1}{\pi} \int_0^x \frac{\phi(a)da}{\sqrt{x - a}}.$$

This equation gives s in terms of the abscissa x, and the curve therefore is completely determined.

We shall aply the expression above to some examples.

I. If

$$\phi(a) = \alpha_0 a^{\mu_0} + \alpha_1 a^{\mu_1} + \ldots + \alpha_m a^{\mu_m} = \Sigma \alpha a^\mu$$

the expression for s will be

$$s = \frac{1}{\pi} \int_0^x \frac{da}{\sqrt{x - a}} \Sigma \alpha a^\mu = \frac{1}{\pi} \Sigma \left(\alpha \int_0^x \frac{a^\mu da}{\sqrt{x - a}} \right).$$

Setting $a = xy$ we have

$$\int_0^x \frac{a^\mu da}{\sqrt{x - a}} = x^{\mu + \frac{1}{2}} \int_0^1 \frac{y^\mu dy}{\sqrt{1 - y}} = x^{\mu + \frac{1}{2}} \frac{\Gamma(\mu + 1)\Gamma(\frac{1}{2})}{\Gamma(\mu + \frac{3}{2})};$$

hence

$$s = \frac{\Gamma(\frac{1}{2})}{\pi} \Sigma \frac{\alpha \Gamma(\mu + 1)}{\Gamma(\mu + \frac{3}{2})} x^{\mu + \frac{1}{2}}$$

or, since $\Gamma(\frac{1}{2}) = \sqrt{\pi}$,

$$s = \sqrt{\frac{x}{\pi}} \left[\alpha_0 \frac{\Gamma(\mu_0 + 1)}{\Gamma(\mu_0 + \frac{3}{2})} x^{\mu_0} + \ldots + \alpha_m \frac{\Gamma(\mu_m + 1)}{\Gamma(\mu_m + \frac{3}{2})} x^{\mu_m} \right].$$

If $m = 0$ and $\mu_0 = 0$, the curve in question is an isochrone, and we find

$$s = \sqrt{\frac{x}{\pi}} \alpha_0 \frac{\Gamma(1)}{\Gamma(\frac{3}{2})} = \frac{2\alpha_0 \sqrt{x}}{\pi}$$

which is known to be equation of a cycloid.[1]

---

[1] [We omit example II, where the function $\phi(a)$ is assumed to be given by different formulas in different intervals.

In the earlier article mentioned above, Abel gives the same final formula for the solution but bases his discussion on the assumption that s can be represented by a sum of terms of the form

$$s = \Sigma \alpha^{(m)} x^m.$$

He then discusses particular cases where the time of descent is proportional to a power of the vertical distance a or is constant (isochrone curve). At the end of the article Abel gives a striking form to his solution by using the notation of derivatives and integrals of non-integral order. We define as the derivative of order α of a function $\psi(x)$ the expression

$$\frac{d^\alpha \psi(x)}{dx^\alpha} = D_z{}^\alpha \psi(x) = \begin{cases} \dfrac{1}{\Gamma(-\alpha)} \displaystyle\int_c^x (x-z)^{-\alpha-1} \psi(z)\,dz \text{ if } \alpha < 0, \\[2ex] \dfrac{d^p}{dx^p} D_z{}^{\alpha-p} \psi(x) \text{ if } p \text{ is an integer and } 0 \leq p-1 < \alpha \leq p, \end{cases}$$

c being a constant which equals 0 in Abel's discussion. If we assume without proof that $D^\alpha D^\beta \psi = D^{\alpha+\beta} \psi$, then Abel's integral equation can be written as

$$\phi(x) = \Gamma(1-n) D_z{}^{n-1} D_z s = \Gamma(1-n) D_z^n s,$$

which can be solved immediately by the formula

$$s(x) = \frac{1}{\Gamma(1-n)} D_z^{-n} \phi = \frac{1}{\Gamma(1-n)\Gamma(n)} \int_0^x \phi(a)(x-a)^{n-1}\,da.$$

To justify this operation Abel proves the identity

$$D_z{}^{-n-1} D_z{}^{n+1} f = f,$$

which also can be derived from the identity (1) above.

In the particular case $n = \frac{1}{2}$ Abel writes the equation and its solution respectively as

$$\psi(x) = \sqrt{\pi}\,\frac{d^{\frac{1}{2}}s}{dx^{\frac{1}{2}}}; \quad s = \frac{1}{\sqrt{\pi}}\frac{d^{-\frac{1}{2}}\psi}{dx^{-\frac{1}{2}}} = \frac{1}{\sqrt{\pi}} \int^{\frac{1}{2}} \psi(x)\,dx^{\frac{1}{2}}.$$

At the end of the article Abel remarks: "In the same fashion as I have found s from the equation

$$\psi(a) = \int_{x=0}^{x=a} \frac{}{(a-x)^n}$$

I have determined the function ϕ from the equation

$$\psi(a) = \int \phi(xa)f(x)\,dx$$

where ψ and f are given functions and the integral is taken between any limits [constant?]; but the solution of this problem is too long to be given here." This solution was never published by Abel.

It should be noted finally that Abel's equation and several others analogous to it were solved by Liouville by using the notion of derivatives and integrals of non-integral order (loc. cit.). Liouville's procedure is purely formal, and he seems to be unaware of Abel's results. It was also Liouville who solved the equation of Poisson mentioned above (Note sur la détermination d'une fonction arbitraire placée sous un signe d'intégration définie, Journal de l'École Polytechnique, Cahier 24, Vol. 15, 1835, pp. 55–60). Poisson's equation is

$$F(r) = \frac{1}{2}\sqrt{\pi}\, r^{n+1} \int_0^\pi \psi(r\cos\omega)\sin^{2n+1}\omega\,d\omega$$

where $F(r)$ is a given function and the unknown function $\psi(a)$ is assumed to be even, $\psi(-u) = \psi(u)$. Poisson's equation is reduced to Abel's type by using $(0, \pi/2)$ as the interval of integration and by making the substitution $\cos\omega = (s/x)^{\frac{1}{2}}$, $r^2 = x$.]

BESSEL

On His Functions

(Translated from the German by Professor H. Bateman, California Institute of Technology, Pasadena, Calif.)

Friedrich Wilhelm Bessel was born on July 22, 1784, and died on March 17, 1846. His father was Regierungssecretär and finally obtained the title of Justizrath. His mother was the daughter of pastor Schrader of Rehme. He married Johanna Hagen of a Königsberg family and had two sons and three daughters. At Olbers's desire and proposal Bessel took the position of inspector to the private Observatory of the Oberamtmann Schröter in Lilienthal. This was early in 1806 and from this date Bessel was an astronomer by profession and worked with great zeal. In his observational work he paid much attention to the planet Saturn. The portion of his work that seems to be best known is the experimental work on pendulums but his name has become famous on account of the work or the functions which now bear his name.

He was not the first to use these functions but he was certainly the first to give a systematic development of their properties and some tables for the functions of lowest order. Bessel considered only the functions of order i where i is an integer, but similar functions of non-integral order have been found to be of importance in applied mathematics. The literature of the subject is now quite vast and many differential equations have been solved in terms of Bessel functions. The values of numerous definite integrals can also be expressed in terms of these functions; indeed, the functions have been found to be so useful that the tables of Bessel have been greatly extended and books have been devoted entirely to the development of the properties. These books have given most mathematicians all the formulas they require and I turn to that very few men turn to the original memoir. This, however, is still of much interest and well deserves a place among the most important contributions to the progress of mathematics. On account of its length the memoir is not given in full, the extracts consist of the preface and some portions relating to the properties of the functions.

The translation (pp. 667–669) is from his "Untersuchungen des Theils der planetarischen Störungen, welcher aus der Bewegung der Sonne entsteht" (Investigation of the portion of the planetary perturbations which arises from the motion of the sun) which appeared in the *Berlin Abbandlungen* (1824), and in his *Werke*, Bd. 1, pp. 84–109. The translation was checked by Morgan Ward, Research fellow in mathematics of California Institute of Technology, Pasadena.

The disturbance of the elliptic motion of one planet by another consists of two parts: one arises from the attraction of the disturbing planet on the disturbed planet; the other arises from the motion

663

of the sun which the disturbing planet produces. The two parts are combined in previous calculations of planetary disturbances but it is worth while to try to separate them. The latter can, in fact, as I shall show in the present memoir, be directly and completely evaluated and so deserves to be separated from the first for which the evaluation has so far not been made; the separation is indeed necessary if we wish to subject to a test the assumption, generally made so far, that the disturbing planet acts on the disturbed and the sun with equal mass.[1]

The two integrals $\int \cos i\mu . \cos \epsilon . d\epsilon$ and $\int \sin i\mu . \sin \epsilon . d\epsilon$ occurring in the first six formulae can easily be reduced to

$$\int \cos (b\epsilon - k \sin \epsilon)d\epsilon$$

where b denotes an integer; this last integer I shall denote by $2\pi I_k{}^h$. We have in fact

$$\int \cos i\mu . \cos \epsilon . d\epsilon = \int \cos i\mu[1 - (1 - e \cos \epsilon)]\frac{d\epsilon}{e}$$

$$= \frac{1}{e}\int \cos i\mu . d\epsilon - \frac{1}{e}\int \cos i\mu . d\mu$$

where the last part vanishes when taken between $\mu = 0$ and $\mu = 2\pi$ thus

$$\int \cos i\mu . \cos \epsilon . d\epsilon = \frac{2\pi}{e} I_{i\circ}^{i}$$

Furthermore

$$\int \sin i\mu . \sin \epsilon . d\epsilon = \int \cos i\mu . \cos \epsilon . d\epsilon - \int \cos (\epsilon + i\mu)d\epsilon.$$

or

$$\int \sin i\mu . \sin \epsilon . d\epsilon = \frac{2\pi}{e} I_{i\circ}^{i} - 2\pi I_{j\circ}^{i+1}$$

The series expansion for $I_h{}^k$ is obtained in the way used in my memoir on Kepler's problem,[2] it is

$$I_k{}^h = \frac{(k/2)^h}{\pi(b)}\left\{1 - \frac{1}{b + 1}\left(\frac{k}{2}\right)^2 + \frac{1}{1.2(b + 1)(b + 2)}\left(\frac{k}{2}\right)^4 - \cdots\right\}$$

---

[1] [Bessel's fundamental equations are

$$\frac{r}{a} \cos \phi = \cos \epsilon - e \qquad\qquad (1)$$

$$\frac{r}{a} \sin \phi = \sqrt{1 - e^2}. \quad \sin \epsilon \qquad\qquad (2)$$

$$u = \acute{e} - e \sin \epsilon, \qquad\qquad r = a(1 - e \cos \epsilon).]$$

[2] ["Analytische Auflösung der Kepler'sche Aufgabe." *Abbandlungen der Berliner Akademie der Wissenscbaften, matb. Cl.* (1816–17), p. 49; *Werke,* Bd. I, p. 17. The paper was read July 2, 1818. It was also communicated to Lindenau in a letter written in June, 1818. See *Zeitscbr. für Astron.,* V., p. 367.]

where

$$\pi(b) = b!$$

Not only the equation for the center and the quantities cos ϕ, sin ϕ, r cos ϕ, r sin ϕ, $\frac{1}{r^2}$ cos ϕ, $\frac{1}{r^2}$ sin ϕ lead on expansion to these definite integrals but this is also the case for

$$\log r, \; r^n, \; r^n \cos m\phi, \; r^n \sin m\phi, \; r^n \cos m\epsilon, \; r^n \sin m\epsilon,$$

whenever n and m are integers which may be positive, negative, or zero. Since most problems of physical astronomy lead to such expansions in series, a fuller knowledge of these integrals is desirable.

For brevity the four integrals, taken between 0 and 2π, will be denoted by symbols as follows:—

$$\frac{2\pi}{e}L = \int \cos i\mu . \cos \epsilon. d\epsilon ; \frac{2\pi}{e}L' = \int \sin i\mu . \sin \epsilon. d\epsilon ;$$

$$\frac{2\pi}{e}M = \int \frac{\cos i\mu . \cos \epsilon . d\epsilon}{1 - e \cos \epsilon} ; \frac{2\pi}{e}M' = \int \frac{\sin i\mu . \sin \epsilon. d\epsilon}{1 - e \cos \epsilon} ,$$

and we must first show that the expansions of the quantities mentioned involve these quantities.

We denote the coefficient of cos $i\mu$ in the expansion of log r by H^i; the expansion being made so that the series runs over both positive and negative values of i. We have

$$2\pi H^i = \int \log r . \cos i\mu d\mu = \frac{1}{i} \log r . \sin i\mu - \frac{e}{i} \int \frac{\sin i\mu . \sin \epsilon. d\epsilon}{1 - e \cos \epsilon}$$

thus, with the exception of $i = 0$,

$$H^i = -\frac{1}{i}M'.$$

For $i = 0$ we obtain a logarithmic expansion; in fact, if we denote $\dfrac{e}{1 + \sqrt{1 - e^2}}$ by λ and take the semi-major axis to be unity we have

$$\frac{1}{r} = \frac{1}{\sqrt{1 - e^2}} \left\{ 1 + 2\lambda \cos \epsilon + 2\lambda^2 \cos 2\epsilon + 2\lambda^3 \cos 3\epsilon + \dots \right\};$$

and, if we multiply by $dr = e \sin \epsilon d\epsilon$ and integrate,

$$\log r = c - 2\{\lambda \cos \epsilon + \tfrac{1}{2}\lambda^2 \cos 2\epsilon + \tfrac{1}{3}\lambda^3 \cos 3\epsilon + \dots \}.$$

For the determination of the constant c we have, for $\epsilon = 0$,

$$\log (1 - e) = c - 2\{\lambda + \tfrac{1}{2}\lambda^2 + \tfrac{1}{3}\lambda^3 + \ldots\}$$
$$= c + 2 \log (1 - \lambda).$$

$\therefore \log r =$

$$\log\frac{1 - e}{(1 - \lambda)^2} - 2\left\{\lambda \,\cos\, \epsilon + \frac{1}{2}\lambda^2 \cos 2\epsilon + \frac{1}{3}\lambda^3 \cos 3\epsilon + \ldots\right\};$$

and, if we multiply this by $d\mu = (1 - e \cos \epsilon)d\epsilon$ and integrate from 0 to 2π,

$$H^0 = \log \frac{1 - e}{(1 - \lambda)^2} + \lambda e = \log\frac{1 + \sqrt{1 - e^2}}{2} + \frac{e^2}{1 + \sqrt{1 - e^2}}$$

Bessel's Recurrence Formulas.

The following recurrence formulae are given in Bessel's paper and will be quoted here without proof

$$o = kI_k{}^{i-1} - 2iI_k{}^i + kI_k{}^{i+1} \qquad (40)$$
$$I_k{}^{-i} = (-)^i I_k{}^i \qquad (41)$$

$$\frac{I_k{}^i}{I_k{}^{i-1}} = \cfrac{\dfrac{k}{2i}}{1 - \cfrac{k^2}{(2i)(2i + 2)}{\Large/}\left(1 - \cfrac{k^2}{(2i + 2)(2i + 4)}\right)}$$

$$\cdots\cdots\cdots\cdots$$

$$\frac{dI_k{}^i}{dk} = \frac{i}{k}I_k{}^i - I_k{}^{i+1}$$

$$\frac{I_k{}^{i+h}}{\left(\dfrac{k}{2}\right)^{ith}} = (-)^h\frac{d^h\left\{\dfrac{I_k{}^i}{\left(\dfrac{k}{2}\right)^i}\right\}}{\left(d\dfrac{k^2}{4}\right)^h},$$

$$\frac{I_k{}^i}{\left(\dfrac{k}{2}\right)^i} = (-)^i\frac{d^i\{I_k{}^0\}}{\left\{d\dfrac{k^2}{4}\right\}^i}$$

$$o = \frac{d^2I_k{}^i}{dk^2} + \frac{1}{k}\frac{dI_k{}^i}{dk} + I_k{}^i\left(1 - \frac{i^2}{k^2}\right)$$

Other integrals can be reduced to the function $I_k{}^i$ as the following examples will show[1].

$$\frac{1}{2\pi}\int \cos(i\epsilon - m\cos\epsilon - n\sin\epsilon)\,d\epsilon = \cos i\alpha . I^i{}_{\sqrt{m^2+n^2}} \qquad (50)$$

$$\frac{1}{2\pi}\int \cos i\epsilon . \cos(m\cos\epsilon + n\sin\epsilon)\,d\epsilon = \cos i\alpha . I^i{}_{\sqrt{m^2+n^2}} \quad i \text{ even} \ (51)$$
$$= 0 \qquad i \text{ odd}$$

$$\frac{1}{2\pi}\int \sin i\epsilon . \sin(m\cos\epsilon + n\sin\epsilon)\,d\epsilon = \cos i\alpha . I^i{}_{\sqrt{m^2+n^2}} \quad i \text{ odd} \ (52)$$
$$= 0 \qquad i \text{ even}$$

Bessel then proves the following relations

$$\frac{1}{2\pi}\int (\cos\epsilon)^{2i}\cos(k\sin\epsilon)\,d\epsilon = \frac{1.3\ldots(2i-1)}{k^i}I_k{}^i \qquad (53)$$

$$\frac{1}{2\pi}\int_0^1 \cos kz.(1-z^2)^{i-\frac{1}{2}}\,dz = \frac{1.3\ldots(2i-1)}{4k^i}I_k{}^i \qquad (54)$$

$$\frac{1}{2\pi}\int e^{n\cos\epsilon}\cos(m\sin\epsilon)\,d\epsilon = I^o{}_{\sqrt{m^2-n^2}} \qquad (55)$$

$$\cos k.I_k{}^o = 1 - \frac{3}{(\pi 2)^2}k^2 + \frac{3.5.7}{(\pi 4)^2}k^4 - \ldots$$

where $(\pi 2)$ is written for $2!$, and so for similar uses of π.

$$\sin k.I_k{}^o = k - \frac{3.5}{(\pi 3)^2}k^3 + \frac{3.5.7.9}{(\pi 5)^2}k^5 - \ldots \qquad (56)$$

$$I_{k+z}{}^i = \left[\left(1+\frac{z}{k}\right)^i\left\{I_k{}^i - I_k{}^{i+1}z\left(\left(1+\frac{z}{2k}\right)+\right.\right.\right.$$
$$\left.\left.\left.\frac{I_k{}^{i+2}}{1.2}z^2\left(1+\frac{z}{2k}\right)^2 - \ldots\right] \qquad (49)$$

The last series can be used for the calculation and interpolation of a table of the functions and was actually used by Bessel to form tables of $I_k{}^o$ and $I_k{}^1$ from $k = 0$ to $k = 3.2$.

The Roots of the Bessel Functions.

"The function $I_k{}^o$ has in common with the sine and cosine the remarkable property of vanishing twice and changing sign when its argument k increases from $2n\pi$ to $(2n+2)\pi$. I shall show that if

m is even $I_k{}^o > 0$ from $k = m\pi$ to $(m + \frac{1}{2})\pi$
" " odd $I_k{}^o < 0$ "   "   "   "   "

---

[1] In these equations $\tan\alpha = n/m$.

If we put $\sin \epsilon = z$ and $k = \dfrac{2m + m'}{2}\pi$ where m denotes a proper fraction, we have, according to the remark (54)

$$I_{k^o} = \frac{2}{\pi}\int_0^1 \cos\left(\frac{2m + m'}{2}\pi z\right)\frac{dz}{\sqrt{1 - z^2}}$$

Writing v for $(2m + m')z$ this expression becomes

$$I_{k^o} = \frac{2}{\pi}\int_0^{2m+m'} \cos\frac{\pi v}{2}\frac{dv}{\sqrt{(2m + m')^2 - v^2}}$$

The integral taken from $v = a$ to $v = b$, when we write $b + u$ for v, is

$$\int_{a-b}^{b-b} \cos\left(\frac{b\pi}{2} + \frac{\pi u}{2}\right)\frac{du}{\sqrt{(2m + m')^2 - (b + u)^2}}$$

Taking successively $b = 1, 3, \ldots 2m - 1$ and a, b always $b - 1$, $b + 1$ respectively, the last expression gives

$$I_{h^0} = \frac{2}{\pi}\int_{-1}^1 \sin\frac{\pi u}{2}\, du\left[-\frac{1}{\sqrt{\mu^2 - (1 + u)^2}} + \frac{1}{\sqrt{\mu^2 - (3 + u)^2}}\right.$$

$$\left. - \ldots + \frac{(-)^{m-1}}{\sqrt{\mu^2 - (2m - 3 + u)^2}} + \frac{(-)^m}{\sqrt{\mu^2 - (2m - 1 + u)^2}}\right\}$$

$$+ \frac{2}{\pi}(-)^m\int_0^{m'} \frac{\cos\frac{\pi u}{2}\,du}{\sqrt{(\mu^2 - 2m + u)^2}}(\mu = 2m + m')$$

The individual terms of this expression are $+$, the last clearly so because $\pi u/2$ is always $< \pi/2$, the other because their part is greater than the negative; for we have

$$\int_{-1}^1 \frac{\sin\frac{\pi\mu}{2}du}{\sqrt{\mu^2 - (b+u)^2}} = \int_0^1 \left\{ \frac{\sin\frac{\pi u}{2}du}{\sqrt{\mu^2 - (b+u)^2}} - \frac{\sin\frac{\pi u}{2}du}{\sqrt{\mu^2 - (b-u)^2}}\right\}$$

where the denominator of the positive part is always smaller than that of the negative. Furthermore, each following term is greater than the preceding on account of the continually decreasing denominator; the sum of two successive terms has therefore the sign of the last. If m is even, the last term in the bracket is positive and therefore the sum of all terms positive; if m is odd the last term is negative and therefore the sum of all terms up to the second negative and the first term as well as the term outside the bracket is negative. This property does not belong to I_{k^o} alone but all the

$I_k{}^i$ possess a similar property. In fact from (46) if for brevity we write $I_k{}^i = \frac{k}{2}R^{(i)}$ and $\frac{k^2}{4} = k$

$$R^{(i+1)} = -\frac{dR^i}{dx}$$

Therefore R^{i+1} vanishes when R_i has a maximum or minimum; but between two values of k or x, for which R^i vanishes there is necessarily a maximum or minimum, thus also a vanishing R^{i+1}. It is therefore clear that $I_k{}^1$ is zero just as often as $I_k{}^0$ is a maximum or minimum; between two values of k for which $I_k{}^1 = 0$ there lies always a maximum or minimum of R^1, therefore a root of $I_k{}^2$ and so on."

MÖBIUS

On the Barycentric Calculus

(Translated from the German by J. P. Kormes, Hunter College, New York City.)

August Ferdinand Möbius (1790–1868) was professor of astronomy in Leipzig and wrote several papers on this subject. His researches in celestial mechanics led him to an extensive study of geometry. In 1827 he published his most important contribution to science under the title: *Der barycentrische Calcul ein neues Huelfsmittel zur analytischen Bebandlung der Geometrie dargestellt und insbesondere auf die Bildung neuer Classen von Aufgaben und die Entwickelung mehrerer Eigenschaften der Kegelschnitte angewendet*, Leipzig, Verlag von Johann Ambrosius Barth," pp. 1–454.

In this work Möbius introduces for the first time homogeneous coordinates into analytic geometry. With the aid of the barycentric calculus the treatment of various problems and in particular those relating to conic sections becomes simple and uniform. He introduces the remarkable classification of the properties of geometric figures according to the transformations (similar, affine, collinear) under which these properties remain invariant. Möbius arrives at the characteristic invariant of the collinear group, the anharmonic ratio. He also succeeds in establishing the most general principle of duality of points and straight lines without the use of a conic section.

§2. Through two given points A and B parallel lines are drawn. If a and b are any two numbers in a given ratio such that $a + b$ is different from zezo, find a straight line intersecting the two parallel lines in A' and B' respectively such that

$$a.AA' + b.BB' = 0.$$

Draw the line AB and find on it a point P such that $AP:PB = b:a$. Every line through P (and no other line) intersecting the two parallel lines will have the required property. From the similarity of the triangles $AA'P$ and $BB'P$ we have

$$AA':BB' = AP:BP = AP:-PB^1 = b: -a;$$

---

[1] [Möbius considers directed segments and triangles. Thus if A and B are any two points on a straight line, $AB + BA = 0$; and if B, C, D are three points on a straight line and A is a point not on the line, then the sum of the areas of the triangles

$$ACD + ADB + ABC = 0.]$$

670

hence
$$aAA' + bBB' = 0.$$
. . . .

§3, c. If we place in A and B weights proportional to a and b respectively, P may be considered as the centroid of the points A and B with the coefficients a and b.

. . . .

§8...THEOREM.—Given a system of n points A, B, C,...N with the coefficients a, b, c,...n respectively where the sum $a + b + c + ... + n$ is different from zero, there can always be found one point and only one point, the centroid S, having the following property: If parallel lines be drawn through the given points and through the point S in any direction and these lines be intersected by any plane in the points A', B', C',...S' respectively, we always have

$$a.AA' + b.BB' + ... + n.NN' = (a + b + ... + n).SS'.$$

In particular if the plane passes through S we have

$$a.AA' + b.BB' + ... + n.NN' = 0.$$

§9... If $a + b + c + ... + n = 0$, the centroid is infinitely remote in the direction determined by the parallel lines.

. . . .

§13... In place of the segments AA', BB',...their endpoints A, B,...shall be used. Thus if S is the centroid of A, B, C with the coefficients a, b, $-c$, we write

$$a.A + b.B - cC = (a + b - c).S.$$

. . . .

§14. The operations with such abbreviated formulas form the barycentric calculus or a calculus based upon the notion of the centroid...**§15.** (1) In barycentric calculus points and their coefficients are considered. The points are denoted by capital letters, their coefficients by small letters...(2) The fact that S is the centroid of the points A, B, C,...with the coefficients a, b, c,...is expressed as follows:

I. $aA + bB + cC + ... = (a + b + c + ...)S^1$...

(3) The fact that the system A, B, C,...with the coefficients a, b, c,...has the same centroid as the system F, G, H,...with the

---

[1] $[aA + bB + cC + ... \equiv S$ is used in place of I.]

coefficients f, g, b,...provided the sum of the coefficients a, b, c,... equals the sum of the coefficients f, g, b,...is expressed as follows:

II. $aA + bB + cC + \ldots = fF + gG + bH + \ldots$

(4) The equation III. $aA + bB + \ldots = 0$ indicates that the system A, B,...with the coefficients a, b,...has no finite centroid ...[1]

§21. THEOREM.—If $aA + bB \equiv C$ then C is on the line through A and B and we have:

$$a:b = BC:CA\ldots$$

§23. THEOREM.—If $aA + bB + cC \equiv D$ and A, B, C are not on a straight line, the point D is in the plane A, B, C, and so

$$a:b:c = DBC:DCA:DAB^2\ldots$$

§24...If $aA + bB + cC + dD \equiv 0$ then A, B, C, D are in one plane and we have:

$$a:b:c:d = BCD:-CDA:DAB:-ABC\ldots$$

§25. THEOREM.—If $aA + bB + cC + dD \equiv E$ and A, B, C, D are not in one plane then we have:

$$a:b:c:d = \text{pyramids } BCDE:CDEA:DEAB:EABC\ldots$$

§28. In order to determine the position of a point be it on a straight line, plane or space quantities of two kinds are essential; the ones of the first kind remain the same for all points like the axes of the usual system of coordinates, the others, the coordinates in the most general sense, depend upon the position of the various points with respect to the quantities of the first kind. By the method under consideration points shall be determined as follows: The quantities of the first kind shall be points and we shall call them "fundamental points" and the point whose position is to be determined shall be considered as their centroid. These fundamental points are taken as the system of coordinates. The coordinates of any point P with respect to these fundamental points are given by the relations which must exist among the coefficients of the fundamental points in order that the point P should be the centroid of these points.

. . . .

---

[1] [All equations in the barycentric calculus assume one of the forms I, II or III and retain this form throughout all transformations.]

[2] [DBC means the area of $\triangle DBC$, etc.]

§36. The change from one system of coordinates to another is very simple. If A', B',..., the new fundamental points, are given in terms of A, B,..., then it is sufficient to express the old fundamental points A, B,...in terms of A', B',...If these values be substituted in the expression for P, P is given in terms of the new coordinates. The simplicity of this process is illustrated by the following example: Let A', B', C', the new fundamental points, be the midpoints of the sides of the fundamental triangle ABC, A' the midpoint of BC,[1] B' the midpoint of CA, and C' the midpoint of AB. We have then

$$2A' = B + C, 2B' = C + A, 2C' = A + B,$$

and therefore

$$A = B' + C' - A', B = C' + A' - B', C = A' + B' - C'.$$

If the expression for P with respect to the system ABC is

$$P \equiv pA + qB + rC,$$

then the expression for P with respect to the new system $A'B'C'$ becomes

$$P \equiv p(B' + C' - A') + q(C' + A' - B') + r(A' + B' - C'),$$

or

$$P \equiv (q + r - p)A' + (r + p - q)B' + (p + q - r)C'.$$

§144...Given a system of points A, B, C,..., three of these may be taken as the fundamental points and any other point in the plane will be determined if the ratios of the coefficients $a:b:c$ are given. If in another system of points A', B', C'...the fundamental triangle formed by A', B', C' has the same sides as the triangle ABC and the ratios $a':b':c'$ are equal to $a:b:c$ for every point, then any figure formed by the points in the second system will be equal and similar to the figure formed by the corresponding points in the first system. If the sides of the fundamental triangle $A'B'C'$ are not equal but are proportional to the sides of the triangle ABC, the corresponding figures are similar. Now let us assume that the ratios of the coefficients of corresponding points are equal but the choice of the fundamental triangle arbitrary.

§145...In order to study the relationship between corresponding figures take any three points A', B', C' as the new fundamental points corresponding to the points A, B, C respectively. Should

---

[1] [See page 671. According to I. we have: $bB + bC = (b + b)S = 2bS$; put $S = A'$.]

any point D' correspond to the point $D \equiv aA + bB + cC$ the expression for D' must be:

$$D' \equiv aA' + bB' + cC'$$

that is the area of the triangles formed by the points A', B', C', D' must be in the same proportion as the areas of the triangles formed by the points A, B, C, D...It follows that $A'B'C' = m.ABC$. This holds true for any two corresponding triangles...Since every figure may be considered as an aggregate of triangles, the nature of the relations under consideration is revealed in the fact that the areas of any two parts in one figure are to each other as the areas of the two corresponding parts in the other figure...

§147...The two figures are then said to be affine...

§153...In general all relations and properties of a figure which are expressed in terms of the coefficients of the fundamental points remain the same in all affine figures. Thus if in one figure two lines are parallel or if they intersect in a given point, the corresponding lines in the affine figure will be parallel or will intersect in the corresponding point. On the other hand all relations which cannot be expressed in terms of the coefficients of the fundamental points are different in affine figures...

§217...Consider now a relationship under the sole condition that straight lines correspond to straight lines and planes to planes...This relationship may be characterized as follows: A correspondence is set up between the points of two planes such that if in one plane a set of points coincides with a straight line [collineantur] the corresponding points in the second plane lie on a straight line. Hence the name for this relation is "collineation"
...

§200...Connect any four points A, B, C, D in a plane by straight lines. The resulting three points of intersection A', B', C' (Fig. 1) connect again by straight lines thus obtaining six new points of intersection: A'', B'', C'', F, G and H which in turn may be again connected with each other and with the seven points previously obtained by straight lines etc. The system of lines thus obtained from any four points A, B, C, D shall be called a *plane net* and the points A, B, C, D shall be called the *four fundamental points of the net*.

§201...THEOREM.—If A, B, C, and $D = aA + bB + cC$ are the four fundamental points of the plane net, every point P of the net can be represented as follows:

$$P \equiv \varphi aA + \chi bB + \psi cC$$

where φ, χ, ψ are rational numbers including zero, which depend only upon the construction of the point P and not upon the four fundamental points.

. . . .

§202...THEOREM.—Every anharmonic ratio formed in a plane net is rational and depends only upon the construction of the straight lines and not upon the four fundamental points.

§219. Every point P of the net in the plane A, B, C, and $D = aA + bB + cC$ can be written in the form

$$P \equiv \varphi aA + \chi bB + \psi cC,$$

where φ, χ, ψ do not depend upon the coefficients a, b, c.   Therefore every point P' of the net formed from the four fundamental points A', B', C', and $D' \equiv a'A' + b'B' + c'C'$ may be expressed as

$$P' \equiv \varphi a'A' + \chi b'B' + \psi c'C',$$

where φ, χ, ψ are the same as in the expression for the point P.

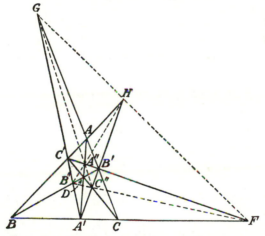

Since on the other hand every point of the plane ABC can be expressed in the form: $\varphi aA + \chi bB + \psi cC$ and when the values of φ, χ, and ψ are given it is always possible to find the point by mere drawing of straight lines, the relationship of collineation may now be defined as follows: Let any four points A', B', C', D', no three of which are on a straight line, correspond to four given points A, B, C, D, no three of which are on a straight line.   To every point P in the first plane there will correspond a point P' in the second plane such that if

$$D \equiv aA + bB + cC, \qquad P \equiv pA + qB + rC,$$
$$D' \equiv a'A' + b'B' + c'C', \quad P' \equiv p'A' + q'B' + r'C'.$$

Then,

$$\frac{p}{a}:\frac{q}{b}:\frac{r}{c} = \frac{p'}{a'}:\frac{q'}{b'}:\frac{r'}{c'}(= \varphi:\chi:\psi)\ldots$$

§220. If we determine the four pairs of the corresponding points A and A', B and B', C and C', D and D' then there corresponds one and only one point P' in the plane $A'B'C'$ to a point P in the plane ABC. The points A, B, C, D, P determine the ratios $a:b:c$ and $\varphi:\chi:\psi$, and the points A', B', C', D' determine the ratios $a':b':c'$. From these, the ratios $\varphi a':\chi b':\psi c'$ can be found and thus P' is uniquely determined. In the expression of two corresponding points

$$P \equiv \varphi aA + \chi bB + \psi cC, \text{ and } P' \equiv \varphi a'A' + \chi b'B' + \chi c'C',$$
if $\qquad \varphi a + \chi b + \psi c = 0, \varphi a' + \chi b' + \psi c'$

may be different from zero. Thus a finite point may correspond to a point at infinity...

§221, 4. The collineation is characterized by the consistency of the ratios $\varphi:\chi:\psi$ for each pair of corresponding points. These ratios $\varphi:\chi:\psi$ can be expressed geometrically as anharmonic ratios. From

$$D \equiv aA + bB + cC \text{ and } P \equiv \varphi aA + \chi bB + \psi cC$$

it follows

$$a:b = -BCD:CDA$$

and

$$\varphi a:\chi b = -BCP:CPA$$

hence

$$\varphi:\chi = (B, A, CP, CD) = (A, B, CD, CP)$$

and similarly

$$\chi:\psi = (B, C, AD, AP)$$

(§190...where (A, B, CE, DE) is the anharmonic ratio the four points of which are A, B and the intersections of the line AB with the lines CE and DE respectively...). Therefore collineation may be defined directly by means of the equality of anharmonic ratios: Two figures are said to be collinear if every expression of the form:

$$\frac{ACD}{CDB}\cdot\frac{AEF}{EFB}$$

is equal to the same expression formed by the corresponding points in the second figure.

SIR WILLIAM ROWAN HAMILTON

On Quaternions

(Selections Edited by Dr. Marguerite D. Darkow, Hunter College, New York City.)

William Rowan Hamilton was born in Dublin in 1805. His early training was in languages. At the age of thirteen, a copy of Newton's *Universal Arithmetic* fell into his hands and turned his thoughts to mathematics. In 1827, although an undergraduate, he was appointed to the chair of Astronomy in Trinity College, Dublin.

After producing a number of papers on various subjects, Hamilton concentrated upon the calculus of directed line-segments in space, and the meaning to be assigned to their product and their quotient. In 1835 and 1843, he wrote on this and allied matters in the *Transactions* of the Royal Irish Academy, and in 1844 in the *Philosophical Magazine*. In 1853, he published his *Lectures on Quaternions*. In 1866, his *Elements of Quaternions* (from which several extracts are to be quoted) appeared posthumously.

Although Hamilton expected his quaternions to prove a tool powerful for the progress of physics, his expectation has not been completely fulfilled, perhaps on account of a loss of naturalness in taking the square of a vector to be a negative scalar. The importance of his quaternions is due rather to the extension through them of the concept of number and the possibility of a variation from the hitherto unchallenged "Laws of Algebra."

Hamilton entered upon the mathematical scene at a time when mathematicians were not yet satisfied as to the sense in which $\sqrt{-1}$ was to be taken as a number, and had not quite divested it of all stigma for being "imaginary." Hamilton, too, was preoccupied with negative and imaginary, and rationalized these concepts by his view of algebra "as being *primarily* the science of ORDER (in time and space), and *not* primarily the science of MAGNITUDE." In his paper of 1835,[1] he introduces couples of moments in time (A, B), the difference A − B of two such moments as being a step a in time, couples of such steps (a, b), and couples of real numbers (a, b) which may be regarded as operators upon couples of steps. He defines

$$(a, b)(a, b) = (aa - bb, ab + ba),$$
$$a(a, b) = (a, 0)(a, b) = (aa, ab).$$

Hence $(0, 1)^2 (a, b) = (0, 1) (-b, a) = (-a, -b) = -(a, b)$, whence $(0, 1)^2 = -1$.

From this beginning, there followed in the paper of 1848[2] a generalization to n-tuples (and in particular to quadruples) of moments and of steps in time.

---

[1] *Transactions* of the Royal Irish Academy, XVII, p. 293.
[2] [*Ibid XXI*, p. *199*.]

He introduced the "momental quaternion" (A, B, C, D), A, B, C, D being moments of time,—the "ordinal quaternion" (a, b, c, d) = (a_0, a_1, a_2, a_3) with time-steps as elements, and the 16×24 operators $R_{\pm\pi,\pm\rho,\pm\sigma,\pm\tau}$ (π, ρ, σ, τ being some permutation of the integers, 0, 1, 2, 3) such that $R_{\pi\rho\sigma\tau}$ (a_0, a_1, a_2, a_3) = $(a_\pi, a_\rho, a_\sigma, a_\tau)$;

e. g. R_{3012} (a, b, c, d) = (d, a, b, c), $R_{-3, 0, 1, -2}$ (a, b, c, d) = $(-d, a, b, -c)$

He defined i as $R_{-1, 0 -3, 2}$, j as R $_{-2, 3, 0, -1}$, k as R $_{-3, -2, 1, 0}$, whence followed the equations in operators:

$$i^2 = j^2 = k^2 = -1, ij = k = -ji, jk = i = -kj, ki = j = -ik,$$

with their non-commutative multiplication. Then came geometric interpretations and applications.

In the same paper, Hamilton laid a foundation for the modern work in linear algebras, regarded as a study of the properties of the set of linear combinations of n linearly independent elements with real or complex coefficients, subject to a multiplication table containing n^3 arbitrary constants, which closed the set under multiplication. Peacock (1791–1858) and DeMorgan (1806–1871) had already recognized the possibility of algebras which differ from ordinary algebra. Such algebras had appeared, but had found little recognition. It was the geometrical application of Hamilton's quaternions that led to a general appreciation of new algebras, in which the laws of combinations of the elements need not retain the classical commutativity, associativity, etc.

Hamilton was much concerned with ensuring respectability for his quaternions. The method of approach in his previous publications did not satisfy him. In his last work, he elected from the outset the geometric point of view. He defined a vector as a directed line-segment in space, and set himself the task of interpreting the quotient of two vectors, making the following assumptions (in which Greek letters denote vectors):

1. β/α = q implies β = qα, in the sense that q, operating upon α, produces β.
2. $\beta'/\alpha' = \beta/\alpha$, $\alpha' = \alpha$, imply $\beta' = \beta$.
3. $q' = q$, $q'' = q$, imply $q' = q''$.
4. $\dfrac{\gamma}{\alpha} \pm \dfrac{\beta}{\alpha} = \dfrac{(\gamma \pm \beta)}{\alpha}$; $\dfrac{\gamma/\alpha}{\beta/\alpha} = \dfrac{\gamma}{\beta}$.
5. $\dfrac{\gamma}{\beta}.\dfrac{\beta}{\alpha} = \dfrac{\gamma}{\alpha}$.

The quotient of two parallel vectors is plus or minus the ratio of their lengths, according to whether they are similarly or oppositely directed.

Three extracts from the *Elements of Quaternions*, London, 1866, are appended. The first (pp. 106–110) examines, on the basis of the preceding assumptions, the nature of the quotient of two vectors, and justifies the use of "quaternion" for such a quotient. The second (pp. 157–160) defines i, j, k and derives their multiplication table; and the third (pp. 149–150) comments on the source of the multiplicational non-commutativity.

"108. Already we may see grounds for the application of the *name* QUATERNION, to such a *Quotient of two Vectors* as has been spoken of in recent articles. In the first place, such a quotient cannot *generally* be what we have called a SCALAR: or in other words, it cannot generally be equal to any of the (so-called) *reals of algebra,* whether of the *positive* or of the *negative* kind. For let *x* denote any such (actual)[1] scalar, and let α denote any (actual) vector; then we have seen that the product $x\alpha$ denotes *another* (actual) *vector,* say β', which is either *similar* or *opposite* in direction to α, according as the scalar coefficient, or *factor, x,* is positive or negative; in *neither* case, then, can it represent any vector, such as β, which is *inclined* to α, at any actual *angle,* whether acute, or right, or obtuse: or in other words, the *equation* $\beta' = \beta$, or $x\alpha = \beta$, is impossible, under the conditions here supposed. But we have agreed to write, as in algebra, $(x\alpha)/\alpha = x$; we must therefore[2]...*abstain* from writing *also* $\beta/\alpha = x$, under the same conditions: *x* still denoting a *scalar.* Whatever *else* a *quotient of two inclined vectors* may be found to be, it is thus, at least, a NON-SCALAR.

"109. Now, in forming the conception of the *scalar itself,* as the *quotient of two parallel*[3] *vectors,* we took into account not only *relative length,* or *ratio* of the usual kind, but also *relative direction,* under the form of *similarity or opposition.* In passing from α to $x\alpha$, we *altered* generally the *length* of the line α, in the ratio of $\pm x$ to 1; and we *preserved or reversed* the *direction* of that line, according as the *scalar coefficient x* was *positive* or *negative,* and, in like manner, in proceeding to form, more definitely than we have yet done, the conception of the *non-scalar quotient,* q = β: α = OB:OA, of *two inclined vectors,* which for simplicity may be supposed to be *co-initial,* we have *still* to take account both of the *relative length* and of the *relative direction,* of the two lines compared. But while the *former* element of the *complex relation* here considered, between these two lines or vectors, is *still* represented by a simple RATIO (of the kind commonly considered in geometry), or by a *number*[4] expressing that ratio; the *latter element* of the same complex relation is *now* represented by an ANGLE, AOB: and not simply (as it was before) by an *algebraical sign* + or −.

"110. Again, in estimating this *angle,* for the purpose of *distinguishing* one quotient of vectors from another, we must consider

---

[1] [*Non zero.*]

[2] [*By the second assumption.*]

[3] [Or collinear.]

[4] ["The tensor of the quotient."]

not only its *magnitude* (or *quantity*), but also its PLANE: since otherwise, in violation of the principle[1]..., we should have $OB':OA = OB:OA$, if OB and OB' were *two distinct rays* or sides of a *cone* of revolution, with OA for its *axis;* in which case...they would necessarily be *unequal vectors*.   For a similar reason, we must attend also to the *contrast* between two *opposite angles*, of equal magnitudes, and in one *common plane*.   In short, for the purpose of knowing *fully* the *relative direction* of two co-initial lines OA, OB in *space*, we ought to know not only *how many degrees*...the *angle* AOB contains; but also...the *direction of the rotation* from OA to OB: including a knowledge of the *plane*, *in which* the rotation is performed; and of the *hand* (as *right* or *left*, when *viewed* from a known *side* of the plane), *towards which* the rotation is *directed*.

"111. Or, if we agree to *select* some one *fixed hand* (suppose the *right* hand), and to call all *rotations positive* when they are directed towards *this* selected hand, but all rotations *negative* when they are directed towards the *other hand*, then, for *any given angle* AOB, supposed for simplicity to be less than two right angles, and considered as representing a *rotation in a given plane* from OA to OB, we may speak of *one perpendicular* OC *to that plane* AOB as being the *positive axis* of that rotation; and of the *opposite perpendicular* OC' to the same plane as being the *negative axis* thereof: the rotation around the positive axis being *itself* positive, and *vice-versa*.   And then the *rotation* AOB may be considered to be entirely *known*, if we know, Ist, its *quantity*, or the *ratio* which it bears to a *right rotation;* and IInd, the *direction* of its *positive axis*, OC, but not without knowledge of these *two* things, or of some data equivalent to them.   But whether we consider the *direction of an* AXIS, or the *aspect of a* PLANE, we find (as indeed is well known) that the *determination* of such a *direction*, or of such an *aspect*, depend on TWO *polar coordinates*, or other *angular elements*.

"112. It appears, then, from the foregoing discussion, that *for the complete determination*, of what we have called the *geometrical* QUOTIENT *of two coinitial Vectors, a System of Four Elements*, admitting each separately of numerical expression, *is generally required*.   Of these four elements, *one* serves to determine the *relative length* of the two lines compared; and the other *three* are in general necessary, in order to determine *fully* their *relative direction*.   Again, of these three latter elements, *one* represents the

---

[1] [Assumption 2.]

mutual *inclination,* or *elongation,* of the two lines; or the *magnitude* (or quantity) of the *angle* between them; while the *two others* serve to determine the *direction* of the *axis,* perpendicular to their common *plane,* round which a *rotation* through that angle is to be performed, in a *sense* previously selected as the *positive* one (or towards a fixed and previously selected *hand*), for the purpose of *passing* (in the simplest way, and therefore in the plane of the two lines) from the *direction* of the *divisor-line, to* the direction of the *dividend-line.* And *no more than four* numerical *elements* are necessary for our present purpose: because the *relative length* of two lines is not changed when their lengths are altered proportionally, nor is their *relative direction* changed, when the *angle* which they form is merely *turned* about, *in its own plane.* On account, then, of this *essential connexion* of that *complex relation* between two lines, which is *compounded* of a *relation of lengths,* and of a *relation of directions,* and to which we have given (by an *extension* from the theory of *scalars*), the name of a *geometrical quotient,* with a *System of* FOUR *numerical Elements,* we have already a *motive* for saying that '*The Quotient of two Vectors is generally a Quaternion'.*"[1]

"181. Suppose that OI, OJ, OK are any three given and coinitial but rectangular unit lines, the rotation around the first from the second to the third being positive; and let OI', OJ', OK' be the three unit vectors respectively opposite to these, so that

$$OI' = -OI, \; OJ' = -OJ, \; OK' = -OK.$$

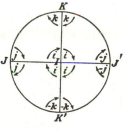

Let the three new symbols i, j, k denote a *system of three right versors,*[2] *in three mutually rectangular planes,* . . . ; so that . . . i = OK:OJ, j = OI:OK, k = OJ:OI, as the figure may serve to illustrate. We shall then have these other expressions for the same three versors.

$$i = OJ':OK = OK':OJ' = OJ:OK';$$
$$j = OK':OI = OI':OK' = OK:OI';$$
$$k = OI':OJ = OJ':OI' = OI:OJ';$$

---

[1] ["Quaternion . . . signifies . . . a Set of Four."]

[2] [A right versor is an operator which produces a rotation of a right angle about a given axis in a given direction.]

while the three respectively *opposite* versors may be thus expressed:

$$-i = OJ:OK = OK':OJ = OJ':OK' = OK:OJ'$$
$$-j = OK:OI = OI':OK = OK':OI' = OI:OK'$$
$$-k = OI:OJ = OJ':OI = OI':OJ' = OJ:OI'$$

and from the comparison of these different expressions several important symbolical consequences follow...

"182. In the *first* place, since

$$i^2 = (OJ':OK).(OK:OJ) = OJ':OJ, \text{ etc.,}$$

we deduce the following equal values for the *squares* of the new symbols:

I $i^2 = -1; j^2 = -1; k^2 = -1,$

. . . .

In the *second* place, since

$$ij = (OJ:OK').(OK':OI) = OJ:OI, \text{ etc.,}$$

we have the following values for the *products* of the same three symbols, or versors, when taken *two by two*, and in a certain *order* of *succession*...:

II $ij = k; jk = i; ki = j.$

But in the *third* place..., since

$$ji = (OI:OK).(OK:OJ) = OI:OJ, \text{ etc.,}$$

we have these other and *contrasted* formulae, for the *binary products* of the *same* three right versors, when taken as factors with an *opposite order:*

III $ji = -k; kj = -i; ik = -j.$

Hence, while the *square of each* of the *three right versors,* denoted by these *three new symbols*, i, j, k, is equal to *negative unity,* the *product of any two* of them is equal either to the *third itself,* or to the *opposite* of that third versor, according as the *multiplier precedes* or *follows* the *multiplicand,* in the *cyclical succession*

i, j, k, i, j, . . .

which the annexed figure may give some help towards remembering.

"183. Since we have thus $ji = -ij, \ldots$ we see that the *laws of combination of the new symbols,* i, j, k, are *not in all respects the same* as the corresponding laws in *algebra;* since the *Commutative Property of Multiplication,* or the *convertibility* of the places of the *factors* without change of value of the *product*, does *not here* hold good; which arises from the circumstance that the factors to be combined are here diplanar versors. It is therefore important to observe that there *is* a respect in which the *laws of* i, j, k *agree* with usual and *algebraic laws:* namely, in the Associative Property of Multiplication; or in the property that the new symbols always obey the associative formula

$$\iota\kappa\lambda = \iota\kappa\lambda,$$

whichever of them may be substituted for ι, for κ, and for λ; in virtue of which equality of values we may *omit the point* in any such symbol of a *ternary product* (whether of equal or unequal factors), and write it simply as $\iota\kappa\lambda$.   In particular, we have thus,

$$\text{i.jk} = \text{i.i} = \text{i}^2 = -1; \qquad \text{ij.k} = \text{k.k} = \text{k}^2 = -1;$$

or briefly

$$\text{ijk} = -1.$$

We may, therefore,...establish the following important *Formula:*

$$\text{i}^2 = \text{j}^2 = \text{k}^2 = \text{ijk} = -1;$$

...which we shall find to contain (virtually) *all the laws of the symbols* i, j, k, and therefore to be a *sufficient symbolical basis* for the whole *Calculus of Quaternions:* because it will be shown that *every quaternion can be reduced to the Quadrinomial Form,*

$$\text{q} = \text{w} + \text{ix} + \text{jy} + \text{kz},$$

where w, x, y, z compose a *system of four scalars*, while i, j, k are the same *three right versors as above."*

"*If two right versors in two mutually rectangular planes, be multiplied together in two opposite orders, the two resulting products will be two opposite right versors, in a third plane, rectangular to the two former;* or in symbols...

$$\text{q'q} = -\text{qq'}$$

...In *this* case, therefore, we have what would be in algebra a *paradox*...When we come to examine what, in the last analysis, may be said to be the *meaning* of this last equation, we find it to be simply this: that *any two quadrantal* or *right rotations, in planes perpendicular to each other, compound themselves into a third right rotation, as their resultant in a plane perpendicular to each of them:* and that this *third* or *resultant* rotation has one or other of two *opposite directions,* according to the *order* in which the two *component rotations* are taken, so that one shall be *successive* to the other."

GRASSMANN

ON THE AUSDEHNUNGSLEHRE

(Translated from the German by Dr. Mark Kormes, New York City.)

Hermann Günther Grassmann (1809–1877) was professor at the gymnasium in Stettin. In 1844 he published *Die lineale Ausdehnungslehre, ein neuer Zweig der Mathematik...*, which was not generally understood on account of its abstract philosophical form. Grassmann therefore rewrote his book many years later and published it in Berlin (1862) under the title *Die Ausdehnungslehre, Vollständig und in strenger Form bearbeitet.* Besides many other important contributions to mathematics he distinguished himself as a scholar of Sanskrit literature.

In his *Ausdehnungslehre* Grassmann created a symbolic calculus so general that its definitions and theorems can be easily applied not only to geometry of n dimensions but also to almost every branch of mathematics. This calculus forms the basis of vector analysis. By its aid Grassmann derived fundamental theorems on determinants and solved many elimination problems in a most elegant manner. In connection with the problem of Pfaff and the theory of partial differential equations, his theorems are of great importance.

The following translation is limited to (1) the development of the idea of non-commutative multiplication—the combinatory (outer) and the inner products; and (2) certain passages relating to geometry. It is taken from the 1862 edition,—the one which made Grassmann's influence felt.

1. We say that a quantity a is derived from the quantities b, c...by means of numbers β, γ,...if

$$a = \beta b + \gamma c + \ldots$$

where β, γ...are real numbers, rational or irrational, and may be equal to zero. We also say in such a case that a is derived numerically from b, c...

2. We further say that two or more quantities a, b, c...are numerically related if one of them can be derived numerically from the others; for example:

$$a = \beta b + \gamma c + \ldots$$

where β, γ,...are real numbers...

3. A quantity from which a set of other quantities may be derived numerically is called a *unit* and in particular a unit which cannot be derived numerically from any other unit is called a *primitive unit*. The unit of numbers is called the *absolute unit*,

all other units are *relative*. Zero should never be regarded as a unit.

5. An *extensive* quantity is any expression derived by means of numbers from a system of units[1] (which system should not consist solely of the absolute unit). The numbers used are called *derivation-coefficients*. For example the polynominal

$$\alpha_1 e_1 + \alpha_2 e_2 + \ldots \text{ or } \Sigma \alpha e \text{ or } \Sigma \alpha_r e_r$$

is 'an extensive quantity when α_1, α_2,...are real numbers and e_1, e_2...is a system of units. Only if the system of units consists solely of the absolute unit (1) the derived quantity is not an extensive but a numerical quantity...

6. To add two extensive quantities derived from the same system of units we add the derivation-coefficients of the corresponding units:

$$\Sigma \alpha_r e_r + \Sigma \beta_r e_r = \Sigma(\alpha_r + \beta_r)e_r$$
$$\ldots\ldots\ldots\ldots\ldots 2$$

9. All laws of algebraic addition and subtraction hold for the extensive quantities...

10. To multiply an extensive quantity by a number we multiply all its derivation-coefficients by this number:

$$\Sigma \alpha_i e_i . \beta = \beta . \Sigma \alpha_i e_i = \Sigma(\alpha_i \beta)e_i$$
$$\ldots\ldots\ldots\ldots\ldots 3$$

13. All laws of algebraic multiplication and division hold for the multiplication and division of an extensive quantity by a number.

37. A product [*ab*] of two extensive quantities *a* and *b* is defined as an extensive quantity (or a numerical quantity) obtained in the following way: Multiply each of the units from which the first quantity *a* is derived by each of the units from which the second quantity *b* is derived so that the unit of the first quantity is always the first factor and the unit of the second quantity is the second factor of the product; multiply then every such product by the product of the corresponding derivation-coefficients and add all the products so obtained:

$$[\Sigma \alpha_r e_r \Sigma \beta_s e_s] = \Sigma \alpha_r \beta_s[e_r e_s] \ldots$$

---

[1] [A set of units not related numerically (See 4 in the original text).]

[2] [Here follows a similar definition for subtraction. (See 7 in original text.)]

[3] [Here follows a similar definition for division by a number. (See 11 in the original text.)]

Remark. Inasmuch as according to the above definition the product of extensive quantities is either an extensive or a numerical quantity, we must be able (see 5) to derive it numerically from a system of units. What is this system of units and how are we to derive numerically the products e_re_s from it, is not explained in the definition. Therefore if we are to determine exactly the concept of a particular product we must agree upon certain necessary rules...Consider, e.g., the product $P = [(x_1e_1 + x_2e_2)(y_1e_1 + y_2e_2)]$ $= x_1y_1[e_1e_1] + x_1y_2[e_1e_2] + x_2y_1[e_2e_1] + x_2y_2[e_2e_2]$...We could then agree that the four products $[e_1e_1]$, $[e_1e_2]$, $[e_2e_1]$ and $[e_2e_2]$ constitute the system of units from which P is to be derived numerically so that the numbers x_1y_1, x_1y_2, x_2y_1, and x_2y_2 are the derivation-coefficients. We would have then a particular product characterized by the fact that no equations are necessary for its determination. We could on the other hand select three of them; $[e_1e_1]$, $[e_1e_2]$ and $[e_2e_2]$ as units and agree that $[e_2e_1] = [e_1e_2]$, the derivation-coefficients of P would be then: x_1y_1, $(x_1y_2 + x_2y_1)$ and x_2y_2; this kind of a product is characterized by the fact that the laws which govern it are identical with those of algebraic multiplication. We could also select $[e_1e_2]$ as a unit and agree that $[e_1e_1] = 0$, $[e_2e_1] = -[e_1e_2]$, and $[e_2e_2] = 0$; in this case the product P would have only one derivation-coefficient, namely $x_1y_2 - x_2y_1$. Such products are subsequently designated as *combinatory*. We may finally agree to select a system of units not containing any one of the products $[e_1e_1]$, $[e_1e_2]$, $[e_2e_1]$, $[e_2e_2]$, and then to determine how to derive these four products from this system; e.g., we may choose as the system of units the absolute unit and agree that $[e_1e_1] = 1$, $[e_1e_2] = 0$, $[e_2e_1] = 0$, $[e_2e_2] = 1$. Under such conditions P becomes a numerical quantity namely $P = x_1y_1 + x_2y_2$. These products are subsequently designated as *inner products*.

50. Every multiplication for which the determining equations[1] remain true if we substitute in place of the units quantities numerically derived from them is said to be a *linear* multiplication.

51. Besides the multiplication which has no determining equation or the multiplication for which all products are zero there exist only two kinds of linear multiplication: the determining equation for the one is

(1) $$[e_re_s] + [e_se_r] = 0$$

---

[1] [A numerical relation between the products of units. (See 48 in the original text.)]

and for the other

(2) $[e_r e_s] = [e_s e_r] \ldots$

52. A *combinatory product* is defined as a product the factors of which are derived from a system of units, provided that the sum of any two products of units obtained from each other by interchanging the last two factors is equal to zero, while every product consisting solely of different units is not zero. The factors of this product are called *simple factors*. If b and c are units and A is an arbitrary set of units the above condition is expressed in the following form:

$$[Abc] + [Acb] = 0.$$

53. In every combinatory product we may interchange the two last factors provided that we change the sign of the product, or $[Abc] + [Acb] = 0$, in the case when A is an arbitrary set of factors and b and c are simple factors.

Proof.—1. Suppose that b and c are units. Since A is a set of arbitrary factors and since these factors may be derived numerically from units we obtain after substitution an expression for A of the form: $A = \Sigma \alpha_r E_r$, where E_r are products of units. Thus we have

$$
\begin{aligned}
[Abc] + [Acb] &= [\overline{\Sigma \alpha_r E_r} bc] + [\overline{\Sigma \alpha_r E_r} cb] \\
&= \Sigma \alpha_r [E_r bc] + \Sigma \alpha_r [E_r cb)] \\
&= \Sigma \alpha_r ([E_r bc] + [E_r cb]) \\
&= \Sigma \alpha_r 0 \qquad\qquad\qquad (52) \\
&= 0.
\end{aligned}
$$

2. Suppose now that b and c are numerically derived from the units $e_1, e_2, \ldots, e$. *e. g.*, $b = \Sigma \beta_r e_r$, $c = \Sigma \gamma_r e_r$ we have then:

$$
\begin{aligned}
[Abc] + [Acb] &= [A \Sigma \beta_r e_r \Sigma \gamma_r e_r] + [A \Sigma \gamma_r e_r \Sigma \beta_r e_r] \\
&= \Sigma \beta_r \gamma_s [A e_r e_s] + \Sigma \gamma_s \beta_r [A e_s e_r] \\
&= \Sigma \beta_r \gamma_s ([A e_r e_s] + [A e_s e_r]) \\
&= \Sigma \beta_r \gamma_s 0 \qquad\qquad\qquad \text{(proof 1.)} \\
&= 0.
\end{aligned}
$$

55. Any two factors of a combinatory product may be interchanged provided that we change the sign of the product:

$$P_{a,b} = -P_{b,a} \text{ or } P_{a,b} + P_{b,a} = 0.$$

. . . .

60. If two simple factors of a combinatory product are equal the product is equal to zero,

$$P_{a,a} = 0.$$

Proof...According to 55 we have: $P_{a,b} + P_{b,a} = 0$ hence if we put $a = b$,

$$P_{a,a} + P_{a,a} = 0 \text{ or } 2P_{a,a} = 0;$$

thus

$$P_{a,a} = 0.$$

61. A combinatory product is equal to zero if its simple factors are numerically related,—i. e., $[a_1a_2a_3\ldots a_m] = 0$,—if one of the quantities $a_1a_2\ldots a_m$ can be derived numerically from the others; e.g., if

$$a_1 = \alpha_2 a_2 + \alpha_3 a_3 + \ldots + \alpha_m a_m.$$

Proof: If we substitute in the product for a_1 the above expression we obtain

$$[a_1a_2a_3\ldots a_m] = [(\alpha_2 a_2 + \alpha_3 a_3 + \ldots + \alpha_m a_m)a_2a_3\ldots a_m]$$
$$= \alpha_2[a_2a_2a_3\ldots a_m] + \alpha_3[a_3a_2a_3\ldots a_m] + \ldots$$
$$+ \alpha_m[a_m a_2 a_3 \ldots a_m]$$
$$= \alpha_2 0 + \alpha_3.0 + \ldots + \alpha_m.0$$
$$= 0.$$

63. In order to obtain a combinatory product of n simple factors which are derived numerically from the n quantities $a_1, a_2, \ldots a_n$ we form the determinant of the derivation-coefficients where the coefficients of the first factor form the first row, etc., and we multiply this determinant by the combinatory product of the quantities $a_1, a_2 \ldots a_n$:

$$[(\alpha_1^{(1)}a_1 + \ldots + \alpha_n^{(1)}a_n)(\alpha_1^{(2)}a_1 + \ldots + \alpha_n^{(2)}a_n)\ldots(\alpha_1^{(n)}a_1 + \ldots + \alpha_n^{(n)}a_n)] = \Sigma \mp \alpha_1^{(1)}\alpha_2^{(2)}\ldots\alpha_n^{(n)}[a_1a_2\ldots a_n]^1\ldots$$

68. All laws of combinatory multiplication hold if we substitute for the n primitive units an arbitrary set of n quantities derived numerically from these units provided that these n quantities are not related numerically.

---

[1] [The symbol $\Sigma \mp \alpha_1^{(1)}\alpha_2^{(2)}\ldots\alpha_n^{(n)}$ is used by Grassmann for the determinant

$$\begin{vmatrix} \alpha_1^{(1)}\alpha_2^{(1)}\ldots\alpha_n^{(1)} \\ \alpha_1^{(2)}\alpha_2^{(2)}\ldots\alpha_n^{(2)} \\ \cdots\cdots\cdots\cdots \\ \alpha_1^{(n)}\alpha_2^{(n)}\ldots\alpha_n^{(n)} \end{vmatrix}$$

64. Multiplicative combinations from a set of quantities are defined as combinations without repetition of these quantities whereby each combination is a combinatory product, the factors of which are the elements of the combination; *e. g.*, the multiplicative combinations of second class from the three quantities a, b, c are: $[ab]$, $[ac]$ and $[bc]$.

77. The multiplicative combinations of class m of the primitive units shall be called *units of order m*. A quantity derived numerically from units of order m shall be called a *quantity of order m*. Such quantity is said to be *simple* if it may be represented as a combinatory product of quantities of the first order, otherwise it is called a *composite* quantity. The aggregate of all quantities which can be derived numerically from the simple factors of a simple quantity is called the *domain* of this quantity.

77b...Remark. As an example of a composite quantity let us consider the sum $(ab) + (cd)$ where a, b, c, d are quantities not related numerically. Suppose $(ab) + (cd)$ were a simple quantity, *e.g.*, equal to $(p.q)$; we would have $[(ab + cd)(ab + cd)] = [pqpq] = 0$. But $[(ab + cd)(ab + cd)] = [abcd] + [cdab]$ on account of $[abab]$ and $[cdcd]$ being equal to zero. It is however $[abcd] = [cdab]$. Hence $[(ab + cd).(ab + cd)] = 2.[abcd]$. If $(ab) + (cd)$ were a simple quantity $[abcd]$ would be equal to zero and a, b, c, d would be related numerically which contradicts the assumption.

78. The *outer* product of two units of higher order is defined as the combinatory product of the simple factors of those quantities, whereby the arrangement of these factors remains undisturbed:

$$[(e_1e_2\ldots e_m)(e_{m+1}\ldots e_n)] = [e_1e_2\ldots e_n]$$

NOTE.—The name *outer multiplication* is used to emphasize the fact that this product holds if and only if one factor is entirely outside of the domain of the other factor.

79. In order to obtain the outer product of two quantities A and B we form the combinatory product of the simple factors of the first quantity with those of the second quantity:

$$[(ab\ldots).(cd\ldots)] = [ab\ldots cd\ldots].$$

80. The parentheses have no effect on the outer product:

$$[A(BC)] = [ABC]\ldots$$

83. Given a sum of simple quantities S and a set of m quantities of the first order $a_1a_2\ldots a_m$ which are not related numerically.

If the outer product of S with every one of the quantities $a_1, a_2 \ldots$ a_m is equal to zero, S may be represented as an outer product in which $a_1, a_2 \ldots a_m$ are factors; *i. e.*,

$$S = [a_1, a_2 \ldots a_m S_m)$$

if

$$0 = [a_1 S] = [a_2 S] = \ldots = [a_m S] \ldots$$

86. The *principal* domain is the domain of the primitive units from which the quantities under consideration were derived...

89. Let us consider a principal domain of order n and let us assume that the product of the primitive units is equal to 1. If E is a unit of any arbitrary order (*i. e.*, either one of the primitive units or a product of a number of them) the *complement* of E is defined as a quantity which is equal to the combinatory product E' of all units which are not in E. The complement is positive or negative according to whether $[EE']$ is equal to $+1$ or to -1. We will denote the complement by a vertical line before the given quantity. The complement of a number shall be this number itself. Thus we have: $|E = [EE']\, E'$ if E and E' contain all the units $e_1 e_2 \ldots e_n$, and if $[e_1 e_2 \ldots e_n] = 1$; also $|\alpha = \alpha$ when α is a number.

90. The complement of an arbitrary quantity A is the quantity $|A$ obtained if in the expression for A we substitute the complements in the place of the units; *i. e.*,

$$|(\alpha_1 E_1 + \alpha_2 E_2 + \ldots) = \alpha_1 |E_1 + \alpha_2 |E_2 + \ldots$$

where $E_1 E_2 \ldots$ are units of any order whatsoever.

91. The outer product of a unit and its complement is equal to 1:
$$[E|E] = 1.$$

Proof.—If E' is the combinatory product of all primitive units not contained in E we have (according to 89):

$$|E = \mp E' \text{ according to whether } [EE'] = \mp 1.$$

In the case of the lower sign we have: $[E|E] = [EE'] = 1$ and in the case of the upper sign: $[E|E] = -[EE'] = -(-1) = 1$.

92. The complement of a complement of a quantity A is either equal to A or to $-A$ according to whether the product of the orders of A and of its complement is even or odd; *i. e.*,

$$||A = (-1)^{qr} A$$

if q and r are the orders of A and $|A$ respectively...

NOTE.—If both r and q are odd, as for example in the case of a domain of order two and complements of quantities of order one, we have $\|A = -A$ so that in such a case the symbol $|$ obeys the same laws as $i = \sqrt{-1}$, which gives an interpretation of the imaginary in the real domain...

93. If the principal domain is of order n and if n is odd we have

$$\|A = A;$$

if n is even

$$\|A = (-1)^q A$$

where q is the order of A...

94. If the sum of the orders of two units is equal to or less than n,—i. e., the order of the principal domain,—the outer product of these units is called a *progressive* product provided that the progressive product of the primitive units is equal to 1. If on the other hand the sum of the orders of two units is greater than n, the *regressive* product of these units is given by a quantity whose complement is equal to the progressive product of the complements of these units. We shall refer to both the regressive and progressive products as *relative products*...

97. The product of the complements of two quantities is equal to the complement of the product of these quantities:

$$[|A|B] = |[AB]...$$

...If the product of two quantities is progressive, the product of their complements is regressive provided that we agree to consider the product of order zero as a progressive and as a regressive product at the same time...

122. A *mixed product*[1] of three quantities $[ABC]$ is equal to zero if and only if either $[AB] = 0$, or all the quantities A, B, and C are contained in a domain of order less than n, or the quantities A, B, and C have a domain of order more than 0 in common...

137. The *inner product* of two units of an arbitrary order is defined as the relative product of the first unit and the complement of the second unit; i. e., if E and F are units of an arbitrary order the inner product is given by $[E|F]$.

138. The inner product of two quantities A and B is equal to the relative product of the first quantity and the complement of the second quantity, i. e., $[A|B]...$

139. If the factors of an inner product are of orders α and β and if the principal domain is of order n, the inner product is of order

---

[1] [That is, a product in which both the progressive and the regressive multiplication is used.]

$n + \alpha - \beta$ or $\alpha - \beta$ according to whether β is greater than α or not...

141. The inner product of two quantities of equal order is a number.

Proof. The difference between the orders is then zero, hence the inner product is of order zero,—*i. e.*, a number.

142. The inner product of two equal units is 1, the inner product of two different units of equal order is zero; *i. e.*,

$$[E_r|E_r] = 1, [E_r|E_s] = 0...$$

143. If $E_1 E_2 \ldots E_m$ are units of an arbitrary but equal order we have:

$$[(\alpha_1 E_1 + \alpha_2 E_2 + \ldots + \alpha_m E_m)|(\beta_1 E_1 + \ldots + \beta_m E_m)] =$$
$$\alpha_1 \beta_1 + \alpha_2 \beta_2 + \ldots + \alpha_m \beta_m \ldots$$

144. The two factors of an inner product may be interchanged provided they are of the same order; *i. e.*,

$$[A|B] = [B|A].$$

Proof.—If $E_1 \ldots E_m$ are the units and $A = \Sigma \alpha_r E_r$, $B = \Sigma \beta_s E_s$, we have from 143,

$$[A|B] = \Sigma \alpha_r \beta_r = \Sigma \beta_r \alpha_r = [B|A].$$

145. For the sake of simplicity we write

$$[A|A] = A^2,$$

and we call it the *inner square* of A...

147. The inner product of two units E and F is not equal to zero if and only if one of the units is incident[1] with the other...

148. If E and F are units and $[EF] \neq 0$, we have

$$[EF|E] = F \text{ and } [F|EF] = |E...$$

151. The *numerical value* of a quantity A is defined as the *positive square root* of the inner square of that quantity. Two quantities are said to be numerically equal if their numerical values, *i. e.*, their inner squares,—are equal.

152. Two quantities different from zero are said to be *orthogonal* if their inner product is zero...

153. A set of n numerically equal quantities of the first order which are orthogonal to each other is called an *orthogonal system*

---

[1] [A quantity is said to be *incident* with another if its domain is incident; that is all quantities of the domain of the first quantity are also quantities of the domain of the second quantity, but not vice versa.]

of order n; in the case that the domain is also of order n we speak
of a *complete orthogonal system*. The numerical value of the given
quantities is said to be the numerical value of the orthogonal
system. Every orthogonal system having the numerical value 1
is said to be *simple*...

157. The quantities of an orthogonal system are not related
numerically and every quantity of the first order can be derived
numerically from any arbitrary complete orthogonal system...

162. The system of the primitive units is a complete orthogonal
system, whose numerical value is 1.

Proof.—If $e_1 e_2 \ldots e_n$ are the primitive units, we have

$$1 = e_1{}^2 = e_2{}^2 = \ldots = e_n{}^2$$
$$0 = [e_1|e_2] = \ldots$$

163. In every domain of order m we can establish an orthogonal
system of order m having an arbitrary numerical value so that this
system is a part of the complete orthogonal system...

168. All previous theorems[1] remain true if we replace the system
of the primitive units by an arbitrary complete orthogonal system
which has the numerical value 1...

175. Given two quantities A and B of order m each of which is
composed of m simple factors. The inner product of these two
quantities is equal to the determinant of m rows and m columns
which is obtained by forming inner products of every simple
factor of one quantity with every simple factor of the other
quantity; *i.e.*

$$[abc\ldots|a'b'c'\ldots] = \text{Determ.} \begin{vmatrix} [a|a'], & [a|b'], & [a|c'], & \ldots \\ [b|a'], & [b|b'], & [b|c'], & \ldots \\ [c|a'], & [c|b'], & [c|c'], & \ldots \\ \cdots\cdots\cdots\cdots\cdots\cdots \end{vmatrix}$$

. . . .

216. Given a point E and let us assume that three lines of equal
length and perpendicular to each other are the principal units.
If α_1, α_2, α_3 are arbitrary numbers the expression:

(a) $E + \alpha_1 e_1 + \alpha_2 e_2 + \alpha_3 e_3$

defines the point A, obtained in the following manner: From E
we proceed along the segment EB which is equal to $\alpha_1 e_1$, that is,
which has the same direction as e_1 or the opposite direction according to whether α_1 is positive or negative and the distance EB is

---

[1] [Relative to the inner product.]

in the same ratio to e_1 as α_1 is to 1. We then proceed from B along the segment BC equal to α_2e_2 in the above sense and finally from C we proceed along the segment CA which is equal to α_3e_3 in the same manner. Furthermore the expression:

(b) $\alpha_1e_1 + \alpha_2e_2 + \alpha_3e_3$

defines a segment, that is, a straight line of a given length and directions and namely such a particular segment which has the same length and direction as the line connecting the point E with the point

$$E + \alpha_1e_1 + \alpha_2e_2 + \alpha_3e_3\ldots$$

229. Every segment of the space may be derived numerically from three arbitrary segments which are not parallel to a plane...

231. If three segments are related numerically they are parallel to a plane...

232. All points of the space can be derived numerically from four arbitrary points which do not lie in one plane...

234. Every point of a straight line may be derived numerically from two arbitrary points of this line...

235. If three points are related numerically they lie in a straight line...

236. If four points are related numerically they lie in a plane...

237. In the space a domain of first order is a point, the domain of second order the unlimited straight line, that of third order the unlimited plane and that of fourth order the unlimited space.

245. The combinatory product of two points vanishes if and only if the two points coincide; the combinatory product of three points vanishes if the points lie in a straight line, that of four points if they lie in a plane and the combinatory product of five points always vanishes...

249. The product [AB] shall be called a *segment* and we shall say that it is a part of the unlimited line AB and that it is of equal length and direction as the segment AB...

273. The sum of two finite segments the lines of which intersect is a segment and its line passes through the point of intersection of the other two lines; the direction and length of this segment are the same as those of the diagonal of the parallelogram formed by segments of the same length and direction as the two original segments...

288. Planimetric multiplication is defined as the relative multiplication with respect to the plane; stereometric multiplication

as relative multiplication with respect to the space (as domain of order four)...

306.[1] The equation of a point x which lies in a straight line with the points a and b is given by:

$$[xab] = 0.$$

Proof.—$[xab]$ vanishes (according to 245) if and only if x lies in a straight line with a and b...

307. The equation of a straight line X which passes through the same point as the straight lines A and B is given by

$$[XAB] = 0...$$

309. If $P_{n,x}$ is a planimetric product of order zero in which the point x is contained n times and if the other factors are only fixed points or lines, the equation:

$$P_{n,x} = 0$$

is then the point-equation of an algebraic curve of order n, provided that it is not satisfied by every point x...

310. If $P(n, X)$ is a planimetric product of order zero in which the line X is contained n times and as the other factors are only fixed points or lines, the equation

$$P(n, X) = 0$$

is then the line-equation of an algebraic curve of class n...

311. If $P_{n,x}$ is a stereometric product of order zero, which contains the point x n times and as other factors has only fixed points, lines or planes, the equation:

$$P_{n,x} = 0$$

is the point-equation of an algebraic surface of order n...

323. The equation of a conic section, passing through the five points a, b, c, d, e, no three of which lie in a straight line is given by...

$$[xaBc_1.Dex] = 0$$

where $B = [cd]$ $c_1 = [ab.de]$ $D = [b.c]$...

324. If A, B, C are three straight lines in space no two of which intersect then:

$$[xABCx] = 0$$

---

[1] [From now on Grassmann uses small letters to denote points and capital letters to denote lines.]

is the equation of the surface of second order which contains the three straight lines A, B, C...

330. For the purpose of inner multiplication we shall always assume as principal units three segments of equal length and perpendicular to each other $(e_1e_2e_3)$, in the plane two such segments (e_1e_2) and we shall assume that the length of these segments shall be the unit of length, $[e_1e_1e_2]$ the unit of volume and $[e_1e_2]$ the unit of area.

331. For the plane[1] the concept of length coincides with the concept of numerical value, orthogonal means perpendicular...

---

[1] [Also for the space (see 333 of original text).]

Index

A CATALOG OF SELECTED
DOVER BOOKS
IN SCIENCE AND MATHEMATICS

Astronomy

BURNHAM'S CELESTIAL HANDBOOK, Robert Burnham, Jr. Thorough guide to the stars beyond our solar system. Exhaustive treatment. Alphabetical by constellation: Andromeda to Cetus in Vol. 1; Chamaeleon to Orion in Vol. 2; and Pavo to Vulpecula in Vol. 3. Hundreds of illustrations. Index in Vol. 3. 2,000pp. 6⅛ x 9¼.
Vol. I: 23567-X
Vol. II: 23568-8
Vol. III: 23673-0

EXPLORING THE MOON THROUGH BINOCULARS AND SMALL TELE-SCOPES, Ernest H. Cherrington, Jr. Informative, profusely illustrated guide to locating and identifying craters, rills, seas, mountains, other lunar features. Newly revised and updated with special section of new photos. Over 100 photos and diagrams. 240pp. 8¼ x 11.
24491-1

THE EXTRATERRESTRIAL LIFE DEBATE, 1750–1900, Michael J. Crowe. First detailed, scholarly study in English of the many ideas that developed from 1750 to 1900 regarding the existence of intelligent extraterrestrial life. Examines ideas of Kant, Herschel, Voltaire, Percival Lowell, many other scientists and thinkers. 16 illustrations. 704pp. 5⅜ x 8½.
40675-X

THEORIES OF THE WORLD FROM ANTIQUITY TO THE COPERNICAN REVOLUTION, Michael J. Crowe. Newly revised edition of an accessible, enlightening book recreates the change from an earth-centered to a sun-centered conception of the solar system. 242pp. 5⅜ x 8½.
41444-2

A HISTORY OF ASTRONOMY, A. Pannekoek. Well-balanced, carefully reasoned study covers such topics as Ptolemaic theory, work of Copernicus, Kepler, Newton, Eddington's work on stars, much more. Illustrated. References. 521pp. 5⅜ x 8½.
65994-1

A COMPLETE MANUAL OF AMATEUR ASTRONOMY: Tools and Techniques for Astronomical Observations, P. Clay Sherrod with Thomas L. Koed. Concise, highly readable book discusses: selecting, setting up and maintaining a telescope; amateur studies of the sun; lunar topography and occultations; observations of Mars, Jupiter, Saturn, the minor planets and the stars; an introduction to photoelectric photometry; more. 1981 ed. 124 figures. 26 halftones. 37 tables. 335pp. 6½ x 9¼.
42820-6

AMATEUR ASTRONOMER'S HANDBOOK, J. B. Sidgwick. Timeless, comprehensive coverage of telescopes, mirrors, lenses, mountings, telescope drives, micrometers, spectroscopes, more. 189 illustrations. 576pp. 5⅝ x 8¼. (Available in U.S. only.)
24034-7

STARS AND RELATIVITY, Ya. B. Zel'dovich and I. D. Novikov. Vol. 1 of *Relativistic Astrophysics* by famed Russian scientists. General relativity, properties of matter under astrophysical conditions, stars, and stellar systems. Deep physical insights, clear presentation. 1971 edition. References. 544pp. 5⅝ x 8¼.
69424-0

Chemistry

THE SCEPTICAL CHYMIST: The Classic 1661 Text, Robert Boyle. Boyle defines the term "element," asserting that all natural phenomena can be explained by the motion and organization of primary particles. 1911 ed. viii+232pp. 5⅜ x 8½.
42825-7

RADIOACTIVE SUBSTANCES, Marie Curie. Here is the celebrated scientist's doctoral thesis, the prelude to her receipt of the 1903 Nobel Prize. Curie discusses establishing atomic character of radioactivity found in compounds of uranium and thorium; extraction from pitchblende of polonium and radium; isolation of pure radium chloride; determination of atomic weight of radium; plus electric, photographic, luminous, heat, color effects of radioactivity. ii+94pp. 5⅜ x 8½.
42550-9

CHEMICAL MAGIC, Leonard A. Ford. Second Edition, Revised by E. Winston Grundmeier. Over 100 unusual stunts demonstrating cold fire, dust explosions, much more. Text explains scientific principles and stresses safety precautions. 128pp. 5⅜ x 8½.
67628-5

THE DEVELOPMENT OF MODERN CHEMISTRY, Aaron J. Ihde. Authoritative history of chemistry from ancient Greek theory to 20th-century innovation. Covers major chemists and their discoveries. 209 illustrations. 14 tables. Bibliographies. Indices. Appendices. 851pp. 5⅜ x 8½.
64235-6

CATALYSIS IN CHEMISTRY AND ENZYMOLOGY, William P. Jencks. Exceptionally clear coverage of mechanisms for catalysis, forces in aqueous solution, carbonyl- and acyl-group reactions, practical kinetics, more. 864pp. 5⅜ x 8½.
65460-5

ELEMENTS OF CHEMISTRY, Antoine Lavoisier. Monumental classic by founder of modern chemistry in remarkable reprint of rare 1790 Kerr translation. A must for every student of chemistry or the history of science. 539pp. 5⅜ x 8½.
64624-6

THE HISTORICAL BACKGROUND OF CHEMISTRY, Henry M. Leicester. Evolution of ideas, not individual biography. Concentrates on formulation of a coherent set of chemical laws. 260pp. 5⅜ x 8½.
61053-5

A SHORT HISTORY OF CHEMISTRY, J. R. Partington. Classic exposition explores origins of chemistry, alchemy, early medical chemistry, nature of atmosphere, theory of valency, laws and structure of atomic theory, much more. 428pp. 5⅜ x 8½. (Available in U.S. only.)
65977-1

GENERAL CHEMISTRY, Linus Pauling. Revised 3rd edition of classic first-year text by Nobel laureate. Atomic and molecular structure, quantum mechanics, statistical mechanics, thermodynamics correlated with descriptive chemistry. Problems. 992pp. 5⅜ x 8½.
65622-5

FROM ALCHEMY TO CHEMISTRY, John Read. Broad, humanistic treatment focuses on great figures of chemistry and ideas that revolutionized the science. 50 illustrations. 240pp. 5⅜ x 8½.
28690-8

Engineering

DE RE METALLICA, Georgius Agricola. The famous Hoover translation of greatest treatise on technological chemistry, engineering, geology, mining of early modern times (1556). All 289 original woodcuts. 638pp. 6¾ x 11. 60006-8

FUNDAMENTALS OF ASTRODYNAMICS, Roger Bate et al. Modern approach developed by U.S. Air Force Academy. Designed as a first course. Problems, exercises. Numerous illustrations. 455pp. 5⅜ x 8½. 60061-0

DYNAMICS OF FLUIDS IN POROUS MEDIA, Jacob Bear. For advanced students of ground water hydrology, soil mechanics and physics, drainage and irrigation engineering, and more. 335 illustrations. Exercises, with answers. 784pp. 6⅛ x 9¼. 65675-6

THEORY OF VISCOELASTICITY (Second Edition), Richard M. Christensen. Complete, consistent description of the linear theory of the viscoelastic behavior of materials. Problem-solving techniques discussed. 1982 edition. 29 figures. xiv+364pp. 6⅛ x 9¼. 42880-X

MECHANICS, J. P. Den Hartog. A classic introductory text or refresher. Hundreds of applications and design problems illuminate fundamentals of trusses, loaded beams and cables, etc. 334 answered problems. 462pp. 5⅜ x 8½. 60754-2

MECHANICAL VIBRATIONS, J. P. Den Hartog. Classic textbook offers lucid explanations and illustrative models, applying theories of vibrations to a variety of practical industrial engineering problems. Numerous figures. 233 problems, solutions. Appendix. Index. Preface. 436pp. 5⅜ x 8½. 64785-4

STRENGTH OF MATERIALS, J. P. Den Hartog. Full, clear treatment of basic material (tension, torsion, bending, etc.) plus advanced material on engineering methods, applications. 350 answered problems. 323pp. 5⅜ x 8½. 60755-0

A HISTORY OF MECHANICS, René Dugas. Monumental study of mechanical principles from antiquity to quantum mechanics. Contributions of ancient Greeks, Galileo, Leonardo, Kepler, Lagrange, many others. 671pp. 5⅜ x 8½. 65632-2

STABILITY THEORY AND ITS APPLICATIONS TO STRUCTURAL MECHANICS, Clive L. Dym. Self-contained text focuses on Koiter postbuckling analyses, with mathematical notions of stability of motion. Basing minimum energy principles for static stability upon dynamic concepts of stability of motion, it develops asymptotic buckling and postbuckling analyses from potential energy considerations, with applications to columns, plates, and arches. 1974 ed. 208pp. 5⅜ x 8½. 42541-X

METAL FATIGUE, N. E. Frost, K. J. Marsh, and L. P. Pook. Definitive, clearly written, and well-illustrated volume addresses all aspects of the subject, from the historical development of understanding metal fatigue to vital concepts of the cyclic stress that causes a crack to grow. Includes 7 appendixes. 544pp. 5⅜ x 8½. 40927-9

ROCKETS, Robert Goddard. Two of the most significant publications in the history of rocketry and jet propulsion: "A Method of Reaching Extreme Altitudes" (1919) and "Liquid Propellant Rocket Development" (1936). 128pp. 5⅜ x 8½.   42537-1

STATISTICAL MECHANICS: Principles and Applications, Terrell L. Hill. Standard text covers fundamentals of statistical mechanics, applications to fluctuation theory, imperfect gases, distribution functions, more. 448pp. 5⅜ x 8½.   65390-0

ENGINEERING AND TECHNOLOGY 1650–1750: Illustrations and Texts from Original Sources, Martin Jensen. Highly readable text with more than 200 contemporary drawings and detailed engravings of engineering projects dealing with surveying, leveling, materials, hand tools, lifting equipment, transport and erection, piling, bailing, water supply, hydraulic engineering, and more. Among the specific projects outlined–transporting a 50-ton stone to the Louvre, erecting an obelisk, building timber locks, and dredging canals. 207pp. 8⅜ x 11¼.   42232-1

THE VARIATIONAL PRINCIPLES OF MECHANICS, Cornelius Lanczos. Graduate level coverage of calculus of variations, equations of motion, relativistic mechanics, more. First inexpensive paperbound edition of classic treatise. Index. Bibliography. 418pp. 5⅜ x 8½.   65067-7

PROTECTION OF ELECTRONIC CIRCUITS FROM OVERVOLTAGES, Ronald B. Standler. Five-part treatment presents practical rules and strategies for circuits designed to protect electronic systems from damage by transient overvoltages. 1989 ed. xxiv+434pp. 6⅛ x 9¼.   42552-5

ROTARY WING AERODYNAMICS, W. Z. Stepniewski. Clear, concise text covers aerodynamic phenomena of the rotor and offers guidelines for helicopter performance evaluation. Originally prepared for NASA. 537 figures. 640pp. 6⅛ x 9¼.
          64647-5

INTRODUCTION TO SPACE DYNAMICS, William Tyrrell Thomson. Comprehensive, classic introduction to space-flight engineering for advanced undergraduate and graduate students. Includes vector algebra, kinematics, transformation of coordinates. Bibliography. Index. 352pp. 5⅜ x 8½.   65113-4

HISTORY OF STRENGTH OF MATERIALS, Stephen P. Timoshenko. Excellent historical survey of the strength of materials with many references to the theories of elasticity and structure. 245 figures. 452pp. 5⅜ x 8½.   61187-6

ANALYTICAL FRACTURE MECHANICS, David J. Unger. Self-contained text supplements standard fracture mechanics texts by focusing on analytical methods for determining crack-tip stress and strain fields. 336pp. 6⅛ x 9¼.   41737-9

STATISTICAL MECHANICS OF ELASTICITY, J. H. Weiner. Advanced, self-contained treatment illustrates general principles and elastic behavior of solids. Part 1, based on classical mechanics, studies thermoelastic behavior of crystalline and polymeric solids. Part 2, based on quantum mechanics, focuses on interatomic force laws, behavior of solids, and thermally activated processes. For students of physics and chemistry and for polymer physicists. 1983 ed. 96 figures. 496pp. 5⅜ x 8½.   42260-7

Mathematics

FUNCTIONAL ANALYSIS (Second Corrected Edition), George Bachman and Lawrence Narici. Excellent treatment of subject geared toward students with background in linear algebra, advanced calculus, physics, and engineering. Text covers introduction to inner-product spaces, normed, metric spaces, and topological spaces; complete orthonormal sets, the Hahn-Banach Theorem and its consequences, and many other related subjects. 1966 ed. 544pp. 6⅛ x 9¼. 40251-7

ASYMPTOTIC EXPANSIONS OF INTEGRALS, Norman Bleistein & Richard A. Handelsman. Best introduction to important field with applications in a variety of scientific disciplines. New preface. Problems. Diagrams. Tables. Bibliography. Index. 448pp. 5⅜ x 8½. 65082-0

VECTOR AND TENSOR ANALYSIS WITH APPLICATIONS, A. I. Borisenko and I. E. Tarapov. Concise introduction. Worked-out problems, solutions, exercises. 257pp. 5⅝ x 8¼. 63833-2

THE ABSOLUTE DIFFERENTIAL CALCULUS (CALCULUS OF TENSORS), Tullio Levi-Civita. Great 20th-century mathematician's classic work on material necessary for mathematical grasp of theory of relativity. 452pp. 5⅝ x 8¼. 63401-9

AN INTRODUCTION TO ORDINARY DIFFERENTIAL EQUATIONS, Earl A. Coddington. A thorough and systematic first course in elementary differential equations for undergraduates in mathematics and science, with many exercises and problems (with answers). Index. 304pp. 5⅜ x 8½. 65942-9

FOURIER SERIES AND ORTHOGONAL FUNCTIONS, Harry F. Davis. An incisive text combining theory and practical example to introduce Fourier series, orthogonal functions and applications of the Fourier method to boundary-value problems. 570 exercises. Answers and notes. 416pp. 5⅜ x 8½. 65973-9

COMPUTABILITY AND UNSOLVABILITY, Martin Davis. Classic graduate-level introduction to theory of computability, usually referred to as theory of recurrent functions. New preface and appendix. 288pp. 5⅜ x 8½. 61471-9

ASYMPTOTIC METHODS IN ANALYSIS, N. G. de Bruijn. An inexpensive, comprehensive guide to asymptotic methods–the pioneering work that teaches by explaining worked examples in detail. Index. 224pp. 5⅜ x 8½ 64221-6

APPLIED COMPLEX VARIABLES, John W. Dettman. Step-by-step coverage of fundamentals of analytic function theory–plus lucid exposition of five important applications: Potential Theory; Ordinary Differential Equations; Fourier Transforms; Laplace Transforms; Asymptotic Expansions. 66 figures. Exercises at chapter ends. 512pp. 5⅜ x 8½. 64670-X

INTRODUCTION TO LINEAR ALGEBRA AND DIFFERENTIAL EQUATIONS, John W. Dettman. Excellent text covers complex numbers, determinants, orthonormal bases, Laplace transforms, much more. Exercises with solutions. Undergraduate level. 416pp. 5⅜ x 8½. 65191-6

CALCULUS OF VARIATIONS WITH APPLICATIONS, George M. Ewing. Applications-oriented introduction to variational theory develops insight and promotes understanding of specialized books, research papers. Suitable for advanced undergraduate/graduate students as primary, supplementary text. 352pp. 5⅜ x 8½.
64856-7

COMPLEX VARIABLES, Francis J. Flanigan. Unusual approach, delaying complex algebra till harmonic functions have been analyzed from real variable viewpoint. Includes problems with answers. 364pp. 5⅜ x 8½.
61388-7

AN INTRODUCTION TO THE CALCULUS OF VARIATIONS, Charles Fox. Graduate-level text covers variations of an integral, isoperimetrical problems, least action, special relativity, approximations, more. References. 279pp. 5⅜ x 8½.
65499-0

COUNTEREXAMPLES IN ANALYSIS, Bernard R. Gelbaum and John M. H. Olmsted. These counterexamples deal mostly with the part of analysis known as "real variables." The first half covers the real number system, and the second half encompasses higher dimensions. 1962 edition. xxiv+198pp. 5⅜ x 8½.
42875-3

CATASTROPHE THEORY FOR SCIENTISTS AND ENGINEERS, Robert Gilmore. Advanced-level treatment describes mathematics of theory grounded in the work of Poincaré, R. Thom, other mathematicians. Also important applications to problems in mathematics, physics, chemistry, and engineering. 1981 edition. References. 28 tables. 397 black-and-white illustrations. xvii+666pp. 6⅛ x 9¼.
67539-4

INTRODUCTION TO DIFFERENCE EQUATIONS, Samuel Goldberg. Exceptionally clear exposition of important discipline with applications to sociology, psychology, economics. Many illustrative examples; over 250 problems. 260pp. 5⅜ x 8½.
65084-7

NUMERICAL METHODS FOR SCIENTISTS AND ENGINEERS, Richard Hamming. Classic text stresses frequency approach in coverage of algorithms, polynomial approximation, Fourier approximation, exponential approximation, other topics. Revised and enlarged 2nd edition. 721pp. 5⅜ x 8½.
65241-6

INTRODUCTION TO NUMERICAL ANALYSIS (2nd Edition), F. B. Hildebrand. Classic, fundamental treatment covers computation, approximation, interpolation, numerical differentiation and integration, other topics. 150 new problems. 669pp. 5⅜ x 8½.
65363-3

THREE PEARLS OF NUMBER THEORY, A. Y. Khinchin. Three compelling puzzles require proof of a basic law governing the world of numbers. Challenges concern van der Waerden's theorem, the Landau-Schnirelmann hypothesis and Mann's theorem, and a solution to Waring's problem. Solutions included. 64pp. 5⅜ x 8½.
40026-3

THE PHILOSOPHY OF MATHEMATICS: An Introductory Essay, Stephan Körner. Surveys the views of Plato, Aristotle, Leibniz & Kant concerning propositions and theories of applied and pure mathematics. Introduction. Two appendices. Index. 198pp. 5⅜ x 8½.
25048-2

INTRODUCTORY REAL ANALYSIS, A.N. Kolmogorov, S. V. Fomin. Translated by Richard A. Silverman. Self-contained, evenly paced introduction to real and functional analysis. Some 350 problems. 403pp. 5⅜ x 8½. 61226-0

APPLIED ANALYSIS, Cornelius Lanczos. Classic work on analysis and design of finite processes for approximating solution of analytical problems. Algebraic equations, matrices, harmonic analysis, quadrature methods, more. 559pp. 5⅜ x 8½. 65656-X

AN INTRODUCTION TO ALGEBRAIC STRUCTURES, Joseph Landin. Superb self-contained text covers "abstract algebra": sets and numbers, theory of groups, theory of rings, much more. Numerous well-chosen examples, exercises. 247pp. 5⅜ x 8½. 65940-2

QUALITATIVE THEORY OF DIFFERENTIAL EQUATIONS, V. V. Nemytskii and V.V. Stepanov. Classic graduate-level text by two prominent Soviet mathematicians covers classical differential equations as well as topological dynamics and ergodic theory. Bibliographies. 523pp. 5⅜ x 8½. 65954-2

THEORY OF MATRICES, Sam Perlis. Outstanding text covering rank, nonsingularity and inverses in connection with the development of canonical matrices under the relation of equivalence, and without the intervention of determinants. Includes exercises. 237pp. 5⅜ x 8½. 66810-X

INTRODUCTION TO ANALYSIS, Maxwell Rosenlicht. Unusually clear, accessible coverage of set theory, real number system, metric spaces, continuous functions, Riemann integration, multiple integrals, more. Wide range of problems. Undergraduate level. Bibliography. 254pp. 5⅜ x 8½. 65038-3

MODERN NONLINEAR EQUATIONS, Thomas L. Saaty. Emphasizes practical solution of problems; covers seven types of equations. ". . . a welcome contribution to the existing literature. . . . "–*Math Reviews.* 490pp. 5⅜ x 8½. 64232-1

MATRICES AND LINEAR ALGEBRA, Hans Schneider and George Phillip Barker. Basic textbook covers theory of matrices and its applications to systems of linear equations and related topics such as determinants, eigenvalues, and differential equations. Numerous exercises. 432pp. 5⅜ x 8½. 66014-1

MATHEMATICS APPLIED TO CONTINUUM MECHANICS, Lee A. Segel. Analyzes models of fluid flow and solid deformation. For upper-level math, science, and engineering students. 608pp. 5⅜ x 8½. 65369-2

ELEMENTS OF REAL ANALYSIS, David A. Sprecher. Classic text covers fundamental concepts, real number system, point sets, functions of a real variable, Fourier series, much more. Over 500 exercises. 352pp. 5⅜ x 8½. 65385-4

SET THEORY AND LOGIC, Robert R. Stoll. Lucid introduction to unified theory of mathematical concepts. Set theory and logic seen as tools for conceptual understanding of real number system. 496pp. 5⅜ x 8¼. 63829-4

CATALOG OF DOVER BOOKS

TENSOR CALCULUS, J.L. Synge and A. Schild. Widely used introductory text covers spaces and tensors, basic operations in Riemannian space, non-Riemannian spaces, etc. 324pp. 5⅜ x 8¼. 63612-7

ORDINARY DIFFERENTIAL EQUATIONS, Morris Tenenbaum and Harry Pollard. Exhaustive survey of ordinary differential equations for undergraduates in mathematics, engineering, science. Thorough analysis of theorems. Diagrams. Bibliography. Index. 818pp. 5⅜ x 8½. 64940-7

INTEGRAL EQUATIONS, F. G. Tricomi. Authoritative, well-written treatment of extremely useful mathematical tool with wide applications. Volterra Equations, Fredholm Equations, much more. Advanced undergraduate to graduate level. Exercises. Bibliography. 238pp. 5⅜ x 8½. 64828-1

FOURIER SERIES, Georgi P. Tolstov. Translated by Richard A. Silverman. A valuable addition to the literature on the subject, moving clearly from subject to subject and theorem to theorem. 107 problems, answers. 336pp. 5⅜ x 8½. 63317-9

INTRODUCTION TO MATHEMATICAL THINKING, Friedrich Waismann. Examinations of arithmetic, geometry, and theory of integers; rational and natural numbers; complete induction; limit and point of accumulation; remarkable curves; complex and hypercomplex numbers, more. 1959 ed. 27 figures. xii+260pp. 5⅜ x 8½. 42804-4

POPULAR LECTURES ON MATHEMATICAL LOGIC, Hao Wang. Noted logician's lucid treatment of historical developments, set theory, model theory, recursion theory and constructivism, proof theory, more. 3 appendixes. Bibliography. 1981 ed. ix+283pp. 5⅜ x 8½. 67632-3

CALCULUS OF VARIATIONS, Robert Weinstock. Basic introduction covering isoperimetric problems, theory of elasticity, quantum mechanics, electrostatics, etc. Exercises throughout. 326pp. 5⅜ x 8½. 63069-2

THE CONTINUUM: A Critical Examination of the Foundation of Analysis, Hermann Weyl. Classic of 20th-century foundational research deals with the conceptual problem posed by the continuum. 156pp. 5⅜ x 8½. 67982-9

CHALLENGING MATHEMATICAL PROBLEMS WITH ELEMENTARY SOLUTIONS, A. M. Yaglom and I. M. Yaglom. Over 170 challenging problems on probability theory, combinatorial analysis, points and lines, topology, convex polygons, many other topics. Solutions. Total of 445pp. 5⅜ x 8½. Two-vol. set.
Vol. I: 65536-9 Vol. II: 65537-7

INTRODUCTION TO PARTIAL DIFFERENTIAL EQUATIONS WITH APPLICATIONS, E. C. Zachmanoglou and Dale W. Thoe. Essentials of partial differential equations applied to common problems in engineering and the physical sciences. Problems and answers. 416pp. 5⅜ x 8½. 65251-3

THE THEORY OF GROUPS, Hans J. Zassenhaus. Well-written graduate-level text acquaints reader with group-theoretic methods and demonstrates their usefulness in mathematics. Axioms, the calculus of complexes, homomorphic mapping, p-group theory, more. 276pp. 5⅜ x 8½. 40922-8

Math–Decision Theory, Statistics, Probability

ELEMENTARY DECISION THEORY, Herman Chernoff and Lincoln E. Moses. Clear introduction to statistics and statistical theory covers data processing, probability and random variables, testing hypotheses, much more. Exercises. 364pp. 5⅜ x 8½. 65218-1

STATISTICS MANUAL, Edwin L. Crow et al. Comprehensive, practical collection of classical and modern methods prepared by U.S. Naval Ordnance Test Station. Stress on use. Basics of statistics assumed. 288pp. 5⅜ x 8½. 60599-X

SOME THEORY OF SAMPLING, William Edwards Deming. Analysis of the problems, theory, and design of sampling techniques for social scientists, industrial managers, and others who find statistics important at work. 61 tables. 90 figures. xvii +602pp. 5⅜ x 8½. 64684-X

LINEAR PROGRAMMING AND ECONOMIC ANALYSIS, Robert Dorfman, Paul A. Samuelson and Robert M. Solow. First comprehensive treatment of linear programming in standard economic analysis. Game theory, modern welfare economics, Leontief input-output, more. 525pp. 5⅜ x 8½. 65491-5

PROBABILITY: An Introduction, Samuel Goldberg. Excellent basic text covers set theory, probability theory for finite sample spaces, binomial theorem, much more. 360 problems. Bibliographies. 322pp. 5⅜ x 8½. 65252-1

GAMES AND DECISIONS: Introduction and Critical Survey, R. Duncan Luce and Howard Raiffa. Superb nontechnical introduction to game theory, primarily applied to social sciences. Utility theory, zero-sum games, n-person games, decision-making, much more. Bibliography. 509pp. 5⅜ x 8½. 65943-7

INTRODUCTION TO THE THEORY OF GAMES, J. C. C. McKinsey. This comprehensive overview of the mathematical theory of games illustrates applications to situations involving conflicts of interest, including economic, social, political, and military contexts. Appropriate for advanced undergraduate and graduate courses; advanced calculus a prerequisite. 1952 ed. x+372pp. 5⅜ x 8½. 42811-7

FIFTY CHALLENGING PROBLEMS IN PROBABILITY WITH SOLUTIONS, Frederick Mosteller. Remarkable puzzlers, graded in difficulty, illustrate elementary and advanced aspects of probability. Detailed solutions. 88pp. 5⅜ x 8½. 65355-2

PROBABILITY THEORY: A Concise Course, Y. A. Rozanov. Highly readable, self-contained introduction covers combination of events, dependent events, Bernoulli trials, etc. 148pp. 5⅜ x 8¼. 63544-9

STATISTICAL METHOD FROM THE VIEWPOINT OF QUALITY CONTROL, Walter A. Shewhart. Important text explains regulation of variables, uses of statistical control to achieve quality control in industry, agriculture, other areas. 192pp. 5⅜ x 8½. 65232-7

Math–Geometry and Topology

ELEMENTARY CONCEPTS OF TOPOLOGY, Paul Alexandroff. Elegant, intuitive approach to topology from set-theoretic topology to Betti groups; how concepts of topology are useful in math and physics. 25 figures. 57pp. 5⅜ x 8½. 60747-X

COMBINATORIAL TOPOLOGY, P. S. Alexandrov. Clearly written, well-organized, three-part text begins by dealing with certain classic problems without using the formal techniques of homology theory and advances to the central concept, the Betti groups. Numerous detailed examples. 654pp. 5⅜ x 8½. 40179-0

EXPERIMENTS IN TOPOLOGY, Stephen Barr. Classic, lively explanation of one of the byways of mathematics. Klein bottles, Moebius strips, projective planes, map coloring, problem of the Koenigsberg bridges, much more, described with clarity and wit. 43 figures. 210pp. 5⅜ x 8½. 25933-1

CONFORMAL MAPPING ON RIEMANN SURFACES, Harvey Cohn. Lucid, insightful book presents ideal coverage of subject. 334 exercises make book perfect for self-study. 55 figures. 352pp. 5⅝ x 8¼. 64025-6

THE GEOMETRY OF RENÉ DESCARTES, René Descartes. The great work founded analytical geometry. Original French text, Descartes's own diagrams, together with definitive Smith-Latham translation. 244pp. 5⅜ x 8½. 60068-8

PRACTICAL CONIC SECTIONS: The Geometric Properties of Ellipses, Parabolas and Hyperbolas, J. W. Downs. This text shows how to create ellipses, parabolas, and hyperbolas. It also presents historical background on their ancient origins and describes the reflective properties and roles of curves in design applications. 1993 ed. 98 figures. xii+100pp. 6½ x 9¼. 42876-1

THE THIRTEEN BOOKS OF EUCLID'S ELEMENTS, translated with introduction and commentary by Thomas L. Heath. Definitive edition. Textual and linguistic notes, mathematical analysis. 2,500 years of critical commentary. Unabridged. 1,414pp. 5⅜ x 8½. Three-vol. set. Vol. I: 60088-2 Vol. II: 60089-0 Vol. III: 60090-4

GEOMETRY OF COMPLEX NUMBERS, Hans Schwerdtfeger. Illuminating, widely praised book on analytic geometry of circles, the Moebius transformation, and two-dimensional non-Euclidean geometries. 200pp. 5⅝ x 8¼. 63830-8

DIFFERENTIAL GEOMETRY, Heinrich W. Guggenheimer. Local differential geometry as an application of advanced calculus and linear algebra. Curvature, transformation groups, surfaces, more. Exercises. 62 figures. 378pp. 5⅜ x 8½. 63433-7

CURVATURE AND HOMOLOGY: Enlarged Edition, Samuel I. Goldberg. Revised edition examines topology of differentiable manifolds; curvature, homology of Riemannian manifolds; compact Lie groups; complex manifolds; curvature, homology of Kaehler manifolds. New Preface. Four new appendixes. 416pp. 5⅜ x 8½. 40207-X

History of Math

THE WORKS OF ARCHIMEDES, Archimedes (T. L. Heath, ed.). Topics include the famous problems of the ratio of the areas of a cylinder and an inscribed sphere; the measurement of a circle; the properties of conoids, spheroids, and spirals; and the quadrature of the parabola. Informative introduction. clxxxvi+326pp; supplement, 52pp. 5⅜ x 8½. 42084-1

A SHORT ACCOUNT OF THE HISTORY OF MATHEMATICS, W. W. Rouse Ball. One of clearest, most authoritative surveys from the Egyptians and Phoenicians through 19th-century figures such as Grassman, Galois, Riemann. Fourth edition. 522pp. 5⅜ x 8½. 20630-0

THE HISTORY OF THE CALCULUS AND ITS CONCEPTUAL DEVELOP-MENT, Carl B. Boyer. Origins in antiquity, medieval contributions, work of Newton, Leibniz, rigorous formulation. Treatment is verbal. 346pp. 5⅜ x 8½. 60509-4

THE HISTORICAL ROOTS OF ELEMENTARY MATHEMATICS, Lucas N. H. Bunt, Phillip S. Jones, and Jack D. Bedient. Fundamental underpinnings of modern arithmetic, algebra, geometry, and number systems derived from ancient civilizations. 320pp. 5⅜ x 8½. 25563-8

A HISTORY OF MATHEMATICAL NOTATIONS, Florian Cajori. This classic study notes the first appearance of a mathematical symbol and its origin, the competition it encountered, its spread among writers in different countries, its rise to popularity, its eventual decline or ultimate survival. Original 1929 two-volume edition presented here in one volume. xxviii+820pp. 5⅜ x 8½. 67766-4

GAMES, GODS & GAMBLING: A History of Probability and Statistical Ideas, F. N. David. Episodes from the lives of Galileo, Fermat, Pascal, and others illustrate this fascinating account of the roots of mathematics. Features thought-provoking references to classics, archaeology, biography, poetry. 1962 edition. 304pp. 5⅜ x 8½. (Available in U.S. only.) 40023-9

OF MEN AND NUMBERS: The Story of the Great Mathematicians, Jane Muir. Fascinating accounts of the lives and accomplishments of history's greatest mathematical minds—Pythagoras, Descartes, Euler, Pascal, Cantor, many more. Anecdotal, illuminating. 30 diagrams. Bibliography. 256pp. 5⅜ x 8½. 28973-7

HISTORY OF MATHEMATICS, David E. Smith. Nontechnical survey from ancient Greece and Orient to late 19th century; evolution of arithmetic, geometry, trigonometry, calculating devices, algebra, the calculus. 362 illustrations. 1,355pp. 5⅜ x 8½. Two-vol. set. Vol. I: 20429-4 Vol. II: 20430-8

A CONCISE HISTORY OF MATHEMATICS, Dirk J. Struik. The best brief history of mathematics. Stresses origins and covers every major figure from ancient Near East to 19th century. 41 illustrations. 195pp. 5⅜ x 8½. 60255-9

Physics

OPTICAL RESONANCE AND TWO-LEVEL ATOMS, L. Allen and J. H. Eberly. Clear, comprehensive introduction to basic principles behind all quantum optical resonance phenomena. 53 illustrations. Preface. Index. 256pp. 5⅜ x 8½.   65533-4

QUANTUM THEORY, David Bohm. This advanced undergraduate-level text presents the quantum theory in terms of qualitative and imaginative concepts, followed by specific applications worked out in mathematical detail. Preface. Index. 655pp. 5⅜ x 8½.   65969-0

ATOMIC PHYSICS: 8th edition, Max Born. Nobel laureate's lucid treatment of kinetic theory of gases, elementary particles, nuclear atom, wave-corpuscles, atomic structure and spectral lines, much more. Over 40 appendices, bibliography. 495pp. 5⅜ x 8½.   65984-4

A SOPHISTICATE'S PRIMER OF RELATIVITY, P. W. Bridgman. Geared toward readers already acquainted with special relativity, this book transcends the view of theory as a working tool to answer natural questions: What is a frame of reference? What is a "law of nature"? What is the role of the "observer"? Extensive treatment, written in terms accessible to those without a scientific background. 1983 ed. xlviii+172pp. 5⅜ x 8½.   42549-5

AN INTRODUCTION TO HAMILTONIAN OPTICS, H. A. Buchdahl. Detailed account of the Hamiltonian treatment of aberration theory in geometrical optics. Many classes of optical systems defined in terms of the symmetries they possess. Problems with detailed solutions. 1970 edition. xv+360pp. 5⅜ x 8½.   67597-1

PRIMER OF QUANTUM MECHANICS, Marvin Chester. Introductory text examines the classical quantum bead on a track: its state and representations; operator eigenvalues; harmonic oscillator and bound bead in a symmetric force field; and bead in a spherical shell. Other topics include spin, matrices, and the structure of quantum mechanics; the simplest atom; indistinguishable particles; and stationary-state perturbation theory. 1992 ed. xiv+314pp. 6⅛ x 9¼.   42878-8

LECTURES ON QUANTUM MECHANICS, Paul A. M. Dirac. Four concise, brilliant lectures on mathematical methods in quantum mechanics from Nobel Prize–winning quantum pioneer build on idea of visualizing quantum theory through the use of classical mechanics. 96pp. 5⅜ x 8½.   41713-1

THIRTY YEARS THAT SHOOK PHYSICS: The Story of Quantum Theory, George Gamow. Lucid, accessible introduction to influential theory of energy and matter. Careful explanations of Dirac's anti-particles, Bohr's model of the atom, much more. 12 plates. Numerous drawings. 240pp. 5⅜ x 8½.   24895-X

ELECTRONIC STRUCTURE AND THE PROPERTIES OF SOLIDS: The Physics of the Chemical Bond, Walter A. Harrison. Innovative text offers basic understanding of the electronic structure of covalent and ionic solids, simple metals, transition metals and their compounds. Problems. 1980 edition. 582pp. 6⅛ x 9¼.   66021-4

HYDRODYNAMIC AND HYDROMAGNETIC STABILITY, S. Chandrasekhar. Lucid examination of the Rayleigh-Benard problem; clear coverage of the theory of instabilities causing convection. 704pp. 5⅜ x 8¼. 64071-X

INVESTIGATIONS ON THE THEORY OF THE BROWNIAN MOVEMENT, Albert Einstein. Five papers (1905–8) investigating dynamics of Brownian motion and evolving elementary theory. Notes by R. Fürth. 122pp. 5⅜ x 8½. 60304-0

THE PHYSICS OF WAVES, William C. Elmore and Mark A. Heald. Unique overview of classical wave theory. Acoustics, optics, electromagnetic radiation, more. Ideal as classroom text or for self-study. Problems. 477pp. 5⅜ x 8½. 64926-1

PHYSICAL PRINCIPLES OF THE QUANTUM THEORY, Werner Heisenberg. Nobel Laureate discusses quantum theory, uncertainty, wave mechanics, work of Dirac, Schroedinger, Compton, Wilson, Einstein, etc. 184pp. 5⅜ x 8½. 60113-7

ATOMIC SPECTRA AND ATOMIC STRUCTURE, Gerhard Herzberg. One of best introductions; especially for specialist in other fields. Treatment is physical rather than mathematical. 80 illustrations. 257pp. 5⅜ x 8½. 60115-3

AN INTRODUCTION TO STATISTICAL THERMODYNAMICS, Terrell L. Hill. Excellent basic text offers wide-ranging coverage of quantum statistical mechanics, systems of interacting molecules, quantum statistics, more. 523pp. 5⅜ x 8½. 65242-4

THEORETICAL PHYSICS, Georg Joos, with Ira M. Freeman. Classic overview covers essential math, mechanics, electromagnetic theory, thermodynamics, quantum mechanics, nuclear physics, other topics. xxiii+885pp. 5⅜ x 8½. 65227-0

PROBLEMS AND SOLUTIONS IN QUANTUM CHEMISTRY AND PHYSICS, Charles S. Johnson, Jr. and Lee G. Pedersen. Unusually varied problems, detailed solutions in coverage of quantum mechanics, wave mechanics, angular momentum, molecular spectroscopy, more. 280 problems, 139 supplementary exercises. 430pp. 6½ x 9¼. 65236-X

THEORETICAL SOLID STATE PHYSICS, Vol. I: Perfect Lattices in Equilibrium; Vol. II: Non-Equilibrium and Disorder, William Jones and Norman H. March. Monumental reference work covers fundamental theory of equilibrium properties of perfect crystalline solids, non-equilibrium properties, defects and disordered systems. Total of 1,301pp. 5⅜ x 8½. Vol. I: 65015-4 Vol. II: 65016-2

WHAT IS RELATIVITY? L. D. Landau and G. B. Rumer. Written by a Nobel Prize physicist and his distinguished colleague, this compelling book explains the special theory of relativity to readers with no scientific background, using such familiar objects as trains, rulers, and clocks. 1960 ed. vi+72pp. 23 b/w illustrations. 5⅜ x 8½. 42806-0 $6.95

A TREATISE ON ELECTRICITY AND MAGNETISM, James Clerk Maxwell. Important foundation work of modern physics. Brings to final form Maxwell's theory of electromagnetism and rigorously derives his general equations of field theory. 1,084pp. 5⅜ x 8½. Two-vol. set. Vol. I: 60636-8 Vol. II: 60637-6

QUANTUM MECHANICS: Principles and Formalism, Roy McWeeny. Graduate student-oriented volume develops subject as fundamental discipline, opening with review of origins of Schrödinger's equations and vector spaces. Focusing on main principles of quantum mechanics and their immediate consequences, it concludes with final generalizations covering alternative "languages" or representations. 1972 ed. 15 figures. xi+155pp. 5⅜ x 8½. 42829-X

INTRODUCTION TO QUANTUM MECHANICS WITH APPLICATIONS TO CHEMISTRY, Linus Pauling & E. Bright Wilson, Jr. Classic undergraduate text by Nobel Prize winner applies quantum mechanics to chemical and physical problems. Numerous tables and figures enhance the text. Chapter bibliographies. Appendices. Index. 468pp. 5⅜ x 8½. 64871-0

METHODS OF THERMODYNAMICS, Howard Reiss. Outstanding text focuses on physical technique of thermodynamics, typical problem areas of understanding, and significance and use of thermodynamic potential. 1965 edition. 238pp. 5⅜ x 8½. 69445-3

TENSOR ANALYSIS FOR PHYSICISTS, J. A. Schouten. Concise exposition of the mathematical basis of tensor analysis, integrated with well-chosen physical examples of the theory. Exercises. Index. Bibliography. 289pp. 5⅜ x 8½. 65582-2

THE ELECTROMAGNETIC FIELD, Albert Shadowitz. Comprehensive undergraduate text covers basics of electric and magnetic fields, builds up to electromagnetic theory. Also related topics, including relativity. Over 900 problems. 768pp. 5⅝ x 8¼. 65660-8

GREAT EXPERIMENTS IN PHYSICS: Firsthand Accounts from Galileo to Einstein, Morris H. Shamos (ed.). 25 crucial discoveries: Newton's laws of motion, Chadwick's study of the neutron, Hertz on electromagnetic waves, more. Original accounts clearly annotated. 370pp. 5⅜ x 8½. 25346-5

RELATIVITY, THERMODYNAMICS AND COSMOLOGY, Richard C. Tolman. Landmark study extends thermodynamics to special, general relativity; also applications of relativistic mechanics, thermodynamics to cosmological models. 501pp. 5⅜ x 8½. 65383-8

STATISTICAL PHYSICS, Gregory H. Wannier. Classic text combines thermodynamics, statistical mechanics, and kinetic theory in one unified presentation of thermal physics. Problems with solutions. Bibliography. 532pp. 5⅜ x 8½. 65401-X

Paperbound unless otherwise indicated. Available at your book dealer, online at **www.doverpublications.com**, or by writing to Dept. GI, Dover Publications, Inc., 31 East 2nd Street, Mineola, NY 11501. For current price information or for free catalogs (please indicate field of interest), write to Dover Publications or log on to **www.doverpublications.com** and see every Dover book in print. Dover publishes more than 500 books each year on science, elementary and advanced mathematics, biology, music, art, literary history, social sciences, and other areas.